Selected Titles in This Series

49 **Ronald L. Graham, Jan Kratochvíl, Jaroslav Nešetřil, and Fred S. Roberts, Editors,** Contemporary Trends in Discrete Mathematics

48 **Harvey Rubin and David Harlan Wood, Editors,** DNA Based Computers III

47 **Martin Farach-Colton, Fred S. Roberts, Martin Vingron, and Michael Waterman, Editors,** Mathematical Support for Molecular Biology

46 **Peng-Jun Wan, Ding-Zhu Du, and Panos M. Pardalos, Editors,** Multichannel Optical Networks: Theory and Practice

45 **Marios Mavronicolas, Michael Merritt, and Nir Shavit, Editors,** Networks in Distributed Computing

44 **Laura F. Landweber and Eric B. Baum, Editors,** DNA Based Computers II

43 **Panos Pardalos, Sanguthevar Rajasekaran, and José Rolim, Editors,** Randomization Methods in Algorithm Design

42 **Ding-Zhu Du and Frank K. Hwang, Editors,** Advances in Switching Networks

41 **David Aldous and James Propp, Editors,** Microsurveys in Discrete Probability

40 **Panos M. Pardalos and Dingzhu Du, Editors,** Network Design: Connectivity and Facilities Location

39 **Paul W. Beame and Samuel R Buss, Editors,** Proof Complexity and Feasible Arithmetics

38 **Rebecca N. Wright and Peter G. Neumann, Editors,** Network Threats

37 **Boris Mirkin, F. R. McMorris, Fred S. Roberts, and Andrey Rzhetsky, Editors,** Mathematical Hierarchies and Biology

36 **Joseph G. Rosenstein, Deborah S. Franzblau, and Fred S. Roberts, Editors,** Discrete Mathematics in the Schools

35 **Dingzhu Du, Jun Gu, and Panos M. Pardalos, Editors,** Satisfiability Problem: Theory and Applications

34 **Nathaniel Dean, Editor,** African Americans in Mathematics

33 **Ravi B. Boppana and James F. Lynch, Editors,** Logic and Random Structures

32 **Jean-Charles Grégoire, Gerard J. Holzmann, and Doron A. Peled, Editors,** The SPIN Verification System

31 **Neil Immerman and Phokion G. Kolaitis, Editors,** Descriptive Complexity and Finite Models

30 **Sandeep N. Bhatt, Editor,** Parallel Algorithms: Third DIMACS Implementation Challenge

29 **Doron A. Peled, Vaughan R. Pratt, and Gerard J. Holzmann, Editors,** Partial Order Methods in Verification

28 **Larry Finkelstein and William M. Kantor, Editors,** Groups and Computation II

27 **Richard J. Lipton and Eric B. Baum, Editors,** DNA Based Computers

26 **David S. Johnson and Michael A. Trick, Editors,** Cliques, Coloring, and Satisfiability: Second DIMACS Implementation Challenge

25 **Gilbert Baumslag, David Epstein, Robert Gilman, Hamish Short, and Charles Sims, Editors,** Geometric and Computational Perspectives on Infinite Groups

24 **Louis J. Billera, Curtis Greene, Rodica Simion, and Richard P. Stanley, Editors,** Formal Power Series and Algebraic Combinatorics/Séries Formelles et Combinatoire Algébrique, 1994

23 **Panos M. Pardalos, David I. Shalloway, and Guoliang Xue, Editors,** Global Minimization of Nonconvex Energy Functions: Molecular Conformation and Protein Folding

22 **Panos M. Pardalos, Mauricio G. C. Resende, and K. G. Ramakrishnan, Editors,** Parallel Processing of Discrete Optimization Problems

21 **D. Frank Hsu, Arnold L. Rosenberg, and Dominique Sotteau, Editors,** Interconnection Networks and Mapping and Scheduling Parallel Computations

(See the AMS catalog for earlier titles)

Contemporary Trends in Discrete Mathematics

From DIMACS and DIMATIA to the Future

Some participants of the DIMATIA-DIMACS conference in the park of the Štiřín Castle. Photographs taken by A. Kotěšovcová on May 20, 1997.

DIMACS

Series in Discrete Mathematics
and Theoretical Computer Science

Volume 49

Contemporary Trends in Discrete Mathematics

From DIMACS and DIMATIA to the Future

DIMATIA-DIMACS Conference
May 19–25, 1997
Štiřín Castle, Czech Republic

Ronald L. Graham
Jan Kratochvíl
Jaroslav Nešetřil
Fred S. Roberts
Editors

NSF Science and Technology Center
in Discrete Mathematics and Theoretical Computer Science
A consortium of Rutgers University, Princeton University,
Bell Labs, Telcordia Technologies (formerly Bellcore), and
NEC Research Institute

American Mathematical Society

This DIMACS volume contains papers from a DIMATIA-DIMACS conference on The Future of Discrete Mathematics. The conference was held May 19–25, 1997, at Štiřín Castle, Czech Republic.

1991 *Mathematics Subject Classification.* Primary 05–06, 05Cxx, 05Dxx, 06A07, 11Bxx, 60C05, 68Rxx, 68Q15, 92C40.

Library of Congress Cataloging-in-Publication Data

Contemporary trends in discrete mathematics : from DIMACS and DIMATIA to the future : DIMATIA-DIMACS conference, May 19–25, 1997, Štiřín Castle, Czech Republic / Ronald L. Graham, editors... [et al.].

p. cm. — (DIMACS series in discrete mathematics and theoretical computer science ; v. 49)

Papers from the DIMACS/DIMATIA Workshop on the Future of Discrete Mathematics.

Includes bibliographical references.

ISBN 0-8218-0963-6 (alk. paper)

1. Graph theory—Congresses. 2. Combinatorial analysis—Congresses. 3. Computer science—Mathematics—Congresses. I. Graham, Ronald L., 1935– . II. DIMATIA (Group) III. DIMACS (Group) IV. DIMACS/DIMATIA Workshop on the Future of Discrete Mathematics (1997 : Štiřín Castle) V. Series.

QA166.C6355 1999

511′.5—dc21 99-19657

CIP

∞ The paper used in this book is acid-free and falls within the guidelines
established to ensure permanence and durability.
Visit the AMS home page at URL: http://www.ams.org/

10 9 8 7 6 5 4 3 2 1   04 03 02 01 00 99

Contents

Foreword

The DIMACS-DIMATIA Workshop on The Future of Discrete Mathematics was held on May 19–25, 1997 at the Štiřín Castle in the Czech Republic. We would like to express our appreciation to Fred S. Roberts, Ronald Graham, Jaroslav Nešetřil and Jan Kratochvíl for their efforts to organize and plan this successful workshop.

The main focus of this workshop was on presenting current trends in discrete mathematics in all its versatility.

The workshop was part of the international cooperation between two centers, DIMACS and DIMATIA, which is supported by grants from the Czech and US National Science Foundations.

DIMACS gratefully acknowledges the generous support that makes its programs possible: The National Science Foundation, through its Science and Technology Center program, the National Security Agency, the New Jersey Commission on Science and Technology, DIMACS' partners at Rutgers, Princeton, AT&T Labs Research, Bell Labs, Telcordia Technologies (formerly Bellcore), and NEC Research Institute.

Fred S. Roberts
Director

Robert Sedgewick
Co-Director for Princeton

Preface

The following book is an outgrowth of the DIMATIA-DIMACS conference "The Future of Discrete Mathematics" which was held at Štiřín Castle in the vicinity of Prague in May 19-25, 1997. This was the inaugural conference under a program of official cooperation of the two centers, which is being supported by joint Czech Academy - National Science Foundation grants 055/1997 and NSF-INT-96-05174. The conference was further supported by Czech grants GAČR 201/1996/0194, GAUK 194/1996, FR MŠMT, and Czech-US Science and Technology grant 94051.

The ambitious aim of the meeting was to present current trends of Discrete Mathematics in all its versatility, width and depth. Discrete Mathematics has taken its place among the leading disciplines of mathematics and theoretical computer science. This is accompanied on one side by its growing importance in university curriculae and on the other side by its growing role in various applications which range from optimization to molecular biology.

More than 50 participants (including students from several participating universities) took part in the conference, and the enclosed program mirrors the quality and the scope of the meeting.

The present book is not just a conference proceedings. We have tried to assemble contributions to the volume that feature some of the most attractive fields of contemporary Discrete Mathematics. We believe that we have managed to put together a sample of both research papers and surveys that reflects the scope and excitement of the field. The delay between the conference and collection of final versions enabled the authors to reflect new developments and include new results from the relevant areas. All contributions were refereed and we take this opportunity to thank our referees for their help.

As expected, the largest number of papers deal with graph theory. Among them are surveys by W. Mader on topological minors and M. Simonovits on extremal graph theory. Research papers in this area include papers on path systems (A. Frank) and on the generalized matching problem and bounded tree-width (A. Gupta, D. Kaller, S. Mahajan, and T. Shermer). Several research papers are devoted to various versions of the classical coloring and recently introduced oriented chromatic number (A. Kostochka, T. Łuczak, G. Simonyi and E. Sopena; P. Boiron, E. Sopena, and L. Vignal; A. Sali and G. Simonyi). To the area of graph coloring problems also belongs a survey of choosability and list colorings by J. Kratochvíl, Zs. Tuza and M. Voigt. Most of these papers also investigate computational complexity of the various graph theoretical problems and investigate various problems of theoretical computer science.

A more structural (algebraic) approach is taken in papers by P. Mihók (a survey of reducibility of graph properties) and by J. Nešetřil and C. Tardif who survey the recent solution of density problems in full generality. A survey of a recent solution of the problem of representations of partially ordered sets by circular orders is presented by W. T. Trotter. This paper also deals to a large extent with new developments in Ramsey theory. Geometrical Ramsey theorems related to the classical Erdős-Szekerés theorem appear in the paper by J. Nešetřil, J. Solymosi, and P. Valtr. Further geometrical problems are studied in an elegant paper by P. Ossona de Mendez and H. de Fraysseix on planar representations of graphs. Geometric methods are also used in the paper by G. Elekes on large homothetic subsets and by J. Griggs and G. Rote on distribution of vector sums.

Combinatorial questions of an algebraic nature are represented by articles by W. Hochstättler and M. Loebl on bases of lattices and M. Rosenfeld on matchings in finite fields, a work related to Hadamard matrices. Two papers are devoted to problems related to Davenport-Schinzel sequences intensively studied in Prague (P. Valtr surveying geometrical applications of this combinatorial structure, and M. Klazar considering several generalizations).

Problems of a purely computer science nature are studied in papers by S. Fekete, S. Kromberg, Ch. Moll, and W. Hochstättler on the complexity of the inverse shortest path problem and by J. Díaz, M. Serna, and P. Spirakis surveying linear and nonlinear systems, and in an extensive survey by R. G. Downey and M. R. Fellows devoted to parametrized complexity analysis.

Finally, applications are represented by a survey of F. S. Roberts on problems stemming from the social sciences and related to limitations on conclusions about optimal solutions or decision making imposed by considerations of measurement scales, and by a paper of V. Janota, P. Pančoška, and J. Nešetřil which is a further contribution to the literature on protein folding problems.

This book is the result of one of the first activities of the DIMATIA Center of Charles University. The Center for Discrete Mathematics, Theoretical Computer Science and Applications was established in September 1996 at Charles University in Prague as a joint project of Charles University, Academy of Sciences of the Czech Republic and Institute of Chemical Technology, Prague. The center is open and actively seeks partners from all parts of the world. The purpose of the center is to foster research in all fields of discrete mathematics and its modern applications and relationship to computer science, operations research and fields as diverse as biology, chemistry and the social sciences. Towards this end the center organizes a continuing program of workshops, conferences and research visits both in Prague and other places together with cooperating institutions. The project concentrates on research topics that are at the forefront of current research in discrete mathematics and theoretical computer science.

DIMATIA is a corporate structure founded by three leading Prague research groups working in discrete mathematics, theoretical computer science and their applications. The research team includes 14 permanent members, mostly mathematicians and computer scientists, though some members of the research team are physicists and chemists. All these members are staff members of the founding institutions – Faculty of Mathematics and Physics of the Charles University (V. Janota, M. Klazar, J. Kratochvíl, L. Kučera, M. Loebl, J. Matoušek, J. Nešetřil, P. Pančoška, P. Valtr), Mathematical Institute of the Academy of Sciences of the Czech Republic (I. Havel, J. Krajíček, P. Pudlák, J. Sgall) and Faculty of Chemical Engineering of the Institute of Chemical Technology, Prague (D. Turzík). This number does not include associated researchers, foreign associates and graduate students.

The center has a Director (J. Nešetřil) to provide scientific leadership and day-to-day operation of the center. In that he is assisted by Deputy Director (J. Kratochvíl) and Scientific Secretary (V. Janota). Charles University provides the center with a part time secretary position.

The center has an International Scientific Advisory Board consisting of representatives of DIMATIA participating organizations. DIMATIA is an open institution, directed to cooperation on the whole European scale and open to further collaboration. This is a very special feature of the DIMATIA concept and key to its intended activities. Presently the foreign associates are Sobolev Institute of Mathematics, Novosibirsk; LaBRI Institute, Bordeaux; Pacific Institute of Mathematical Sciences; Institute of Computational Mathematics, Pisa; Operations Research Institute, Bonn; University Bielefeld; Mathematical Institute, Budapest; University Barcelona; and DIMACS.

DIMACS, the Center for Discrete Mathematics and Theoretical Computer Science, is a consortium of Rutgers and Princeton Universities and AT&T Labs, Bell Labs, Bellcore, and NEC Research Institute (a newly added partner). DIMACS has entered into the partnership with DIMATIA in order to enhance and stimulate international collaborations and direct the attention of the international mathematical sciences community toward the fields of discrete mathematics and theoretical computer science and their applications. Such an international partnership also offers outstanding young people from both sides the opportunity to be exposed to the world's leading mathematical scientists and to gain early international experiences. DIMACS is delighted to celebrate the opening of the DIMATIA Center and is enthusiastic about the partnership with such a leading scientific organization. The Štiřín conference was the first joint project within the newly founded DIMATIA-DIMACS framework, which also includes more focused workshops, exchanges of visitors, and programs for postdoctoral fellows, graduate students, and undergraduates. We all believe and hope this was the first of many mutually beneficial and exciting scientific collaborations.

Ronald L. Graham
Jan Kratochvíl
Jaroslav Nešetřil
Fred S. Roberts

Program

Monday May 19, 1997

4 - 6 p.m.

M. Rosenfeld: Three Problems in Combinatorial Number Theory

Robin Thomas: Permanents, Pfaffian Orientations, and Even Directed Circuits

Tuesday May 20, 1997

9 a.m.–12 noon

William T. Trotter: Ramsey Theory and Sequences of Random Variables

Vera Sós: On Combinatorial Number Theory

Fred S. Roberts: Meaningless Statements

2–6 p.m.

Jiří Matoušek: Upper Bounds in Geometric and Combinatorial Discrepancy

Günter Rote: Convex 3-polytopes Can Be Realized with Linearly Many Bits per Coordinate

Pavel Valtr: Several Recent Results on Geometric Graphs

Mary Inaba: Geometric Clustering Problem

Peter Brass: On Lattice-like Extremal Sets for Combinatorial Geometry Problems

Wednesday May 21, 1997

9 a.m.–12 noon

Joel Spencer: Sophisticated Probability

Tomasz Łuczak: Random Trees and Random Graphs

Krystyna T. Balińska: An Algorithmic Approach to a Random Process for Graphs with Bounded Degree

2–6 p.m.

Eugene Luks: Algorithmic Applications of the Simple Groups Classification

Luděk Kučera: Fast Deflection Routing of Long Messages

Josep Díaz: Approximation of #P Problems in RNC

Jiří Sgall: On Pairs of Intersecting Families

Conference banquet

Thursday May 22, 1997

9 a.m.–12 noon

Jaroslav Nešetřil: The Structure of Graph Homomorphisms

Jan Kratochvíl: List Colorings and Choosability of Graphs

Alexander Kostochka: On Colour-critical Graphs and Hypergraphs with Few Edges

Afternoon excursion to České Budějovice (exhibition of paintings by Načeradský and Nešetřil) and to Velké Popovice (brewery)

Friday May 23, 1997

9 a.m.–12 noon

Lászlo Lovász: Random Walks, Mixing, and Sampling

András Frank: Algorithms for Increasing the Connectivity of Graphs

Eric Sopena: Acyclic Improper Colorings

2–6 p.m.

Jan Krajíček: Proof Complexity and Some Related Algebraico - combinatorial Problems

Arvind Gupta: Monadic Second Order Logic and Complement Problems

Mike Fellows: The Shortest Vector Problem and the Prospects for Cryptosystems Based on NP-Hard Problems

Gyorgy Elekes: Metric Combinatorics and Combinatorial Algebra

Gábor Simonyi: Information Theory in Combinatorics

William F. Smyth: Periodicity in Strings

Saturday May 24, 1997

9 a.m.–1 p.m.

Walter Deuber: Ramsey Theory and Linear Algebra

Graham Brightwell: Graph Homomorphisms and Branching Random Walks

Wolfgang Mader: An Extremal Problem for Subdivisions of K_5 and Topological Subgraphs in Graphs of Large Girth

Sándor P. Fekete: A New Exact Algorithm for General Orthogonal d-dimensional Knapsack Problems and a New Lower Bound for Bin Packing Problems

Shiyu Zhou: Computing undirected connectivity in small space

3–6 p.m.

Martin Klazar: Extremal Problems for Colored Trees

Miklós Simonovits: General Methods to Prove Exact Results in Extremal Graph Theory

Peter Mihók: The Structure of Hereditary Properties of Graphs and Minimal Reducible Bounds

Petr Pančoška: Proteins, Graphs and Spectra - Selected Problems

Sunday May 25, 1997

9–11 a.m.

Jerry Griggs: The Distribution of Subset Sums in R^m and in Finite Abelian Groups

Pavel Pudlák : New algorithms for satisfiability

List of participants

R. Babilon, DIMATIA
K. Balińska, Poznań
J. Balogh, Szeged
P. Brass, Germany
G. Brightwell, London
T. Dacic, PIMS
J. Díaz, Barcelona
W. Deuber, Bielefeld
G. Elekes, Budapest
S. Fekete, Köln
M. Fellows, PIMS
A. Frank, Budapest
H. de Fraysseix, Paris
J. Griggs, Columbia
A. Gupta, PIMS
P. Haxell, Waterloo
P. Hliněný, DIMATIA
M. Inaba, Tokyo
V. Janota, DIMATIA
T. Kaiser, DIMATIA
M. Klazar, DIMATIA
A. Kostochka, Novosibirsk
J. Krajíček, DIMATIA
J. Kratochvíl, DIMATIA
L. Kučera, DIMATIA
L. Lovász, Budapest
T. Łuczak, Poznań
E. Luks, Eugene
W. Mader, Hannover
J. Matoušek, DIMATIA
F. Matúš, Bielefeld
P. Mihók, Košice
J. Nešetřil, DIMATIA
P. Ossona de Mendez, Paris
P. Pančoška, DIMATIA
G. Pete, Szeged
P. Pudlák, DIMATIA
F. Roberts, DIMACS

M. Rosenfeld, Seattle
G. Rote, Germany
J. Sgall, DIMATIA
M. Simonovits, Budapest
G. Simonyi, Budapest
W. Smyth, Ottawa
E. Sopena, Bordeaux
V. T. Sós, Budapest
J. Spencer, DIMACS
C. Tardif, Bielefeld
R. Thomas, Atlanta
W. Trotter, Bonn
D. Turzík, DIMATIA
P. Valtr, DIMATIA
S. Zhou, DIMACS

DIMACS Series in Discrete Mathematics
and Theoretical Computer Science
Volume **49**, 1999

Acyclic Improper Colourings of Graphs with Bounded Degree

Pierre Boiron, Eric Sopena, and Laurence Vignal

ABSTRACT. In this paper, we continue the study of acyclic improper colourings of graphs introduced in a previous work. An improper colouring of a graph G is a mapping c from the set of vertices of G to a set of colours such that for every colour i, the subgraph induced by the vertices with colour i satisfies some property depending on i. Such an improper colouring is acyclic if for every two distinct colours i and j, the subgraph induced by all the edges linking an i-coloured vertex and a j-coloured vertex is acyclic.

We consider in this paper the case of graphs with bounded degree. We prove some positive and negative results for graphs with maximum degree three and generalize some of the negative results to graphs with maximum degree k.

1. Introduction

Let G be a graph. We denote by $V(G)$ the vertex set of G and by $E(G)$ the edge set of G. A (proper) *k-colouring* of G is a mapping c from $V(G)$ to a set of k colours such that $c(x) \neq c(y)$ whenever xy is an edge in G. We say that G is *k-colourable* if G admits a k-colouring. The *chromatic number* $\chi(G)$ of G is defined as the minimum k such that G is k-colourable. A k-colouring of a graph G may be equivalently regarded as a partition $V_1, V_2, \ldots, V_k$ of $V(G)$ where for every i, $1 \leq i \leq k$, V_i is the set of vertices with colour i. Elements of such a partition are usually called *colour classes*.

Many variations and extensions of graph colourings have been considered in the literature (see **[8]** for a general overview). In particular *improper* colourings (sometimes called *generalized*, *defective* or *relaxed* colourings) have extensively been studied **[6, 15, 16]**. A colouring of a graph G is said to be improper whenever two adjacent vertices may be assigned the same colour. In such colourings, the adjacency constraint is replaced by some constraint on the structure of *monochromatic subgraphs*, that is subgraphs induced by colour classes.

Following Borowiecki and Mihók **[4]**, improper colourings may be defined as follows. Let $P_1, P_2, \ldots, P_k$ be graph properties. A *$(P_1, P_2, \ldots, P_k)$-partition* of a graph G is a partition $(V_1, V_2, \ldots, V_k)$ of $V(G)$ such that for every i, $1 \leq i \leq k$, the

1991 *Mathematics Subject Classification.* Primary 05C15.

This research was supported by the Barrande Grant #97137.

subgraph of G induced by V_i, denoted by $G[V_i]$, has property P_i. If a graph G has such a partition we say that G has property $P_1 \circ P_2 \circ \ldots \circ P_k$. If $P_1 = P_2 = \ldots = P_k = P$ we will denote by P^k the property $P_1 \circ P_2 \circ \ldots \circ P_k$. If $\mathcal{D}_0$ denotes the property to be without edges, then $\mathcal{D}_0^k$ is the property to be k-colourable.

In the following we will call $(P_1, P_2, \ldots, P_k)$-*colouring* of a graph G a mapping c from $V(G)$ to the set of colours $C = \{1, 2, \ldots, k\}$ such that the partition $(V_1, V_2, \ldots, V_k)$ given by $V_i = \{x \in V(G) : c(x) = i\}$, $1 \leq i \leq k$, is a $(P_1, P_2, \ldots, P_k)$-partition of G. Moreover, for every graph property P we will denote by $F(P)$ the family of all graphs satisfying P.

A proper k-colouring of a graph G is *acyclic* if for every two colours i and j the subgraph induced by $V_i \cup V_j$ has no cycle [**3, 7**]. The *acyclic chromatic number* of a graph G is then defined as the minimum k such that G has an acyclic proper k-colouring. This notion can be extended to the case of improper colourings as follows. Let $V_1, V_2, \ldots, V_k$ be a $(P_1, P_2, \ldots, P_k)$-partition of a graph G. Let us call a (i, j)-*edge* an edge linking a vertex in V_i to a vertex in V_j. Such a partition is said to be *acyclic* if for every i, j, $1 \leq i < j \leq k$, the subgraph induced by all the (i, j)-edges has no cycle. The reader should note here that the subgraph induced by any two colour classes may contain bichromatic cycles but that any such cycle has to contain at least one monochromatic edge. A bichromatic cycle with no monochromatic edge will be called an *alternating bichromatic cycle.* If a graph H has an acyclic $(P_1, P_2, \ldots, P_k)$-partition we say that G has property $P_1 \odot P_2 \odot \ldots \odot P_k$. If $P_1 = P_2 = \ldots = P_k = P$ we will denote by $P^{(k)}$ the property $P_1 \odot P_2 \odot \ldots \odot P_k$. Observe that $\mathcal{D}_0^{(k)}$ is the property of being acyclically (properly) k-colourable in the usual sense.

Acyclic improper colourings originally arose in the study of oriented colourings. An *orientation* of a graph G is a digraph obtained from G by giving to every edge of G one of the two possible orientations. A digraph is called an *oriented graph* if it is an orientation of some undirected graph. An *oriented k-colouring* of an oriented graph $\vec{G}$ is a mapping c from $V(\vec{G})$ to a set of k colours such that (i) for every arc $\vec{xy}$ in $E(\vec{G})$, $c(x) \neq c(y)$ (an oriented colouring is thus a proper colouring) and (ii) for every two arcs $\vec{xy}$ and $\vec{zt}$ in $E(\vec{G})$, $c(x) = c(t) \Longrightarrow c(y) \neq c(z)$. The *oriented chromatic number* $\vec{\chi}(\vec{G})$ of an oriented graph $\vec{G}$ is then defined as the minimum k such that $\vec{G}$ has an oriented k-colouring.

Oriented chromatic numbers have been studied in [**9, 11, 12, 13**]. In [**2**] we have proved the following result which gives a link between acyclic improper colourings and oriented colourings:

THEOREM 1.1 (Boiron *et al.*, 1997). *Let $P_1, P_2, \ldots, P_k$ be graph properties such that for every i, $1 \leq i \leq k$, every orientation of every graph in the family $F(P_i)$ has oriented chromatic number at most α_i. Then every orientation of every graph in $F(P_1 \odot P_2 \odot \ldots \odot P_k)$ has oriented chromatic number at most $2^{k-1} \times \sum_{i=1}^{k} \alpha_i$.*

Acyclic improper colourings of outerplanar and planar graphs have been considered in [**2**]. We continue here the study of acyclic improper colourings by considering the case of graphs with bounded degree. We will denote by $\mathcal{D}_k$ the property of having degree at most k and by $\mathcal{T}_1$ the property of being a forest. The main result concerning improper colourings of graphs with bounded degree is due to Lovàsz [**10**]:

THEOREM 1.2 (Lovàsz, 1966). *For every $p, q \geq 0$, $F(\mathcal{D}_{p+q+1}) \subseteq F(\mathcal{D}_p \circ \mathcal{D}_q)$.*

In the following, this result will be referred to as Lovàsz's Theorem.

The problem of determining the acyclic chromatic number of graphs with bounded degree appears to be more difficult and it is still an open question [**8**]. Alon *et al.* [**1**] proved the following result:

THEOREM 1.3 (Alon *et al.*, 1991). *For every k, there exist positive constants a_k and b_k such that:*
1. Every graph with maximum degree k has acyclic chromatic number at most $a_k \times k^{4/3}$.
2. There exist graphs with maximum degree k and acyclic chromatic number at least $b_k \times k^{4/3} \times (\log k)^{-1/3}$.

For small values of k, Grünbaum proved in [**7**] that every graph with maximum degree three has acyclic chromatic number at most four (this result is optimal). In [**5**] Burstein proved that every graph with maximum degree four has acyclic chromatic number at most five (he also mentionned that this result was independently obtained by Kostochka). Using our notation we thus get that $F(\mathcal{D}_3) \subseteq F(\mathcal{D}_0^{(4)})$ and $F(\mathcal{D}_4) \subseteq F(\mathcal{D}_0^{(5)})$.

We will consider acyclic improper colourings of graphs with bounded degree whose colour classes induce either forests or graphs with bounded degree. In the case of graphs with maximum degree three we know from Grünbaum's result that we only have to consider acyclic improper colourings using at most three colours. Moreover, since there exist infinitely many graphs with maximum degree three and acyclic chromatic number four we clearly get that $F(\mathcal{D}_3) \not\subseteq F(\mathcal{T}_1 \odot \mathcal{D}_0)$ (and thus $F(\mathcal{D}_3) \not\subseteq F(\mathcal{D}_1 \odot \mathcal{D}_0)$). In addition, we will prove the following negative results:

THEOREM 1.4.
1. $F(\mathcal{D}_3) \not\subseteq F(\mathcal{D}_0 \odot \mathcal{D}_2)$,
2. $F(\mathcal{D}_3) \not\subseteq F(\mathcal{D}_1 \odot \mathcal{D}_1)$.

Our second theorem gives some positive results:

THEOREM 1.5.
1. $F(\mathcal{D}_3) \setminus \{K_4, K_{3,3}\} \subseteq F(\mathcal{T}_1 \odot \mathcal{T}_1)$,
2. $F(\mathcal{D}_3) \subseteq F(\mathcal{T}_1 \odot \mathcal{D}_2)$,
3. $F(\mathcal{D}_3) \subseteq F(\mathcal{D}_1 \odot \mathcal{D}_1 \odot \mathcal{D}_1)$.

The general case of graph with maximum degree k seems to be quite difficult to handle. Nevertheless, we can prove the following negative results:

THEOREM 1.6.
1. For every $k \geq 2$, $F(\mathcal{D}_k) \not\subseteq F(\mathcal{D}_{k-2} \odot \mathcal{D}_{k-2})$,
2. For every p, q, $pq \neq 0$, $p + q \geq 4$, $F(\mathcal{D}_{p+q}) \not\subseteq F(\mathcal{D}_p \odot \mathcal{D}_q)$.

In the following sections we will prove our three theorems and we will end this paper by stating some open problems.

DRAWING CONVENTION. In all the forthcomming figures, grey (resp. black) vertices will represent vertices with colour 1 (resp. 2). Uncoloured vertices will be drawn as white vertices.

2. Negative results for graphs with maximum degree three

We prove in this section the negative statements of Theorem 1.4. In every case we will exhibit infinitely many graphs having no corresponding acyclic colouring.

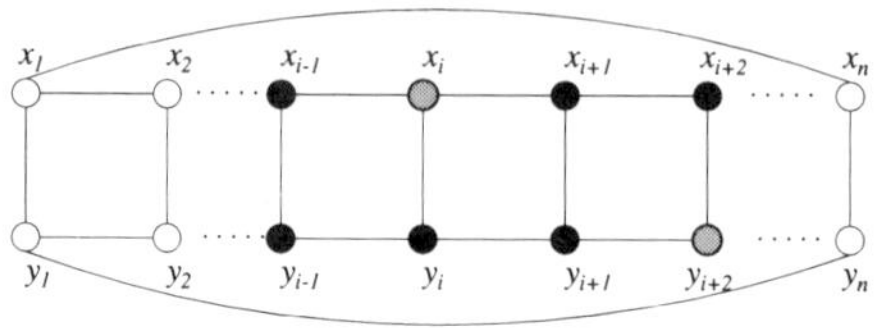

FIGURE 1. The circular grid $T(2,n)$

PROOF OF FIRST STATEMENT. Let $T(2\times n)$, $n\geq 3$, be the circular grid of dimension two with 2 rows of n vertices each (see Figure 1). Let $\{x_1,x_2,\dots,x_n\}$ (resp. $\{y_1,y_2,\dots,y_n\}$) denote the vertices of row 1 (resp. of row 2). Since every vertex has degree 3, at least one of them, say x_i, must get colour 1. Vertices x_{i-1},x_{i+1},y_i then necessarily get colour 2. In order to avoid alternating bichromatic cycles, vertices y_{i-1} and y_{i+1} must also get colour 2. Since they both have two neighbours already coloured with colour 2, vertices y_{i-2} and y_{i+2} must get colour 1 and so on. It is not difficult to observe that the pattern $\left(\begin{smallmatrix}2122\\2221\end{smallmatrix}\right)$ must be repeated all along the grid. Therefore, if $n\equiv 0 \pmod 4$, the colouring thus obtained is an acyclic $(\mathcal{D}_0,\mathcal{D}_2)$-colouring, if $n\equiv 1 \pmod 4$, the colour 1 does not induce a graph with no edges, if $n\equiv 2\pmod 4$, the colour 2 does not induce a graph with maximum degree two and if $n\equiv 3 \pmod 4$, we get an alternating bichromatic cycle. □

PROOF OF SECOND STATEMENT. Let G be a 3-regular graph and c a $(\mathcal{D}_1,\mathcal{D}_1)$-colouring of G (we know from Lovàsz's Theorem that such a colouring exists). We denote by B the bipartite subgraph of G defined by $V(B)=V(G)$ and $E(B)=\{xy\in E(G),\ c(x)\neq c(y)\}$. Since every vertex in B has at least two neighbours, B contains a cycle. Thus the colouring c cannot be acyclic. □

3. Positive results for graphs with maximum degree three

We prove in this section the positive statements of Theorem 1.5. Observe first that it is sufficient to prove these statements for 3-regular graphs. From Lovàsz's Theorem, we know that $F(\mathcal{D}_3)\subseteq F(\mathcal{T}_1\circ\mathcal{T}_1)$ and $F(\mathcal{D}_3)\subseteq F(\mathcal{T}_1\circ\mathcal{D}_2)$. In the following we will prove that we can modify every (not acyclic) such improper colouring in order to strictly decrease the number of alternating bichromatic cycles. By repeating this step we will get in each case an acyclic improper colouring.

Let $C=(x_0,x_1,\dots,x_{2n-1})$ be an alternating bichromatic cycle. For every i, $0\leq i\leq 2n-1$, let y_i be the neighbour of x_i distinct from x_{i-1} and x_{i+1} (the subscripts are always taken modulo $2n$). We say that C is of type I if there exists some i such that y_i and y_{i+1} have the same colour. We say that C is of type II if it is not of type I and for every i, x_i and y_i have distinct colours. Otherwise, we say that C is of type III (in that case, for every i, x_i and y_i have the same colour).

Unless otherwise specified, in the following pictures vertices not belonging to the alternating bichromatic cycle may be identified with vertices of the cycle provided they have the same color.

PROOF OF FIRST STATEMENT. Let G be a 3-regular graph, c a (not acyclic) $(\mathcal{T}_1,\mathcal{T}_1)$-colouring of G and C an alternating bichromatic cycle of smallest type in G. We have three cases to consider:

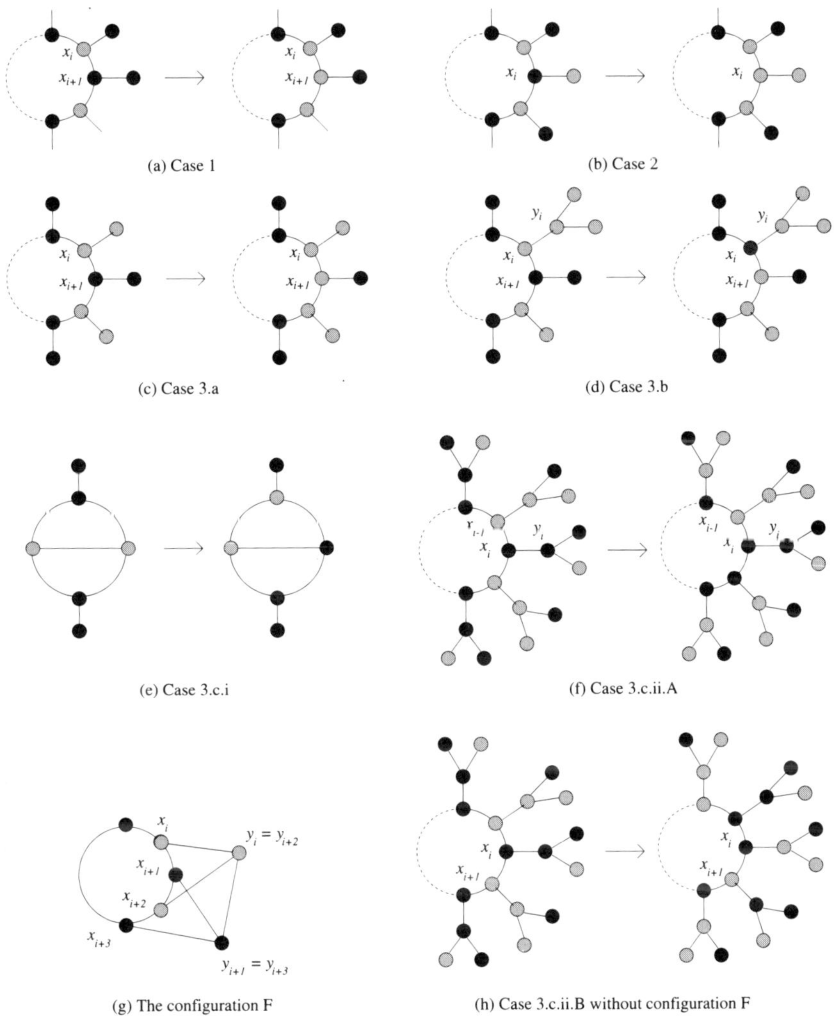

FIGURE 2. Deleting an alternating bichromatic cycle in a $(\mathcal{T}_1, \mathcal{T}_1)$-colouring

1. *C is of type I.* Let i be such that y_i and y_{i+1} have the same colour. If x_i and y_i have the same colour it suffices to change the colour of x_i, otherwise it suffices to change the colour of x_{i+1} (see Figure 2(a)).
2. *C is of type II.* It suffices to change the colour of any of the $x_i's$ (see Figure 2(b)).
3. *C is of type III.* We have three subcases to consider:
   (a) if there exists an index i such that x_i and x_{i+2} does not belong to the same tree (induced by their colour), it suffices to change the colour of x_{i+1} (see Figure 2(c)).
   (b) Otherwise, if there exists a vertex y_i with the same colour as its three neighbours, it suffices to change the colour of x_i and x_{i+1} (see Figure 2(d)).

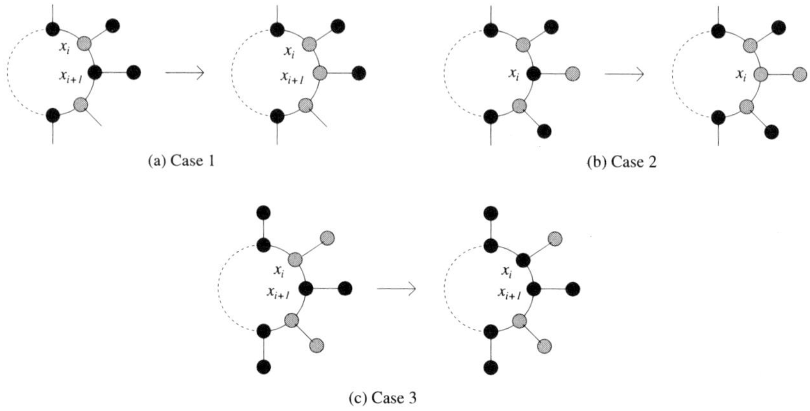

FIGURE 3. Deleting an alternating bichromatic cycle in a $(\mathcal{T}_1, \mathcal{D}_2)$-colouring

(c) if none of the previous subcases happens, we still have two cases to consider:
   (i) if some y_i belongs to C then C has length exactly 4, since x_i and y_i belong to the same tree (induced by their colour) and we are not in subcase 3(a). Since the graph G is not K_4, we necessarily have the configuration depicted by Figure 2(e) and it suffices to change the colours of two consecutive vertices of the cycle.
   (ii) if none of the y_i's belongs to C we again have two cases to consider:
      (A) if there exists some y_i such that y_i is distinct from all other y_j's and y_i has colour, say 2, it suffices then to change the colours of all the 1-coloured x_j's except x_{i-1} and of all the 2-coloured y_j's except y_i (see Figure 2(f)).
      (B) if every y_i is identified with some y_j, $j \neq i$, we denote by F the configuration depicted in Figure 2(g). If C does not contain the configuration F, it suffices to change the colours of all the y_j's and of all the x_j's except two consecutive ones, say x_i and x_{i+1} (see Figure 2(h)). If C contains the configuration F then, since we are not in subcase 3(a), the cycle C has exactly 4 vertices and the graph G is isomorphic to $K_{3,3}$, which is not possible.

□

PROOF OF SECOND STATEMENT. Let G be a 3-regular graph, c a (not acyclic) $(\mathcal{T}_1, \mathcal{D}_2)$-colouring of G and C an alternating bichromatic cycle of smallest type in G. We have three cases to consider:

1. *C is of type I.* Let i be such that y_i and y_{i+1} have the same colour. If x_i and y_i have the same colour it suffices to change the colour of x_i, otherwise it suffices to change the colour of x_{i+1} (see Figure 3(a)).
2. *C is of type II.* It suffices to change the colour of any of the x_i's having colour 2 (see Figure 3(b)).

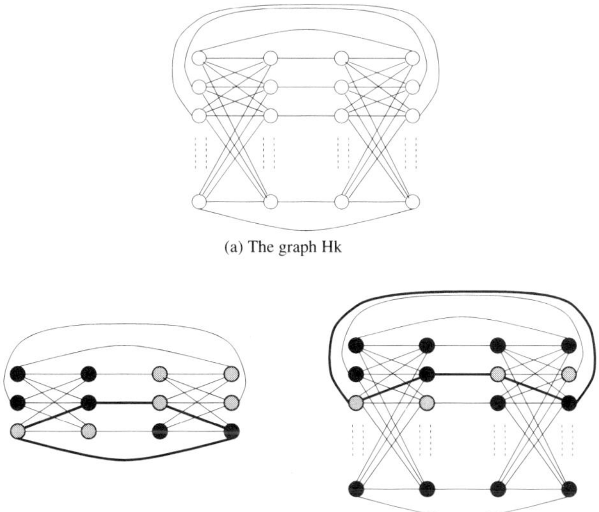

(a) The graph Hk

(b) The case p = q = 2

(c) The case p = 1, q = k-1

FIGURE 4. Construction for second statement of Theorem 1.6

3. *C is of type III.* It suffices to change the colour of any of the $x_i's$ having colour 1 (see Figure 3(c)).

□

PROOF OF THIRD STATEMENT. This statement can be proved by induction on the number of vertices of the graph. Due to the lack of space, we will not report the proof here. The complete proof can be found in [**14**]. □

4. Negative results for graphs with maximum degree k

We prove in this section the negative statements of Theorem 1.6. In every case we will exhibit infinitely many graphs having no corresponding acyclic colouring.

PROOF OF FIRST STATEMENT. Using the same technique we used in the proof of Theorem 1.4 (second statement), it is easy to show that every k-regular graph has no acyclic $(\mathcal{D}_{k-2}, \mathcal{D}_{k-2})$-colouring. □

PROOF OF SECOND STATEMENT. Let H_k^2, $k \geq 4$, be the graph obtained by cyclically connecting two copies of $K_{k-1,k-1}$, as depicted by Figure 4(a). Let us denote by $x_0^1, \dots, x_k^1$ and $x_0^2, \dots, x_k^2$ (resp. by $x_0^3, \dots, x_k^3$ and $x_0^4, \dots, x_k^4$) the vertices of the first (resp. the second) copy of $K_{k-1,k-1}$. A vertex x_j^i will be said to be of type i. Let c be any $(\mathcal{D}_p, \mathcal{D}_q)$-colouring of H_k^2, with $p+q=k$, $pq \neq 0$ (we know by Lovàsz's Theorem that such a colouring exists).

We first claim that each copy of $K_{k-1,k-1}$ contains at least two vertices of each colour. If all vertices of a given copy have the same colour, say 1, then $p = k-1$. All their neighbours in the other copy are thus 2-coloured and we get $p+q = 2k-2 > k$, a contradiction since $k \geq 4$. If all vertices but one, say x_0^1, in the first copy, have the same colour, say 1, then $p = k-1$ (all the x_j^1's, $j \neq 0$, have $k-1$ neighbours of type 2 with the same colour) and thus $q = 1$. But in that case, vertices $x_1^4, \dots, x_k^4$

must have colour 2 and thus all the x_j^3's must have colour 1. It is not difficult to check that we then necessarily have an alternating bichromatic cycle.

Our second claim is that for every type of vertex, one colour is used at most once. This is obvious if $k = 4$. If $k > 4$, suppose on the contrary that two vertices of type 1 are 1-coloured and two vertices of type 1 are 2-coloured. In that case, no two vertices of type 2 may have the same colour in order to avoid alternating bichromatic cycles. We thus get $k = 3$, a contradiction.

From these two claims we get that each copy of $K_{k-1,k-1}$ contains exactly two vertices of one colour, say 1. Colour 2 thus induces a graph of degree at least $q = k - 2$. Therefore, we have two cases to consider:

1. $p = 2$ and $q = k - 2$. The subgraph induced by the 2-coloured vertices in the first copy of $K_{k-1,k-1}$ has degree $k-2$. The neighbours of these vertices in the second copy are thus 1-coloured and we get $p = q = 2$ and thus $k = 4$. In that case, we necessarily have an alternating bichromatic cycle (see Figure 4(b)).
2. $p = 1$ and $q = k - 1$. Every 1-coloured vertex in a copy of $K_{k-1,k-1}$ is adjacent to a 2-coloured vertex in the second copy and we also get an alternating bichromatic cycle (see Figure 4(c)).

Thus H_k^2 has no acyclic $(\mathcal{D}_p, \mathcal{D}_q)$-colouring. It is not difficult to check that this is also the case for every graph H_k^p obtained by cyclically connecting p copies of $K_{k-1,k-1}$. □

5. Discussion and open problems

None of the positive results given in Theorem 1.5 improves the best known upper bound for the oriented chromatic number of oriented graphs with maximum degree three (this bound has been settled to 11 in [**14**] and it has been conjectured in [**13**] that every *connected* oriented graph with maximum degree three has oriented chromatic number at most 7).

We do not know whether every (or almost every) graph with maximum degree three admits an acyclic $(\mathcal{D}_0, \mathcal{D}_0, \mathcal{D}_1)$-, $(\mathcal{D}_0, \mathcal{D}_1, \mathcal{D}_1)$-, $(\mathcal{D}_1, \mathcal{D}_2)$- or $(\mathcal{T}_1, \mathcal{D}_1)$-colouring. If the answer to this last question is YES, this would decrease to 10 the upper bound for the oriented chromatic number of oriented graphs with maximum degree three. We know that at least three graphs do not admit an acyclic $(\mathcal{T}_1, \mathcal{D}_1)$-colouring, namely the Petersen graph and the cyclic grids $T(2 \times 3)$ and $T(2 \times 4)$.

We know by Lovàsz's Theorem that every graph in $\mathcal{D}_3$ admits a $(\mathcal{D}_2, \mathcal{D}_2)$-colouring. Concerning acyclic colourings, we have the following conjecture:

CONJECTURE 5.1. $F(\mathcal{D}_3) \subseteq F(\mathcal{D}_2 \odot \mathcal{D}_2)$.

The problem of acyclic improper colourings of graphs with degree at most k seems really difficult to handle. We have no idea of whether there exist or not such colourings whose colour classes induce graphs with bounded degree or forests.

ACKNOWLEDGMENT. The authors would like to thank Jean-Luc Fouquet and Bruce Reed for their helpful comments on the first version of this paper.

References

[1] N. Alon, C. McDiarmid and B. Reed, *Acyclic colorings of graphs.* Random Structures and Algorithms **2** (1991), 277–289.

[2] P. Boiron, E. Sopena and L. Vignal, *Acyclic improper colorings of graphs.* Manuscript, 1997.

[3] O.V. Borodin, *On acyclic colorings of planar graphs.* Discrete Math. **25** (1979), 211–236.
[4] M. Borowiecki and P. Mihók, *Hereditary properties of graphs.* In: V. R. Kulli, editor, Advances in Graph Theory. Vishwa International Publishers, 1991.
[5] M.I. Burstein, *Every 4-valent graph has an acyclic 5-coloring.* (in Russian), Soobšč. Akad. Nauk Gruzin. SSR **93** (1979), 21–24.
[6] L.J. Cowen, R.H. Cowen and D.R. Woodall, *Defective colorings of graphs in surfaces: Partitions into subgraphs of bounded valency.* J. Graph Theory **10** (1986), 187–195.
[7] B. Grünbaum, *Acyclic coloring of planar graphs.* Israel J. Math. **14** (1973), 390–412.
[8] T.R. Jensen and B. Toft, *Graph coloring problems.* Wiley Interscience, 1995.
[9] A.V. Kostochka, E. Sopena and X. Zhu, *Acyclic and oriented chromatic numbers of graphs.* J. Graph Theory **24** (1997), 331–340.
[10] L. Lovász, *On decomposition of graphs.* Studia Sci. Math. Hungar. **1** (1966), 237–238.
[11] J. Nešetřil, A. Raspaud and E. Sopena, *Colorings and girth of oriented planar graphs.* Discrete Math. **165/166** (1997), 519–530.
[12] A. Raspaud and E. Sopena, *Good and semi-strong colorings of oriented planar graphs.* Inf. Processing Letters **51** (1994), 171–174.
[13] E. Sopena, *The chromatic number of oriented graphs.* J. Graph Theory **25** (1997), 191–205.
[14] L. Vignal, *Homomorphismes et colorations de graphes orientés.* PhD Thesis (in French), Univ. Bordeaux I, December 1997.
[15] M.L. Weaver and D.B. West, *Relaxed chromatic numbers of graphs.* Graphs Combin. **10** (1994), 75–93.
[16] D.R. Woodall, *Improper colourings of graphs.* In: R. Nelson and R. J. Wilson, editors: Graph Colourings, vol. 218 of Pitman Research Notes in Mathematics Series, 45–63, Longman, 1990.

LaBRI, Université Bordeaux I, 351 cours de la Libération, 33405 Talence, France
E-mail address: `sopena@labri.u-bordeaux.fr`
URL: `http://dept-info.labri.u-bordeaux.fr/~sopena/`

DIMACS Series in Discrete Mathematics
and Theoretical Computer Science
Volume **49**, 1999

Intersection Graphs of Jordan Arcs

P. Ossona de Mendez and H. de Fraysseix

ABSTRACT. A family of Jordan arcs, such that two arcs are nowhere tangent, defines a hypergraph whose vertices are the arcs and whose edges are the intersection points. We shall say that the hypergraph has a *strong intersection representation* and, if each intersection point is interior to at most one arc, we shall say that the hypergraph has a *strong contact representation.*

We first characterize those hypergraphs which have a strong contact representation and deduce some sufficient conditions for a simple planar graph to have a strong intersection representation.

Then, using the Four Color Theorem, we prove that a large class of simple planar graphs have a strong intersection representation.

1. Introduction

A family of Jordan arcs, such that two arcs are nowhere tangent, defines a hypergraph whose vertices are the arcs and whose edges are the intersection points. We shall say that the hypergraph has a *strong intersection representation* and, if each intersection point is interior to at most one arc, we shall say that the hypergraph has a *strong contact representation.*

Such a family classically defines a *string graph*, which is a simple graph, whose vertices are the arcs, two vertices being adjacent if the corresponding arcs intersect at least once [**8**]. It is not known whether deciding whether a graph is a string graph is algorithmically decidable [**15**]. However, all planar graphs are obviously string graphs.

If two arcs may only intersect once, we shall say that the intersection graph has an *intersection representation.* Deciding whether a given graph has an intersection representation is known to belong to the NP-complete class [**13**], even if strong restrictions are requested on the family of arcs [**14**]. It has been a challenge for long to prove or disprove that all planar graphs have an intersection representation. It is clear that a simple graph has an intersection representation if and only if it has a strong intersection representation.

First, we shall introduce a general framework on strong contact representation.

Then, the introduction of a 4-coloration will lead to prove that a large class of planar graphs have an intersection representation.

Most of the results proved here were first presented in [**3**] and [**6**].

1991 *Mathematics Subject Classification*: Primary 05C10; Secondary 05B30.

This work was partially supported by the **Esprit LTR Project no 20244-ALCOM IT**.

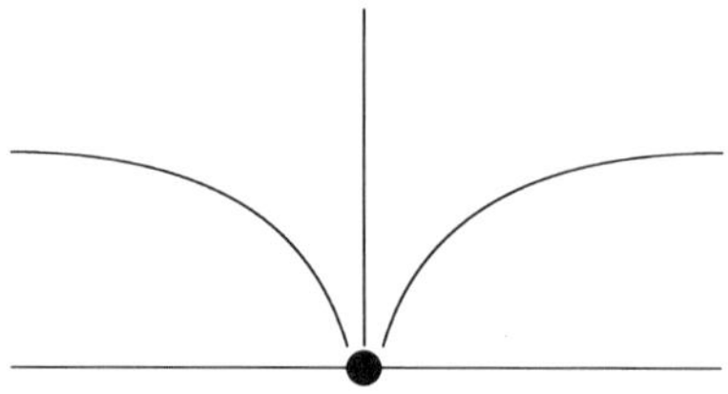

FIGURE 1. Locally one-sided contact

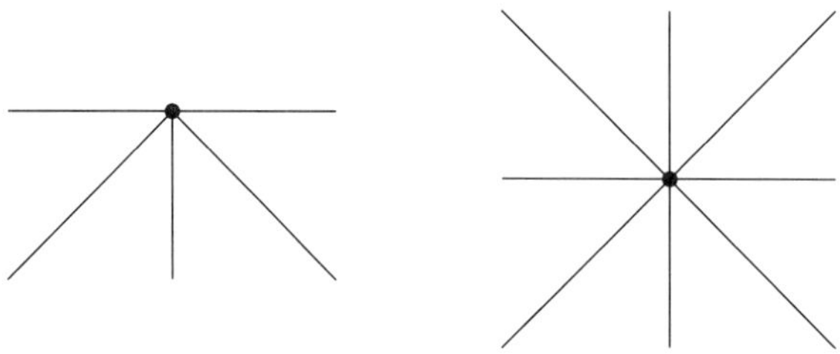

FIGURE 2. A contact and an intersection representations of K_4

GENERAL FRAMEWORK

Arc intersection problems arise topological difficulties that vanish when arcs are only in contact, as the arc-point incidence then defines a plane graph (i.e. a planar graph together with an embedding).

A *contact family* $\mathcal{F} = (\mathcal{V}, \mathcal{P})$ of Jordan arcs is a finite family $\mathcal{V}$ of Jordan arcs in the plane and a finite point set $\mathcal{P}$ on these arcs, such that two arcs share at most one point (called *contact point*), and such that all the contact points belong to $\mathcal{P}$ and are interior to at most one arc. Furthermore, at each point p interior to an arc ν, the arc ν appears twice consecutively in a clockwise traversal around p (i.e. we consider only locally one-sided contacts). This family defines a particular type of intersection graph, called *contact graph* (which may be not planar) and also a colored planar bipartite *arc-point incidence graph* $\text{Incid}(\mathcal{F})$, whose vertex set is the union of the arc set $\mathcal{V}$ (colored white) and the contact point set $\mathcal{P}$ (colored black), and whose incidence is the inclusion relation of a point in an arc.

2. Graphs, Hypergraphs and Frames

2.1. Graphs. Given a graph G, we use the following notations:

- $V(G)$ is the vertex set of G,
- $E(G)$ is the edge set of G,
- $N(x)$ is the neighbor set of the vertex x,
- $N(A) = \bigcup_{x \in A} N(x)$ is the neighbor set of a subset A of vertices,
- $d_G(x)$ is the degree of the vertex x,
- G_A is the subgraph induced by the subset A of vertices,

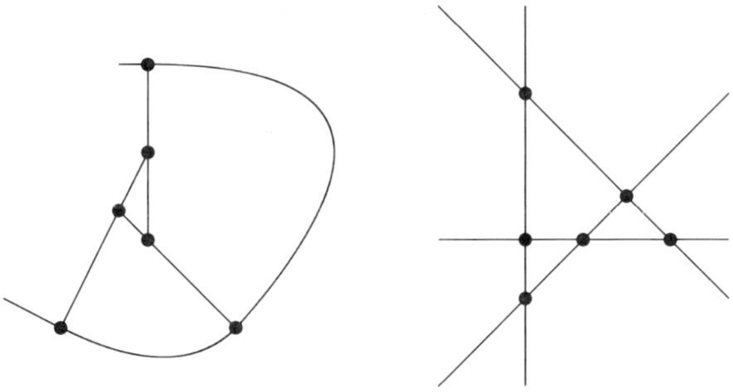

FIGURE 3. A strong contact and a strong intersection representations of K_4

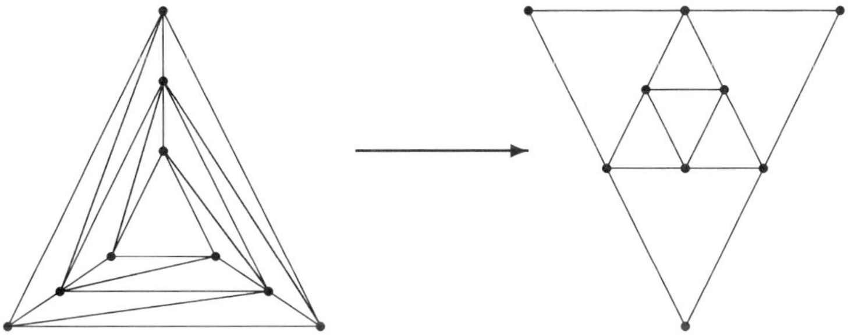

FIGURE 4. A planar graph, which has a contact representation, but no strong contact representation

- $T(G)$ is the set of bounded triangular faces of G, if G is a plane graph (i.e. a planar graph embedded in the plane).

  We shall also need the following definition, which is useful when dealing with indegree constrained plane orientations [**4**]:

DEFINITION 2.1. An *embracing cycle* is an elementary cycle with a non-empty interior and without interior chords.

DEFINITION 2.2. A graph G has a *strong contact representation* if there exists a contact family $\mathcal{F}$ such that the arc-point incidence of $\mathcal{F}$ is the vertex-edge incidence of G

REMARK 2.3. In a strong contact representation, the points represent the edges and thus belong to exactly 2 arcs, unlike the contact points of a contact representation which represent cliques.

2.2. Hypergraphs. A *hypergraph* is a pair $\mathcal{H} = (X, \mathcal{E})$, where X is a finite set and $\mathcal{E}$ is a family $(E_i, i \in I)$ of subsets of X, such that: $E_i \neq \emptyset (\forall i \in I)$ and $\bigcup_{i \in I} E_i = X$. The elements of X and $\mathcal{E}$ are respectively the *vertices* and the *edges* (or *hyperedges*) of the hypergraph. Two vertices $x, y \in X$ are *adjacent* if they both belong to some edge of $\mathcal{H}$; two edges E_i, E_j are *adjacent* if their intersection is not empty. A vertex $x \in X$ is *incident* to an edge $E_i \in \mathcal{E}$ if x belongs to E_i (see [**1**]). A *loop* is a single element edge.

A hypergraph is *linear* if any two edges have at most one common element:

$$\forall i \neq j, \quad \left|E_i \cap E_j\right| \leq 1$$

The *sub-hypergraph* of $\mathcal{H}$ *induced* by a subset $Y \subseteq X$ is the hypergraph $\mathcal{H}_Y = (Y, \mathcal{E}_Y)$, where

$$\mathcal{E}_Y = \{E_i \cap Y, \quad E_i \in \mathcal{E}; E_i \cap Y \neq \emptyset\}$$

DEFINITION 2.4. The *equivalent edge number* of $\mathcal{H}$ is the sum

$$\mu(\mathcal{H}) = \sum_{i \in I} \Big(\left|E_i\right| - 1\Big)$$

DEFINITION 2.5. The *incidence graph* $\operatorname{Incid}(\mathcal{H})$ of $\mathcal{H}$ is the colored bipartite graph on $(X, \mathcal{E})$ defined by the vertex-edge incidence, a vertex being colored white (resp. black) if it belongs to X (resp. $\mathcal{E}$).

As a special case, if G is a graph, $\operatorname{Incid}(G)$ is the bicolored vertex-edge incidence graph of G.

LEMMA 2.6. *A hypergraph $\mathcal{H}$ is linear if and only if* $\operatorname{Incid}(\mathcal{H})$ *is C_4-free.*

PROOF. The hypergraph $\mathcal{H}$ is linear if and only if no two edges of $\mathcal{H}$ have two common adjacent vertices, that is, if and only if $\operatorname{Incid}(\mathcal{H})$ is C_4-free. □

2.3. Constrained Orientations of Hypergraphs. Basic results on graph orientation may easily be extended to hypergraphs.

DEFINITION 2.7. An *out-orientation* of a hypergraph $\mathcal{H}$ is a function $\mathrm{h} : \mathcal{E} \to V$ such that, for any edge E of $\mathcal{H}$, $h(E)$ belongs to E. The vertex $\mathrm{h}(E)$ is the *head* of E. The complement $\operatorname{Tail}(E) = E \setminus \{\mathrm{h}(E)\}$ is the *tail set* of E.

An out-orientation of a hypergraph is obviously equivalent to an orientation of its incidence graph in which the vertices corresponding to edges have indegree 1.

DEFINITION 2.8. Let $\mathcal{H}$ be a hypergraph and let h be an out-orientation of it. The *indegree* $\mathrm{d}^-_{\mathcal{H}}(x)$ of a vertex x is the number of edges having x in their tail set.

DEFINITION 2.9. Let $\mathcal{H}$ be a hypergraph and h an out-orientation of it. A *directed path* P of $\mathcal{H}$ with respect to the out-orientation h is a sequence

$$P = (v_1, E_1, \ldots, v_{k-1}, E_{k-1}, v_k)$$

where v_i are vertices and E_i edges of $\mathcal{H}$ such that $\forall 1 \leq i < k, \quad v_i = \mathrm{h}(E_i)$ and $v_{i+1} \in \operatorname{Tail}(E_i)$.

LEMMA 2.10. *Let $\mathcal{H}$ be a hypergraph and let* h *be an out-orientation of it. Then:*

$$\mu(\mathcal{H}) = \sum_{x \in V} \mathrm{d}^-_{\mathcal{H}}(x) \tag{2.1}$$

PROOF. This follows from the following equality:

$$\sum_{x \in V} \mathrm{d}^-_{\mathcal{H}}(x) = \sum_{(x,E) \in V \times \mathcal{E}, x \in \operatorname{Tail}(E)} 1 = \sum_{E \in \mathcal{E}} |\operatorname{Tail}(E)| = \sum_{E \in \mathcal{E}} (|E| - 1) = \mu(\mathcal{H})$$

□

We shall extend a classical result on indegree bounded orientations (see [**10, 16**] or [**17**, pp 79–80]) to out-oriented hypergraphs:

LEMMA 2.11. *Let $\mathcal{H}$ be a hypergraph and $\lambda : V \to \mathbb{N}$ be any integral function on the vertices of $\mathcal{H}$. Then, there exists an out-orientation* h *of $\mathcal{H}$ such that*

$$\forall x \in V, \qquad d^-_{\mathcal{H}}(x) \leq \lambda(x) \tag{2.2}$$

if and only if

$$\forall A \subseteq V, \qquad \mu(\mathcal{H}_A) \leq \sum_{x \in A} \lambda(x) \tag{2.3}$$

PROOF. If the indegree of any vertex is bounded by λ then, for any induced sub-hypergraph $\mathcal{H}_A$, we get:

$$\mu(\mathcal{H}_A) = \sum_{E \in \mathcal{E}, E \cap A \neq \emptyset} (|E \cap A| - 1) \leq \sum_{E \in \mathcal{E}} |Tail(E) \cap A| = \sum_{x \in A} d^-_{\mathcal{H}}(x) \leq \sum_{x \in A} \lambda(x)$$

Conversely, assume that $\mu(\mathcal{H}_A) \leq \sum_{x \in A} \lambda(x)$ for all subsets A of vertices. Consider an out-orientation h, such that the value

$$C(\mathrm{h}) = \sum_{x \in V, d^-_{\mathcal{H}}(x) > \lambda(x)} (d^-_{\mathcal{H}}(x) - \lambda(x))$$

is minimal. If $C(\mathrm{h}) = 0$, then the out-orientation satisfies $d^-_{\mathcal{H}}(x) \leq \lambda(x)$ for all $x \in V$. Otherwise, there exists a vertex v such that $d^-_{\mathcal{H}}(v) > \lambda(v)$. Consider the set A of the vertices from which there exists a directed path to v. Then, for any edge E, if $\mathrm{Tail}(E)$ intersects A, then $\mathrm{h}(E)$ belongs to A and thus the restriction of h to A defines an out-orientation of $\mathcal{H}_A$ and the functions $d^-_{\mathcal{H}_A}$ and $d^-_{\mathcal{H}}$ take the same values over A. According to Lemma 2.3, $\mu(\mathcal{H}_A) = \sum_{x \in A} d^-_{\mathcal{H}_A}(x)$. By hypothesis, this value is bounded by $\sum_{x \in A} \lambda(x)$. As $d^-_{\mathcal{H}_A}(v) = d^-_{\mathcal{H}}(v) > \lambda(v)$, there exists at least a vertex w in V such that $d^-_{\mathcal{H}}(w) = d^-_{\mathcal{H}_A}(w) < \lambda(w)$. By hypothesis there exists a directed path $(x_1 = w, E_1, \ldots, E_{k_1}, x_k = v)$ from w to v. We define a new out-orientation h' equal to h except for the edges E_i in the directed path, for which we define $\mathrm{h}'(E_i) = x_i$ With respect to this new out-orientation, the only indegrees which have changed are those of v and w that respectively decreased and increased by 1. Thus, $C(\mathrm{h}') - C(\mathrm{h}) - 1$, what contradicts the minimality of $C(\mathrm{h})$. □

2.4. Frames and Hypergraph Planarity. Different generalizations of the concept of planarity to hypergraphs have been proposed (See [**11**], for instance). The first generalization is due to Zykov [**20**]. He proposed to represent the edges of a hypergraph by the faces of a planar map :

DEFINITION 2.12. A hypergraph $\mathcal{H} = (X, \mathcal{E})$ is *planar* if there is a multigraph $G = (X, E)$ (parallel edges allowed), a planar embedding M of G and an injection from $\mathcal{E}$ to the faces of M such that an edge of $\mathcal{H}$ is mapped into a face of M with the same set of incident vertices.

Walsh [**18**] has shown that Zykov's definition (as well as another definition due to Cori [**2**]) is equivalent to the following :

LEMMA 2.13 ([**18**]). *A hypergraph $\mathcal{H}$ is* planar *if and only if* $\mathrm{Incid}(\mathcal{H})$ *is planar.*

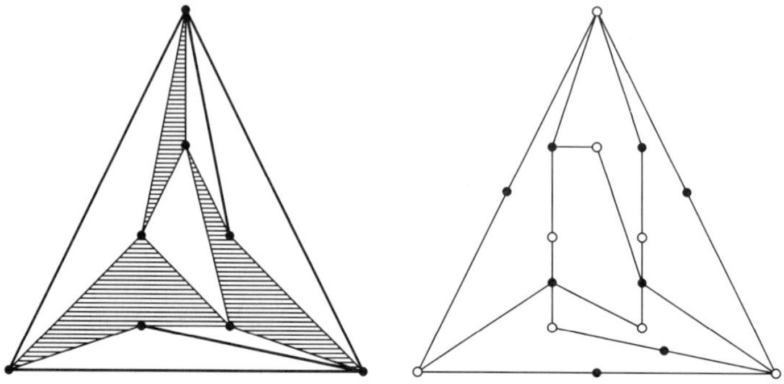

FIGURE 5. A frame representation of a planar hypergraph $\mathcal{H}$ and the incidence graph of $\mathcal{H}$ (the edges of $\mathcal{H}$ are the dashed faces and the thick line segments of the frame)

In [**19**], White shows that a hypergraph is planar if and only if there exists a face-bipartite plane graph whose "white" faces represents the edges of the hypergraph (see also [**12**]). The frame representation we propose here is nothing but this representation in which the "white" faces of length 2 are collapsed into a single edge. Then, the representation of a graph (as a special case of a hypergraph) is the graph itself.

DEFINITION 2.14. A *frame* is a pair $\Phi = (\Gamma, \Pi)$, where Γ is a plane multigraph and $\Pi = (F_1, \ldots, F_p)$ is a partition of the edge set $E(\Gamma)$, such that each class is either a 2-connected face of Γ of length at least 3 or a single edge (which may be a loop).

A vertex $x \in V(\Gamma)$ is incident to a class $F_i \in \Pi$ if it is incident to at least one edge in F_i.

A frame Φ defines a hypergraph $\mathcal{H}_\Phi = (V(\Gamma), \mathcal{E})$, where E_i is the set of the vertices incident to the class F_i of Π. We shall say that Φ is a *frame representation* of $\mathcal{H}_\Phi$ (see Fig. 5).

PROPOSITION 2.1. *A hypergraph $\mathcal{H} = (X, \mathcal{E})$ is planar if and only if it has a frame representation.*

PROOF. The bijection between the embedded incidence graph of a planar hypergraph and frame representations is shown on Fig. 6:

- From the incidence graph of a planar hypergraph $\mathcal{H}$, each black vertex (corresponding to some E_i) is split into a C_k (if its degree is $k > 2$), a K_2 (if its degree is 2), or a loop (if its degree is 1), which edges forms a set F_i. Then, the black-white edges are contracted. The resulting graph Γ together with the partition $\Pi = (F_1, \ldots, F_m)$ is a frame representation of $\mathcal{H}$. By construction, an edge of Γ belongs at most to a face in Π and Γ may be embedded in such a way that its unbounded face does not belong to Π.
- From a frame $\Phi = (\Gamma, \Pi)$, we add a vertex in each face in Π (including the faces defined by a loop) and join it to the vertices of the face, we bisect the non-loop edges that forms a single element class in Π and then delete all the original edges of Γ that have not been bisected.

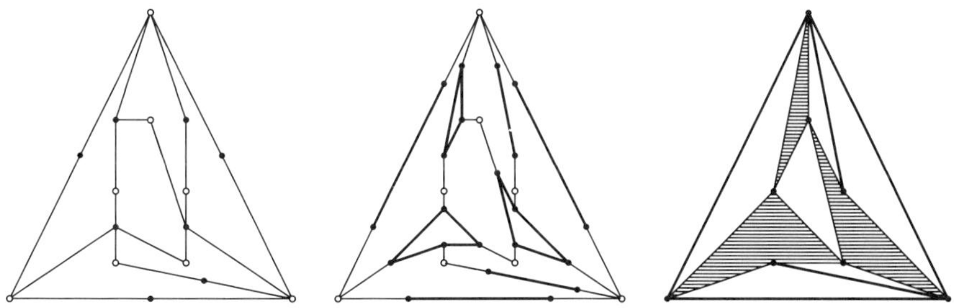

FIGURE 6. Bijection between incidence graphs of planar hypergraphs and frames

□

LEMMA 2.15. *If* (Γ, Π) *is a frame representation of a loop-less linear hypergraph* $\mathcal{H}$, *then* Γ *is a simple graph.*

PROOF. The loops of Γ corresponds exactly to the loops of $\mathcal{H}$. Hence, if $\mathcal{H}$ is loop-less, so is Γ.

Assume Γ has two parallel edges adjacent to vertices x and y. As these edges do not belong to a same class (they would define a face of length 2), $\mathcal{H}$ has two edges incident to both x and y what contradicts the linearity of $\mathcal{H}$. □

2.5. Frame Triangulations. Triangulations of frames allow to give an characterization of loop-less linear hypergraphs and a useful alternate expression for the equivalent edge number $\mu(\mathcal{H})$ of a hypergraph.

First, we introduce some definitions and notations related to triangulation:

Let $\mathcal{H}$ be a planar hypergraph and $\Phi = (\Gamma, \Pi)$ be a frame representation of $\mathcal{H}$.

- Let C be a polygon, then $\mathrm{Triang}(C)$ is the set of the maximal outer-planar graphs which may be obtained by triangulating the inside of C.
- Let E be an edge of $\mathcal{H}$. Then $\mathrm{Triang}(E) = \{E\}$ if E as cardinality at most two and $\mathrm{Triang}(E) = \mathrm{Triang}(C)$ if C is the polygon of Γ induced by the edges of the class of Π corresponding to the edge E of $\mathcal{H}$.
- $\mathrm{Triang}(\Phi)$ is the set of the plane multigraphs which may be obtained from Γ by triangulating the faces belonging to Π.
- For any plane graph $G \in \mathrm{Triang}(\Phi)$, $\mathrm{T}_\Phi(G)$ denotes the set of all the triangular faces of G which either are not faces of Γ or belong to Π.

  For any induced plane subgraph H of a plane graph $G \in \mathrm{Triang}(\Phi)$, $\mathrm{T}_\Phi(H)$ is the set of all the triangular faces of H which belong to $\mathrm{T}_\Phi(G)$. Notice that this set does not depend on the choice of G in $\mathrm{Triang}(\Phi)$.

LEMMA 2.16. $\mathcal{H}$ *is a loop-less linear hypergraph if and only if all the graphs in* $\mathrm{Triang}(\Phi)$ *are simple.*

PROOF. As the loops of $\mathcal{H}$ corresponds exactly of the loops of Γ, we only have to pay attention to possible multiple edges.

- If $\mathcal{H}$ is not linear, there exists two edges E, E' of $\mathcal{H}$ having at least two vertices x and y in common. Hence, there exists a multigraph in $\mathrm{Triang}(\Phi)$

such that x an y are linked by an edge in $\text{Triang}(E)$ and by an edge induced in $\text{Triang}(E')$.

- If $\mathcal{H}$ is linear, two vertices may simultaneously belong to at most one edge of $\mathcal{H}$ and hence are linked by at most one edge in a graph belonging to $\text{Triang}(\Phi)$.

□

LEMMA 2.17. *Let C be a polygon. Then, for any subset $W \subseteq V(C)$, we have:*

$$\max_{G\in\text{Triang}(C)} \left|E(G_W)\right| - \left|T(G_W)\right| = \begin{cases} \left|W\right| - 1 & \text{if } W \neq \emptyset \\ 0 & \text{otherwise} \end{cases} \tag{2.4}$$

PROOF. The number $\left|T(G_W)\right|$ of bounded triangular faces of G_W is one less than the number of faces of G_W. Hence, $\left|E(G_W)\right| - \left|T(G_W)\right|$ is equal to $\left|W\right| - c(G_W)$, where $c(G_W)$ is the number of connected components of G_W. If W is nonempty, this number is at most $\left|W\right| - 1$, and this value is achieved by any triangulation G of C, such that G_W is connected. □

LEMMA 2.18. *Let $\Phi = (\Gamma, \Pi)$ be a frame representation of a hypergraph $\mathcal{H}$. Let $W \subseteq V$ be a subset of the vertex set of Γ. Then,*

$$\mu(\mathcal{H}_W) = \max_{G\in\text{Triang}(\Phi)} \left(\left|E(G_W)\right| - \left|\mathrm{T}_\Phi(G_W)\right|\right) \tag{2.5}$$

PROOF. Let Γ_i be the partial graph of Γ induced by (edges of) $F_i \in \Pi$. Let W_i be the subset of W of the vertices incident to F_i. Any graph $G \in \text{Triang}(\Phi)$ is uniquely defined by the list of the graphs $G^{(i)} \in \text{Triang}(\Gamma_i)$.

The number $\left|E(G_W)\right|$ of edges of the subgraph of G induced by W is the sum of the numbers $\left|E(G^{(i)}_{W_i})\right|$ of edges of the subgraphs of the $G^{(i)}$ induced by W_i.

Similarly, the number $\left|\mathrm{T}_\Phi(G_W)\right|$ is the sum of the numbers $\left|T(G^{(i)}_{W_i})\right|$ of triangles of the subgraphs of the $G^{(i)}$ induced by W_i, as a triangle belongs to $\mathrm{T}_\Phi(G_W)$ if and only if its vertices belongs to the same Γ_i.

Hence,

$$\mu(\mathcal{H}_W) = \sum_{i, W_i\neq\emptyset} \left(\left|W_i\right| - 1\right) \tag{2.6}$$

$$= \sum_i \left(\max_{H\in\text{Triang}(\Gamma_i)} \left(\left|E(H_{W_i})\right| - \left|T(H_{W_i})\right|\right)\right) \tag{2.7}$$

$$= \max\left\{\sum_i \left(\left|E(G^{(i)}_{W_i})\right| - \left|T(G^{(i)}_{W_i})\right|\right), \quad \forall j, G^{(j)} \in \text{Triang}(\Gamma_j)\right\} \tag{2.8}$$

$$= \max_{G\in\text{Triang}(\Phi)} \left(\left|E(G_W)\right| - \left|\mathrm{T}_\Phi(G_W)\right|\right) \tag{2.9}$$

□

3. Strong Contact Representation

3.1. Strong contact representation of graphs.

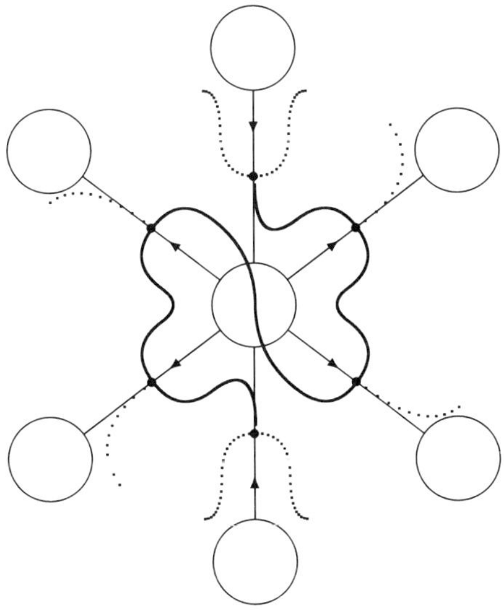

FIGURE 7. A contact representation induced by a 2-orientation

THEOREM 3.1. *A graph G has a strong contact representation if and only if it is planar and satisfies:*

$$\forall A \subseteq V, \quad \left|E(G_A)\right| \leq 2\left|A\right| \tag{3.1}$$

PROOF. Equation (3.1) is equivalent to the existence of a 2-orientation of G, that is an orientation of G such that each vertex has at most 2 incoming edges.

- A strong contact representation of a graph G induces a planar embedding of G and a 2-orientation of it: an edge incident to a vertex v is oriented from v if it corresponds to an internal point of the arc ν representing v; edges corresponding to point which are internal to no arc are oriented arbitrarily.
- Conversely, an embedding and a 2-orientation of G defines a strong contact representation of G (see Fig. 7).

□

REMARK 3.2. This theorem implies that outer-planar graphs and triangle free planar graphs have a strong contact representation. Actually, bipartite planar graphs have a strong contact representation using segments in two directions [**7**].

3.2. Strong contact representation of hypergraphs. The definition of strong contact representation of graphs extends naturally to hypergraphs:

DEFINITION 3.3. A hypergraph $\mathcal{H}$ has a *strong contact representation* if there exists a contact family $\mathcal{F}$, whose arc-point incidence is the vertex-edge incidence of $\mathcal{H}$, so that arcs represents vertices and points represent edges (see Fig. 8).

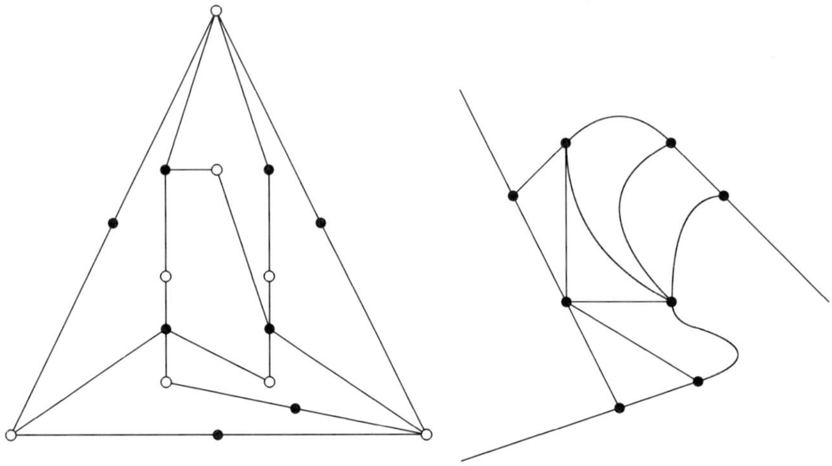

FIGURE 8. The incidence graph Incid($\mathcal{H}$) of a planar hypergraph $\mathcal{H}$ and a strong contact representation of $\mathcal{H}$

THEOREM 3.4. *A hypergraph $\mathcal{H}$ has a strong contact representation if and only if $\mathcal{H}$ is planar and satisfies*

$$\forall A \subseteq X, \quad \mu(\mathcal{H}_A) \leq 2|A| \tag{3.2}$$

PROOF. According to Lemma 2.11, Equation (3.2) is equivalent to the existence of an out-orientation of $\mathcal{H}$ with indegrees bounded by 2, which in turn is equivalent to the existence of an orientation of Incid($\mathcal{H}$) in which the vertices corresponding to vertices of $\mathcal{H}$ have indegree at most 2 and those corresponding to edges of $\mathcal{H}$ have indegree 1.

The theorem follows from the following arguments:

- A strong contact representation of a hypergraph $\mathcal{H}$ induces a planar embedding of Incid($\mathcal{H}$) and an out-orientation h of $\mathcal{H}$ with indegrees bounded by 2: h(E) is any mapping such that if the point representing E is interior to the arc representing v then h(E) $= v$.
- Conversely, if $\mathcal{H}$ is a planar hypergraph, an embedding and an orientation of Incid($\mathcal{H}$) in which the vertices corresponding to vertices of $\mathcal{H}$ have indegree at most 2 and those corresponding to edges of $\mathcal{H}$ have indegree 1 defines a strong contact representation of $\mathcal{H}$ (see Fig. 9).

□

According to Lemma 2.18, we deduce the following corollary:

COROLLARY 3.4.1. *Let $\Phi = (\Gamma, \Pi)$ be a frame representation of a planar hypergraph $\mathcal{H}$. Then, $\mathcal{H}$ has a strong contact representation if and only if the following inequality holds:*

$$\forall G \in \mathrm{Triang}(\Phi), \forall W \subseteq V(\Gamma) \qquad |E(G_W)| - |\mathrm{T}_\Phi(G_W)| - 2|W| \leq 0 \tag{3.3}$$

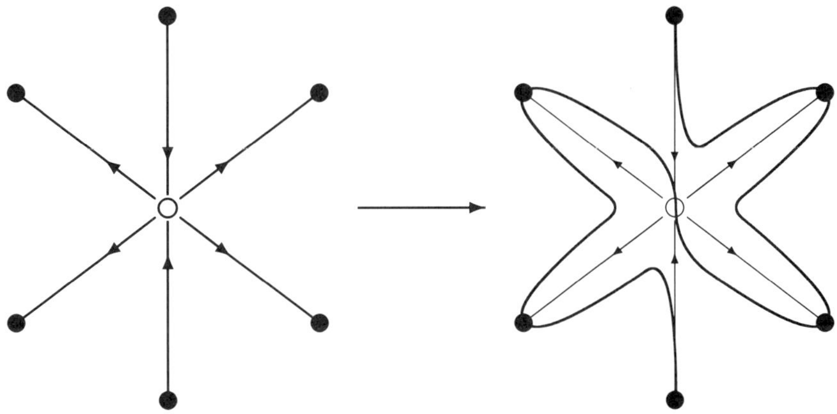

FIGURE 9. Strong contact representation of a 2-oriented planar hypergraph

4. Local Intersections

Let H be a maximal outer-planar graph and let $\mathcal{F}$ be a family of $\left|V(H)\right|$ arcs which share a single contact point p.

The graph H is a *local intersection graph* if, for any vertex v of H, there exists a local deformation of $\mathcal{F}$ at p which represents H, and such that

- the arc to which p is interior represents v,
- the circular order of the arcs is the same as the circular order of the vertices they represent.

A *zig-zag graph* is a maximal outer-planar graph which interior edges defines a path.

LEMMA 4.1. *A zig-zag graph is a local intersection graph.*

PROOF. The construction of a local deformation of a contact point into a zig-zag graph is shown on Fig. 10. □

REMARK 4.2. One can prove the reverse, that is: a maximal outer-planar graph is a local intersection graph if and only if it is a zig-zag graph.

Let $\mathcal{Z}(\Phi)$ denotes the set of all possible graphs obtained from Γ by triangulating the faces in Π using paths, and let $K(\Phi)$ be the graph obtained from Γ by replacing each face in Π by a clique.

THEOREM 4.3. *Let $\Phi = (\Gamma, \Pi)$ be a frame.*
If the following condition holds:

$$\forall G \in \mathrm{Triang}(\Phi), \forall W \subseteq V(\Gamma) \qquad \left|E(G_W)\right| - \left|\mathrm{T}_\Phi(G_W)\right| - 2\left|W\right| \leq 0 \tag{4.1}$$

then:

- *$K(\Phi)$ has a contact representation,*
- *any graph in $\mathcal{Z}(\Phi)$ has an intersection representation.*

PROOF. The conclusion follows from the previous lemma and Corollary 3.4.1. □

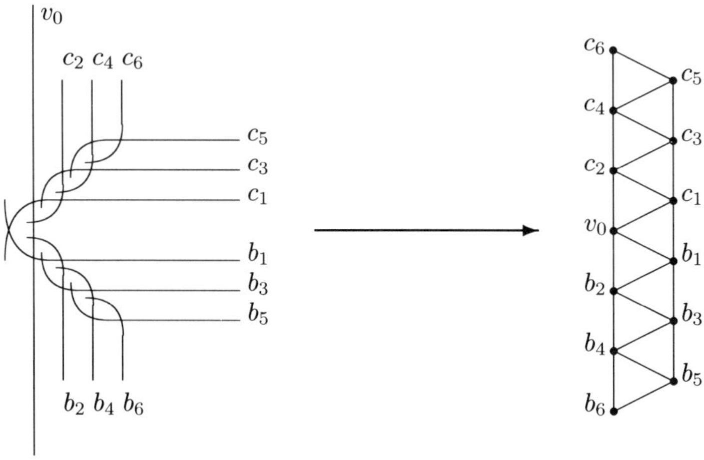

FIGURE 10. The local deformation of a contact point representing a zig-zag graph

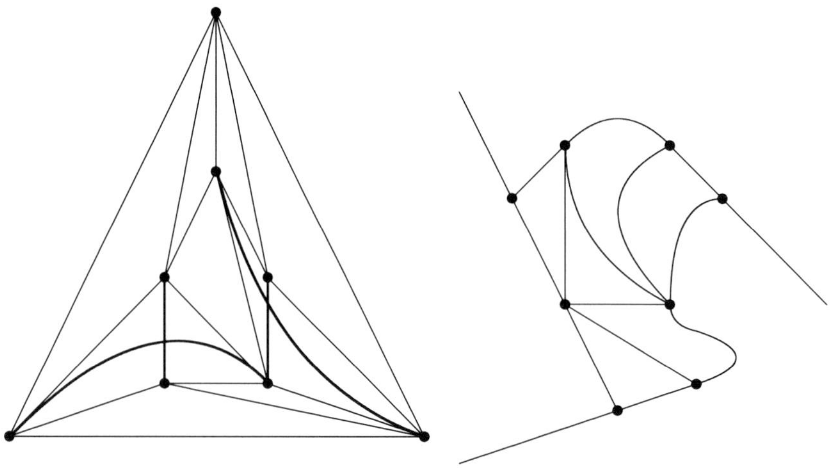

FIGURE 11. A contact representation of $K(\Phi)$

By considering the hypergraph represented by the frame $\Phi = (\Gamma, \Pi)$ where Π is derived from an independent (i.e. edge-disjoint) set of triangular faces of a plane graph Γ by $\Pi = \mathcal{I} \cup \big\{\{e\}, \quad e \in E \setminus (\bigcup_{T\in\mathcal{I}} T)\big\})$, we get:

COROLLARY 4.3.1. *Let $\Gamma(V, E)$ be a plane graph and let $\mathcal{I}$ be an independent (i.e. edge-disjoint) set of triangular faces. If the following condition holds:*

$$\forall W \subseteq V \qquad \big|E(\Gamma_W)\big| - \big|T(\Gamma_W) \cap \mathcal{I}\big| - 2\big|W\big| \leq 0 \tag{4.2}$$

then Γ has a contact representation.

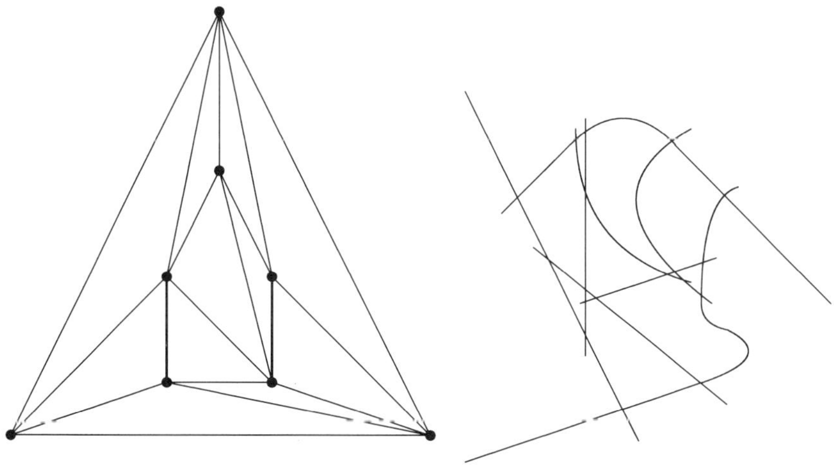

FIGURE 12. A strong intersection representation of a graph $G \in \mathcal{Z}(\Phi)$

REPRESENTATION OF PLANAR GRAPHS

5. The Coloration Method

Let G be a 2-connected 4-colored plane graph. Let $\mathrm{T}^+(G)$ be the set of all the bounded triangular faces of G in which two consecutive colors appear clockwise.

First notice that a triangle belonging to $\mathrm{T}^+(G)$ is adjacent to at most one other triangle belonging to $\mathrm{T}^+(G)$. More precisely, two bounded adjacent triangles both belong to $\mathrm{T}^+(G)$ if and only if the vertices of their union are colored $(1, 2, 3, 4)$ in clockwise order.

A frame $\Phi = (\Gamma, \Pi)$ is obtained from G and $\mathrm{T}^+(G)$:

- The graph Γ is the graph obtained from G by erasing the edges belonging to two triangles belonging to $\mathrm{T}^+(G)$.
- The partition Π has the following classes:
  - $\{e\}$, if the edge e belongs to no triangle in $\mathrm{T}^+(G)$,
  - T, if the triangle T belongs to $\mathrm{T}^+(G)$ and no triangle adjacent to it belongs to $\mathrm{T}^+(G)$,
  - $(T_1 \cup T_2) \setminus (T_1 \cap T_2)$, if the triangles T_1 and T_2 are adjacent and both belong to $\mathrm{T}^+(G)$.

LEMMA 5.1. *If G has no induced C_4 colored $(1, 2, 3, 4)$ and no embracing C_3 (see Definition 2.1) with two consecutive colors appearing clockwise, then all the multigraphs in* Triang(Φ) *are simple graphs.*

PROOF. Assume the hypergraph $\mathcal{H}$ represented by Φ is not linear. According to Lemma 2.16, there exists a graph H in Triang(Φ) with multiple edges. These edges either belong to G or come from the triangulation of a face of length 4 colored $(1, 2, 3, 4)$ in clockwise order. As G is simple, at least one edge corresponds to the later case. If there exists in G a face of length 4 colored $(1, 2, 3, 4)$ in clockwise order and an external chord of this cycle, then, G has an embracing C_3 with two consecutive colors appearing clockwise, which contradicts the hypothesis. Otherwise, there exists in G two faces of length 4 (u, v, w, x) and (u', v, w', x) colored

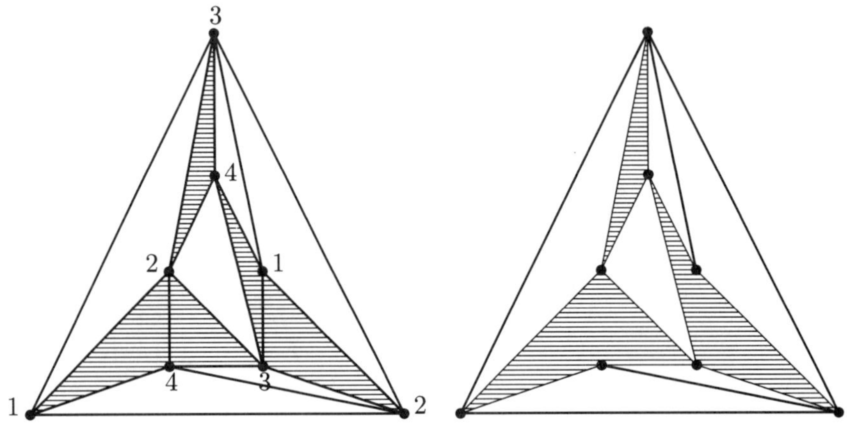

FIGURE 13. The 4-colored plane graph G (the dashed triangles are the members of $\mathrm{T}^+(G)$) and the associated frame

$(1,2,3,4)$ in clockwise order, where v and x are not adjacent in G. Then, either (u,v,w',x) or (u',v,w,x) is an embracing C_4 colored $(1,2,3,4)$ in clockwise order. We may assume it is (u,v,w',x). By hypothesis, this cycle may not be an induced cycle. Moreover, v and x are not adjacent. Hence u and w' are adjacent and define with v or x an embracing C_3 with two consecutive colors appearing clockwise, what contradicts the hypothesis. □

LEMMA 5.2. *If G has no induced C_4 colored $(1,2,3,4)$ and no embracing C_3 with two consecutive colors appearing clockwise, then all the graphs in* Triang(Φ) *have the same property.*

PROOF. All the graphs in Triang(Φ) may be obtained from G by a sequence of edge-switchings performed on edges not in $E(G)$. These edges are adjacent to vertices colored 1 and 3, or 2 and 4. Hence, such a switching cannot create an induced C_4 colored $(1,2,3,4)$.

If such a switching creates an embracing C_3 with two consecutive colors appearing clockwise (say v_3, v_1, v_2, where the edge (v_3, v_1) comes from the switching of an edge (x_2, x_4)), then the original graph had (v_3, x_4, v_1, v_2) as an induced C_4. □

LEMMA 5.3. *Assume G has no induced C_4 colored $(1,2,3,4)$ and no embracing C_3 with two consecutive colors appearing clockwise. Then, for any $G \in$* Triang(Φ), *and for any induced subgraph $H \subseteq G$, we have:*

$$\mathrm{T}^+(H) = \mathrm{T}_\Phi(H) \tag{5.1}$$

PROOF. This is a direct consequence of the fact that G has no embracing C_3 with two consecutive colors appearing clockwise. □

LEMMA 5.4. *If G has no induced C_4 colored $(1,2,3,4)$ in the clockwise order, we have:*

$$\big|E(G)\big| - \big|\mathrm{T}^+(G)\big| - 2\big|V(G)\big| \leq -3 \tag{5.2}$$

PROOF. Let us prove the lemma by induction on the size sequence of the faces.

A face-sequence is minimal if each face has length 3 and hence corresponds to a maximal planar graph G. Then, any triangle of G includes exactly one edge colored $(1,2)$ or $(3,4)$ and such an edge belongs to two adjacent triangles. This matching shows that $\mathrm{T}^+(G)$ includes either the half of the triangles of G or the half the triangles of G minus 1 (depending on the coloration of the outer face which never belongs to $T^+(G)$). Hence:

$$\begin{aligned}|E(G)| - |\mathrm{T}^+(G)| - 2|V(G)| &\le \Big(3|V(G)| - 6\Big) - \Big(\frac{1}{2}\Big(2|V(G)| - 4\Big) - 1\Big) - 2|V(G)| \\ &\le -3\end{aligned}$$

Otherwise, we shall construct a 4-colored 2-connected plane graph G' with no face of length 4 colored $(1,2,3,4)$ in the clockwise order that satisfies (together with the associated subset $\mathrm{T}^+(G')$ of its face set):

$$|E(G')| - |\mathrm{T}^+(G')| - 2|V(G')| \ge |E(G)| - |\mathrm{T}^+(G)| - 2|V(G)| \tag{5.3}$$

Let F be a non-triangular face of G of maximal size. We shall consider several cases:

- There exists a color c, such that 4 vertices of F at least are not colored c.

  Then G' is obtained from G by adding a vertex x in F colored c, which is linked to all the $k \ge 4$ vertices of F, which are not colored c. We get:

$$\begin{aligned}|V(G')| &= |V(G)| + 1\\ |E(G')| &= |E(G)| + k\\ |\mathrm{T}^+(G')| &\le |T(G)| + \left\lfloor \frac{2k}{3} \right\rfloor\\ |E(G')| - |\mathrm{T}^+(G')| - 2|V(G')| &\ge \left\lceil \frac{k}{3} \right\rceil - 2 + |E(G)| - |\mathrm{T}^+(G)| - 2|V(G)|\\ &\ge |E(G)| - |\mathrm{T}^+(G)| - 2|V(G)|\end{aligned}$$

  Remark that no face of length 4 colored $(1,2,3,4)$ in clockwise order may be created that way.

- Otherwise, no color is present twice on F. If $V_i(F)$ denotes the set of the i-colored vertices of F and l denotes the length of F, then:

$$4 \le l = \Big|\bigcup_i V_i(F)\Big| = \frac{1}{3}\sum_i \Big|\bigcup_{j\ne i} V_j(F)\Big| \le 4$$

  Hence, F has length 4.

  We triangulate F by adding an edge and, as F is not colored $(1,2,3,4)$ in clockwise order, at most one additional triangle is added to $\mathrm{T}^+(G')$. Hence,

$$|E(G')| - |\mathrm{T}^+(G')| - 2|V(G')| \ge |E(G)| - |\mathrm{T}^+(G)| - 2|V(G)| \tag{5.4}$$

□

LEMMA 5.5. *If G has no induced C_4 colored $(1,2,3,4)$ in the clockwise order and no embracing C_3 with two consecutive colors appearing clockwise, we have:*

$$\max_{W\subseteq V(G)} |E(G_W)| - |\mathrm{T}^+(G_W)| - 2|V(G_W)| \le -3 \tag{5.5}$$

PROOF. Let us first prove that the inequality

$$|E(H)| - |\mathrm{T}^+(H)| - 2|V(H)| \leq -3 \tag{5.6}$$

holds for any induced 2-connected subgraph $H \subseteq G$

Actually, let $\mathrm{T}^+(H)$ be the set of all the bounded triangles of H in which two consecutive colors appear clockwise. As G has no embracing C_3 with two consecutive colors appearing clockwise, $\mathrm{T}^+(H) = T(H) \cap \mathrm{T}^+(G)$. Moreover, H has no face of length 4 colored $(1, 2, 3, 4)$ in clockwise order as it would be an induced 4-cycle of G.

Hence, by the preceding lemma, any 2-connected induced subgraph H of G satisfies (5.6). From this inequality, we deduce that the same holds for any induced subgraph H of G (two components share at most one vertex). □

THEOREM 5.6. *Let G be a 4-colored 2-connected plane graph with no induced 4-cycle colored $(1, 2, 3, 4)$ in the clockwise order and no embracing C_3 with two consecutive colors appearing clockwise.*

Then, G has an intersection representation.

PROOF. From the previous lemmas, we get

$$\max_{G \in \mathrm{Triang}(\Phi), W \subseteq V(\Gamma)} |E(G_W)| - |\mathrm{T}_\Phi(G_W)| - 2|W| \leq -3 \tag{5.7}$$

and, as G belongs to $\mathcal{Z}(\Phi) = \mathrm{Triang}(\Phi)$, the results follows. □

COROLLARY 5.6.1. *Let G be an internally 5-connected planar graph. Then, G has an intersection representation.*

THEOREM 5.7. *Let G be a 3-colored 2-connected plane graph with no embracing C_3 having two consecutive colors appearing clockwise,*

Then, G has a contact representation.

PROOF. As previously, we get

$$\max_{G \in \mathrm{Triang}(\Phi), W \subseteq V(\Gamma)} |E(G_W)| - |\mathrm{T}_\Phi(G_W)| - 2|W| \leq -3 \tag{5.8}$$

and, as G is equal to $K(\Phi)$, the results follows. □

COROLLARY 5.7.1. *Let G be a 4-connected 3-colorable planar graph. Then, G has a contact representation.*

6. Cut and Paste

By cutting C_3-separated components, reversing the color order if necessary to avoid an external triangle with two consecutive colors appearing clockwise and pasting back their representations we obtain the following:

THEOREM 6.1. *Let G be a 4-colored plane graph with no induced C_4 having vertices using all the 4 colors.*

Then, G has an intersection representation.

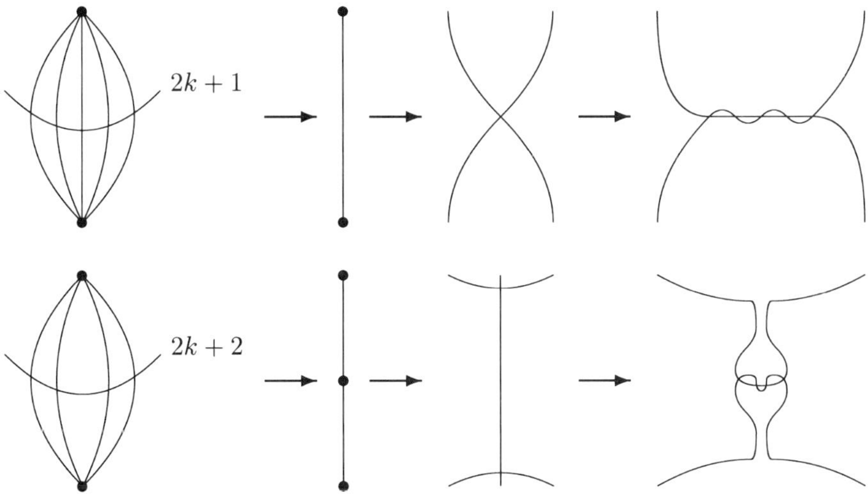

FIGURE 14. The existence of a strong intersection representation for all simple planar graphs is equivalent to the existence of a strong intersection representation for all planar multigraphs

7. Conclusion

In this paper, we have shown that a large class a planar graphs have an intersection representation. However, the general problem is not solved:

PROBLEM 1. Has any planar simple graph an intersection representation?

DEFINITION 7.1. A hypergraph $\mathcal{H}$ (or a multigraph G) has a *strong intersection representation* if there exists a family $\mathcal{F}$ of Jordan arcs, whose arc-point incidence is the vertex-edge incidence of $\mathcal{H}$ (or G): arcs represents vertices and points represent edges.

PROBLEM 2. Has any planar multigraph a strong intersection representation?

Actually, the problems 1 and 2 are equivalent:

PROPOSITION 7.1. *Any planar simple graph has an intersection representation if and only if any planar multigraph has a strong intersection representation.*

PROOF. If any planar multigraph has a strong intersection representation, any planar simple graph has indeed an intersection representation.

Conversely, if any planar simple graph has an intersection representation, any planar simple graph has also a strong intersection representation obtained by a trivial local deformation of the arcs at the intersection points. If a planar multigraph G has multiple edges, we consider the planar simple graph G' obtained from G by replacing any edge set E_i linking two vertices x and y by:

- a single edge $\{x, y\}$ if $|E_i|$ is odd,
- a bisected edge $\{x, z\}, \{z, y\}$ if $|E_i|$ is even.

Then, a strong intersection representation of G' induces a strong intersection representation of G (see Fig. 14). □

We shall conclude with an extension of Problems 1 and 2 to planar hypergraphs:

PROBLEM 3. Has any planar (linear) hypergraph a strong intersection representation?

ACKNOWLEDGMENT 1. The authors would like thank R. Šámal for his helpful comments and remarks.

References

[1] C. Berge, *Graphes et hypergraphes*, second ed., Dunod, Paris, 1973.
[2] Robert Cori, *Un code pour les graphes planaires et ses applications*, vol. 27, Société mathématique de france, Paris, 1975.
[3] H. de Fraysseix, T. Matsumoto, P. Ossona de Mendez, and P. Rosenstiehl, *Regular orientations and graph drawing*, Abstracts for the Third Slovenian international conference in graph theory, 1995, pp. 12–13.
[4] H. de Fraysseix and P. Ossona de Mendez, *On topological aspects of orientations*, in Fifth Czech-Slovak Symposium on Combinatorics, Graph Theory, Algorithms and Applications (to appear in Disc. Math.)
[5] ———, *On regular orientations*, Prague Midsummer Combinatorial Workshop, 1994, pp. 9–13.
[6] ———, *Intersection graphs of jordan arcs*, Discrete and Computational Geometry : Ten Yeras Later, 1995, p. 14.
[7] H. de Fraysseix, P. Ossona de Mendez, and J. Pach, *Representation of planar graphs by segments*, Intuitive Geometry **63** (1991), 109–117.
[8] G. Ehrlich, S. Even, and R.E. Tarjan, *Intersection graphs of curves in the plane*, Journal of Combinatorial Theory **21(B)** (1986), 8–20.
[9] András Frank, *On the orientation of graphs*, Journal of Combinatorial Theory, Series B **28** (1980), 251–261.
[10] S.L. Hakimi, *On the degree of the vertices of a directed graph*, J. Franklin Inst. **279** (1965), 4.
[11] D.S. Johnson and H.O. Pollak, *Hypergraph planarity and the complexity of drawing venn diagrams*, Journal of Graph Theory **11** (1987), no. 3, 309–325.
[12] R.P. Jones, *Colourings of hypergraphs*, Ph.D. thesis, Royal Holloway College, Egham, 1976, p. 209.
[13] J. Kratochvíl, *A special planar satisfiability problem and a consequence of its NP-completeness*, Discrete Applied Mathematics **52** (1994), 233–252.
[14] J. Kratochvíl and J. Matoušek, *Intersection graphs of segments*, Journal of Combinatorial Theory **62(B)** (1994), no. 2, 289–315.
[15] ———, *Are string graph algorithmically recognizable ?*, Problem Session of the conference Graph and Geometry, Prague, 1997.
[16] T Matsumoto, *Orientations contraintes*, Ph.D. thesis, Ecole des Hautes Etudes en Sciences Sociales, Paris, 1997.
[17] P. Ossona de Mendez, *Orientations bipolaires*, Ph.D. thesis, Ecole des Hautes Etudes en Sciences Sociales, Paris, 1994.
[18] T.R.S. Walsh, *Hypermaps versus bipartite maps*, J. Combinatorial Theory **18(B)** (1975), 155–163.
[19] A.T. White, Graphs, Groups and Surfaces, Mathematics Studies, vol. 8, ch. 13, pp. 205–210, Mathematics Studies, North-Holland, Amsterdam, revised ed., 1984, pp. 205–210.
[20] A.A. Zykov, *Hypergraphs*, Uspeki Mat. Nauk (1974), no. 6, 89–154.

PATRICE OSSONA DE MENDEZ, CNRS UMR 0017, E.H.E.S.S., 54 BD RASPAIL, 75006 PARIS, FRANCE
E-mail address: `pom@ehess.fr`

HUBERT DE FRAYSSEIX, CNRS UMR 0017, E.H.E.S.S., 54 BD RASPAIL, 75006 PARIS, FRANCE
E-mail address: `hf@ehess.fr`

DIMACS Series in Discrete Mathematics
and Theoretical Computer Science
Volume **49**, 1999

Linear and Non-Linear Systems: A Survey

J. Díaz, M. Serna, and P. Spirakis

ABSTRACT. In this paper we present the research that has been done with Linear Dynamical Systems to generate almost uniformly elements from a given set, and thus approximate some hard counting problems. We also indicate how non-linear systems can help to parallelize the computation. Finally we outline possible applications of linear systems to formalize heuristics.

1. Introduction

Many problems involving counting solutions of combinatorial structures are well known to be difficult. Valiant defined the class $\#P$ of computationally equivalent counting problems ([**Val79b**]). For many problems in this class, their decision counterpart is in P. It is known that, unless the polynomial hierarchy collapses, $P \neq \#P$. This fact implies that for any $\#P$-complete problem, exact counting is apparently intractable ([**Pap94**]). The most notorious of these problems is to compute the permanent of a dense matrix. That problem turns out to be equivalent to counting the number of perfect matchings in a dense bipartite graph ([**Val79a**]). The hardness of these counting problems motivated research on approximate counting. Pioneering work in this line was the paper [**KLM89**] where a Randomized Fully Approximation Scheme is constructed, for some difficult counting problems. Later, it was discovered that for the problems which are self-reducible, approximate counting is equivalent to almost uniform generation ([**JVV86**]) The almost uniform generation problem consist in picking at random an element of a finite set according to a distribution. Such distribution must be guarantee to be within a relative error ϵ of the uniform distribution.

Karmarkar, Karp, Lipton, Lovasz and Luby give a Monte-Carlo algorithm for approximating the permanent of dense positive matrices based in computing an unbiased estimator ([**KKL$^+$93**]). Their estimator can be easily computed in polylog parallel time with a polynomial number of processors ussing a Randomized PRAM, thus we get a parallel RNC algorithm to approximate the permanent of the adjacency matrix of a bipartite graph with minimum degree $(1/2+\alpha)n$ where $\alpha > 0$. Therefore there exists a RNC approximation to the number of perfect matchings for "quite dense" bipartite graphs.

1991 *Mathematics Subject Classification.* 60J15,68R99.

This research was partially supported by the ESPRIT Long Term Research Project no. 20244 – ALCOM-IT and CICYT project TIC97-1475-CE.

A technique that has proved to be very useful for solving the almost uniform generation problem, is the Markov Chain technique. Given a problem, define a Markov chain where the states are all possible solutions, plus possibly a small fraction of "non-solutions", and the transitions are certain probabilistic rules that allow us to remain in the same state or move to a new state. Under certain properties of the underlying graph, it can be proved that a polynomial (in the input size) random walk on the chain gives a random element, almost distributed according to the stationary distribution of the chain. The difficulty of this method is to prove convergence in a polynomial number of steps to the stationary distribution, usually refereed to as the "rapid mixing" property. Broder used the Markov chain technique to approximate the value of the permanent of a dense matrix ([**Bro86**]). The rapid mixing property of his chains was shown by Jerrum and Sinclair ([**JS89**]). Over the past years, a large body of literature has been devoted to the subject of almost uniform generation through Markov chains and methods of proving rapid mixing. Excellent surveys can be found in [**Sin93, Vaz91, Kan94, MR95, JS95**].

A question of interest is the possibility of parallelizing the almost uniform generation and approximate counting problems. Consider the Markov Chain defined by Broder for almost uniform generation of perfect matchings in dense bipartite graphs ([**Bro89**]). Teng has proved that the problem of computing the final node m' of a sequential walk, starting at a node m is P-complete ([**Ten95**]). This result does not exclude the possibility of generating in parallel an almost uniform perfect matching. The Teng result excludes the possibility of the computation of a given sequential random walk, in NC that is; given the walk and the initial state, compute the final state using a PRAM in polylogarithmic steps using a polynomial number of processors.

To obtain a parallel generator instead of using a Markov chain, we define a "genetic system". Such a system starts from a set S of objects with a given initial distribution Π_0, this will be the initial generation at time $t = 0$. From that initial population, new generations are grown by mating two randomly selected parents. Define a **mating** rule to crossover objects: our rule will be defined in such a way that for any two objects sampled according to distribution Π_t at time t, form a new object that will be an element of the next population. Formally, if u and v are the objects sampled from Π_t mate them with probability $p(u, v, w)$ to outcome the new element w. Then the probability distribution of the population at time $t + 1$ follows the non-linear dynamic equation

$$\Pi_{t+1}(x) = \sum_u \Pi_t(u) \sum_v p(u, v, x)\Pi_t(v).$$

We shall show that the system evolves towards a unique stationary distribution. In general quadratic dynamic systems are difficult and no too much is known about their behavior. For instance, it is known that a quadratic dynamic system can solve any problem in PSPACE, using a polynomial amount of time [**Pud94, ARV94**]. Therefore unless P=NP there is no polynomial time simulation of a general quadratic dynamic system. There are results for some particular non-liner dynamic systems, for example in the work of Rabani et al. [**RSW92, RRS95**].

The next section in this survey presents some general results on Markov chain theory and the Markov chain method. We survey the sequential approach to solve a the monomer-dimer system. Section 3 presents the non-linear approach to parallelize the almost uniform sampling. We show the convergence of the system and how to implement in parallel the defined mating rule. Finally section 4 surveys some of the work done trying to formalize hillclimbing heuristics, together with some final remarks.

2. Linear Systems

Recall that a **Markov chain** μ is an stochastic process, defined on a set of states S, in terms of a transition matrix $P = (p_{ij})_{i,j \in S}$, where each p_{ij} denotes the probability of going from i to j. Therefore, $\forall i \in S$, $\sum_{j \in S} p_{ij} = 1$. Moreover, at $t = k$, we define π_k as

$$\pi_k(i) = \sum_{j \in S} p_{ji} \cdot \pi_{k-1}(j) \tag{1}$$

Let X_t be a stochastic variable such that, at time t, it denotes the state where μ is. Also let the initial distribution $\pi_0(i)$ (at $t = 0$) defined as follows: $\forall i \in S$, $\pi_0(i) = Pr\{X_0 = i\}$. If $S = \{1, 2, 3, \ldots, m\}$ and $\overline{\pi}_k = (\pi_k(1), \ldots, \pi_k(m))$ is the distribution at time $t = k$, then

$$\overline{\pi}_k = \overline{\pi}_{k-1} \cdot P = \overline{\pi}_0 \cdot P^k. \tag{2}$$

So every Markov chain defines a linear system and reciprocally any linear system can be viewed as a Markov chain.

Let p_{ij}^t denote the probability of going from state i to state j in t steps. A Markov chain is **irreducible** if $\forall i, j \in S$, $\exists t$ such that $p_{ij}^t > 0$. A Markov chain is **aperiodic** if $\forall i, j \in S$, $gcd\left\{t | p_{ij}^t > 0\right\} = 1$. A Markov chain is said to be **ergodic** if $\forall j \in S$, $\lim_{t \to \infty} p_{ij}^t = \pi_\infty(j) > 0$. If μ is ergodic then $\overline{\pi}_\infty = (\pi_\infty(1), \ldots, \pi_\infty(n))$ is called the **stationary distribution**.

Let $\overline{1} = (1, 1, \ldots, 1)^\top$. As P is stochastic, then $P \cdot \overline{1} = \overline{1}$. Notice

$$P^\infty = \lim_{t \to \infty} [P_{ij}^t] = \begin{pmatrix} \pi_\infty(1) & \ldots & \pi_\infty(n) \\ \vdots & \ddots & \vdots \\ \pi_\infty(1) & \ldots & \pi_\infty(n) \end{pmatrix}$$

The following results could be found in any basic book of Markov chains (see for ex. [**Nor97, Sin93**])

THEOREM 1. *A Markov chain μ is ergodic if and only if it is irreducible and aperiodic. Moreover, if a Markov chain μ is ergodic then its stationary distribution is the unique distribution that satisfies:*

$$\overline{\pi}_\infty \cdot P = \overline{\pi}_\infty$$
$$\sum_{i \in S} \pi_\infty(i) = 1.$$

To assure that the stationary distribution is the uniform distribution, we need further conditions on μ. A Markov chain is **symmetric** if $\forall i, j \in S$, $p_{ij} = p_{ji}$. An ergodic Markov chain is **reversible** if $\forall i, j \in S$ satisfy the Balance Equation

$$\pi_\infty(i)\, p_{ij} = \pi_\infty(j)\, p_{ji}.$$

The next two results can be found in any textbook on Markov chains,

PROPOSITION 1. *Let μ be an ergodic Markov chain. If $\exists \overline{\pi}_*$ such that $\forall i, j \in S:\ \pi_*(i) p_{ij} = \pi_*(j) p_{ji}$ and $\sum \pi_*(i) = 1$, then μ is reversible and $\overline{\pi}_\infty = \overline{\pi}_*$.*

THEOREM 2. *If μ is a symmetric, ergodic Markov chain with $|S| = n$ then*

$$\forall i \in S, \ \pi_\infty(i) = \frac{1}{n}.$$

Let us consider the problem of given a large and finite set S, and a probability distribution π on S, sample an element in S according to π. The **Markov Chain Technique** gives an approximate solution of the previous problem, and consists in the following steps: Construct a Markov chain μ with states S and stationary distribution π. Starting from an arbitrary state $s \in S$, perform a random walk in the chain large enough to set a *closed* point to equilibrium distribution. In the light of our previous comments μ must be ergodic and if μ is ergodic and symmetric, then

$$\overline{\pi}_\infty = \left(\frac{1}{|S|}, \dots, \frac{1}{|S|}\right)$$

in other words, $\overline{\pi}_\infty$ is uniform.

Therefore, once we have a Markov chain μ, to approximate sample from π_∞, simulate the Markov chain for a finite number of steps and get *close* to π_∞. The question is, what is the *rate of convergence*? How long should the random walk be to sample close enough to the limit distribution? Recall that for all k, $\overline{\pi}_k = \overline{\pi}_0 P^k$. We need to control powers of P, hence, we need to look at the eigenvalues of P. The basic idea is to use spectral theory as it is done in Graph Theory (see for example [**Chu96**]).

Recall from linear algebra that any $n \times n$ matrix M over K could be considered as a linear operator $V \to V$. Moreover, if M has $\lambda_1, \lambda_2, \dots, \lambda_n$ real eigenvalues, not necessarily all different, then $M = \sum_{i=1}^{n} \lambda_i H^i$ with

$$H^i \cdot H^j = \begin{cases} \neq 0 & \text{if } i = j \\ 0 & \text{if } i \neq j. \end{cases}$$

Moreover, if M diagonalizes, then $H^i \cdot H^i = H^i$.

Let P be the transition matrix of μ with eigenvalues $\lambda_1, \dots, \lambda_n$, and assume that P diagonalizes. Then as P is an stochastic matrix $\lambda_1 = 1$. We will assume that eigenvalues are sorted by absolute value, so that $1 = |\lambda_1| \geq |\lambda_2| \geq \dots \geq |\lambda_n|$. So,

$$P^m = \sum_{i \geq 1} \lambda_i^m H^i = \lambda_1^m H^1 + \sum_{i \geq 2} \lambda_i^m H^i.$$

Let $\overline{\pi}_\infty = \overline{\pi}$ be the left eigenvector of $\lambda_1 = 1$ and let

$$H^1 = \begin{pmatrix} \pi(1) & \dots & \pi(n) \\ \vdots & \ddots & \vdots \\ \pi(1) & \dots & \pi(n) \end{pmatrix}$$

then

$$\lim_{m \to \infty} P^m = H^1 + \lim_{m \to \infty} \sum_{i \geq 2} \lambda_i^m H^i \approx H^1 + \lim_{m \to \infty} |\lambda_2| H^2.$$

Therefore if μ is an ergodic Markov chain with stationary distribution π_∞, then we must have $|\lambda_2| < 1$.

To avoid that negative eigenvalues can delay the ratio of convergence, we can increase the value of the self-loop in P and make all eigenvalues positive. The following theorem is proved in [**Sin93**],

THEOREM 3. *If P is the transition matrix of an ergodic and reversible Markov chain with eigenvalues $\lambda_1 = 1 > \lambda_2 \geq \lambda_3 \geq \ldots \geq \lambda_n > -1$, then the Markov chain with matrix $P' = \frac{1}{2}(I + P)$ is also ergodic and reversible, and it has the same limit distribution $\overline{\pi}_\infty$. The eigenvalues $\{\lambda'_i\}_{i=1}^n$ are similarly ordered and $\lambda'_i = \frac{1}{2}(1 + \lambda_i) > 0$.*

We wish to approach the stationary distribution in a random walk of polynomial length. Define the **relative pointwise distance** at time t is

$$\Delta(t) = \max_{i,j \in S} \frac{|p_{ij}^t - \pi_\infty(j)|}{\pi_\infty(j)}.$$

To see how fast $\Delta(t) \to 0$ and get some bounds on $\Delta(t)$, the following bound is useful for an ergodic and reversible Markov chain, ([**Sin93**])

$$|\lambda_2|^t \leq \Delta(t) \leq \frac{|\lambda_2^t|}{\min_{j \in S} \pi_\infty(j)}$$

historically the way to bound the convergence of Markov chains was coupling ([**Lin92**]). The big breakthrough of Jerrum and Sinclair was to use structural properties of the graphs representing the Markov chains associated to certain counting problems, to bound the convergence. Let us define the concept of rapidly mixing Markov Chain.

The rate of convergence in a Markov chain μ to its stationary distribution $\overline{\pi}_\infty$ is given by its mixing time function defined by $\tau_i(\epsilon) = \min\{t \mid \forall t' \geq t : \; \Delta_i(t') \leq \epsilon \}$.

We say that a Markov chain is **rapidly mixing** if from any state $i \in S$ and $\forall \epsilon : 0 < \epsilon \leq 1$ we have

$$\tau_i(\epsilon) \leq \text{poly}\left(|i|, \log \frac{1}{\epsilon}\right).$$

Notice a Markov Chain μ can be considered as a weighted directed graph $G = (V, E, w)$, where V is the set of states, $p_{ij} > 0 \implies (i,j) \in E$, and the weight of an edge $w(i,j)$ is defined as the probability $\pi_\infty(i)p_{ij}$. Notice that when the Markov chain is reversible we have $w(i,j) = w(j,i)$.

Let us give some topological definitions on the underlying graph G of μ. For S' be a nonempty subset of S. The **capacity** of S' measures the probability of being in a state of S' when reaching π_∞, and it is defined $C_{S'} = \sum_{i \in S'} \pi_\infty(i)$. The **ergodic flow** of S' measures the probability of leaving S', and it is defined $F_{S'} = \sum_{i \in S', j \notin S'} \pi_\infty(i)p_{ij}$. The probability of leaving S' once inside S' is given by the formula $\Phi_{S'} = F_{S'}/C_{S'}$. The **conductance** Φ of a Markov chain μ is defined as $\Phi = \min\{\Phi_{S'} \mid S' \subsetneq S \wedge C_{S'} \leq 1\}$. Notice the conductance of μ measures the worst bottle-neck. If the Markov chain μ is reversible we have

$$\forall S' \subsetneq S : F_S = F_{S'} \implies \Phi = \min\left\{\max\{\Phi_{S'}, \Phi_{\overline{S'}}\} \; : \; S' \subsetneq S, S' \neq \emptyset\right\}.$$

It is possible to bound λ_2, and hence the convergence, in terms of a topological characteristic of μ, its conductance Φ. The following result is from [**JS89**],

THEOREM 4 (Jerrum-Sinclair). *Let μ be an ergodic and reversible Markov chain. Then*

$$1 - 2\Phi \leq \lambda_2 \leq 1 - \frac{\Phi^2}{2}.$$

COROLLARY 1. *If μ is ergodic and reversible then*

$$\Delta(t) \leq \frac{(1 - \Phi^2/2)^t}{\min_i \pi_\infty(i)}.$$

Moreover, if $\Phi \leq 1/2$ then $\Delta(t) \geq (1 - 2\Phi)^t$.

The following theorem gives us a characterization of rapid mixing in terms of Φ:

THEOREM 5. *Let μ be an ergodic and reversible Markov chain, that for all $i \in S$ has $p_{ii} \leq 1/2$, and such that if $\pi^*_\infty = \min_{i \in S}\{\pi_\infty(i)\}$ then $\ln pi^{*\,-1}_\infty \leq poly(|i|)$. Then μ is rapidly mixing if and only if $\Phi \geq \frac{1}{poly(|i|)}$ $\forall i \in S$.*

Therefore to prove that an ergodic and reversible Markov chain μ is rapidly mixing, we have to find a polynomial p such that $\Phi \geq 1/p(|i|)$.

Still it is necessary to compute or find bounds for the conductance. For that, Jerrum and Sinclair considered a clever argument to estimate the bottleneck of the Markov chain. Define a unique canonical path, between every pair of states. Given a transition edge, count the number of canonical path going through it: Given any $S' \subseteq S$ let $\delta(S') = \{i \in S' | \ \exists j \in \overline{S'} : \ e = (i,j)\}$. We define the **edge magnification** of the graph of μ as

$$\gamma(\mu) = \min_{0<|S'|<\frac{|S|}{2}} \frac{|\delta(S')|}{|S'|}.$$

Notice that if d is the maximum degree of the graph of μ then $\forall S' \subseteq S: \ |\delta(S')| \geq |S'| \cdot d$. So $0 < \delta(\mu) \leq d$.

Many Markov chains can be considered as a random walk in the graph of μ, with maximum degree d, where transitions from i to j are made with probability β/d for some constant β $(0 < \beta \leq 1)$. In addition, $\forall i \in S$, i has a self loop with probability $1 - \beta\text{deg}(i)/d$. In such a situation, the conductance of the corresponding graph verifies $\Phi = \frac{\beta\gamma(\mu)}{d}$. Therefore, in this kinds of Markov chains, to prove rapid mixing, it is enough to find a polynomial p such that $\Phi = \frac{\beta\gamma(\mu)}{d} \geq \frac{1}{p(|i|)}$. If $d \leq \text{poly}(|i|)$, we just have to find a polynomial lower bound on the edge magnification γ.

2.1. Monomer-Dimer Systems. Let us see an generic example taking from [**Sin93**]. Given a graph $G = (V, E)$ with $|V| = n$ and $|E| = \mu$. For $k \in \{0, \cdots, \lfloor n/2 \rfloor\}$, let $M_k(G)$ denote the set of matchings of size k in G, and M be the set of all its matchings, that is $M = \cup_k M_k$. Let $N = |M|$. From now on, G will denote the input graph.

We start by defining a Markov chain $\mathcal{D}$ for a given weighted graph G, where $c(e)$ denotes the weight of edge e. The chain $\mathcal{D}$ contains as states the set M of all matchings, and the transitions are defined as follows,

Definition of transitions in $\mathcal{D}$:
Given a matching $m \in M$,
(0) Sample uniformly a random edge $e = (u, v)$.
(1) With probability $1/2$ stay in m
otherwise

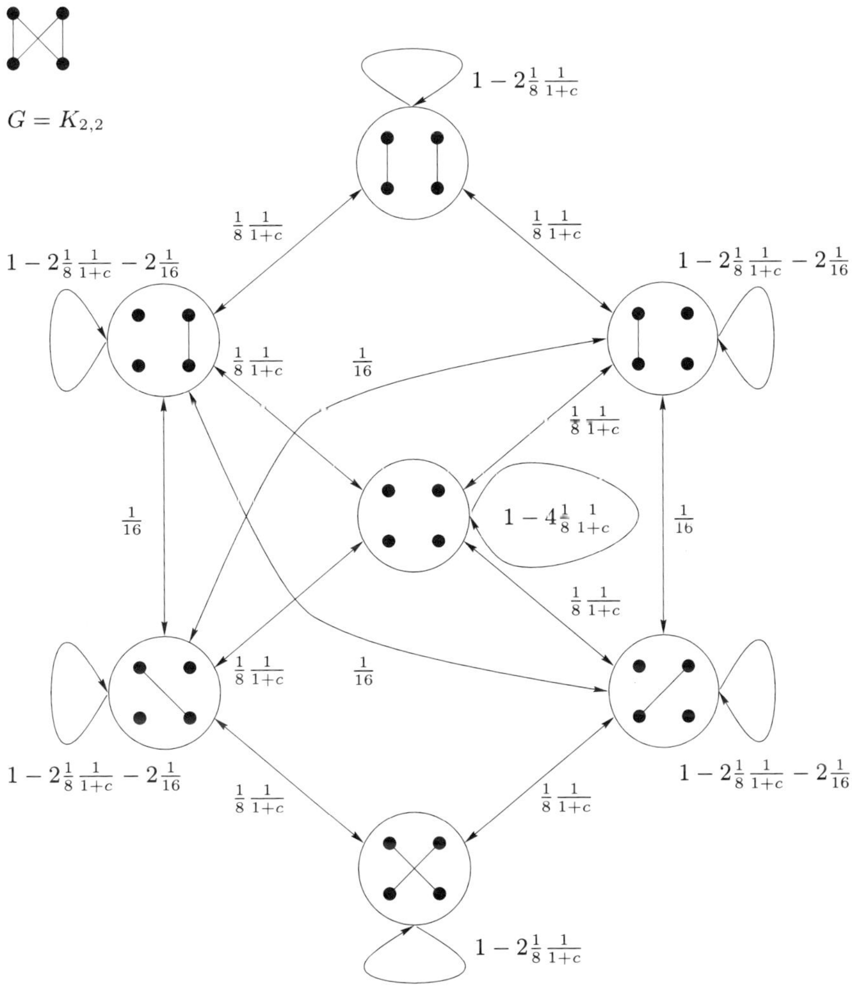

FIGURE 1. Example with $G = K_{2,2}$ and $|E| = 4$

(1.1) (Deletion) If e is in m then with probability $1/(1+c(e))$ go to matching $m - \{e\}$, otherwise stay in m
(1.2) (Augmentation) If $m \cup \{e\}$ is a matching then go to new matching $m \cup \{e\}$ with probability $c(e)/(1+c(e))$.
(1.3) (Rotation) If u is unmatched in m and v is matched in m by edge $e' = (v, w)$, then with probability $c(e)/(c(e')+c(e))$ the new matching is $m - \{e'\} \cup \{e\}$, and with probability $c(e')/(c(e')+c(e))$ keep m.
(1.4) Otherwise stay in m.

Figure 2.1 shows an example of the Markov chain corresponding to a monomer-dimer system in which all edge weights are equal to a constant c. The convergence of the monomer-dimer system can be found in [**Sin93**]

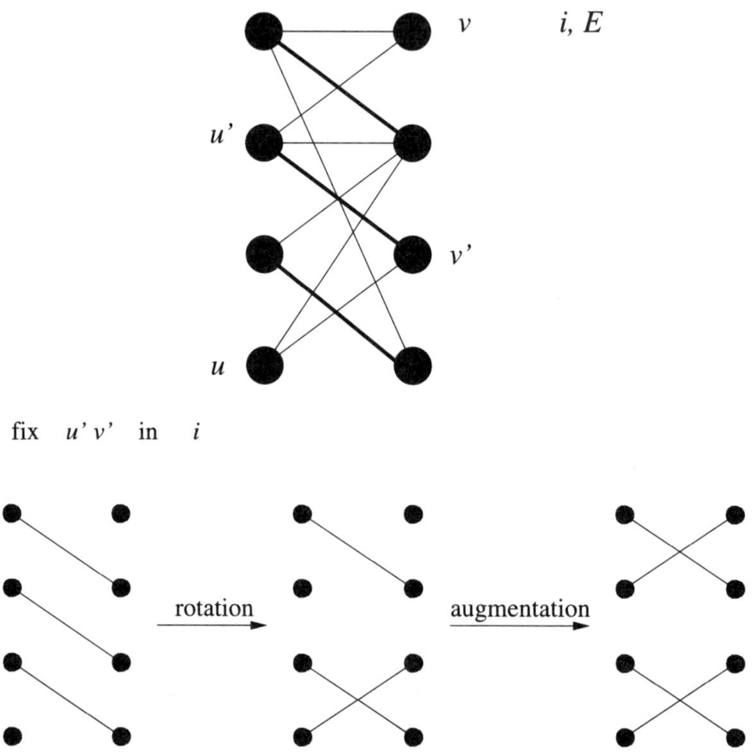

FIGURE 2. An example of canonical path.

THEOREM 6 (Sinclair). *Given a weighted graph G, the Markov chain $\mathcal{D}$ is reversible and ergodic. So it converges, and the stationary distribution is*

$$\Pi_\infty(i) = \frac{\prod_{e \in m_i} c(e)}{\sum_{j \in M} \prod_{e \in m_j} c(e)}.$$

In order to analyze its mixing time, using the conductance argument, we suppose that there is and underlying order on all simple paths in G (including cycles). For any path, fix a start vertex, it must be an endpoint if the path is not a cycle. For any two given matchings m_1 and m_2 their symmetric difference is a set of disjoint cycles and paths. We begin by sorting this set of paths according to the fixed order. The canonical path from m_1 to m_2 involves the transformation of the initial matching into the final one, by modifying in order all the paths and cycles, starting from the corresponding start vertices.

To unwind a path that is not a cycle, we have two cases: first one, the path starts with an edge of m_2, in such a case we change edge by edge and finish by adding the last if necessary. Second, the path starts with an edge of m_1 remove that edge and continue as in the previous case. An example of such unwinding is given in figure 2.

To unwind a cycle, fix a direction to traverse its edges in such a way that the first edge from the start vertex is in m_1. Remove this edge and then proceed as in the corresponding path using as start vertex the endpoint that is not the start vertex in the actual path.

We also need to define an injective mapping from the set of canonical paths that pass through a given transition t. Let t be a transition from matching m to matching m', and let $P(t)$ be the set of canonical paths that contain t. For any pair of matchings such that the canonical path from m_1 to m_2 goes through t define $\sigma_t(m_1, m_2) = m_1 \oplus m_2 \oplus (m \cup m')$, and remove the edge e of m_1 adjacent to the start vertex of the path currently treated in case the resulting set of edges is not a matching. The difference $m_1 \oplus m_2$ can be recovered from $\sigma_t(m_1, m_2)$ using

$$m_1 \oplus m_2 = \begin{cases} \sigma_t(m_1, m_2) \oplus (m \cup m') \oplus e & \text{if } t \text{ is an augmentation} \\ & \text{the current path is a cycle} \\ & \text{and } e \text{ is the removed edge} \\ \sigma_t(m_1, m_2) \oplus (m \cup m') & \text{otherwise} \end{cases}$$

It is possible to tell whether the path is a cycle or is not, because we are unwinding cycles in different direction than paths. Therefore we can recover the original matchings using the path ordering. Hence σ_t is injective. Furthermore it can be shown (see [**Sin93**])

LEMMA 1. *For any transition t and any $\langle m_1, m_2 \rangle \in P(t)$ we have*

$$\pi_\infty(m_1)\pi_\infty(m_2) \leq 4|E|c_{\max}^2 w_t \pi_\infty(\sigma_t(m_1, m_2)),$$

where $c_{\max} = \max\{1, \max_{e \in E} c(e)\}$ and w_t and w_t is $\pi_\infty(m)$ multiplied by probability of the transition t.

3. Genetic System

In order to give a parallel implementation of the methods described in section 2 we introduce the non-linear systems.

To simplify the presentation, we consider only the monomer-dimer Markov chain $\mathcal{M}$ in the case that all weights are equal to a given fixed parameter $c > 0$. Such chain $\mathcal{M}$ contains as states the set M of all matchings, and the transitions are defined as follows,

Definition of transitions in $\mathcal{M}$:
Given a matching $m \in M$,
(0) Sample uniformly a random edge $e = (u, v)$.
(1) With probability $1/2$ stay in m
otherwise
 (1.1) If e is in m then with probability $1/(1 + c)$ go to matching $m - \{e\}$, otherwise stay in m.
 (1.2) If $m \cup \{e\}$ is a matching then with probability $c/(1 + c)$ go to new matching $m \cup \{e\}$.
 (1.3) If u is unmatched in m and v is matched in m by edge $e' = (v, w)$, then with probability $1/2$ the new matching is $m - \{e'\} \cup \{e\}$, and with probability $1/2$ keep m.
 (1.4) Otherwise stay in m.

As $\mathcal{M}$ is a restricted version of the monomer-dimer system we have

THEOREM 7 (Sinclair). *Given a fixed c, the Markov chain $\mathcal{M}$ is reversible and ergodic. Moreover the stationary probability $\Pi_\infty(i) = c^{|m_i|} / \sum_{j \in M} c^{|m_j|}$.*

In [**Sin93**] it is also shown that for graphs verifying

$$\frac{|M_n(G)|}{|M_{n-1}(G)|} \leq q(n) \tag{3}$$

with q a polynomial function, then taking $c = 2q(n)$, the chain $\mathcal{M}$ converges to an uniform stationary distribution on the subset of perfect matchings. Furthermore in the limit distribution, the probability of getting a perfect matching is bigger than $1/2$. As every dense graph satisfies equation (3), then the class of bipartite dense graphs is a subset of the class of graphs satisfying (3). Notice that the stationary distribution is non-uniform on the set M.

We define a *genetic system* $\mathcal{G}$ over the population of all matchings M that will produce the next generation according to a mating rule grounded in the transitions of $\mathcal{M}$.

DEFINITION 1 (Mating Rule). *From parents m_l and m_r, sort randomly the edges of m_r. The offspring m_k is the matching resulting of applying the following procedure:*

(1) *With probability $1/2$, $m_k = m_l$.*
Otherwise,
 (2.1) *For every edge in $m_r \cap m_l$ with probability $1/(1+c)$ choose that the edge that does not belong to m_k.*
 (2.2) *For every edge $e = (u, v) \in m_r$ such that u and v are unmatched in m_l, with probability $c/(1+c)$ choose e to be in m_k.*
 (2.3) *For every maximal increasing path in $m_r \cap m_l$, starting in a node unmatched in m_l, and having even length, see Figure 3, For each edge in the path coming from m_r, with probability $1/2$ choose label 1, otherwise choose label 0. Beginning with the first edge in the path, compute the longest prefix formed with edges labeled 1 (if any). Then m_k consists of the edges from m_r in the prefix, and the edges from m_l after the first edge labeled 0.*

Given three matchings m_i, m_j and m_k, let $P(i, j, k)$ denote the probability of getting m_k as an offspring of m_i and m_j.

To define a system evolving in time t, start from a given initial generation Π_0 over M at $t = 0$. The generation at time $t + 1$ is obtained from the generation Π_t at time t, by sampling two matchings m_l and m_r according to Π_t, and applying the mating rule to m_l and m_r. The system evolves according to the following dynamical equation,

$$\Pi_{t+1}(k) = \sum_{m_l \in M} \Pi_t(l) \cdot \sum_{m_r \in M} P(l, r, k) \cdot \Pi_t(r) \tag{4}$$

Given a probability distribution Π on the set M of all matchings in G, let us define a Markov chain $\mathcal{M}(\Pi)$ on the set of states M, using the mating operation as rule for the transitions. Formally the transitions are defined,

Given a matching $m_l \in M$,
(1) Sample a matching m_r according to distribution Π,
(2) move to the matching m_k defined by the mating of m_l and m_r.

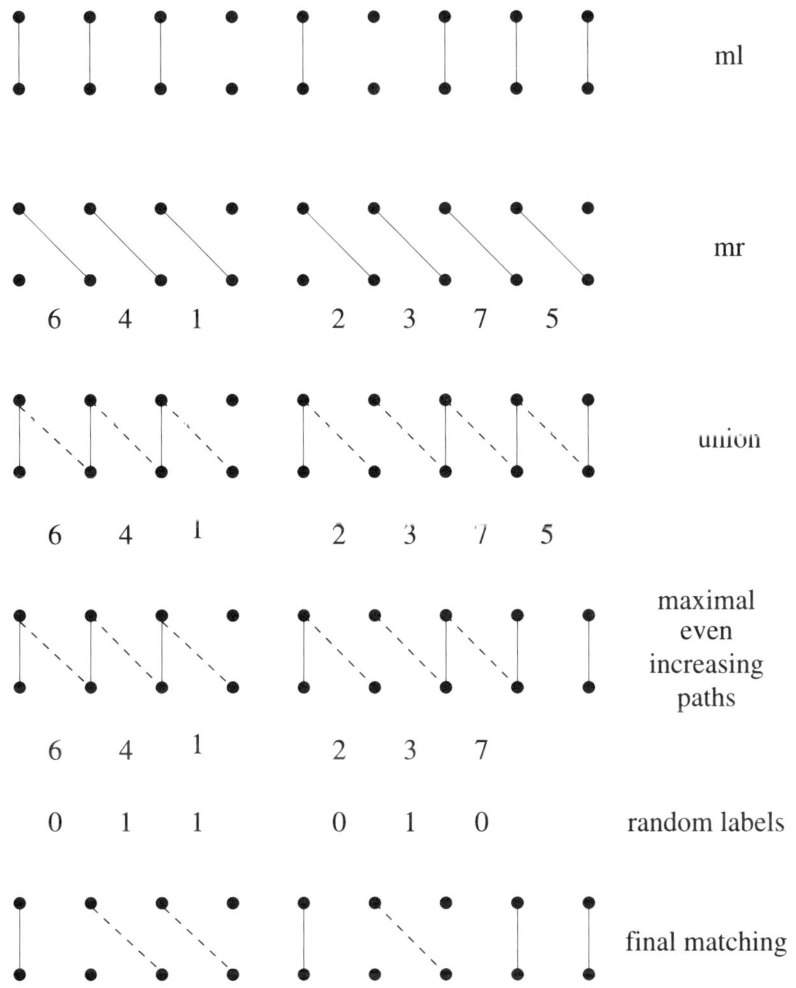

FIGURE 3. Two matchings and the set of maximal increasing paths.

From the way the Markov chain has been defined, it evolves accordingly to the following equation,

$$\Pi_{t+1}(k) = \sum_{m_l \in M} \Pi_t(l) \cdot \sum_{m_r \in M} P(l,r,k) \cdot \Pi(r). \tag{5}$$

Notice that the coefficient $\Pi(l,k)$ in the transition matrix of $\mathcal{M}(\Pi)$ is given by the equation

$$\Pi(l,k) = \sum_{m_r \in M} P(l,r,k) \cdot \Pi(r).$$

3.1. Convergence. We want to study the conditions under which the genetic system and the Markov chain converge.

LEMMA 2. *Given a path $\{m_1, \cdots, m_r\}$ on the underlying graph of $\mathcal{M}$, the stationary distribution Π_∞ verifies,*

$$\Pi_\infty(1) \prod_{i=1}^{r-1} p_{i,i+1} = \Pi_\infty(r) \prod_{i=1}^{r-1} p_{i+1,i}.$$

PROOF. The time reversibility of $\mathcal{M}$ implies $\Pi_\infty(i) \cdot p_{ij} = p_{ji} \cdot \Pi_\infty(j)$, therefore

$$\begin{aligned} \Pi_\infty(1) \prod_{i=1}^{r-1} p_{i,i+1} &= \Pi_\infty(1) \cdot p_{1,2} \prod_{i=2}^{r-1} p_{i,i+1} \\ &= p_{2,1} \cdot \Pi_\infty(2) \prod_{i=2}^{r-1} p_{i,i+1} \\ &\vdots \\ &= \prod_{i=1}^{s-1} p_{i+1,i} \cdot \Pi_\infty(s) \cdot \prod_{i=s}^{r-1} p_{i,i+1} \\ &\vdots \\ &= \prod_{i=1}^{r-1} p_{i+1,i} \cdot \Pi_\infty(r)\,. \end{aligned}$$

□

Given a matching m_i, we denote by $S(i)$ the set of edge sequences obtained by sorting the edges in m_i. Given three matchings m_l, m_r and m_k, and an element $\vec{r} \in S(r)$ let us denote by $P(l, \vec{r}, k)$ the probability of going from m_l to m_k following a sequence given by $\vec{r}$ in $\mathcal{M}$, so we have $P(l, r, k) = \sum_{\vec{r} \in S(r)} P(l, \vec{r}, k)/|S(r)|$. Notice that $P(l, r, k) \in [0, 1]$, with $\sum_k P(l, r, k) = 1$. Using lemma 2 and the fact that, when $P(l, \vec{r}, k) \neq 0$ there is a matching m_s of the same size as m_r such that $P(k, \vec{s}, l)$, we get,

LEMMA 3. *Given three matchings m_l, m_r and m_k, we have*

$$\Pi_\infty(l) \cdot P(l, r, k) = P(k, r, l) \cdot \Pi_\infty(k).$$

This lemma gives us the property we need to prove convergence of both, $\mathcal{M}(\Pi)$ and $\mathcal{G}$ to the same distribution Π_∞.

THEOREM 8. *Given a distribution Π over the set of matchings of a given graph $G = (V, E)$. If for every matching m with exactly one edge we have $\Pi(m) > 0$, then $\mathcal{M}(\Pi)$, and $\mathcal{G}$ converge to the limit distribution of $\mathcal{M}$.*

PROOF. In order to prove the convergence of $\mathcal{M}(\Pi)$, we show the time reversibility of $\mathcal{M}(\Pi)$ with respect to the distribution Π_∞. Recall that the probability of going from m_l to m_k in the Markov chain $\mathcal{M}(\Pi)$ is given by the equation

$\Pi(l,k) = \sum_{m_r \in M} P(l,r,k) \cdot \Pi(r)$, therefore using lemma 3 we have

$$\begin{aligned}\Pi_\infty(l) \cdot \Pi(l,k) &= \Pi_\infty(l) \cdot \sum_{m_r \in M} P(l,r,k) \cdot \Pi(j) \\ &= \sum_{m_r \in M} \Pi_\infty(l) \cdot P(l,r,k) \cdot \Pi(r) \\ &= \sum_{m_r \in M} \Pi_\infty(k) \cdot P(k,r,l) \cdot \Pi(r) \\ &= \Pi(k,l) \cdot \Pi_\infty(k)\end{aligned}$$

Let us prove the convergence of the genetic system $\mathcal{G}$. Recall that the system evolves according to the equation

$$\Pi_{t+1}(k) = \sum_{m_l \in M} \Pi_t(l) \sum_{m_r \in M} P(l,r,k) \cdot \Pi_t(r).$$

Substituting Π_∞ in the previous equation and using again lemma 3 we get,

$$\begin{aligned}\sum_{m_l \in M} \Pi_\infty(l) \cdot \sum_{m_r \in M} & P(l,r,k) \cdot \Pi_\infty(r) \\ &= \sum_{m_l \in M} \sum_{m_r \in M} \Pi_\infty(l) \cdot P(l,r,k) \cdot \Pi_\infty(r) \\ &= \sum_{m_l \in M} \sum_{m_r \in M} \Pi_\infty(k) \cdot P(k,r,l) \cdot \Pi_\infty(r) \\ &= \Pi_\infty(k) \sum_{m_r \in M} \Pi_\infty(r) \sum_{m_l \in M} P(k,r,l) \\ &= \Pi_\infty(k) \sum_{m_r \in M} \Pi_\infty(r) = \Pi_\infty(k)\end{aligned}$$

Therefore Π_∞ is a fix point for the system, let us see that it is the unique fix point of the system. Suppose that Δ is another fix point, by the restriction on the initial distribution, Δ must assign positive probability to any matching. Therefore the Markov chain defined using the mating rule and the Δ distribution converges to Π_∞ and Δ therefore $\Delta = \Pi_\infty$. □

In figure 4 it is given a diagram of the behavior of the genetic system and the sequence of Markov chains. The previous theorem, give us one of the rear cases of convergence for non-linear systems.

3.2. RNC computation of the mating rule. Given two matchings m_l and m_r we wish to compute with a randomized PRAM, in polylogarithmic number of steps and using a polynomial number of processors, the mating operation, that gives birth to child m_k. Consider the following procedure:

(1) With probability $1/2$, $m_k = m_l$
    Otherwise :
(2) In parallel assign an order to the edges in m_r.
(3) For every edge in the graph, check if it is in both matchings. If so with probability $1 - 1/(1+c)$ choose that the edge is in the offspring m_k.
(4) For every edge $(u,v) \in m_r$ check whether u and v are unmatched in m_l. If so, with probability $c/(1+c)$ choose (u,v) to be in m_k.

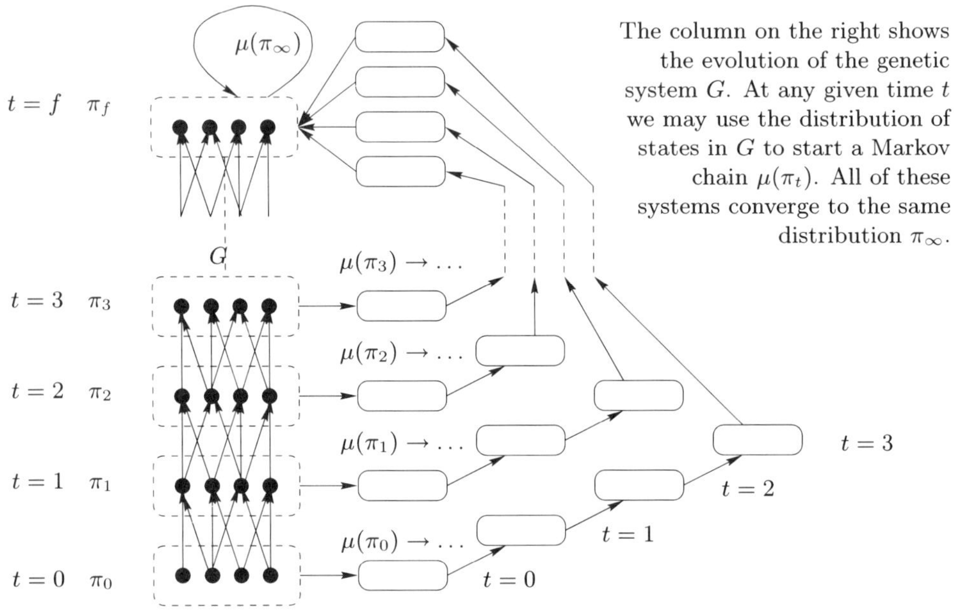

FIGURE 4. The genetic system $\mathcal{G}$and the Markov chains associated to every distribution.

(5) For each free node in m_l, obtain the list of edges in m_r that form a maximal increasing path,
if it has even length.
 (5.1) Toss the coin and assign labels 0/1 with equal probability to the edges.
 (5.2) Obtain the first edge a with label 0. All the edges before a that belong to m_r and all the edges after a that belong to m_l form the matching m_k.
otherwise keep the edges in m_l.

It is easy to implement steps (1) to (4) with a CREW PRAM in $O(\log n)$ steps and μ processors. The data structure we use to represent a matching is an array with $2n$ positions, numbers between 1 and n represent nodes in one bipartition and numbers between $n+1$ and $2n$ the other one. A value j in position i, $j \neq 0$ means that edge (i, j) is in the matching, when $j = 0$ i is unmatched. To implement step (5) we add pointers to the data structure. For an edge $e = (y, z)$ in m_l the number of edges in m_r that touches e may be 0,1, or 2. If this number is 0 we we link the edge to itself. When there is only one e', assuming that $e' = (x', x)$, we link (x', x) with (x, x'). And in the case that there are two edges e', e'', such that e' is previous to e'', suposse the $e' = (x', x)$ and $e'' = (y, y')$, we link (x', x) with (x, y) and (x, y) with (y, y') (see figure 5). Finally each node i unmatched by m_l is linked to the corresponding node position in m_r. The additional pointer structure can be computed in $O(1)$ time with $O(n)$ processors with a CREW PRAM. Starting from the edges leaving free points in m_l, use pointer jumping to obtain the maximal paths and compute their length. Keep those paths which have even length. All this can be done in $O(\log n)$ using $O(n)$ processors in a randomized CREW PRAM,

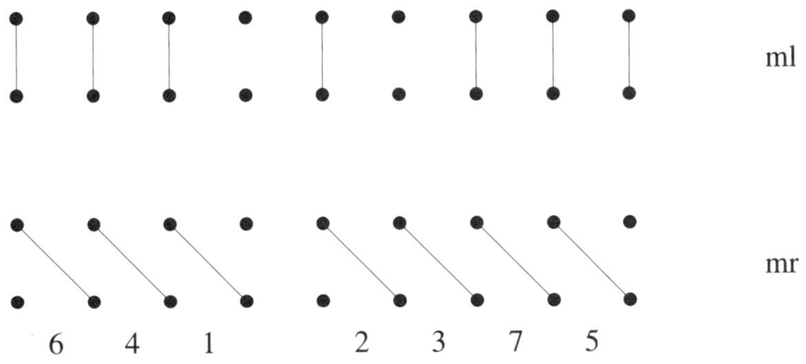

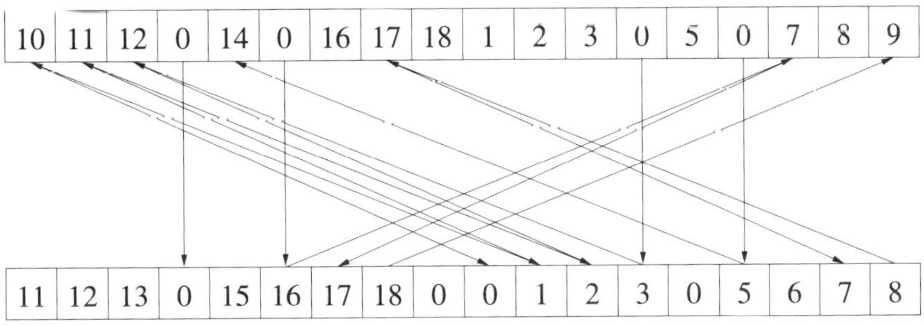

FIGURE 5. Data structure and links

including the label assignment. Finally step (6) can be implemented with the same bounds, using again the pointer jumping technique.

4. Local Search and Optimization Problems

One of the characterization of algorithms in recent time is the use of heuristics greedy type. Those heuristics seems to work quite well in practice for some problems but the theoretical foundations of *why* or *how* they work is an open and difficult topic of research.

Let us recall that given a combinatorial search space S and an objective function $f : S \to R^+$, a **maximization problem** consists on finding $\sigma^* \in S$ such that $\forall \sigma \in S : \ f(\sigma) \leq f(\sigma^*)$, i.e., to find a maximum. A **minimization problem** consists on finding $\sigma^* \in S$ such that $\forall \sigma \in S : \ f(\sigma) \geq f(\sigma^*)$, i.e., to find a minimum.

For example the Graph Bisection problem consists in given $G = (V, E)$ with $|V| = n = 2k$, find $V_1, V_2 \in S$ with $|V_1| = |V_2| = n/2$ and $V_1 \cup V_2 = V$ such that $|\{\{u, v\} : \ u \in V_1, \ v \in V_2\}|$ is minimum.

Here the search space S is the set of all possible bisections and $f(\sigma)$ is the number of crossing edges. This problem is NP-hard [**GJ79**].

A **black-box heuristic** is a randomized search heuristic operating on a connected neighborhood structure H on the vertex set S. Usually, the edge neighbors of a particular state are defined under some measure of distance that is natural to the combinatorial problem under consideration. The term Black-box was coined by Ari Juels in his PhD dissertation [**Jue90**].

The choice of the neighborhood represents a key decisions in the application of an algorithm and affects much of its performance.

An heuristic that has been used to solve efficiently some difficult optimization problems is **hillclimbing**. It has the following generic structure,

```
function HillClimbing(S, N, f)
    Select initial state σ ∈ S
    while movement is possible do
        Randomly select σ' ∈ N(σ)
        if f(σ') > f(σ) then — could also be ≥ —
            σ := σ'
        end if
    end while
    return σ
end
```

The algorithm terminates when it encounters a local maximum (or minimum), i.e. an state σ such that $\forall \sigma' \in N(\sigma): \; f(\sigma) \geq f(\sigma')$. A local maximum can be on a "peak" (its objective function value is strictly greater than all its neighbors) or in a "plateau" (its objective function value is greater or equal than all its neighbors. This is why one must decide to choose a $>$ or a $\geq$ sign in the algorithm and be careful in the last case to avoid cycling.

For instance, in [**KP92**] the authors prove that hillclimbing finds a satisfying truth assignment, if one exists, with high probability.

The problem of the hillclimbing algorithm is that once a local optimum is found the algorithm returns it, but this local optimum can be different of the global optimum. In order to enable the algorithm to accept downhill moves, the Metropolis algorithm is parameterized by a temperature t and proceeds as follows:

```
function Metropolis(S, N, f, t)
    Select initial state σ ∈ S
    while movement is possible do
        Randomly select σ' ∈ N(σ)
        Δ := f(σ) − f(σ')
        with probability min(1, e^{−Δ/t}) do
            σ := σ'
        end with
    end while
    return σ
end
```

Observe that uphill movements will be automatically accepted, whereas downhill movements are accepted randomly in function of the height (Δ) of the movement and the temperature t. With a high temperature the probability of descending is high; with a small temperature, it is low. In the limit, as $t \to \infty$ Metropolis makes a random walk and as $t \to 0$ Metropolis becomes the hillclimbing algorithm.

Formally, let $d(\sigma)$ be the degree of σ and $D = \max_{\sigma \in S}\{d(\sigma)\}$. The Metropolis algorithm can be seen as a Markov chain on H with transitions defined by

1. Self loop with $\mathbf{Pr} = 1/2$ when $\sigma' = \sigma$.
2. Choose: σ' with probability

$$\mathbf{Pr}(\sigma') = \begin{cases} 1/D & \text{if } \sigma' \in N(\sigma) \\ 1 - d(\sigma)/D & \text{if } \sigma' = \sigma \\ 0 & \text{otherwise.} \end{cases}$$

3. With probability $\min(1, e^{-\Delta/t})$ go to σ'.

Let μ_t be the above chain. Its transition probability matrix is $P = [p_{\sigma\sigma'}]$ with $p_{\sigma\sigma'} = \mathbf{Pr}(\sigma') \cdot \min(1, e^{-\Delta/t})$. Define $\pi_t(\sigma) = \frac{e^{f(\sigma)/t}}{z^t}$, where π_t represents the stationary distribution for μ_t. It is straightforward to prove that μ_T is ergodic and reversible. Moreover notice that when $t \to \infty$, we have $e^{1/t} \to 1$ and thus π_t is the uniform distribution $(1/|S|)$. On the other hand, when $t \to 0$, π_t becomes more sharply peaked around optimal solutions in S.

Therefore, fixing t, the Metropolis algorithm is just a "sufficiently long" random walk on the Markov chain μ_t.

The Monomer-Dimer procedure described in Section could be seen as a Metropolis algorithm for the problem of finding a matching of maximum cardinality in a graph. Using that chain, it can be proved the following result [**SH88**],

THEOREM 9. *For graphs such that $M_{n-1}/M_m \leq poly(e^t)$ running Metropolis for $\mathcal{O}(n)$ steps, with probability $1 - \frac{1}{n+1}$ a perfect matching is found.*

Another classical example is the Graph Bisection, that we already mentioned above. Jerrum and Sorkin considered the problem for the following restricted graph model [**JS93**],

In the model $\mathcal{G}_{4n,p,r}$, a graph G has $4n$ nodes, colored half white and half black. Edges between nodes with the same color are included independently with probability p, while those between nodes of different colors are included with probability r with $r < p$. For sufficiently large values of $p-r$, these graph instances will contain a bisection β^* (the one in which white and black nodes are separated) which is very likely to be the unique minimum. This is referred to as the *planted* bisection.

The value $p - r$ characterizes the difficulty. For $p = r = 1/2$, the expected cut of the planted bisection is $n^2/8$. In the case $p = 1/2$ and $r = p - n^{\Delta-2}$ for $3/2 < \Delta < 2$, the expected cut of the planted bisection is $n^2/8 - n^{\Delta}/4$.

Given a bisection β, we define $b(\beta)$ as its cut. We refer $\sigma(l, r)$ as the operation of swapping two vertices $r \in V_1$ and $l \in V_2$. We define $N(\beta)$ as the set of all possible states obtainable from β by a single move σ. We finally define B as the set of all possible bisections of G. The Metropolis algorithm is given:

```
function Metropolis(C) is
    Choose β_0 uniformly from B
    for i := 1 .. t do
        Choose randomly β' ∈ N(β)
        ΔC := b(β') − b(β)
        with probability 1/(1+e^{ΔC/T}) do
            β := β'
        end with
    end for
    return β
end
```

The Markov chain for this problem is given by the set of states B and the following transitions:

1) With probability $1/2$, stay in the same state; 2) Choose $\beta' \in N(\beta)$ and with probability $\frac{1}{1+e^{\Delta C/T}}$ move to it.

The transition matrix is $P = [p_{\beta\beta'}]$ with $p_{\beta\beta'} = \mathbf{Pr}(\beta') \cdot \frac{1}{1+e^{\Delta C/T}}$. As we have an ergodic chain, it converges to a stationary distribution π_0. Jerrum and Sorkin proved the following result,

THEOREM 10. *Let* $\epsilon > 0$, $p - r = n^{1-6+2\epsilon}$, $T = n^{5/6+\epsilon}$. *Select* $G \in \mathcal{G}_{4n,p,r}$. *At constant temperature, the metropolis algorithms reaches the unique* β^* *in* $\mathcal{O}(n^{2+\epsilon})$ *steps with overwhelming probability .*

Ari Juels [**Jue90**] proved that in some measure these results also extrapolate to the hillclimbing algorithm. Jerrum and Sorkin [**JS93**] proved that T is too high to be effective on small instances. There have been other work on proving formalizing convergence and rapid mixing for Metropolis algorithms. For instance Nolte and Schrader use similar ideas to the ones developed by Jerrum and Sorkin, to give a kind of Metropolis algorithm for [**NS97**] 3-colorability, restricted to some particular "coloring planted" kind of graphs [**NS97**].

The authors believe that an important and difficult topic of research is the formalization of hillclimbing type algorithms, and their parallel implementation and formalization. For the parallel implementation, non-linear systems could be of help. But they have the inconvenience that little is know about the theory of non-linear dynamic systems. In fact another important and difficult open area of research is to formalize and study the convergence of genetic algorithms of which the systems studied in section 3 are just an oversimplified model. To give conditions for convergence of genetic algorithms would be of great theoretical and practical relevance.

5. Acknowledgments

We thank Jordi Petit, Xavier Molinero and Marko Rieddel for letting us use some of their Latex macros.

References

[ARV94] S. Arora, Y. Rabani, and U. Vazirani. Simulating quadratic dynamical systems is PSPACE-complete. In *26th ACM Symposium on Theory of Computing*, pages 83–92, 1994.

[Bro86] A. Broder. How hard is to marry at random. In *18th ACM Symposium on Theory of Computing*, pages 50–58, 1986.

[Bro89] A. Broder. Generating random spanning trees. In *30th IEEE Symposium on Foundations of Computer Science*, pages 442–447, 1989.

[Chu96] F.R.K. Chung. *Spectral Graph Theory.* American Mathematical Society, 1996.

[GJ79] M.R. Garey and D.S. Johnson. *Computers and Intractability: A Guide to the Theory of NP-Completeness.* Freeman, San Francisco, 1979.

[JS89] M. Jerrum and A. Sinclair. Approximating the permanent. *SIAM Journal on Computing*, 18:1149–1178, 1989.

[JS93] M. Jerrum and G. Sorkin. Simmulated annealing for graph bisection. In *34th IEEE Symposium on Foundations of Computer Science*, pages 94–103, 1993.

[JS95] M. Jerrum and A. Sinclair. *The Markov Chain Monte Carlo method: An approach to approximate counting and integration*, pages 482–520. PWS, Boston, 1995.

[Jue90] Ari Juels. *Basic in Blax-box Combinatorial Optimization.* Phd thesis, University of California at Berkeley, 1990.

[JVV86] M. Jerrum, L. Valiant, and V. Vazirani. Random generation of combinatorial structures from a uniform distribution. *Theoretical Computer Science*, 43:169–188, 1986.

[Kan94] R.. Kannan. Markov chains and polynomial time algorithms. In *35th IEEE Symposium on Foundations of Computer Science*, pages 656–671, 1994.

[KKL+93] N. Karmarkar, R Karp, R. Lipton, L. Lovasz, and M. Luby. A Monte–Carlo algorithm for estimating the permanent. *SIAM Journal on Complexity*, 22:284–293, 1993.

[KLM89] R.M.. Karp, M. Luby, and N. Madras. Monte-Carlo approximation algorithms for enumeration problems. *Journal of Algorithms*, 18:429–448, 1989.

[KP92] E. Koutsupias and X. Papadimitriu. On the greedy algorithm for satisfiability. *Information Processing Letters*, 43:53–55, 1992.

[Lin92] Torgny Lindvall. *Lectures on the Coupling Method.* Wiley InterScience, 1992.

[MR95] R. Motwani and P. Raghavan. *Randomized Algorithms.* Cambridge University Press, 1995.

[Nor97] J.R. Norris. *Markov Chains.* Cambridge Series in Statistical and Probabilistic Mathematics, 1997.

[NS97] A. Nolte and R. Schrader. Coloring in sublinear time. In R. Burkard and G. Woeginger, editors, *Algorithms-ESA'97*, pages 388–401. Lecture Notes in Computer Science, Springer-Verlag, 1997.

[Pap94] C. Papadimitriou. *Computational Complexity.* Addison-Wesley, Reading, Mass., 1994.

[Pud94] P. Pudlak. Complexity theory and genetics. In *9th. IEEE Symposium on Structur in Complexity Theory*, pages 383–395. IEEE, 1994.

[RRS95] Y. Rabani, Y. Rabinovich, and A. Sinclair. A computational view of population genetics. In *27th ACM Symposium on Theory of Computing*, pages 83–92, 1995.

[RSW92] Y. Rabinovich, A. Sinclair, and A. Wigderson. Quadratic dynamical systems. In *33th IEEE Symposium on Foundations of Computer Science*, pages 304–313, 1992.

[SH88] G.H. Sasaki and B. Hajek. The time complexity of maximum matching by simulate annealing. *Journal of the ACM*, 35:387–403, 1988.

[Sin93] A. Sinclair. *Algorithm for random generation and counting: A Markov chain approach.* Birkhäuser, Boston, 1993.

[Ten95] Shang-Hua Teng. Independent sets versus perfect matchings. *Theoretical Computer Science*, pages 1–10, 1995.

[Val79a] L.G. Valiant. The complexity of computing the permanent. *Theoretical Computer Science*, 8:189–201, 1979.

[Val79b] L.G. Valiant. The complexity of enumeration and reability problems. *SIAM J on Computing*, 8:410–421, 1979.

[Vaz91] V. Vazirani. Rapidly mixing Markov chains. In B. Bollobas, editor, *Probabilistic combinatorics and its applications*, pages 99–121. American Mathematical Society, 1991.

Departament de Llenguatges i Sistemes, Universitat Politècnica Catalunya, Mòdul C5 Campus Nord, Jordi Girona Salgado 1-3, E-08034 Barcelona, Spain
E-mail address: diaz@lsi.upc.es

Departament de Llenguatges i Sistemes, Universitat Politècnica Catalunya, Mòdul C5 Campus Nord, Jordi Girona Salgado 1-3, E-08034 Barcelona, Spain

Computer Technology Institute (C.T.I.), P.O. Box 1122, 26110 Patras, Greece
E-mail address: spirakis@cti.gr

DIMACS Series in Discrete Mathematics
and Theoretical Computer Science
Volume **49**, 1999

Parameterized Complexity: A Framework for Systematically Confronting Computational Intractability

Rodney G. Downey, Michael R. Fellows, and Ulrike Stege

ABSTRACT. In this paper we give an overview of parametcrized computational complexity in the broad context of the problem of coping with computational intractability. We give some examples of how fixed-parameter tractability techniques can deliver practical algorithms in several different ways, including: (1) by providing useful exact algorithms for small parameter ranges, and (2) by providing guidance in the design of heuristic algorithms. In particular, we describe an improved *FPT* algorithm for VERTEX COVER, a practical *FPT* algorithm for the MAXIMUM AGREEMENT SUBTREE (MAST) problem parameterized by the number of species to be deleted, and new heuristics for these problems based on *FPT* techniques. In the course of making this overview, we also investigate some structural and hardness issues. We prove that an important parameterized problem in artificial intelligence, STRIPS PLANNING (where the parameter is the size of the plan) is complete for $W[1]$. As a corollary, this implies that k-STEP REACHABILITY FOR PETRI NETS is also complete for $W[1]$. We describe how the concept of *treewidth* can be applied to STRIPS PLANNING and other problems of logic to obtain *FPT* results. We describe a surprising structural result concerning the top end of the parameterized complexity hierarchy: the naturally parameterized GRAPH k-COLORING problem cannot be resolved with respect to *XP* either by showing membership in *XP*, or by showing hardness for *XP* without settling the $P = NP$ question one way or the other.

1. Introduction

There are basically two different ways that one can view the theory of parameterized complexity. The first way, and the one that is easiest to arrive at, is as a kind of first aid that can sometimes be applied to problems that are *NP* hard, *PSPACE*-hard or undecidable. That is, it can be viewed as a potential means of coping with *classical* intractability.

1991 *Mathematics Subject Classification.* Primary: 68Q15, 68Q25. Secondary: 68Q68, 03D15, 03A05.

Rodney G. Downey. Research supported by a grant from the United States/New Zealand Cooperative Science Foundation and by the New Zealand Marsden Fund for Basic Science.

Michael R. Fellows. Research supported by the National Science and Engineering Research Council of Canada.

The second way that one can view parameterized complexity is as a fundamentally richer and generally more productive *primary framework* for problem analysis and algorithm design, including the design of heuristic and approximation algorithms. For the moment, we will concentrate on the first point of view, and spend some time considering the current situation of our collective efforts to "cope with intractability" in the words of Garey and Johnson [**GJ79**], Chapter 6. In the concluding section we will summarize the arguments for the second point of view.

It is difficult to discuss the current intellectual situation with respect to theoretical computer science without taking note of the unhappiness, in some sense, of the audience of our efforts to understand complexity and deliver useful products. For some time, the practical folks portrayed in the amusing cartoons of Garey and Johnson [**GJ79**], Chapter 1, have been somewhat disappointed with theoretical computer science, and there have been numerous incidents of "restless drums" in the jungles of computer applications. Such grumblings may have something to do with the recent spate of soul-searching among theorists [**HL92, Hart94, PGWRS96, AFGPR96**]. We mention two recent examples.

EXAMPLE 1.1 (Yet Another Call for Reform). One of the plenary addresses at the AAAI meeting in 1996 was concerned with the broad theme of how computer science practice and theory interact [**DKL96**]. The discussion in [**DKL96**] can be summarized as:

1. Pointing to a particular problem, STRIPS PLANNING, as central to the entire field of artificial intelligence.
2. Proposing that practitioners and theorists collaborate in an intense analysis of this one problem, to understand what makes it hard, and to come up with something more useful than a *PSPACE*-completeness classification.
3. Suggesting that the usual framework for concrete complexity analysis is wrong-headed, historically contingent, unnatural (especially worst-case asymptotic analysis), and reflects an unhappy state of interaction between computer science theory and practice.

Regarding the third point, most theorists have probably heard similar charges and complaints from colleagues in applied areas of computer science in their own departments.

EXAMPLE 1.2 (An Encounter With a Computational Biologist). In recent conversations with a biologist who is heavily involved in combinatorial computing [**Fel97**], the following summary was offered of his interaction with theoretical computer scientists.

> *"About ten years ago some computer scientists came by and said they had heard that we have some really cool problems. They showed that the problems are* NP*-complete and went away!"*

We might comment that if this interaction had been more recent, then perhaps the computer scientists would also have proved that the problems are unlikely to have efficient approximation algorithms.

The Pervasive Nature of Computational Intractability and Various Coping Strategies. Arguably the most fundamental discovery of the first decades of theoretical computer science is that most computational problems are *hard* in a variety of mathematically interesting ways. Computer science practitioners quite naturally would like to shoot the messenger who brings so much bad news! In this difficult situation, computer science theory has articulated a few general programs for systematically coping with the ubiquitous phenomena of computational intractability. We list these basic approaches:

- The idea of focusing on average-case as opposed to worst-case analysis of problems.
- The idea of settling for approximate solutions to problems, and of looking for efficient approximation algorithms.
- The idea of using randomization in algorithms.
- The idea of harnessing quantum mechanics, or biomolecular chemistry, to create qualitatively more powerful computational mechanisms.

To this list of basic mathematical strategies for coping with intractability, we argue should be added:

- The idea of devising *FPT* algorithms for parameterizations of a problem.

A list such as this cannot be considered complete without including another coping strategy that mathematical theorists have frequently contributed to. This approach antedates modern complexity-theoretic frameworks and has recently regained some respectability among theorists as reflected in the DIMACS Challenges [**JM93, JT96**] and in new journals such as the *Journal of Heuristics* and the ACM *Journal of Experimental Algorithms*:

- The design of mathematically informed, but perhaps unanalyzable heuristics, that are empirically evaluated by their performance on sets of benchmark instances.

The actual state of the practical world of computing is that (with the exception of some areas) there is not much systematic connection to work in theoretical computer science on algorithms and complexity. Overwhelmingly, in fact, it is heuristic algorithms that are relied on to deal with the problems encountered in most applications.

2. Fixed-Parameter Tractability

The basic concept of the parameterized complexity framework is that of *fixed-parameter tractability*. It is the notion of good complexity behaviour from which all other aspects of the theory follow. The definition is best introduced through concrete examples.

VERTEX COVER
Instance: A graph $G = (V, E)$ and a positive integer k.
Parameter: k
Question: Does G have a vertex cover of size k? That is, is there a subset $V' \subseteq V$ of size at most k such that for every edge $uv \in E$, either $u \in V'$ or $v \in V'$?

DOMINATING SET
Instance: A graph $G = (V, E)$ and a positive integer k.
Parameter: k
Question: Does G have a dominating set of size k? That is, is there a subset $V' \subseteq V$ of size at most k such that every vertex $u \in V$ of G either belongs to V' or has a neighbor in V'?

THE STEINER PROBLEM FOR HYPERCUBES
Instance: A set $S = \{x_i : 1 \leq i \leq k\}$ of binary vectors, $x_i \in \{0,1\}^n$ for $i = 1, ..., k$, and a positive integer m.
Parameter: k
Question: Is there a tree $T = (V, E)$ and a labeling of the vertices of T with elements of $\{0,1\}^n$ such that the following conditions are satisfied? (1) The leaves are labeled 1:1 with the elements of S. (2) The sum over the edges uv of T of the Hamming distance between the labels $l(u) \in \{0,1\}^n$ and $l(v) \in \{0,1\}^n$ is at most m.

THE MAXIMUM AGREEMENT SUBTREE PROBLEM (MAST)
Instance: A set of rooted trees $T_1, ..., T_r$ $(r \geq 3)$ with the leaf set of each T_i labeled 1:1 with a set of species X, and a positive integer k.
Parameter: k
Question: Is there a subset $S \subseteq X$ of size at most k such that T_i restricted to the leaf set $X' = X - S$ is the same (up to label-preserving isomorphism and ignoring vertices of degree 2) for $i = 1, ..., r$?

All of these problems are *NP*-complete ([**GJ79, AK94**]) and are described above in the standard way for the parameterized complexity framework. Part of the input (which may be some aggregate of various aspects of the input) is identified as the *parameter* for the problem specification. (In order to consider a parameterized problem classically, just ignore the parameter part of the specification.) All of these problems can be solved in time $O(n^{f(k)})$ by simple brute force algorithms. For example, for VERTEX COVER and DOMINATING SET we can simply try all k-subsets of vertices.

For VERTEX COVER we can do qualitatively better. Papadimitriou and Yannakakis showed that VERTEX COVER can be solved in time $O(3^k n)$ [**PY96**]. Balasubramanian, Fellows and Raman gave an algorithm with running time $O((53/40)^k k^2 + kn)$ [**BFR98**]. In §4 we describe a new and relatively simple *FPT* algorithm that improves on this. Note that since the exponential dependence on the parameter k in the last expression is *additive*, VERTEX COVER is *well-solved for input of any size so long as k is no more than around 60.* The difference between the complexities of DOMINATING SET and VERTEX COVER is displayed in Table 1.

	$n = 50$	$n = 100$	$n = 150$
$k = 2$	625	2,500	5,625
$k = 3$	15,625	125,000	421,875
$k = 5$	390,625	6,250,000	31,640,625
$k = 10$	1.9×10^{12}	9.8×10^{14}	3.7×10^{16}
$k = 20$	1.8×10^{26}	9.5×10^{31}	2.1×10^{35}

TABLE 1. The Ratio $\frac{n^{k+1}}{2^k n}$ for Various Values of n and k.

In parameterized complexity, the process of interviewing practice in order to formulate the fundamental object of study is enriched. Besides specifying the input to the problem, we specify distributional aspects that may belong to some limited range of values for which an exponential contribution to overall problem complexity may be acceptable. This distributional information, which may be an aggregate of factors, is codified as the *parameter*. A single classical problem may thus shatter into many different associated *parameterized problems*.

EXAMPLE 2.1 (THE STEINER PROBLEM FOR HYPERCUBES). This problem is of interest to biologists in the computation of phylogenetic trees under the criterion of minimum evolution / maximum parsimony [**SOWH96**]. The set S corresponds to a set of species, and the binary vectors correspond to information about the species, each component recording the answer to some question (as 0 or 1), such as: "Does it have wings?" or "Is there a thymine at a certain position in the DNA sequence?" Each such bit of information is termed a *character* of the species. In realistic applications the number k of species may usefully be around 40 or 60, while the number of characters n may be very large.

How Parameters Naturally Arise. Most computational problems involve several pieces of input, one or more of which may be a relevant parameter for various applications.

- *Graph linear layout width metrics* are of interest in VLSI layout and routing problems and have important applications for width values of $k \leq 10$. Interval graphs of pathwidth $k \leq 10$ have applications in DNA sequence reconstruction problems [**BDFHW95**].
- *Logic and database problems* frequently are defined as having input consisting a formula (which may be small and relatively invariant), and some other structure (such as a database) which is typically quite large and changeable. Formula size, or other aspects of formula structure may be a relevant parameter [**Yan95**].
- *Hardware constraints* are a common source of natural parameters. The number of processors or machines to be scheduled may be bounded by a value such as $k \leq 20$. In [**AMOV91**] it was proposed to streamline chip implementations of cryptosystems by bounding the Hamming weight of keys. The number of degrees of freedom in a robot motion-planning problem is commonly in the range $k \leq 10$ [**CW95**]. The number of wiring layers in VLSI chip manufacture is typically bounded by $k \leq 30$ [**Len90**].
- *Network problems* may be naturally concerned with optimally locating a small number of facilities.

These few examples are only suggestive and by no means exhaustive. There are myriad ways in which numbers that are small or moderately large (e.g., $k \leq 40$) arise naturally in problem specifications. Important distributional parameters may also arise in ways that are not at all obvious. For an example of this sort, Thorup has recently shown that the flow graphs of structured programs for the major computer languages have treewidth $k \leq 7$ [**Th97**]. Graphs of pathwidth bounded by a similar number have been used in modeling dependencies in sentences of natural languages [**KT92**]. Attention to hidden parameters can sometimes explain why a problem is easier to solve in practice than *NP* or *PSPACE* hardness results would suggest.

The Basic Definitions. The basic definitions of parameterized complexity are as follows.

DEFINITION 2.1. A *parameterized language* L is a subset $L \subseteq \Sigma^* \times \Sigma^*$. If L is a parameterized language and $(x, y) \in L$ then we will refer to x as the *main part*, and refer to y as the *parameter*. It makes no difference to the theory and is occasionally more convenient to consider that y is an integer, or equivalently to define a parameterized language to be a subset of $\Sigma^* \times \mathbb{N}$.

DEFINITION 2.2. A parameterized language L is *fixed-parameter tractable* if it can be determined in time $f(k)n^\alpha$ whether $(x, k) \in L$, where $|x| = n$, α is a constant independent of both n and k and f is an arbitrary function. The family of fixed-parameter tractable parameterized languages is denoted *FPT*.

It is somewhat surprising, although the argument is not hard, that FPT is unchanged if the definition above is modified by replacing $f(k)n^\alpha$ by $f(k) + n^\alpha$ [**CCDF97**].

About half of the naturally parameterized problems cataloged as *NP*-complete in the book by Garey and Johnson [**GJ79**] are in *FPT*, including three of the six basic problems singled out for attention in Chapter 3.

It is always possible to parameterize a problem in various ways that are fixed-parameter tractable, yet it is not surprising that many parameterized problems apparently do not belong to *FPT*. The naturally parameterized DOMINATING SET problem defined above is one of these. Just as with the issue of polynomial-time complexity, we can find evidence for fixed-parameter *intractability* by studying the appropriate notion of problem transformation.

DEFINITION 2.3. A *parametric transformation* from a parameterized language L to a parameterized language L' is an algorithm that computes from input consisting of a pair (x, k), a pair (x', k') such that:

1. $(x, k) \in L$ if and only if $(x', k') \in L'$,
2. $k' = g(k)$ is a function only of k, and
3. the computation is accomplished in time $f(k)n^\alpha$, where $n = |x|$, α is a constant independent of both n and k, and f is an arbitrary function.

EXAMPLE 2.2 (An Illustrative Non-Example). In first examining the notion of a parametric transformation it can be helpful to see how they differ from ordinary polynomial-time reductions. Recall that for a graph $G = (V, E)$ on n vertices, a set of vertices $V' \subseteq V$ is a k-clique in G if and only if $V - V'$ is a vertex cover in the complementary graph G' where vertices are adjacent if and only if they are not adjacent in G. This gives an easy polynomial-time reduction of the naturally parameterized CLIQUE problem to the naturally parameterized VERTEX COVER problem, transforming the instance (G, k) of CLIQUE into the instance (G', k') of VERTEX COVER. But this is not a parametric transformation, since $k' = n - k$ is not purely a function of k. The evidence is that there is no parametric transformation in this direction between these two problems (although there is a parametric transformation in the reverse direction, either trivially, since VERTEX COVER is in *FPT*, or nontrivially by the construction described in [**DF95b**]).

EXAMPLE 2.3 (An Illustrative Example). There is a fairly elaborate parametric transformation from the naturally parameterized CLIQUE problem to the naturally parameterized DOMINATING SET problem, mapping (G, k) to (G', k') where $k' = 2k$ [**DF95a, DF98**]. The evidence is that there is no parametric transformation in the other direction.

The essential property of parametric transformations is that if L transforms to L' and $L' \in FPT$, then $L \in FPT$. This naturally leads to a completeness program based on a hierarchy of parameterized problem classes:

$$FPT \subseteq W[1] \subseteq W[2] \subseteq \cdots \subseteq W[SAT] \subseteq W[P] \subseteq AW[P] \subseteq XP$$

The parameterized analog of *NP* is $W[1]$, and $W[1]$-hardness is the basic evidence that a parameterized problem is likely not to be fixed-parameter tractable. The k-STEP HALTING PROBLEM FOR NONDETERMINISTIC TURING MACHINES is $W[1]$-complete [**CCDF97**] (where the amount of nondeterminism at each step of the Turing machine computation is unbounded). Since the $q(n)$-STEP HALTING PROBLEM is essentially the defining problem for *NP*, the analogy between *NP* and $W[1]$ is quite strong. The parameterized complexity class $W[P]$ is an analog of *PSPACE* [**ADF95**]. The class *XP* is the set of all parameterized languages L for which there are functions f and g and an algorithm to determine whether $(x, k) \in L$ in time $f(k)|x|^{g(k)}$. The reader will find more about parametric intractability in §6.

How Compelling is Fixed-Parameter Tractability? The notion of *fixed-parameter tractability* is the basic concept of the theory — but how good, really, is this notion of *good complexity behaviour*? It might be objected that Table 1 is misleading, unless FPT parameter functions such as 2^k are typical. Certainly functions such as $2^{2^{2^k}}$ are allowed by the definition, and would be impractical for $k = 3$, which suggests that the basic definition allows too much pathology.

There are two main responses. First of all, we are already used to some risk-taking in definitions, since the notion of polynomial time allows for, e.g., $O(n^{12})$, which is thoroughly impractical. A parameterized problem is just an ordinary problem for which some aspect of the input has been designated as the parameter. Ignoring the parameter, if the problem can be solved in polynomial time, that is, in time polynomial in both the total input size n and the parameter k, then trivially this is an FPT algorithm. In other words, considered classically, FPT is a superset of *P*, and it is intended to be a generalization that allows us to do something for problems that are not in *P* and that may even be *PSPACE* hard or undecidable. We have to expect to risk something in formulating such a general attack on intractability. The definition simply asks whether the difficulty of the problem can be confined to a function of the parameter, while the other costs are polynomial. How else would one naturally formulate a generalization of *P* having such ambitions for a general attack on intractability?

The second response is that there are many good FPT examples other than VERTEX COVER suggesting that "reasonable" (e.g., single exponential) parameter functions are obtainable for natural problems (possibly after some rounds of improvement). For example, consider the problem MAXIMUM SATISFIABILITY where the parameter k denotes the number of clauses to be satisfied. This was shown by Cai and Chen [**CC97**] to be in FPT with the parameter function $2^{2^{ck}}$, when the clause size is bounded by c. The parameter function was improved by Mahajan and Raman [**MR98**] to ϕ^k (without assuming a bound on the clause size), where ϕ is the golden ratio $(1+\sqrt{5})/2$. Franco and others in [**Fr97**] have shown that the falsifiability problem for pure implicational formulas having k negations is in FPT with parameter function k^k. (Can this be improved to 2^k?) Although the type checking problem for the programming language ML is *PSPACE*-complete [**HM91**], this is handled in implementations in linear time with a parameter function of 2^k, where k is the nesting depth of *let*'s, a very natural parameter for this problem (one that explains why the problem did not seem hard in practice). In §4 we give an FPT

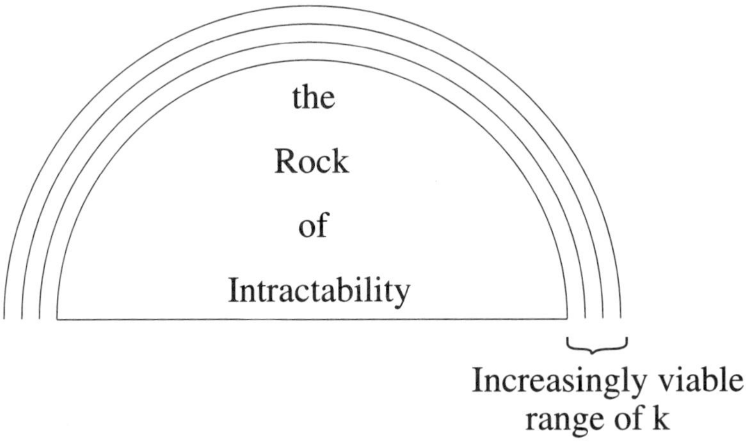

FIGURE 1. The point of view of parameterized complexity is *lichenism.*

algorithm for a natural parameterization of the MAXIMUM AGREEMENT SUBTREE problem having the parameter function 3^k. Many more examples can be found in [**DF98**]. The improvement of parameter functions for *FPT* problems seems to be a productive area for research, where many different ideas and techniques can be employed.

The point of view that parameterized complexity adopts can be summarized in a metaphorical picture. There is an assumption that most interesting problems are hard, so we can picture them as stones, or perhaps planets. The trick is to identify and develop thin zones of computational viability, as suggested in Figure 1.

3. A Review of the Major Coping Strategies

In §1 we noted the fundamental problem for concrete computational complexity that has emerged in the first decades of computer science:

> *The need to deal in some systematic way with the pervasive phenomena of computational intractability.*

We also noted that it is possible to point to five general *mathematical* strategies that have been proposed so far: (1) average-case analysis, (2) approximation, (3) randomization, (4) fundamentally new kinds of computing devices, and (5) parameterization, as well as another, quasi-mathematical coping strategy, (6) heuristics.

In this section, we review the accomplishments and prospects of these programs.

Average-Case Analysis. In many applications practitioners would be happy with algorithms having good average-case performance. This criterion is implicit in the common practice of evaluating heuristic algorithms on sets of representative benchmarks. The idea that average-case analysis is more realistic than worst-case analysis has been around since the beginnings of theoretical computer science, and its potential role as a method of coping with intractability is discussed by Garey and Johnson in their chapter on this subject in [**GJ79**]. Classic examples of such analysis include the results of Grimmet and McDiarmid [**GM75**] who described a simple graph coloring algorithm with an average performance ratio of 2 for the uniform distribution (where all n vertex graphs are equally likely), and the surprising theorem of Wilf that the average number of steps required by the straightforward backtracking algorithm to determine whether a graph is 3-colorable is 197 for the uniform distribution [**Wilf85**]. [1]

Obtaining theorems concerning average-case complexity is usually mathematically challenging, even for the simplest of algorithms and distributions. As a theoretical program for coping with intractability, average-case analysis seems to be too difficult to carry out for typical hard problems, realistic distributions and sophisticated algorithms. It is also frequently unclear what constitutes a reasonable assumption about the distribution of problem instances, apart from the analytical difficulties of the program.

A completeness notion for average-case complexity has been introduced by Levin [**Lev86**]. This also seems to be difficult to apply, and has only been demonstrated for a few problems. The main weaknesses of average-case analysis as a mathematical program for coping with intractability seem to be:

- In general, it seems to be too difficult to prove the mathematical results that the program calls for.
- The positive toolkit has not developed well. Few general methods are known for designing algorithms with provably good average-case performance. (One interesting example of a general method has been described by Gustedt and Steger [**GS94**].)

The real strength of average-case analysis as a means of coping is that for most applications of computing it is the *right idea* for how complexity should (usually) be measured, and it is what practitioners generally continue to *do* about complexity measurement in practice, although informally.

Approximation. Perhaps the greatest hopes for a general program for coping with intractability have been pinned on approximation (and also on the average-case analysis of approximation heuristics). Most of the discussion in the chapter on coping with *NP*-completeness in the famous book by Garey and Johnson [**GJ79**] is devoted to explaining the basic ideas of *polynomial time approximation algorithms and schemes.*

Early on, important polynomial time approximation schemes were found for *NP*-complete problems such as BIN PACKING [**JDUGG74**] and KNAPSACK [**IK75**]. However, apart from a few similar results on problems mostly of this same general flavor, it now seems to be clear, on the basis of powerful new proof techniques [**ALMSS92**], that these results are *not* typical for *NP*-hard and otherwise intractable problems. After a long period where the complexity of approximation for most problems remained mysterious, it now seems to be the case that the vast majority of natural *NP*-hard optimization problems probably do not admit efficient approximation schemes.

The study of *the extent to which problems can be approximated* has emerged as a mathematically very rich and productive area of investigation. As pointed out by Hochbaum [**Hoch97**], there is now such an accumulation of interesting and deep results and methods that it is tempting to assess the true difficulty of a hard problem by the degree to which optimal solutions can be approximated in polynomial time. For example, VERTEX COVER (finding a vertex cover that is as small as possible) can be approximated to within a factor of $c = 2$ in polynomial

[1] For the uniform distribution, it is usually easy to discover that the answer is "no". In fact, what Wilf has shown is that it is usually not even necessary to look at the entire graph!

time, and the constant c cannot be improved to $c \leq 16/15$ unless $P = NP$ [**BGS95**]. This contrasts sharply with the CLIQUE problem (find a clique that is as large as possible), which cannot be approximated to better than a factor of $c = n^{0.5-\epsilon}$ for n-vertex graphs, without unlikely complexity-theoretic consequences [**Has96**].

While approximation allows for the clever deployment of mathematics and can be a very effective cure for worst-case intractability when it can be applied, it seems also fair to say that, as with *P versus NP*, most of the results are negative. The main weakness of the program is:

- Most unrestricted classically hard problems cannot be approximated very well.

Randomized Polynomial Time. Randomization is discussed in Chapter 6 of Garey and Johnson [**GJ79**] as a means of avoiding one of the weaknesses of average-case analysis as a coping strategy — the need to have some knowledge in advance of a realistic distribution of problem instances. An algorithm that flips coins as it works may be able to conform to whatever distribution it is given, and either produce an answer in polynomial-time that is correct with high probability (Monte Carlo randomization), or give an answer that is guaranteed to be correct after what is quite likely to be a polynomial amount of time (Las Vegas randomization).

These seemed at first to be potentially very powerful generalizations of polynomial time. Randomized Monte Carlo and Las Vegas algorithms are a workhorse of cryptography [**SS77, GM84**], and have important applications in computational geometry [**Cl87**], pattern matching, on-line algorithms and computer algebra (see [**Karp86**] and [**MR95a**] for surveys), in part because they are often simple to program. Approximate counting is another area of notable success. Randomization is an important new idea that is now applied in many different kinds of algorithms, including approximations and heuristics.

Despite these successes, it now seems that randomized polynomial time is better at delivering good algorithms for difficult problems that "probably" are in P to begin with, than at providing a general means for dealing with intractable problems. There have recently been a number of important results replacing fast probabilistic algorithms with ordinary polynomial time algorithms through the use of sophisticated derandomization techniques [**Rag88, MR95b**]. Even more to the point, Impagliazzo and Wigderson have recently shown that if there merely exists a language in the exponential time class EXP that requires circuits of size $2^{\Omega(n)}$ then $BPP = P$ [**IW97**]. The main weaknesses of randomization (in the sense of algorithms with performance guarantees) as a general program for coping with intractability are:

- With a few exceptions, it does not seem that randomized polynomial time algorithms are any more effective against problems that are truly hard than ordinary polynomial time algorithms.
- Although the positive toolkit of methods for designing and analyzing randomized algorithms is rich, there is no specifically corresponding negative toolkit that can be used in tandem to negotiate problem complexity and guide the design of effective algorithms.

New Forms of Computation: DNA and Quantum Mechanics. Although these programs have been launched with great fanfare, they so far offer much less of substance than the other items on this list in terms of a general mathematically-powered program for coping with intractability. So far, DNA computing essentially amounts to computation by molecular brute force. Combinatorial explosion can quickly force one to contemplate a very large test tube for brute force computations, despite the fact that information can be stored in molecules with a factor of 10^{12} improved efficiency compared to magnetic tape storage. Mathematically, the program seems to come down to the potential for significant constant speed-ups by means of this physical miniaturization of computing.

It is still unclear whether quantum computers useful for any kind of computation can actually be built. The notion of quantum polynomial time is mathematically interesting, but so far appears to be applicable only to a few special kinds of problems.

The main weakness of these approaches are:

- Practical implementation of the new computing devices seems to be far in the future.
- Biomolecular computing is essentially a physical attack on intractability, not a mathematical one.
- It is unclear whether quantum polynomial time is a significant generalization of ordinary polynomial time, except for a few special kinds of problems.

Parameterization. We can trace the idea of coping with intractability through parameterization to early discussions in Garey and Johnson [**GJ79**], particularly Chapter 4, where it is pointed out that parameters associated with different parts of the input to a problem can interact in a wide variety of ways in producing non-polynomial complexity. The internal structure of an intractable problem — the identification of those aspects (parameters) to which the non-polynomial complexity can be confined — is precisely what is at issue in parameterized complexity.

The structure of non-polynomial complexity is addressed again in Chapter 6 of [**GJ79**], in the discussion of efforts to develop exponential algorithms that improve on simple exhaustive search. A classic example is the algorithm of Nešetřil and Poljak that uses fast matrix multiplication to solve the k-CLIQUE problem for n-vertex graphs in time $O(n^{ck})$ where $c \approx 0.792$ [**NP85**].

A weakness of the parameterized complexity program is that some of the most general and spectacular positive methods, such as the celebrated results of Robertson and Seymour [**RS85**], yield algorithms having parameter functions that are supremely impractical (e.g., towers of 2's of height described by towers of 2's ...). If *tractability* has friends like these, who needs enemies?

The main strengths of parameterization as a program are that it does seem to be very generally applicable to hard problems throughout the classical hierarchy of intractable classes, and it supports a rich toolkit of both positive and negative techniques. The crucial strike against the program seems to be:

- The extent to which *FPT* is really useful is unclear.

Heuristics. Since heuristic algorithms that work well in practice are now, and have always been, the workhorses of industrial computing, there is no question about the ultimate significance of this program for dealing with intractability. There has recently been a revival of interest in obtaining systematic empirical performance evaluations of heuristic algorithms for hard problems [**BGKRS95, Hoo95, JM93, JT96**]. There have been vigorous developments of new ideas for designing heuristic algorithms, particularly new ideas employing randomization in various ways. These approaches frequently have colorful and extravagant names based on far-fetched analogies in other sciences, such as *simulated annealing*, *genetic algorithms*, *neural nets*, *great deluge algorithms* [**Due93**], and *roaming ants* [**CDM92**]. Many of these can be considered as variants on the basic technique of local search.

The main problem with considering heuristics as a systematic program for coping with intractability is that it is not a coherent mathematical program. The positive toolkit properly includes voodoo and the kitchen sink. As a program, it doesn't call for any theorems, only empirical performance. The undeniable successes of sometimes "mindless" and generally unanalyzable heuristic algorithms puts computer science theory in an uncomfortable position.

Heuristics based on local search perhaps come the closest to constituting a mathematically articulated general coping strategy for intractability (see the articles in [**AL97**]). There is a negative toolkit (of sorts) based on the notion of *polynomial local search (PLS)* problems and *PLS*-completeness, introduced by Johnson, Papadimitriou and Yannakakis [**JPY88**]. Although a number of local search problems have been shown to be complete, the reductions are quite demanding (so there aren't very many such results), and there is furthermore a notable peculiarity of this framework. For a concrete example, because the TRAVELING SALESMAN PROBLEM is *NP*-hard, one resorts to a local search heuristic based on the k-Opt neighborhood structure, such as the Lin-Kernighan algorithm [**LK73**], and this is considered to be a particularly successful local search heuristic. Yet, Krentel has shown that for $k = 8$, this neighborhood structure is *PLS*-complete [**Kr90, JMc97**]. This seems like a weapon firing in the wrong direction, or perhaps just a kind of reiteration that TSP is a hard problem. It is unclear how *PLS*-completeness provides any guidance in designing local search heuristics.

The main difficulty is summarized:

- Although heuristics are the principal coin of the realm in practical computing, the design of heuristics is not well-organized as a mathematical program.

Some general remarks. If these research programs were the *Knights of Theoretical Computer Science* who would deal with the *Dragon of Intractability*, we would seem to have: one well-equipped for battle but immobilized by the weight of the armaments, one in impressive battlegear with mystic insignia riding backwards away from the fray, one armed with lucky charms effective against small lizards, one in a baby carriage shaking a rattle, and one on horse, going in the right direction, and armed with a lance — but a lance that is effective only after stabbing the Dragon a number of times bounded by

$$2^{2^{2^{2^{2^{2^{2^{k}}}}}}}$$

where k is ... unfortunately, it doesn't matter ... while the townspeople are cheering on another, who is marching towards the Dragon on foot, drunk and waving a wooden cudgel (actually just a hired ruffian).

Leaving fanciful impressions aside, it seems fair to say that the central problems of theoretical computer science, both structural and concrete, have turned out to be *much* harder to crack than was hoped in the early years of the field.

Mathematical computer scientists have so far had little to offer of any general value from a systematic perspective to practitioners who are frequently concerned with very specific hard problems (such as STRIPS PLANNING [**FN71**]). What seems to be happening is that as mathematical science makes no progress on systematically addressing intractability, we are visited by a series of fads based on turning to *some other science* as a source of solutions, often accompanied by a fanfare of "paradigm shifts" such as:

- Turning to Physics for metaphors, and getting algorithms called "simulated annealing".
- Turning to Brain Science for metaphors, and getting "neural nets".
- Turning to Evolutionary Biology for metaphors, and getting "genetic algorithms".
- Turning to Physics for more metaphors, this time about "phase transitions".

Each of these essentially shallow notions has in turn been widely and enthusiastically embraced by practitioners who must deal with their favorite hard problems somehow. From the point of view of computer science theory, each of these algorithmic fads is disappointing — they provide little opportunity to deploy the power of mathematical reasoning. It is frequently unclear if their effectiveness is much more than superstition. (If this summary of the field seems unpleasantly harsh, then look again at papers such as [**DKL96**].)

The present situation of theoretical computer science is that it has *not* been very successful in coming up with a viable systematic program for dealing in a general way with computational intractability in its own native terms (i.e., in terms of productive and interesting mathematical science). No wonder that practitioners turn in other directions, and complain!

4. Industrial Strength *FPT*

There are two main points that we will argue in this section:

1. The notion of *FPT*, in many cases, simply provides a new name and a new perspective on heuristic algorithms already in use. *FPT* algorithms frequently turn out to be *what clever practitioners implemented* after their problems were proved *NP*-hard. Where natural instance distributions exhibit limited parameter ranges, these have often been implicitly exploited in the design of useful heuristics.
2. The parameterized complexity perspective can lead to useful algorithms in several different ways, including:
    - Directly and analytically, when the parameter function is reasonable (i.e., not too fast-growing) and the parameter describes a restriction of the general problem that is still useful.
    - Directly and empirically, in cases where the analytic bound on the running time of the *FPT* algorithm turns out to be too crude or pessimistic, that is, where the algorithm turns out to be useful for larger parameter ranges than the parameter function would indicate.
    - By supplying guidance in the systematic design of heuristic algorithms in the form of explicit general methods based on *FPT* techniques, using the theory to understand "the internal structure of the complexity of the problem" by identifying those parameterizations of the problem that are *FPT* and those that are probably not (because they are hard for $W[1]$, the parameterized analog of *NP*).
    - Via methodological connections between *FPT* and the design of efficient polynomial-time approximation schemes (where the relevant implicit parameter is $k = 1/\epsilon$, for approximations to within a factor of $(1 + \epsilon)$ of optimal) and other approximation heuristics.

4.1. Various *FPT* Algorithms and Heuristics for VERTEX COVER. We can continue to use the VERTEX COVER problem to illustrate many of the main ideas. This is a useful example because it is a simple problem to describe, and because a wide variety of *FPT* algorithmic techniques can be applied to it. We will see how these different techniques can be systematically adapted into a corresponding variety of heuristics.

We begin by describing a new *FPT* algorithm for VERTEX COVER that is currently the best known. It is based on two simple, but standard methods, *reduction to a problem kernel*, and *search trees*. More examples and discussion of these methods can be found in [**DF95c, DF98, KST94, MR98**].

ALGORITHM 4.1. (An Improved Direct *FPT* Algorithm for VERTEX COVER.) The algorithm proceeds in two phases. In the first phase we compute the *kernel* of the given instance (G, k) or answer "no". The kernel is an instance (G', k') where $|G'| \leq k^2$ and $k' \leq k$ such that G' has a vertex cover of size k' if and only if G has a vertex cover of size k. The reduction from (G, k) to (G', k') is computable in time $O(kn)$, where k is the number of vertices in G. (That is, this is a *polynomial time* parametric transformation of VERTEX COVER *to itself*, such that the target instance has size bounded by a function of the parameter k.) Given that we can do this, we have immediately demonstrated membership in *FPT* for VERTEX COVER, since the question for (G', k') can now be answered in time bounded by a function of k simply by an exhaustive analysis of G'.

Phase 1 (Reduction to a Problem Kernel): Starting with (G, k) we apply the following reduction rules until no further applications are possible (the justifications for the reductions are given below):

(0): If G has a vertex v of degree greater than k, then replace (G, k) with $(G - v, k - 1)$.

(1): If G has two nonadjacent vertices u, v such that $|N(u) \cup N(v)| > k$, then replace (G, k) with $(G + uv, k)$.

(2): If G has adjacent vertices u and v such that $N(v) \subseteq N[u]$, then replace (G, k) with $(G - u, k - 1)$.

(3): If G has a pendant edge uv with u having degree 1, then replace (G, k) with $(G - \{u, v\}, k - 1)$.

(4): If G has a vertex x of degree 2, with neighbors a and b, and none of the above cases applies (and thus a and b are not adjacent), then replace (G, k) with (G', k) where G' is obtained from G by:

- Deleting the vertex x.
- Adding the edge ab.
- Adding all possible edges between $\{a, b\}$ and $N(a) \cup N(b)$.

(5): If G has a vertex x of degree 3, with neighbors a, b, c, and none of the above cases applies, then replace (G, k) with (G', k) according to one of the following cases depending on the number of edges between a, b and c.

(5.1): There are no edges between the vertices a, b, c. In this case G' is obtained from G by:

- Deleting vertex x from G.
- Adding edges from c to all the vertices in $N(a)$.
- Adding edges from a to all the vertices in $N(b)$.
- Adding edges from b to all the vertices in $N(c)$.
- Adding edges ab and bc.

(5.2): There is exactly one edge in G between the vertices a, b, c, which we assume to be the edge ab. In this case G' is obtained from G by
- Deleting vertex x from G.
- Adding edges from c to all the vertices in $N(a) \cup N(b)$.
- Adding edges from a to all the vertices in $N(c)$.
- Adding edges from b to all the vertices in $N(c)$.
- Adding edge bc.
- Adding edge ac.

The reduction rules described above are justified as follows:

(0): Any k-element vertex cover in G must contain v, since otherwise it would be forced to contain $N(v)$, which is impossible.

(1): It is impossible for a k-element vertex cover of G not to contain at least one of u, v, since otherwise it would be forced to contain all of the vertices of $N(u) \cup N(v)$. This allows us to add uv.

(2): If a vertex cover C did not contain u then it would be forced to contain $N[v] - u$. But then there would be no harm in exchanging v for u.

(3): If G has a k-element vertex cover C that does not contain v, then it must contain u. But then $C - u + v$ is also a k-element vertex cover. Thus G has a k-element vertex cover if and only if it has one that contains v.

(4): We first argue that if G has a k-element vertex cover C, then it must have one with one of the following forms:
1. C contains a and b, or
2. C contains x but neither of a, b, and therefore also contains $N(a) \cup N(b)$.

If C did not have either of these forms, then it must contain exactly one of a, b, and therefore also x. But in this C can be modified to form 1. If G has a k-element vertex cover of the form 1, then this also forms a k-element vertex cover in G'. If G has a k-element vertex cover of the form 2, then this same set of vertices with x replaced by either a or b is a k-element vertex cover in G'. Conversely, suppose G' has a k-element vertex cover C. If C contains both a and b, then it is also a k-element vertex cover in G. Otherwise, it must contain at least one of a, b, suppose a. But then the edges from b to all of the vertices of $N(a) \cup N(b)$ in G' force C to contain $N(a) \cup N(b)$. So $C - a + x$ is a k-element vertex cover in G.

(5.1): Let C denote a k-element vertex cover in G. If C does not contain x, then necessarily C contains $\{a, b, c\}$. In this case C is also a k-element vertex cover in G'. Assume that C contains x. We can assume (easy to check) that at most one of the vertices of $\{a, b, c\}$ belongs to C. If none of the vertices of $\{a, b, c\}$ belongs to C, then b belongs to the vertex cover of G'. If $a \in C$ then $C - x + c$ is a vertex cover for G', if $b \in C$ then $C - x + a$ is a vertex cover for G', and if $c \in C$ then $C - x + b$ is a vertex cover for G'.

(5.2): Let C denote a k-element vertex cover in G. If C does not contain x, then necessarily C contains $\{a, b, c\}$. In this case C is also a k-element vertex cover in G'. Assume that C contains x. We can assume (easy to check) that exactly one of the vertices of $\{a, b\}$ belongs to C. W.l.o.g we assume $a \in C$. Then $C - x + c$ is a vertex cover for G'.

It is easy to see that at the end of Phase 1 we have reduced (G, k) to (G', k') where G' has minimum degree 4, if we have not already answered the question.

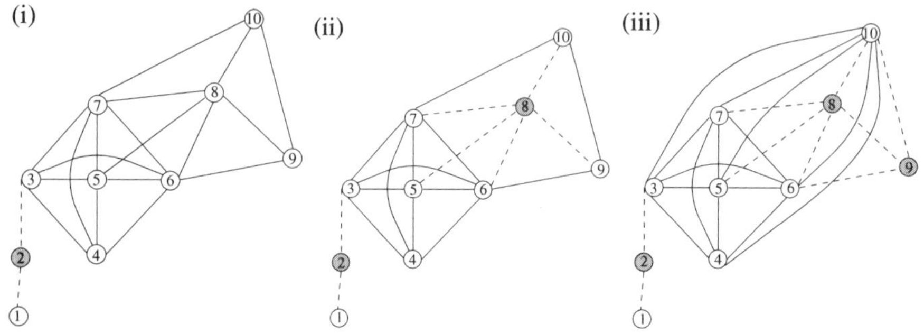

FIGURE 2. Let $k = 8$. (i) In the given graph $(1,2)$ is a pendant edge (case (3)). $k' = 7$, $G' = G - \{1,2\}$. 2 is added to the vertex cover. (ii) Vertices 8 and 10 are adjacent and $N(10) \subseteq N[8]$ (case (2)). (iii) The degree of vertex 9 is 2 (case (4)). $k''' = 7$, G''' is received by adding the following edges to $G'' - \{9\}$: $(10,3)$, $(10,4)$, $(10,5)$, $(10,6)$.

Furthermore, simply because of the reduction rules (1) and (2) we can conclude that the answer is "no" if the number of vertices in G' is more than k^2. Phase 1 is a good example of what is meant by the *FPT* technique of *reduction to a problem kernel.* If we still have no answer about the original input (G,k), then we are left with considering (G',k') where $|G'| \leq k^2$ and $k' \leq k$. A demonstration that the problem is in *FPT* is now trivial, since we can just (in the absence of further ideas) *exhaustively* answer the question for the kernel (G',k'). However, we can do a little better by analyzing the kernel instance by means of a search tree.

Figure 2 shows an example of the kernelization procedure of Phase 1.

Phase 2 (Search Tree): In this phase of the algorithm we build a search tree of height at most k. The root of the tree is labeled with the output (G',k') of Phase 1. We will describe various rules for deriving the children of a node in the search tree. For example, we can note that for a vertex v in G', and for any vertex cover C, either $v \in C$ or $N(v) \subseteq C$. Consequently we could create two children, one labeled with $(G'-v, k'-1)$, and the other labeled with $(G'-N[v], k'-\deg(v))$. In our algorithm, we perform this branching if there is a vertex of degree at least 6. By repeating this branching procedure, at each step reapplying the reductions of Phase 1, we can assume that that at each leaf of the resulting search tree we are left with considering a graph where every vertex has degree 4 or 5. If there is a vertex x of degree 4, then the following branching rules are applied. Suppose that the neighbors of a vertex x are $\{a,b,c,d\}$. We consider various cases according to the number of edges present between the vertices a,b,c,d.

Note that if not all of $\{a,b,c,d\}$ are in a vertex cover, then we can assume that at most two of them are.

Case 1. The subgraph induced by the vertices a,b,c,d has an edge, say ab.

Then c and d together cannot be in a vertex cover unless all four of a,b,c,d are there. We can conclude that one of the following is necessarily a subset of the vertex cover C and branch accordingly:

1. $\{a,b,c,d\} \subseteq C$
2. $N(c) \subseteq C$

3. $\{c\} \cup N(d) \subseteq C$.

Case 2. The subgraph induced by the vertices a, b, c, d is empty. We consider three subcases.

Subcase 2.1 Three of the vertices (say a, b, c) have a common neighbor y other than x.

Then when not all of a, b, c, d are in a vertex cover, x and y must be. We can conclude that one of the following is a subset of the vertex cover C and branch accordingly:

1. $\{a, b, c, d\} \subseteq C$
2. $\{x, y\} \subseteq C$.

Subcase 2.2 If Subcase 2.1 does not hold, then there may be a pair of vertices who have a total of six neighbors other than x, suppose a and b. If all of a, b, c, d are not in the vertex cover C then $c \notin C$, or $c \in C$ and $d \notin C$, or both $c \in C$ and $d \in C$ (in which case $a \notin C$ and $b \notin C$). We can conclude that one of the following is a subset of the vertex cover C and branch accordingly:

1. $\{a, b, c, d\} \subseteq C$
2. $N(c) \subseteq C$
3. $\{c\} \cup N(d) \subseteq C$
4. $\{c, d\} \cup N(\{a, b\}) \subseteq C$.

Subcase 2.3 If Subcases 2.1 and 2.2 do not hold, then the graph must have the following structure in the vicinity of x: (1) x has four neighbors a, b, c, d and each of these has degree four. (2) There is a set E of six vertices such that each vertex in E is adjacent to exactly two vertices in $\{a, b, c, d\}$, and the subgraph induced by $E \cup \{a, b, c, d\}$ is a subdivided K_4 with each edge subdivided once. In this case we can branch according to:

1. $\{a, b, c, d\} \subseteq C$
2. $(E \cup \{x\}) \subseteq C$.

If the graph G is regular of degree 5 (that is, there are no vertices of degree 4 to apply one of the above branching rules to) and none of the reduction rules of Phase 1 can be applied, then we choose a vertex x of degree 5 and do the following. First, we branch from (G, k) to $(G - x, k - 1)$ and $(G - N[x], k - 5)$. Then we choose a vertex u of degree 4 in $G - x$ and branch according to one of the above cases. The net result of these two combined steps is that from (G, k) we have created a subtree where one of the following cases holds:

1. There are four children with parameter values $k - 5$, from Case 1.
2. There are three children with parameter values $k_1 = k - 5$, $k_2 = k - 5$ and $k_3 = k - 3$, from Subcase 2.1.
3. There are five children with parameter values $k_1 = k - 5$, $k_2 = k - 5$, $k_3 = k - 5$, $k_4 = k - 6$ and $k_5 = k - 9$, from Subcase 2.2.

Note that if reduction rule (2) of Phase 1 cannot be applied to $G - x$, then at least one of the neighbors of u has degree 5, and so Subcase 2.3 is impossible.

The bottleneck recurrence comes from the degree 5 situation which produces four children with parameter values $k - 5$. The total running time of the algorithm is therefore $O(r^k k^2 + kn)$, where $r = 4^{1/5}$, or $r = 1.31951$ approximately. This time bound is a slight improvement on the $r = 1.32472$ of [**BFR98**]. (The tiny difference amounts to a 21% improvement in the running time for $k = 60$.)

The parameter function of our analysis of the running time of Algorithm 4.1 indicates that this "exact" algorithm is useful for input graphs of any size, so long as the parameter k is no more than about 60. There are a number of uses of VERTEX COVER in analyzing biological sequence data, and for this reason, essentially the above algorithm has been implemented as part of the DARWIN project at ETH [**HGS98**]. It turns out that the useful parameter range of $k \leq 60$ indicated by the parameter function above is overly pessimistic. In practice, the algorithm seems to be useful for k up to around 200. Note that if you run this algorithm on (G, k) where k is apparently too large, the worst that can happen is that the algorithm fails to finish in a reasonable amount of time. If it terminates, then it does give the correct answer.

The two phases of the above algorithm are independent. If one intended to solve the general VERTEX COVER problem by simulated annealing (or any other method), it would still make sense to apply Phase 1 *before* doing the simulated annealing, since the "simplifications" accomplished by the kernelization require only polynomial time (that is, they in no way depend on k being small). Consequently, Phase 1 is a reasonable first step for *any* general algorithmic attack on the *NP*-complete VERTEX COVER problem. We can codify this discussion by describing the following heuristic algorithm for the *general* VERTEX COVER problem. That is, the following algorithm, although it is based on *FPT* methods, has nothing to do with small parameter ranges, and it runs in polynomial time.

ALGORITHM 4.2. (A General Heuristic Algorithm for VERTEX COVER Based on Kernelization.) The algorithm simply reduces the input graph G to nothing by repeating the following two steps:

1. Apply Phase 1 of Algorithm 4.1.
2. Choose a vertex of maximum degree.

The reduction path gives an approximate minimum vertex cover for G. To see that this works correctly, it is necessary to observe that for each of the rules of Phase 1 reducing (G, k) to (G', k'), and given any vertex cover C' in G', we can reverse the reduction to obtain a vertex cover C in G.

In Algorithm 4.2, we have simply replaced Phase 2 of Algorithm 4.1, which is exponential in k, with a well-known approximation heuristic for this problem. However, we could also *adapt* Phase 2 by *simply not exploring all of the branches of the search tree*, either by making a random selection, or by branch selection heuristics, etc. The following is a natural adaptation of Algorithm 4.1 that exploits this idea for designing heuristics based on the *FPT* search tree technique.

ALGORITHM 4.3. (A General Heuristic Algorithm for VERTEX COVER Based on Kernelization and a Truncated Search Tree.) This is the same as Algorithm 4.1, except that in the second phase of building the the search tree, we do two things. First of all, we decide in advance how large of a search tree we can afford, and we build the tree until we have this number of leaves. Then for each leaf of the search tree, we apply an approximation heuristic such as the greedy method used in Algorithm 4.2, and take the best of all the solutions computed from the leaf instances of the search tree.

The search tree is developed according to the following branching heuristic. Given the instance (G, k) as the label on a node of the search tree to be expanded, we calculate the most valuable branch in the following way:

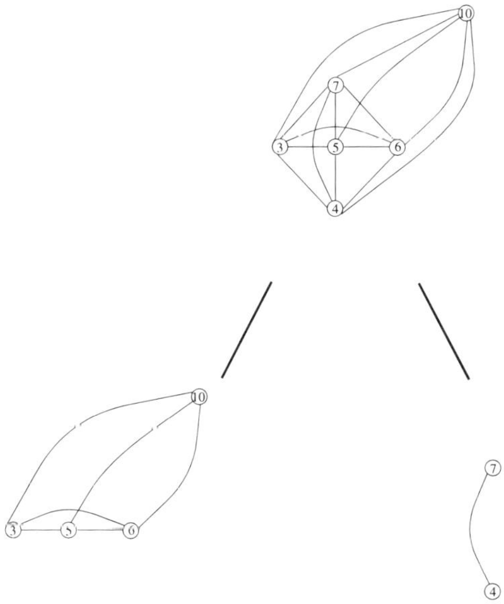

FIGURE 3. The resulting graph in Figure 1 is taken as root of the search tree in Algorithm 3. Here $a = 3$, $b = 4$, and $c = 7$.

1. Find a vertex a of maximum degree r in G.
2. Find a pair of vertices b, c such that $s = |S|$, where $S = N(b) \cap N(c)$ is maximized.
3. Determine which of the two equations $x^r - x^{r-1} - 1 = 0$ and $x^s - x^{s-2} - 1 = 0$ has the smallest positive real root. If the first equation, then branch to $(G - a, k - 1)$ and $(G - N[a], k - r)$. If the second, then branch to $(G - b - c, k - 2)$ and $(G - S, k - s)$. (This information can be precomputed and determined by table lookup.)

The branching heuristic above is justified because it essentially chooses the branching rule to apply according to the heuristic criterion that if it were the only rule applied in building a tree, then this would result in the smaller tree. (This could obviously be generalized by considering 3-element subsets of the vertices, or by employing a different criterion for choosing a branching rule.)

Algorithm 4.3 is illustrated in Figure 3.

In view of the remarkably nice *FPT* algorithms we now have for VERTEX COVER it is both ironic and instructive that the VERTEX COVER problem was first proved to be fixed-parameter tractable by the application of methods that are much more complicated than the problem actually requires. In fact, this happened in print at least three times [**FL87, Jo87, PY96**]. The first of these approaches is based on the Robertson-Seymour theorems, the second on finite-state dynamic programming on graphs of bounded treewidth, and the third on maximum matching. Probably the correct moral to draw from this history is that the design of *FPT* algorithms has its own distinctive issues and opportunities, and is not necessarily well-served by the habits we have developed for the design of polynomial time algorithms.

With respect to the connection between *FPT* and practical heuristics, the correct point of view seems to be that *all of the various FPT methods that can be applied to a problem may be useful in the design of heuristics.* We next describe a new, utterly *inefficient* (in terms of the parameter function) *FPT* algorithm for VERTEX COVER, based on completely different *FPT* techniques, and then describe how these methods can contribute, nevertheless, to the design of an interesting and apparently powerful heuristic.

DEFINITION 4.1. $G \geq H$ if there is a surjective map $h : V(G') \to V(H)$ with the property that for all $uv \in E(H)$, there are vertices $x, y \in V(G)$ such that $h(x) = u$, $h(y) = v$ and $xy \in E(G)$. (It is easy to verify that this defines a partial order on graphs.)

If h is a function that k-colors the vertices of $G = (V, E)$, $h : V \to \{1, ..., k\}$, then we will write $h(G)$ to denote the graph with the vertex set $h(V) \subseteq \{1, ..., k\}$ and the edge set

$$\{ij : \exists x, y \in V \text{ with } h(x) = i, h(y) = j \text{ and } xy \in E\}.$$

LEMMA 4.2. *For each fixed k there is a finite set $\mathcal{O}_k$ of graphs, such that for every graph G, G has a k-element vertex cover if and only if $\forall H \in \mathcal{O}_k :\ G \not\geq H$.*

PROOF. By the Graph Minor Theorem [**RS96**], it is enough to argue that if H is a minor of G then $G \geq H$. But this is essentially trivial, since one way of viewing the fact that H is a minor of G is that there is a *folio representation* of H in G. That is, there is a set of disjoint connected subgraphs $G_v \subseteq G$, one such subgraph for each vertex $v \in V(H)$, such that if $uv \in E(H)$ then there are vertices $x \in G_u$ and $y \in G_v$ with $xy \in E(G)$. From this it is easy to construct a function $h : V(G) \to V(H)$ showing that $G \geq H$. □

The celebrated Graph Minor Theorem used in the above argument simply states that every set of finite graphs has a finite number of minimal elements in the partial ordering of graphs by minors. For a gentle survey of this result with an eye to algorithmic applications, see [**FL88**].

LEMMA 4.3. *For every fixed graph H, it can be determined in time $O(n \log n)$, for an input graph G on n vertices, whether $G \geq H$.*

PROOF. Note that we are essentially proving that the problem of determining whether $G \geq H$ is in *FPT* for the natural parameter H. As a consequence of a theorem of Mader [**Mad72**], there is a constant c_H such that if a graph G on n vertices has more then $c_H n$ edges then necessarily G has H as a minor and therefore $G \geq H$. In linear time we can determine if this provides a reason to answer "yes" and otherwise we can assume a linear bound on the number of edges of G.

We use the powerful and general *FPT* method of *color-coding* introduced by Alon, Yuster and Zwick. In [**AYZ94**] it is shown that for every k there is a family $\mathcal{H}_k$ of functions $h : \{1, ..., n\} \to \{1, ..., k\}$ with the property that if S is any subset of $\{1, ..., n\}$ of size k, then $\exists h \in \mathcal{H}_k$ such that h maps S 1:1 onto $\{1, ..., k\}$. The crucial thing is that the family of functions $\mathcal{H}_k$ is not very large. In [**AYZ94**] it is shown that it is possible to produce such families of hash functions with $|\mathcal{H}_k| \leq 2^{O(k)} \log n$. (Note the reasonable parametric function, in the sense of the discussion in §2.)

Suppose H has m edges. Then $G \geq H$ if and only if there is a set of at most $2m$ vertices of G that can be mapped onto the vertices of H in order to witness this fact.

Let $\mathcal{F}_{2m}$ denote the set of all functions from the set $\{1, ..., 2m\}$ onto the vertex set of H. Then $G \geq H$ if this can be detected by the composition $f = g \circ h$ of some function $h \in \mathcal{H}_{2m}$ and $g \in \mathcal{F}_{2m}$, by having H isomorphic to $f(G)$. Our algorithm thus consists simply of computing $f(G)$ for all such functions f. This can clearly be accomplished in time $O(n \log n)$ with a hidden constant that is exponential in the size of the fixed graph H. □

The two lemmas above allow us to conclude immediately that VERTEX COVER is in *nonuniform FPT*. Nonuniform, because so far we can only conclude that for each fixed k there is a different algorithm for k-VERTEX COVER, based in each case on the obstruction set $\mathcal{O}_k$, which is different for each k, and not in any obvious way computable. In order to combine these separate algorithms into a single (uniform) *FPT* algorithm, we will use the self-reduction technique of [**FL94**]. The subroutine that we need is described by the following lemma.

LEMMA 4.4. *Suppose A is an oracle (black box) for the* VERTEX COVER *decision problem. Then by making $O(k^2)$ calls to the oracle, we can compute a k-element vertex cover for a graph G on n vertices (if it has one).*

PROOF. First note that by attaching pendant edges to vertices of G, we immediately have an easy algorithm that makes at most n calls to the oracle by repeatedly "probing" the graph by attaching a pendant edge to the vertices of G. Note that if uv is a pendant edge where u has degree 1, then if G has any k-element vertex cover, then it has one including the vertex v. We simply keep probing the vertex set until we have attached pendant edges to k vertices.

The operation of *shattering* a vertex x is accomplished by:

- Deleting x.
- Adding pendant edges to each vertex $y \in N(x)$.

If G has a k element vertex cover, then for any partition of the vertex set of G into $k+1$ classes there must be at least one class that is *safe to shatter* in the sense that if G' is obtained from G by shattering every vertex in the class, then G' has a k-element vertex cover that is also a vertex cover for G. Thus by making $k+1$ calls to the oracle, one for each such G', we can discover at least one new vertex that can be tagged with a pendant edge. (If $|G| >> k$ then we would expect to find not just one, but many, and then be able to apply the kernelization subroutine to substantially reduce the graph remaining.) At an intermediate step of the algorithm, there is some set S of $k' < k$ vertices of G that are tagged with pendant edges, and we know these k' vertices belong to some k-element vertex cover. We can delete a vertex $z \notin S$ if $N(z) \subseteq S$ (since the vertices of S are forced to be in the vertex cover, it is safe to assume that z is not). The remaining vertices are partitioned into $k - k' + 1$ approximately equal-sized classes. By inquiring about the graphs G' obtained by shattering these classes, we can discover at least one new vertex to add to S. Thus $O(k^2)$ calls to the oracle are sufficient. □

The astute reader will note that Lemma 4.4 can be improved to $O(k)$ calls to the oracle. We do not pursue this at the moment because the shattering strategy can be used in an interesting way in the design of a heuristic for the problem, and this is our main objective.

ALGORITHM 4.4. (A Direct But Completely Impractical *FPT* Algorithm for VERTEX COVER Based on Well-Quasiordering, Hashing, and Fast Self-Reduction.) The algorithm receives an input instance (G, k) and begins to build a list of the elements of $\mathcal{O}_k$ by systematically generating *all* finite graphs, and checking each

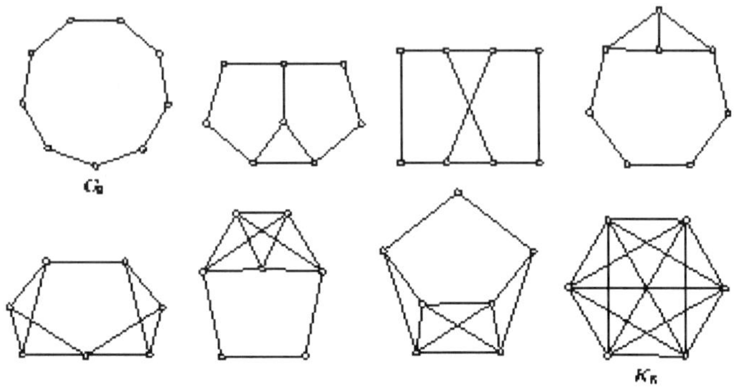

FIGURE 4. The connected obstructions for 4-VERTEX COVER

one to see if it should be added to the list (this is easy to determine). Since by what we have argued so far we have no way of knowing when the list is complete, we interleave this process with another, until we have decided whether to answer "yes" or "no". We call this procedure for finding a new obstruction *Process 1.*

Process 2 is executed whenever Process 1 has added something new to the list of $\mathcal{O}_k$. Process 2 acts under the *assumption* that the list is complete. If this were so, then by the means described by Lemmas 4.2 and 4.3, the list provides an "oracle" for the decision problem, by checking, for each H on the list, whether $G \geq H$.

Each call to this decision problem "oracle" requires time $O(f(k)n \log n)$, where f is some function of the parameter. We use the oracle to obtain either a definite reason to answer "no", by discovering that $G \geq H$ for some $H \in \mathcal{O}_k$, or a definite reason to answer "yes", by actually identifying and checking a k-element vertex cover in G. In order to obtain the latter evidence, we use the self-reduction algorithm of Lemma 4.4. Either of these efforts to give a definite answer may in fact succeed, even though our assumption that the list is complete is incorrect. (If the assumption is correct, then of course we will succeed in giving a definite answer.)

The only remaining possibility is that something weird happens. The incorrect assumption may cause the self-reduction algorithm to take too long or to output an incorrect solution. These malfunctions are easy to recognize. If we discover that the assumption was incorrect in this manner, then we return to Process 1 until we generate a new graph to add to the list of known elements of $\mathcal{O}_k$. Since $\mathcal{O}_k$ is finite, we alternate between the two processes at most $|\mathcal{O}_k|$ times.

The running time can be calculated to be $O(f(k)n \log n)$ for a function $f(k)$ that grows explosively mainly because of the size of the obstruction sets $\mathcal{O}_k$.

The set of connected obstructions for $k = 4$ is shown in Figure 4. In principle, the set $\mathcal{O}_k$ can be mechanically computed for any k. The set shown in Figure 4 was computed by Cattell and Dinneen [**CD94**].

Algorithm 4.4 is a nice example of the use of some very powerful and general *FPT* methods, and it is a good example as well of an *FPT* result that is utterly impractical as a direct algorithm. Nevertheless, we can adapt the *methods* of Algorithm 4.4 in designing new and interesting heuristics for VERTEX COVER. General heuristic design strategies that correspond to some of the main *FPT* methods are displayed in Table 2. The essential theme is to obtain heuristic methods from

FPT algorithms by strategies for *deflating the parametric costs* by truncating or sampling the search trees or obstruction sets, etc.

FPT Technique	**Heuristic Design Strategy**
Reduction to a Problem Kernel	A useful pre-processing subroutine for any heuristic.
Search Tree	Explore only an affordable, heuristically chosen subtree.
Well-Quasiordering	Use a sample of the obstruction set.
Color-Coding	Use a sample of the hash functions.

TABLE 2. Some *FPT* Methods and Heuristic Strategies

The well-informed reader may note the absence from Table 2 of an entry corresponding to the methods of bounded pathwidth and treewidth. This is an important open problem! We have the following heuristic algorithm for VERTEX COVER adapted from Algorithm 4.4.

ALGORITHM 4.5. (A New Family of Heuristics for VERTEX COVER.)

The heuristic algorithm that we describe is *self-analyzing* in the sense that it computes both a vertex cover $C \subseteq V$ for an input graph $G = (V, E)$, and a guarantee r that C is within a factor of r of optimal.

Step 1: An Initial Solution. The first step of the algorithm is to compute a vertex cover that is within a factor of 2 of optimal, using one of the well-known approximation algorithms that provide this performance guarantee.

Step 2: Improving the Upper Bound. Knowing that G has a vertex cover of size k_{upper}, we attempt to show that that it has one of size $k' < k_{\text{upper}}$. To do this, we use a strategy suggested by Lemma 4.4. If $S \subseteq V$ is a set of vertices of G, define the *shattering of G with respect to S* to be the graph obtained by "unplugging" any edges incident on vertices of S. If S consists of a single vertex, then this coincides with the definition given in the proof of Lemma 4.4. If two vertices $u, v \in S$ are adjacent, then this results in a K_2 from uv being unplugged from both endpoints. There are two salient facts that we use concerning this operation:

- If (G, k) is a yes-instance, then for any $(k+1)$-partition of the vertex set, there must be at least one class S of the partition such that (G', k) is also a yes-instance, where G' is obtained by shattering G with respect to S.
- If G' is the result of shattering G with respect to S, and C' is a vertex cover in G', then we can easily compute from C' a vertex cover C for G with $|C| < |C'|$.

Typically, the result G' of shattering G with respect to S will have many vertices of degree 1, and hence G' can be reduced by applying the rules for the kernelization phase of Algorithm 4.1. In this way, we obtain a much smaller instance from which we can possibly compute a vertex cover for G of the targeted size to improve the upper bound. Using the facts above, we develop a search tree based on some randomly or heuristically chosen partitions, and explore some of the branches (where a branch is given by shattering on a class of the partition).

There are various ways to work out the details of this exploration process. We might think of the tree as organized into *lower branches* (closer to the root) and *upper branches*. The lower branches create a search space of affordable size consisting of graphs that are somewhat smaller than G, but a search space that is guaranteed to contain yes-instance under the assumption that G is a yes-instance (because the lower branches include *all* shatterings for some partition). The upper branches are developed by repeating the shattering process enough times, but only exploring some of the branches, until the resulting graphs are completely solved. The exploration of the upper branches could also be based on a greedy heuristic.

Step 3: Improving the Lower Bound. Knowing a lower bound k_{lower} on the minimum size of a vertex cover in G, we attempt to prove a better lower bound $k > k_{\text{lower}}$. To do this, we use a strategy suggested in part by Lemma 4.2. However, before doing anything fancy, we should compute a maximal matching in G to see if there is an easy way to improve the lower bound by detecting the easy obstruction $(k+1)K_2 \in \mathcal{O}_k$.

Note that if (G, k) is a no-instance, then for any $(2k+3)$-coloring of the vertex set, there must be at least one class that can be *pinched* in the following way, to produce a (smaller) graph G' such that (G', k) is also a no-instance. That this is true follows from Lemma 4.2 and the fact that the maximum number of vertices of a graph $H \in \mathcal{O}_k$ is $2k+2$ (attained uniquely by $H = (k+1)K_2$). This can be used to develop the lower branches of a search tree as in Step 2, that is similarly guaranteed to have at least one branch to a simpler graph from which a proof that (G, k) is a no-instance can be obtained (if it is a no-instance).

For the upper branches of this search tree we can employ a partial exploration of the obstruction set for $\mathcal{O}_k$ and a partial exploration of the family of hash functions used in the proof of Lemma 4.3 to see if a reason for answering "no" can be obtained. It is easy to generate elements in $\mathcal{O}_k$ by taking disjoint unions of obstructions for smaller parameter values. For example $C_5 \in \mathcal{O}_2$ and $K_6 \in \mathcal{O}_4$ so $(C_5 \cup K_6) \in \mathcal{O}_7$.

Instead of incrementally applying Steps 2 and 3 separately to close the gap between k_{upper} and k_{lower}, some form of binary search could be adopted. For example, the first round of such a strategy might be to apply Steps 2 and 3 to $k = (k_{\text{upper}} + k_{\text{lower}})/2$.

The main point in describing the above heuristic (which has not yet been implemeted) is to illustrate how various typical *FPT* techniques can be systematically adapted into heuristic algorithms.

4.2. A Useful Parameterized Algorithm for MAST. We next describe a useful direct *FPT* algorithm for the MAXIMUM AGREEMENT SUBTREE (MAST) problem defined in §1, when it is restricted to binary trees, a reasonable restriction for biological applications. Apart from the intrinsic interest of this result, it is a nice example of two important points concerning *FPT* algorithms.

- Our algorithm for MAST ultimately uses an algorithm for VERTEX COVER as a subroutine. Useful *FPT* algorithms lead to other useful *FPT* algorithms, as one might naturally expect.
- There is already a polynomial time algorithm for MAST for binary trees, so why bother with an exponential *FPT* algorithm? The answer is that the polynomial time algorithm for MAST due to [**FPT95**] runs in time $O(rn^3 + n^d)$ for r phylogenetic trees of maximum degree d on a set of n species. The algorithm we describe requires time $O(c^k rn \log n)$ where c is a constant less than 3. Consequently, this is an example of a situation *where a classically exponential FPT algorithm may be preferable to a polynomial time algorithm.*

THEOREM 4.5. *The parameterized MAST problem can be solved in time $O(c^k rn \log n)$ for r trees on n species.*

SKETCH. The input to the problem is a set of rooted binary trees $T_1, ..., T_r$ each having n leaves labeled in 1:1 correspondence with a set X of n species. The problem is to determine if it is possible to delete at most k species from X to obtain a set X' on which all of the trees *agree*. In considering this problem, there is an elegant point of view developed by Bryant [**Bry97**] based on *triples*. If $\{a, b, c\}$ is a set of three species in X, then the restriction of each of the trees T_i to these three species must be one of the three possible alternatives (using parenthetical notation to represent trees): $(a, (b, c))$, $(b, (a, c))$, $(c, (a, b))$, or (a, b, c). If two or more of these three possibilities arise among the T_i, then obviously it will be necessary to eliminate at least one of the species in $\{a, b, c\}$ from X in order to obtain an agreement subtree. In this situation we will refer to $\{a, b, c\}$ as a *conflicted triple* of species.

An argument due to Bryant [**Bry98**] shows that our problem can be reduced in this way to the well-known 3-HITTING SET problem (see [**GJ79**]), which takes as input a collection $\mathcal{C}$ of 3-element subsets of a set X and a positive integer k, and must answer whether there is a subset $X' \subseteq X$ with $|X'| \leq k$ such that for each $A \in \mathcal{C}$, $A \cap X' \neq \emptyset$. Bryant's argument shows that our problem is equivalent to finding a k-element hitting set for the triples of X that are conflicted with respect to the T_i.

The set of conflicted triples could be computed by exhaustively computing the restrictions of the T_i for each 3-element subset of X, but this is not very efficient. In time $O(n \log n)$ it is possible to either determine that two trees are isomorphic, or identify a conflicted triple. Once a conflicted triple is identified, we can branch in a search tree based on 3 possibilities for resolving the conflict. For example, if the conflicted triple is $\{a, b, c\}$ then we create three branches in the search tree by deleting one element (say a) from X and from all of the trees T_i. We now recursively attempt to determine if the modified instance can be solved with $k' = k - 1$.

There will be at most $O(3^k)$ nodes in the search tree, and the running time due to each node is $O(rn \log n)$, which yields the claimed running time. □

The algorithm sketched above uses 3-HITTING SET implicitly as a means of solving the parameterized MAST problem. An improved but more elaborate *FPT* algorithm for this problem is described in [**BFRS98**], where we reduce MAST to 3-HITTING SET, and in turn devise a good algorithm for 3-HITTING SET using a subroutine for VERTEX COVER, to obtain an algorithm with $c = (1 + \sqrt{17})/2$. Some further examples of the systematic adaptation of *FPT* algorithms into useful heuristics for problems in computational biology (where there is an abundance of naturally parameterized problems) can be found in [**FKS98**].

4.3. The Steiner Problem for Generalized Hypercubes. In this section we give another *FPT* result based on the method of reduction to a problem kernel. This leads to an important heuristic algorithm that is *already in use.* We consider a slightly more general problem than the one described in §1.

THE STEINER PROBLEM FOR GENERALIZED HYPERCUBES
Instance: The input the problem consists of the following pieces of information:

1. A set of complete weighted digraphs D_i for $i = 1, ..., n$, each described by a set of vertices V_i and a function

$$t_i : V_i \times V_i \to I\!N$$

(We refer to the vertices of D_i as *character states*, to D_i as the *character state digraph*, and to t_i as the *state transition cost function* for the ith character.)
2. A positive integer k_1 such that $|V_i| \leq k_1$ for $i = 1, ..., n$.
3. A set X of k_2 length n vectors x_j for $j = 1, ..., k_2$, where the ith component $x_j[i] \in V_i$. That is, for $j = 1, ..., k_2$,

$$x_j \in \Omega = \prod_{i=1}^{n} V_i$$

4. A positive integer M.

Parameter: (k_1, k_2)
Question: Is there a rooted tree $T = (V, E)$ and an assignment to each vertex $v \in V$ of T of an element $y_v \in \Omega$, such that:

- X is assigned 1:1 with the set of leaves of T,
- The sum over all parent-child edges uv of T, of the *total transition cost* for the edge, defined to be

$$\sum_{i=1}^{n} t_i(y_u[i], y_v[i])$$

is bounded by M?

THEOREM 4.6. THE STEINER PROBLEM FOR GENERALIZED HYPERCUBES *is fixed-parameter tractable.*

PROOF. We define an equivalence relation $i \sim j$ on the index space $\{1, ..., n\}$ that allows us to combine D_i and D_j and obtain an equivalent smaller instance. In order to define $\sim$ we first define some other equivalences.

Fix $m \leq k_1$ and let l be an integer edge labeling of the complete digraph K_m on m vertices. Let $v_1, ..., v_m$ denote the vertices of K_m. Let T be a rooted tree with k_2 leaves labeled from $v_1, ..., v_m$. Define the *cost* of T with respect to l to be the minimum, over all possible labelings s of the internal vertices of T with labels taken from $\{v_1, ..., v_m\}$, of the sum over the parent-child edges of T of the transition costs given by l on the labels, and write this as

$$\text{cost}(T, l) = \min_s\{\text{cost}(T, s, l)\}$$

If l and l' are integer edge labelings of K_m and T is as above, then define $l \sim_T l'$ if and only if $\exists s$ such that

$$\text{cost}(T, l) = \text{cost}(T, s, l) = \text{cost}(T, s, l') = \text{cost}(T, l')$$

and define $l \sim l'$ if and only $l \sim_T l'$ for all such trees T.

For $i, i' \in \{1, ..., n\}$ define $i \sim i'$ if and only if:

1. $|V_i| = |V_{i'}| = m$ so that the only difference between D_i and $D_{i'}$ is in their arc-labelings l and l', and
2. $l \sim l'$.

The kernelization algorithm can now be described quite simply. Let $\mathcal{I}$ be an instance of the problem. If there are indices $i \neq i'$ for which $i \sim i'$, then modify $\mathcal{I}$ by combining these into one character state digraph with the state transition cost function given by the arc-labeling given $l + l'$, where these are the cost functions for D_i and $D_{i'}$, respectively. Let $\mathcal{I}'$ denote the modified instance.

The correctness of the reduction to the smaller instance is obvious. We need only to note that the equivalence $i \sim i'$ can be determined in time bounded by a function of the parameter and that number of equivalence classes is similarly bounded by a function of the parameter. □

The parameter function for this simple kernelization algorithm is not very good and can probably be much improved. We remark that most of the expense is in determining when two transition digraph indices i and i' are equivalent by testing them on all possible trees with k_2 leaves. This suggests a heuristic algorithm that combines indices when they fail to be distinguished by a (much smaller) random sample of trees and leaf-labelings.

In §1 we reported an encounter with an evolutionary biologist who reported earlier, rather fruitless interactions with theoretical computer scientists who proved that his problems were *NP*-complete and "went away". We claimed that we were different! and that we had a result on one of his computational problems (THE STEINER PROBLEM FOR HYPERCUBES) that might be of interest. After we described the *FPT* algorithm he said simply [**Fel97**]:

"That's what I already do!"

4.4. Kernelization and Heuristics: A Universal Connection. The example problems we have considered above exhibit a profound connection between kernelization algorithms for an *FPT* problem, and heuristics for the general unparameterized problem. The connection goes in both directions:

- An improved algorithm for the general problem, applied to the problem kernel, may extend the useful range of parameter values for which the problem is well solved.
- Because the kernelization phase is a polynomial time algorithm that simplifies and decreases the size of the problem instance, it is a reasonable first step for any general attack on the unparameterized problem.

It is interesting to investigate the question of to what extent good kernelization algorithms are typical of problems in *FPT*. This can be formalized as follows.

DEFINITION 4.7. A parameterized problem L is *kernelizable* if there is there is a parametric transformation of L to itself that satisfies:

1. the running time of the transformation of (x, k) into (x', k'), where $|x| = n$, is bounded a polynomial $q(n, k)$ (so that in fact this is a polynomial-time transformation of L to itself, considered classically, although with the additional structure of a parametric reduction),
2. $k' \leq k$, and
3. $|x'| \leq h(k)$, where h is an arbitrary function.

LEMMA 4.8. *A parameterized problem L is in* FPT *if and only if it is kernelizable.*

PROOF. The "if" direction is trivial. We can derive the "only if" direction from the result proved in [**CCDF97**] that a parameterized language L is in FPT if and only if there is a constant $\alpha \geq 1$ and a function $f(k)$ such that we can determine if $(x, k) \in L$ in time $O(n^\alpha + f(k))$. Our kernelization algorithm for L considers two cases:

1. If $k < f^{-1}(n)$ then in time $O(n^\alpha)$ we can simply answer the question.
2. If $k \geq f^{-1}(n)$ then $n \leq f(k)$ and the instance (x, k) already belongs to the problem kernel.

□

Since we are naturally interested in the *efficiency* with which problem kernels can be computed, it makes sense to consider the following modification of the above definition. This essentially sets up a time hierarchy inside of FPT based on the running time of the kernelization algorithm that every FPT problem has (by the lemma above).

DEFINITION 4.9. Let c be a fixed constant. A parameterized problem L is *strong n^c-kernelizable* if there is a recursive function f, a polynomial q and a parametric transformation of L to itself running in time $O(q(k)n^c)$ for $n = |x|$, taking (x, k) to (x', k') such that:

1. $k' \leq k$, and
2. $|x'| \leq f(k)$.

We say L is *weak n^c-kernelizable* if f is not required to be recursive. Let $K(n^c)$ denote the c-kernelizable subset of FPT (either strong or weak depending on the context).

The following somewhat technical theorem corresponds to the standard time hierarchy theorem, but is more difficult to prove because of the need for a "wait and see" argument to deal with the parameter functions.

THEOREM 4.10. *If $c < c'$ are constants, then*

$$K(n^c) \subset K(n^{c'}) \subset FPT$$

PROOF. It is sufficient to prove that $K(n^c) \subset$ strong $FPT(n^c)$ for all $c \geq 1$, where by strong $FPT(n^c)$ we mean the parameterized problems solvable in time $f(k)n^c$ for some recursive function f. We will only do the case $c = 1$, the more general case follows by the same technique. We consider the strong flavor of the theorem first.

Let φ_e denote the e-th partial recursive function. Let g_e denote the e-th parametric self reduction, so that $g_e : (x, k) \mapsto (h_e(x), q_e(k))$, and is polynomial in k, say in time $|k|^e$, but linear in $|x|$. It suffices to construct a $FPT(n)$ language L to meet the requirements $\mathcal{R}_{e,i}$ below.

$$\begin{aligned} \mathcal{R}_{e,i}: \quad & \text{either } \varphi_i \text{ is not total or} \\ & \exists x, k\ |h_e(x)| \not\leq \varphi_i(k) \text{ or} \\ & \exists x, k\ (g_e((x, k)) \in L \text{ iff } (x, k) \notin L). \end{aligned}$$

We set aside row $\langle e, i \rangle$ for the sake of requirement $\mathcal{R}_{e,i}$. Our language L will be in FPT with each row in time $2^k|x|$. At each stage s we will look only at one requirement, cycling through the requirements so that we see them infinitely often. At stage s we will decide the fate of all strings (x, k) with $|x| = s$. For a single requirement $\mathcal{R}_{e,i}$, our strategy is as follows. We need to do nothing until we arrive at a stage s where $\varphi_i(k)\downarrow$ in fewer than s steps. At this stage $\mathcal{R}_{e,i}$ becomes *active*. Note that if $\mathcal{R}_{e,i}$ never becomes active then φ_i is not total. (Similarly we can now check that $|h_e(x)| \leq \varphi_i(k)$, which will be implicit in the discussion below.)

We next consider an active requirement at a stage t where we can know that $2^{k-1}t > |k|^e t + \varphi_i(k)$. Note that at such a stage, we can, in $2^{k-1}t$ steps, work out the image of $(1^t, k)$ under g_e, namely $(h_e(1^t), q_e(k)) = (x', k')$, say. The diagonalization is then simple. If we see that we have already put (x', k') into L then we declare that (x, k) will be kept out of L. If we see that we have $(x', k') \notin L$ then we put (x, k) into L.

This concludes the strategy for a single requirement $\mathcal{R}_{e,i}$ and there is obviously no problem in dealing with the combinations of requirements.

In the arbitary case, where f is not computable, then we will replace the requirement $\mathcal{R}_{e,i}$ above by the infinitely many requirements $\mathcal{R}_{e,i,j}$ below.

$$\begin{aligned}\mathcal{R}_{e,i,j}: \quad & \text{either } \exists x, k h_e(x) \not\leq j \text{ or} \\ & \exists x, k\ (g_e((x,k)) \in L \text{ iff } (x,k) \notin L).\end{aligned}$$

The idea is that j is the "guess" we will use for $f(k)$. While we think that $f(k) = j$ we regard j as active. We activate $j+1$ when j proves wrong. If all the j prove wrong then there is no value for $f(k)$. We devote a single row $\langle e, i\rangle$ to the collection $\{\mathcal{R}_{e,i,j} : j \in \mathbb{N}\}$. The idea is essentially the same. When we consider j, we will be looking for stages t where $2^{k-1}t > |k|^e t + j$. If we see that the size of $h_e(1^t)$ exceeds j then we cancel j and move on to $j+1$. If not, then we can diagonalize as before. □

Cheeseman et al. [**CKT91**] have discussed (essentially) the idea that the *FPT* method of *reduction to a problem kernel* is a strategy that can and should be applied as a preprocessing step for *any* computational problem. They describe some reduction rules for GRAPH COLORING, and present statistical evidence that the set of reduced instances signal a complexity "phase transition" where the hard instances of the problem are concentrated. (The notion of reduction discussed in [**CKT91**] is technically distinct from our notion of parametric kernelization because they make no assumption about the size of the resulting reduced instances.)

4.5. *FPT* **Methods and Other Means of Coping.** There seem to be fairly strong methodological connections between *FPT* algorithm design methods and methods for devising polynomial time approximation schemes and other approximation heuristics. Such a connection might naturally be expected, since in an approximation scheme for an *NP*-hard problem, one would normally anticipate a running time exponential in $k = 1/\epsilon$ for an algorithm that computes an approximation to within a factor of $(1+\epsilon)$ of optimal. (This is isn't strictly necessary, since the problem may have a *fully polynomial-time approximation scheme* (see [**GJ79**]). However, these seem to be quite rare.) We mention three examples of connections between *FPT* methods and other coping strategies, especially approximation.

EXAMPLE 4.1 (Efficient Approximation Schemes for Planar Graphs). A notable example of efficient approximation is the family of schemes devised by Baker [**Ba94**] for planar graph problems. The basic idea is quite simple and very much based on the standard *FPT* techniques of bounded treewidth and pathwidth

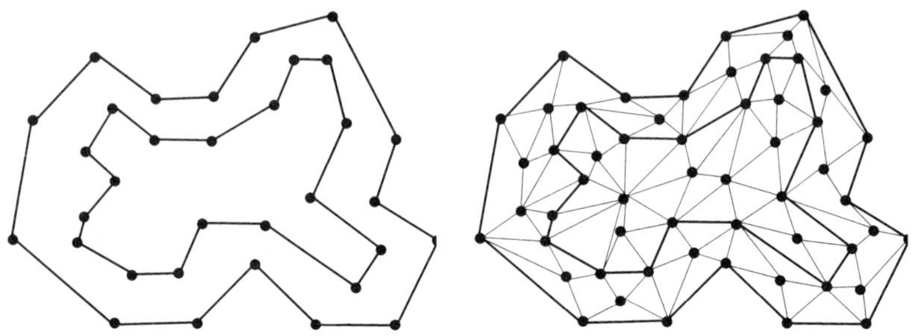

FIGURE 5. Illustration of bounded width onion decomposition for a planar graph. The figure on the left shows the boundaries of the layers.

(thoroughly exposited in [**DF98**]). For concreteness, consider the problem of computing a minimum dominating set for a planar graph. Starting from a planar embedding of the graph G, we can define a kind of "onion" decomposition of G. The layers of this decompostion can be defined in such a way that each has bounded treewidth. Figure 5 illustrates the idea.

The basic idea is to solve the problem exactly for each layer, using bounded treewidth machinery, and then combine the solutions for the layers. There is a small amount of inefficient overlap as the solutions for the layers are combined (by taking the union of the relevant vertex sets, for example). As the width of the layers is increased (at the usual *FPT* costs for bounded treewidth algorithms), the overall effect of the inefficient overlap on the goodness of the approximation progressively decreases, yielding an approximation to within a factor of $(1+\epsilon)$ of optimal for G for an appropriately chosen width of the layers. This same idea can be applied to a number of other graph problems.

EXAMPLE 4.2 (Testing Hereditary Properties Efficiently on Average). One of the few general techniques for devising algorithms that are efficient on average has been described by Gustedt and Steger [**GS94**] based on obstruction sets, one of the most powerful ideas in the *FPT* toolkit. The basic idea of their approach is that for some properties of combinatorial objects that are lower ideals in an appropriate partial order (which implies that the property is characterized by an obstruction set), the probability may be high for "typical" inputs that the answer is "no" (the object does not have the property), and that this sometimes can quickly be detected by checking for obstructions (not necessarily all of them). This strategy is employed as a preface, quickly answering the question for *most* inputs, thus raising the performance on average of whatever algorithm is used to deal with the remaining cases. Arguing along similar lines to our discussion of kernelization, they point out that this is a reasonable preface for *any* algorithmic attack on these kinds of problems. Our Algorithm 4.5 can be viewed in part as an example of this strategy. Langston et al. have similarly used approximate obstruction sets to design heuristics for VLSI layout problems [**LR91, GLR92**].

EXAMPLE 4.3 (Iterative Approximation Heuristics). An important example of a much-used approximation heuristic is the k-Opt heuristic for the TRAVELING SALESMAN PROBLEM [**JMc97**]. This can be viewed as a simple example of the following general recipe for the design of local search heuristics based on *FPT* methods.

First: Identify an *FPT* result for a modification of the problem that can be used as a subroutine to compute a single iteration of the local search. (For the k-Opt heuristic for the TSP, the relevant subroutine computes the best rearrangement of the k pieces into which the current best tour has been cut. This can be accomplished in time $O(k!n)$.)

Second: Iterate the application of this subroutine.

This design paradigm naturally raises the following question about the k-Opt neighborhood structure for TSP: Is there an *FPT* algorithm to compute the best neighbor s' of a solution s? Fisher [**Fis95**] has shown that it is *NP*-complete to determine whether there is a local optimum within k steps of s for the 2-Opt neighborhood structure for TSP. Whether this problem is in *FPT* is also open.

4.6. Some Further Discussion.

REMARK 4.11. The design of non-polynomial algorithms has been an area somewhat neglected by theoretical computer scientists — or at least it is an area in which there are many unexplored opportunities of importance to the applied community. The design of *FPT* algorithms systematically focuses attention in this direction. It is notable how the original, unparameterized problem is resurrected in the problem kernel (e.g., for VERTEX COVER in the second phase of Algorithm 4.1 we are essentially addressing the problem for a graph of size n with $k = \sqrt{n}$). It would be interesting to know if any systematic connection can be made between the "optimum" size of a problem kernel (which requires a definition) and the *complexity threshold* for the problem [**KS94**], or with the structural notion of the *complexity core* of the decision problem.

REMARK 4.12. For *FPT* problems that can be kernelized efficiently we end up in the happy situation that the problem can be solved for input of *any* size, so long as the parameter function $f(k)$ is not too large. This means that our reward for work on an improved algorithm for the general hard problem (with k unrestricted) is *leveraged* in a very significant way for problems that are *FPT* (by applying the improved general algorithm to the kernel). This also suggests measuring our success on an *FPT* problem L in terms of *c-klams*, where the c-klam value of an algorithm A is defined to be the largest k_{max} such that:

1. L can be solved by A in time $f(k) + n^c$, and
2. $f(k_{\max}) \leq U$, where U is some reasonable absolute bound on the maximum number of steps of any computation, e.g., $U = 10^{12}$ as suggested by Yessenin-Volpin [**YV70**] (or perhaps a somewhat larger number).

   If we can achieve for TREEWIDTH (see below) a $c = 3$ klam value of even just 8 or 10, this will be very significant for number of practical applications.

REMARK 4.13. The program of dealing with a hard problem by parameterizing can be compared to the program in Chapter 4 of Garey and Johnson [**GJ79**] of mapping the boundary between *NP*-hardness and polynomial-time subproblems in the following way. Suppose that we have an *FPT* result requiring time $O(2^k n^c)$ about the problem Π with parameter k. There are two different natural "forgetful functors" from the parametric to the classical framework. The first regards this *FPT* result as saying that for each fixed k the classical problem is solvable in polynomial time. The information loss in this is obvious, since it forgets entirely the distinction between *FPT* and *XP* (see Table 1). A second possible mapping of the *FPT* result to the classical framework would regard this as saying that Π restricted to $k = O(\log n)$ is solvable in *P*-time. In general, it would be fruitful to find, for a collection of parameters of a problem Π, some such bounds (as functions of the input size n) such that for parameters less than these bounds, Π can be solved in polynomial time. (I.e., this is a nice canonical kind of subproblem.) This is *precisely* the issue in parameterized complexity: for a given collection of parameters, is there *any* corresponding collection of bounds (as functions of n) such that the canonical subproblem defined by these bounds is in *P*? $W[1]$-hardness shows when the answer is probably "no". Improving the parameter function of an *FPT* problem is equivalently the problem of improving such bounds.

REMARK 4.14. We conjecture that the most promising method that the reader can apply towards developing useful *FPT* algorithms and heuristics for *NP*-hard problems is to purchase the moderately priced and entertaining book [**DF98**].

5. Applications of Treewidth to Problems of Logic

In this section we explore the theme of treewidth as a hidden parameter in many problems, particularly those of logic and programming languages. It is possible that treewidth represents one of the most universal hidden economies of computational tractability. (Parts of this section are unavoidably somewhat technical. For some of the arguments here the reader needs to be familiar with the "Myhill-Nerode" bounded treewidth algorithmic machinery developed in [**FL89, Co90, CL95, AF93**]. A thorough exposition of this machinery can be found in [**DF98**].)

Many computational problems about graphs that are intractable for general graphs can solved efficiently for trees. Graphs of *bounded treewidth* constitute classes of graphs that generalize from trees (which have a treewidth bound of 1) to graphs that are (intuitively) "tree-shaped". A concrete and relatively accessible definition of the family of graphs of treewidth at most k can be given in two steps as follows. We first define k-trees, and then note that a graph has treewidth bounded by k if and only if it is a subgraph of a k-tree. A k-tree is defined recursively by the rules:

- The complete graph on $k+1$ vertices is a k-tree.
- If H is a k-tree and $C \subseteq H$ is k-clique in H, then also the graph H' is a k-tree, where H' is obtained from H by adding a new vertex u and adding edges uv between u and each vertex v of the clique C.

The reader who is unfamiliar with bounded treewidth should sketch a few 2-trees to develop some intuition about this generalization of trees. Given the enormous importance of trees in feasible computing, it is elegant that bounded treewidth should turn out to be such a useful and natural structural generalization arising in many different contexts.

Thorup recently proved the remarkable result that the flow graphs that arise from structured programs have bounded treewidth. The bound on the treewidth is specific to the particular programming language and depends on whether certain language constructions (such as loop-exits, conditional 'or's, etc.) are allowed. For example, the flow graphs of structured C++ programs have a treewidth bound of 7 [**Th97**].

We explore how treewidth can be applied as a parameter to two problems in logic, MINIMUM AXIOM SET and STRIPS PLANNING. These problems are defined classically as follows.

MINIMUM AXIOM SET

Instance: A finite set S of *sentences*, an *implication relation* R consisting of pairs (A, t) where $A \subseteq S$ and $t \in S$, and a positive integer k.

Question: Is there a set $S_0 \subseteq S$ with $|S_0| \leq k$ and a positive integer n such that if we define S_i for $1 \leq i \leq n$ to consist of exactly those $t \in S$ for which either $t \in S_{i-1}$ or there exists a set $U \subseteq S_{i-1}$ such that $(U, t) \in R$, then $S_n = S$?

MINIMUM AXIOM SET with parameter k has been shown to be complete for $W[P]$ in [**DFKHW94**].

PROPOSITIONAL STRIPS PLANNING

Instance: A dimension n representing a number of atomic formulas called *conditions*, an *initial state vector* $S \in \{0,1\}^n$ indicating which of conditions are initially *true*, a goal vector $G \in \{0,1,*\}^n$ expressing the goal in terms of conditions that should be *true* (1), *false* (0) or *don't care* ($*$), and a collection $\mathcal{O}$ of *operators*, where each operator $o \in \mathcal{O}$ is a pair $o = (P, Q)$ with $P \in \{0,1,*\}^n$ expressing the *preconditions* of the operator in the natural way, and $Q \in \{0,1,*\}^n$ expressing the *postconditions* of the action. An operator $o = (P, Q)$ can be applied to the current state vector X if for all i, $1 \leq i \leq n$, $P[i] = a \in \{0,1\}$ implies $X[i] = a$, that is, the preconditions naturally expressed by P are met by the state vector X. The state vector X' resulting from the operation o in this situation is defined by:

1. $X'[i] = X[i]$ if $Q[i] = *$, and
2. $X'[i] = Q[i]$ otherwise.

Question: Is there a sequence of operations that can be applied, starting from the initial state vector S, and resulting in the state vector G.

The PROPOSITIONAL STRIPS PLANNING problem was first described by Fikes and Nilsson [**FN71**] and was proved to be *PSPACE*-complete by Bylander [**Byl94**].

We must first define how the notion of treewidth can be applied to these problems. To do this we define two digraph pebbling problems to which these logic problems can be conveniently translated.

DIGRAPH PEBBLING

Instance: A digraph $D = (V, A)$ for which the vertex set V is partitioned $V = R \cup B$ into two classes, the red vertices of R and the blue vertices of B, and a positive integer k.

Question: Is it possible to pebble each vertex of the digraph according to the following set of rules?

1. Start with k pebbles placed on red vertices.
2. A blue vertex b can be pebbled if there is a pebbled vertex u such that $ub \in A$.
3. A red vertex r can be pebbled if for every vertex u such that $ur \in A$, u is pebbled.

If every vertex is red, then this is the same problem as DIRECTED FEEDBACK VERTEX SET and therefore DIGRAPH PEBBLING is *NP*-complete [**Karp72**]. The parameterized complexity of DIRECTED FEEDBACK VERTEX SET (with parameter k) is semi-famously open.

SIGNED DIGRAPH PEBBLING

Instance: A red/blue bipartite digraph $D = (V, A)$ for which the vertex set V is partitioned $V = R \cup B$ into two classes, and also the arc set A is partitioned into two classes $A = A^+ \cup A^-$.

Question: Starting from the *start state* where there are no pebbles on any of the

red vertices, is it possible to reach the *finish state* where there are pebbles on all of the red vertices, by a series of moves of the following form?
A legal move: If b is a blue vertex for which $\forall u$ such that $ub \in A^+$, u is pebbled, and $\forall u$ such that $ub \in A^-$, u is not pebbled (in which case we say that b is *enabled*), then the set of vertices v such that $bv \in A^+$ are reset by making them all pebbled, and the set of vertices v such that $bv \in A^-$ are reset by making them all unpebbled.

We leave it to the reader to verify that an instance $\mathcal{I}$ of MINIMUM AXIOM SET can be naturally translated into an instance $\mathcal{I}'$ of DIGRAPH PEBBLING, where the set of sentences S of $\mathcal{I}$ becomes the set of blue vertices B of $\mathcal{I}'$. We can define the *treewidth* of $\mathcal{I}$ to be the treewidth of the digraph of $\mathcal{I}'$. The representation of a PROPOSITIONAL STRIPS PLANNING instance as an instance of SIGNED DIGRAPH PEBBLING is equally straightforward, and we adopt a similar notion of the treewidth of an instance.

We now consider the following parameterizations of these problems. By DIGRAPH PEBBLING I we refer to this problem with the parameter k. (This is complete for $W[P]$ by the results on MINIMUM AXIOM SET cited above.) By DIGRAPH PEBBLING II we refer to this problem with parameter (k, w) where w is the treewidth of the instance. By SIGNED DIGRAPH PEBBLING I we refer to this problem with parameter w, where w is the treewidth of the instance, and by SIGNED DIGRAPH PEBBLING II we refer to this problem with parameter (k, w) where w is the treewidth, and where k is a bound on the number of moves in which to reach the finish state. Note that the number of moves is a reasonable parameter for planning problems. The new results that follow all make use of the Myhill-Nerode techniques for bounded treewidth developed in [**MP94, FL89, AF93, BFW92, FHW93, DF98**].

We remark that it appears difficult to express either DIGRAPH PEBBLING II or SIGNED DIGRAPH PEBBLING II in Monadic Second-Order logic, a very powerful but not always applicable method for obtaining tractability results for bounded treewidth (see [**Co90**] and [**DF98**] for expositions).

In order to apply the Myhill-Nerode machinery for bounded treewidth the key notion that is needed is that of a *t-boundaried graph,* which is just a graph with t distinguished vertices. What is ultimately going on is that the structure of a bounded treewidth graph or digraph is represented by a rooted labeled *parse tree.* A graph of treewidth bounded by t can be parsed in linear time (i.e., this problem is linear *FPT*) by a theorem of Bodlaender [**Bod96**]. If $\mathcal{F}$ is a set of rooted labeled trees, then leaf-to-root tree automata provide a notion of finite-state recognizability for $\mathcal{F}$. The Myhill-Nerode theorem for bounded treewidth relates finite-state recognizability of $\mathcal{F}$ to the amount of information that must flow across a t-sized boundary in order to determine membership in $\mathcal{F}$. The machinery is applicable not just to ordinary graphs, but also to finitely edge- and vertex-colored digraphs.

THEOREM 5.1. DIGRAPH PEBBLING II *is in FPT.*

PROOF. It is sufficient to argue that the canonical equivalence relation on t-boundaried (vertex-colored) digraphs, for the family $\mathcal{F}_k$ of k-pebblable digraphs, has a finite number of equivalence classes for each fixed pair (t, k). We first review the essential notions.

By a *t-boundaried digraph* we mean a digraph $D = (V, A)$ equipped with t distinguished vertices labeled $1, ..., t$. We refer to these vertices as constituting the *boundary set* $\partial(D) \subseteq V$. The *interior* vertices of D are those vertices that are not boundary vertices, and we write $\text{int}(D) = V - \partial(D)$.

If D and D' are t-boundaried digraphs, we say that they are *compatible* if their boundaries are colored in the same way, that is, for $i = 1, ..., t$ the ith boundary vertex of D is red (blue) if and only if the ith boundary vertex of D' is red (blue).

If D and D' are compatible t-boundaried digraphs, then by $D \oplus D'$ we denote the digraph obtained by identifying the vertices $i \in \partial(D)$ with $i \in \partial(D')$ for $i = 1, ..., t$. The canonical equivalence relation on the set of all 2-vertex-colored t-boundaried digraphs, for a family of such digraphs $\mathcal{F}$, is defined for compatible digraphs by:

$$D \sim_{\mathcal{F}} D' \text{ if and only if } \forall D'' : D \oplus D'' \in \mathcal{F} \leftrightarrow D' \oplus D'' \in \mathcal{F}$$

Fix k and t. Our approach is to define a different (easier to work with) equivalence relation $\sim$ on t-boundaried digraphs, and then to show two things:
Claim 1. that $\sim$ has finite index, and
Claim 2. that $D \sim D'$ implies $D \sim_{\mathcal{F}} D'$, where $\mathcal{F} = \mathcal{F}_k$.
This allows us to conclude that $\sim_{\mathcal{F}_k}$ has finite index on t-boundaried digraphs.

Our definition of $\sim$ is based on a set of abstract *tests* in the sense of the "method of test sets" developed and exposited in [**FL89, AF93, CDDFL98, DF98**].

A *test* T is specified by the following information:

1. A partition of $\{1, ..., t\}$ into two subsets, the red subset T_R and the blue subset T_B.
2. A *boundary starter set* $S \subseteq \{1, ..., t\}$. (We will use k_0 to denote the size of S.)
3. A positive integer k_1 such that $k_0 + k_1 \leq k$. (We will use k_2 to denote $k - k_0 - k_1$.)
4. A permutation π of $\{1, ..., t\} - S$.
5. A subset $Q \subseteq T_B$.

A test T is *compatible* with a t-boundaried red/blue digraph D if the partition given in (1) accords with the colors of the boundary vertices of D.

Let $\mathcal{T}$ denote the set of tests. We have the bound on the size of the test set

$$|\mathcal{T}| \leq k2^{2t}t!$$

For a digraph $D = (V, A)$ and a vertex $v \in V$, let

$$\text{pred}(v) = \{u \in V : uv \in A\}$$

be the set of *predecessors* of v.

By a *pebbling sequence* for a t-boundaried digraph D we mean a permutation σ of the vertices of D. If σ is a pebbling sequence for D and v is a vertex of D, then v is *legal* with respect to σ if either of the following statements holds.

1. v is a blue vertex, and at least one predecessor of v precedes v in σ
2. v is a red vertex, and every predecessor of v precedes v in σ.

Say that a t-boundaried digraph $D = (V, A)$ *passes* the test $T \in \mathcal{T}$ if T is compatible with D and there is a pebbling sequence σ for D that satisfies the requirements:

1. The prefix σ' consisting of the first $k_0 + k_1$ vertices of σ includes the k_0 vertices of S and k_1 additional interior vertices of D.
2. The order in which the vertices of $\partial(D) - S$ occur in σ is specified by the permutation π of $\partial(D) - S$.
3. Every red vertex not in the prefix σ' is legal.

4. Every blue vertex b not in the prefix σ' is either legal or belongs to Q.

Intuitively, the set Q represents blue vertices that will be pebbled for "reasons on the other side of the boundary".

Let $\mathcal{T}(D)$ denote the set of tests passed by D. Given two t-boundaried digraphs D and D', define $D \sim D'$ if and only if $\mathcal{T}(D) = \mathcal{T}(D')$. Since the relation $\sim$ is defined in terms of equality of sets, it is clearly an equivalence relation.

Now that we have defined a suitable equivalence relation, we have only to prove the two Claims listed above. Claim 1 is trivial, since $\mathcal{T}$ is finite.

Proof of Claim 2. We argue by contraposition. Suppose $D \not\sim_{\mathcal{F}} D'$, and let this be witnessed by a t-boundaried digraph X with $D \oplus X \in \mathcal{F}$ and $D' \oplus X \notin \mathcal{F}$. Let σ denote a permutation of the vertices of $D \oplus X$ corresponding to a successful k-pebbling. Thus every vertex u occuring in σ after the initial prefix of length k is legal with respect to $D \oplus X$. Let S denote the starter set of size at most k for the successful pebbling (consisting of both interior and boundary vertices), and let S_0 denote the subset of S on the boundary

$$S_0 = S \cap \partial(D \oplus X)$$

with $k_0 = |S_0|$. Let k_1 denote the size of $S \cap \text{int}(D)$. Let π denote the permutation of $\partial(D) - S_0$ inherited as a subsequence of σ. Then D passes the test T defined by the information: (1) the color partition of the boundary, (2) the boundary starter set S_0, (3) the positive integer k_1, (4) the permutation π, and (5) the set Q consisting of those blue vertices of the boundary which are illegal for σ restricted to the vertices of D.

We argue that D' fails T. Suppose not, and let σ' be a permutation of the vertices of D' that starts with S_0 followed by k_1 interior vertices of D' that is a valid pebbling sequence for D' showing that it passes the test T. Using σ' we can modify σ to obtain a valid pebbling sequence for $D' \oplus X$ and reach a contradiction. The contradiction will allow us to conclude that D' fails T and thus $D' \not\sim D$, which will establish Claim 2.

The modification of σ is described as follows. The permutation σ can be partitioned into three subsequences σ_0, σ_1 and σ_2 according to membership in the three disjoint sets of vertices $\partial(D \oplus X)$, $\text{int}(D)$ and $\text{int}(X)$. The modification consists of two steps:

1. The subsequence σ_1 is deleted.
2. Substrings of σ' are inserted.

The relevant substrings of σ' are those consisting only of interior vertices of D' that are bounded by either the initial block of starter vertices for the pebbling of D' described by σ' or by vertices of the boundary, or by the end of the sequence. Each such substring of interior vertices of D' has a left boundary and a right boundary. Call the substring whose right boundary is the end of σ' the *last substring.* All of the other substrings have a right boundary consisting of a boundary vertex of D'.

These substrings, other than the last substring, should be inserted into the modified σ just before each boundary vertex in σ, and the last substring should be concatenated to the end of the modified result. The block of interior starter vertices of σ' should be concatenated to the beginning of the modified σ. Call the result of all this σ''.

The point of all these modifications is to produce a pebbling sequence σ'' that works for $D' \oplus X$. What we have done is replace "what happened inside of D"

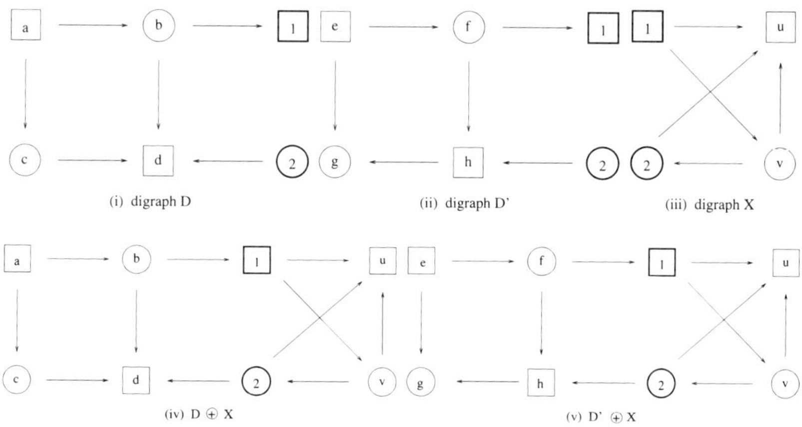

FIGURE 6. Examples for Theorem 5.1.

according to σ, with equivalent bits from σ' describing how D' can pass the test T, which simply records "what D saw of X" in the pebbling described by σ (where D interacts with X only across the boundary). It is not hard to check that σ'' is a valid pebbling sequence for $D' \oplus X$. □

Figure 6 illustrates the argument. The boundary vertices of the two digraphs D and D' are indicated in bold in Figure 6(i) and Figure 6(ii). The boundary size is $t = 2$. We will take $k = 1$ It happens to be a fact that $D \sim D'$. The sequence $\sigma = (a, b, c, 1, v, 2, d, u)$ is a valid pebbling sequence for $D \oplus X$. The test T that corresponds to this is specified by: (1) the obvious partition of the boundary, (2) $S_0 = \emptyset$, (3) $k_1 = 1$, (4) $\pi = (1, 2)$, and (5) $Q = \{2\}$. Note that D' passes the test T with the sequence $\sigma' = (e, f, 1, 2, h, g)$. Using this to modify σ as described we get $\sigma'' = (e, f, 1, v, 2, u, h, g)$, a valid pebbling sequence for $D' \oplus X$.

REMARK 5.2. The algorithm implicitly described by Theorem 5.1 consists of two parts: (1) the computation of a tree decomposition, and (2) the evaluation of the resulting parse tree by a finite-state leaf-to-root tree automaton. The time complexity arising from (1) is bounded by the running time of Bodlaender's algorithm, $O(2^{35w^3} n)$ for a graph on n vertices and treewidth w. Given the tree decomposition, part (2) consists only of making finite-state transitions, which is fast. The problem is that the automaton that represents the algorithm may be very large. Based on the proof of the theorem, the best we can say is that the automaton has $O(2^{k2^{2w}} w!)$ states. This means that the practical difficulty is more likely to be about *space* than *time*, although this bound on the number of states is undoubtedly extremely pessimistic.

THEOREM 5.3. SIGNED DIGRAPH PEBBLING II *is in FPT*.

SKETCH. This is proved in much the same way, the key point being that there is a finite set of abstract tests that define an equivalence relation on the edge- and vertex-colored digraphs that refines the canonical equivalence relation. How to define these tests is the essential matter.

Since only k moves are to be made, the boundary would be in a sequence of at most k different states in the course of a solution. Note that there are at most 2^t

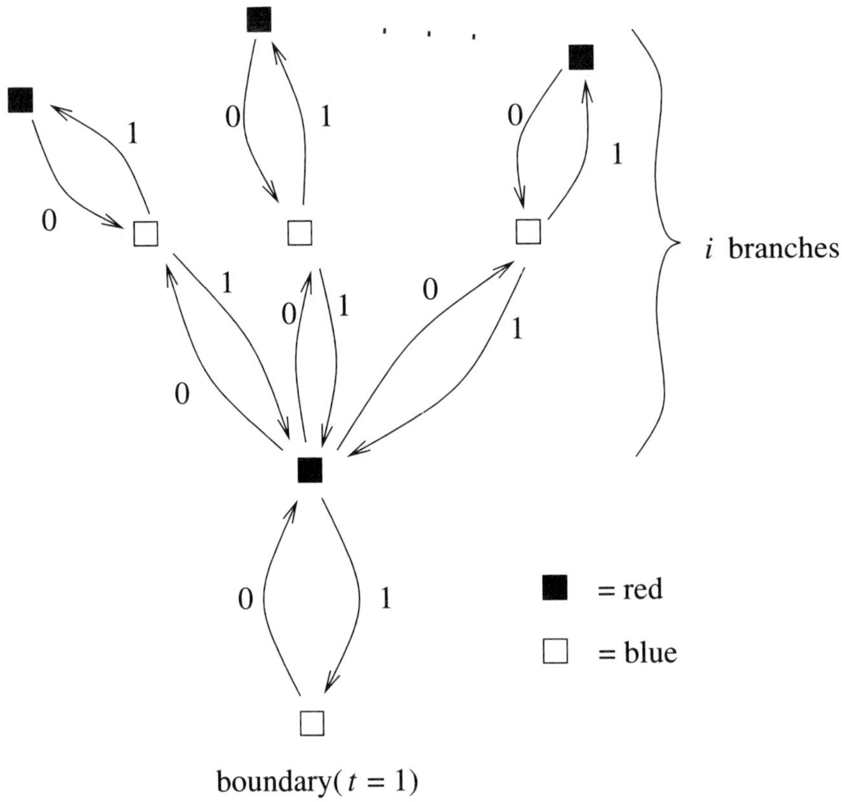

FIGURE 7. The trees for Theorem 5.4

different states that the boundary set can be in, in terms of which vertices of the boundary set are pebbled. A suitable test is specified by the sequence of boundary states (including information on which blue vertices of the boundary are "enabled" for a move by the states of their predecessors in the interior of D), and a budget of some number of moves (always less than k) to get from one state to the next in the sequence, together with a specification of at most k changes of the boundary state due to actions on the other side. Call all this information a *plan* for a k-step pebbling of the digraph. The number of plans is bounded by a function of t and k. The question that defines the test for a digraph D is whether D can realize this plan, ending up with all of its red interior vertices pebbled.

The size of an appropriate set of tests $\mathcal{T}$ can be bounded by $|\mathcal{T}| \leq 3^{2kt}k^k$, which leads to a crude bound on the size of the resulting automaton of $O(2^{3^{2kw}k^k})$ states. $\square$

THEOREM 5.4. SIGNED DIGRAPH PEBBLING *is not finite-state for any fixed treewidth bound* $t \geq 1$.

PROOF. Given the machinery, the argument is quite simple. Recall the usual argument using the Myhill-Nerode theorem that the formal language $L = \{a^n b^n : n \geq 0\}$ is not regular (i.e., finite-state recognizable by a linear automaton). The argument consists simply in exhibiting an infinite set of words x_i such that if $i \neq j$

then there is a word $y_{i,j}$ with $x_i y_{i,j} \in L$ and $x_j y_{i,j} \notin L$. One can take $x_i = a^i$ for $i = 1, 2, ...$ and $y_{i,j} = b^i$. We make here a completely analogous argument. (Other examples for graph problems can be found in [**AF93, BFW92, FHW93, DF98**].) We describe an infinite family of trees T_i for $i = 1, 2, ...$ such that if $i \neq j$, then there is a compatible tree $T_{i,j}$ such that:

1. $T_i \oplus T_{i,j}$ is a "yes" instance for the problem.
2. $T_j \oplus T_{i,j}$ is a "no" instance for the problem.

The trees T_i are shown in Figure 7. Each tree T_i consists a single blue boundary vertex connected to a red vertex to which i pendant paths of length 2 are attached. We can take $T_{i,j} = T_i$.

The reader can easily verify that 1 and 2 hold. □

We suspect that SIGNED DIGRAPH PEBBLING I is $W[1]$-hard. (It may even be that SIGNED DIGRAPH PEBBLING is *NP*-hard for trees, which would imply $W[P]$ hardness.) It seems that the *length of the plan* is an important parameter on which to focus on for the STRIPS PLANNING problem, in order to obtain tractable restrictions of the problem.

6. A Working Guide to Parametric Intractability

The main classes of parametric complexity are described in the tower:

$$FPT \subseteq W[1] \subseteq W[2] \subseteq \cdots \subseteq W[SAT] \subseteq W[P] \subseteq AW[P] \subseteq XP$$

The focus of this survey is on the analysis of concrete computational problems. For this purpose, all of the beautiful structural questions associated with the degrees of parameterized complexity, and even the definitions of most of the classes in the above hierarchy, are not very important. Suffice it to say that the $W[t]$ classes have natural complete problems that are variants of SATISFIABILITY. For concrete problem analysis, only the three classes

$$FPT \subseteq W[1] \subseteq XP$$

have compelling practical significance, where *XP* is the family of parameterized languages that can be recognized in time $O(f(k)n^{g(k)})$ where f and g are arbitrary functions. Knowing that a problem is in *XP* has some practical value and can be difficult to show. For example, for UNIT TASK LENGTH PRECEDENCE CONSTRAINED k-PROCESSOR SCHEDULING it is a famous open problem (discussed in [**GJ79**]) whether this is in *XP* or whether it is *NP*-complete for any fixed k, although it is proved in [**BF95**] to be hard for $W[2]$.

Table 3 exhibits some of the many results that are now known about the complexity of naturally parameterized problems. Those problems which are only known to be hard for a parameterized class (as opposed to complete) are annotated as such. (The problem definitions can be found in Garey and Johnson [**GJ79**] and also [**DF98**] where there is a comprehensive inventory of known results.) The problems in the table are all known to belong to *XP*, with the exception of UNIT LENGTH PRECEDENCE CONSTRAINED SCHEDULING.

6.1. The Nature of the Evidence for Parametric Intractability. As in the theory of *NP*-completeness, there are two kinds of evidence indicating, when a parameterized problem is hard for $W[1]$, that it is unlikely to be fixed-parameter tractable. The first is that given a sufficient amount of unsuccessful effort to demonstrate tractability for various problems in a class, the knowledge that a problem

Class	Problem	Reference
$W[P]$	LINEAR INEQUALITIES	**[ADF95]**
	MINIMUM AXIOM SET	**[DFKHW94]**
	SHORT SATISFIABILITY	**[ADF95]**
	WEIGHTED CIRCUIT SATISFIABILITY	**[ADF95]**
$W[SAT]$	WEIGHTED SATISFIABILITY	**[ADF95]**
$W[t]$, for all t	LONGEST COMMON SUBSEQUENCE (k = NUMBER OF SEQS.,$\|\Sigma\|$) (hard)	**[BDFHW95]**
	FEASIBLE REGISTER ASSIGNMENT (hard)	**[BFH94]**
	TRIANGULATING COLORED GRAPHS (hard)	**[BFH94]**
	BANDWIDTH (hard)	**[BFH94]**
	PROPER INTERVAL GRAPH COMPLETION (hard)	**[BFH94]**
	WEIGHTED t–NORMALIZED SATISFIABILITY	**[DF95a]**
$W[2]$	WEIGHTED $\{0,1\}$ INTEGER PROGRAMMING	**[DF95a]**
	DOMINATING SET	**[DF95a]**
	TOURNAMENT DOMINATING SET	**[DF95c]**
	UNIT LENGTH PRECEDENCE CONSTRAINED SCHEDULING (hard)	**[BF95]**
$W[1]$	SHORTEST COMMON SUPERSEQUENCE (k SEQS.) (hard)	**[FHK95]**
	MAXIMUM LIKELIHOOD DECODING (hard)	**[DFVW98]**
	WEIGHT DISTRIBUTION IN LINEAR CODES (hard)	**[DFVW98]**
	NEAREST VECTOR IN INTEGER LATTICES (hard)	**[DFVW98]**
	SHORT PERMUTATION GROUP FACTORIZATION (hard)	**[CCDF96]**
	SHORT POST CORRESPONDENCE	**[CCDF96]**
	WEIGHTED q–CNF SATISFIABILITY	**[DF95b]**
	VAPNIK–CHERVONENKIS DIMENSION	**[DEF93]**
	LONGEST COMMON SUBSEQUENCE (length m common subseq. for k seqs., parameter (k,m))	**[BDFW95]**
	INDEPENDENT SET	**[DF95b]**
	SQUARE TILING	**[CCDF96]**
	MONOTONE DATA COMPLEXITY FOR RELATIONAL DATABASES	**[DFT96]**
	k-STEP DERIVATION FOR CONTEXT SENSITIVE GRAMMARS	**[CCDF96]**
	CLIQUE	**[DF95b]**
	SHORT NTM COMPUTATION	**[CCDF96]**
FPT	FEEDBACK VERTEX SET	**[DF95c]**
	GRAPH GENUS	**[RS85]**
	MINOR ORDER TEST	**[RS85]**
	TREEWIDTH	**[Bod96]**
	VERTEX COVER	**[BFR98]**

TABLE 3. A Sample of Parametric Complexity Classifications

is hard for the class offers a cautionary sociological message, of the sort depicted in the famous cartoon in the opening pages of [**GJ79**]. As shown by Table 3, the amount of evidence of this sort is now substantial.

A second reason for the belief that $W[1]$-hardness implies parametric intractability, is rooted in the following fundamental theorem [**DFKHW94, CCDF96**].

THEOREM 6.1 (Downey-Fellows). *The k-*STEP HALTING PROBLEM FOR NONDETERMINISTIC TURING MACHINES *is complete for* $W[1]$.

On input consisting of a Turing machine M and a positive integer k (with k being the parameter), the question is whether M can reach a halting configuration in at most k steps. This problem is so generic and opaque that it is hard to imagine that there is any algorithm for it that radically improves on simply exploring the n-branching depth k tree of allowed nondeterministic transitions exhaustively. The theorem can be viewed as essentially a miniaturization of Cook's Theorem.

In general, W-hardness results tend to be harder to prove than theorems about *NP*-completeness, because of the control requirement concerning the parameter. CLIQUE is a typical starting point for $W[1]$-hardness arguments. It is interesting that most natural parameterized problems seem to belong to a small number of degrees (*FPT*, $W[1]$, $W[2]$, $W[P]$, $AW[*]$ and $AW[P]$; for details see [**DF98**]).

6.2. $W[1]$-Hard Means No Good PTAS. One might suspect that parameterized complexity is related to the complexity of approximation. A very good connection is supplied by the following theorem first proved by Bazgan [**Baz95**], and later independently by Cesati and Trevisan [**CT97**], strengthening an earlier result of Cai and Chen [**CC97**].

DEFINITION 6.2. An optimization problem has an *efficient PTAS* if there is an algorithm that computes a solution within a factor of $(1+\epsilon)$ of optimal in time $O(f(\epsilon)n^c)$ for some constant c.

DEFINITION 6.3. For a maximization (resp. minimization) problem A, the induced language L_A is the parameterized language consisting of all pairs (x, k) where x is an instance of A and $\text{opt}(x) \geq k$ (resp. $\text{opt}(x) \leq k$).

THEOREM 6.4 (Bazgan). *If A has an efficient PTAS then $L_A \in FPT$.*

Thus if the parameterized problem naturally associated with an optimization problem A is hard for $W[1]$, then A cannot have an efficient PTAS unless $FPT = W[1]$. It is worth noting that some (but by no means all) *NP*-completeness reductions are serendipitously parametric and thus may provide demonstrations of $W[1]$-hardness and non-approximability "for free". An important optimization problem that has a PTAS but is not known to have an efficient PTAS is the EUCLIDEAN TRAVELING SALESMAN PROBLEM. The PTAS for this problem due to Arora runs in time $O(n^{30/\epsilon})$ [**Ar96**].

6.3. The Complexity of Propositional STRIPS Planning (for Short Plans) and k-Step Petri Net Reachability. We prove two negative complexity results about k-step PROPOSITIONAL STRIPS PLANNING and k-step PETRI NET REACHABILITY (for parameter k) for otherwise unrestricted inputs to these problems. These results should be interpreted as indicating that some sort of structure on the Petri nets and the set of planning options is a vital parameter for tractability of these problems.

THEOREM 6.5. *k-STEP PROPOSITIONAL STRIPS PLANNING is hard for $W[1]$.*

PROOF. We reduce from the k-STEP HALTING PROBLEM FOR NONDETERMINISTIC TURING MACHINES that is proved to be complete for $W[1]$ in [**CCDF96**]. Let the instance of the HALTING PROBLEM consist of the Turing machine $M = (Q, \Sigma, \delta, F)$ where Q is the set of states, Σ is the alphabet, δ is the transition relation, and F is the set of accept states, together with a positive integer k (the number of steps of a computation).

The dimension n of the instance of STRIPS PLANNING to which we reduce (M, k) is given by

$$n = |Q| + 1 + k + k|\Sigma|.$$

We can think of the components of the vectors of the STRIPS PLANNING as belonging to $k+3$ blocks. The first block of size $|Q|$ will be constrained to have exactly one component that is a "1" with all the rest "0", indicating the state of M. The second block of size 1 indicates whether the state indicated by the "1" component of the first block is an accept state of M. The third block of size k indicates the position of the tape head of M. The k remaining blocks, each of size $|\Sigma|$, indicate the symbols on the k tape squares that might be visited during a k-step computation.

Each possible transition of M:

$$(q, a) \vdash (p, b, X) \text{ where } X \in \{L, R\}$$

becomes k different STRIPS operators (one for each of the locations on the tape where the transition may occur).

We leave the remaining details to the reader. □

THEOREM 6.6. *k-STEP PETRI NET REACHABILITY is hard for $W[1]$.*

For the definition of the PETRI NET REACHABILITY problem we refer the reader to [**Pet81**].

PROOF. The reduction from k-STEP STRIPS PLANNING to k-STEP PETRI NET REACHABILITY is essentially trivial, using two places of the Petri net to represent each component of the planning vectors. □

Showing that these problems belong to $W[1]$ is straightforward and can be done in much the same manner as the proof that the k-STEP HALTING PROBLEM FOR NONDETERMINISTIC TURING MACHINES is in $W[1]$ (see [**CCDF96**]).

6.4. The Upper Reaches of Parametric Complexity. There are a number of naturally parameterized problems, such as GRAPH k-COLORING that are *NP*-complete for fixed values of k. Does the theory of parameterized complexity have anything to say about these? Since this is a well-defined parameterized problem, it has to be located *somewhere* in the parametric degree structure. The following theorem shows that the hierarchy of parameterized complexity classes seems to disappear into clouds of unknowing at the top end.

THEOREM 6.7. *The naturally parameterized* GRAPH k-COLORING *problem has an imponderable relationship to* XP *in the following sense:*

1. *If* GRAPH k-COLORING *is in XP then $P = NP$.*
2. *If* GRAPH k-COLORING *is hard for XP then $P \neq NP$.*

PROOF. Clearly, if GRAPH k-COLORING belongs to *XP* then $P = NP$. On the other hand, if it is hard for *XP*, then $P \neq NP$ by the following argument. For each k, let L_k be a problem that is complete for DTIME(n^k) by a polynomial-time reduction where the degree of the polynomial bounding the time required for the reduction is independent of k, such as the k-MICE CAT AND MOUSE GAME problem [**DF98**]. Consider the parameterized language

$$L = \{(y, k) : y \in L_k\}.$$

It is obvious that $L \in XP$. Suppose that there is a parametric reduction of L to GRAPH k-COLORING. Thus from (y, k) we can produce a graph G of size bounded by $f(k)n^c$ and an integer $k' = g(k)$ such that G is k' colorable if and only if $(y, k) \in L$, i.e., $y \in L_k$, where f, g are some fixed but arbitrary functions, c is a constant and $n = |y|$. The time required for the transformation can also be assumed to be bounded by $f(k)n^c$.

If $P = NP$ then there is an algorithm running in time $O(n^{c'})$ that can decide if G is k'-colorable, and thus we have a way of determining whether $y \in L_k$ in time bounded by $h(k)n^{c''}$ for some function h and constant c''. When k and n are sufficiently large this contradicts the DTIME hierarchy theorem. □

Thus we are unable to say anything about the relationship of GRAPH COLORING and *XP* without settling the question of whether $P = NP$.

7. The Role of Parameterized Complexity Analysis

The current approach to the analysis of concrete computational problems is dominated by two kinds of effort:

(1): The search for asymptotic worst-case polynomial-time algorithms.

(2): Alternatively, proofs of classical hardness results, particularly *NP*-hardness.

We expect that these will become substantially supplemented by:

(1'): The design of *FPT* algorithms for various parameterizations of a given problem, and the development of associated heuristics.

(2'): Alternatively, demonstrations of $W[1]$-hardness.

We think this will happen because we are inevitably forced towards something like an *ultrafinitist* [**YV70**] outlook concerning computational complexity because of the nature of the universe of interesting yet feasible computation. The main point of this outlook is that numbers in different ranges of magnitude should be treated in qualitatively different ways.

The pair of notions (1′) and (2′) are actually rather straightforward mutations of (1) and (2), and they inherit many of the properties that have made the framework provided by (1) and (2) so successful. We note the following in support of this position.

- The enrichment of the dialogue between practice and theory that parameterized complexity is based on always makes sense. It *always* makes sense to ask the users of algorithms, "Are there aspects of your problem that may typically belong to limited distributional ranges?"
- Fixed-parameter tractability is a more accurate notion of "the good". If you were concerned with inverting very large matrices and could identify a bounded structural parameter k for your application that allows this to be

done in time $O(2^k n^2)$, then you might well prefer this *classically exponential-time* algorithm to the usual $O(n^3)$ polynomial-time algorithm.

- The "bad", $W[1]$-hardness, is based on a miniaturization of Cook's Theorem in a way that establishes a strong analogy between *NP* and $W[1]$. Proofs of $W[1]$-hardness are generally more challenging than *NP*-completeness, but it is obvious by now (see Table 3) that this is a very applicable complexity measurement.
- Problems that are hard do not just go away. Parameterization allows for several kinds of sustained dialogue with a single problem, in ways that allow finer distinctions about the causes of intractability (and opportunities for practical algorithms, including systematically designed heuristics) to be made than the exploration of the "*NP*-completeness boundary" described in [**GJ79**].
- Polynomial time has thrived because of the empirical circumstance that when polynomial-time algorithms can be devised, one almost always has small exponent polynomials. This is also true for *FPT* algorithms.
- Polynomial time is robust in that it seems to support a strong form of Church's thesis, i.e., that polynomial time on Turing machines is the same as polynomial time on any reasonable computing device. This also seems to be true for *FPT*.
- Polynomial time has thrived because it is a mathematically rich and productive notion allowing for a wide variety of algorithm design techniques. *FPT* seems to offer an even richer field of play, in part because it encompasses polynomial time as (usually) the best kind of *FPT* result. Beyond this, the *FPT* objective encompasses a rich and distinctive positive toolkit, including novel ways of defining and exploiting parameters.
- There is good evidence that not only are small polynomial exponents generally available when problems are *FPT*, but also that simple exponential parameter functions such as 2^k are frequently achievable, and that many of the problems in *FPT* admit kernelization algorithms that provide useful start-ups for *any* algorithmic attack on the problem.
- The complexity of approximation is handled more elegantly than in the classical theory, with $W[1]$-hardness immediately implying that there is no efficient PTAS. Moreover, *FPT* algorithm design techniques appear to be fruitful in the design of approximation algorithms.
- Parameterization is a very broad idea. It is possible to formulate and explore notions such as randomized *FPT* [**FK93**], parameterized parallel complexity [**Ces96**], parameterized learning complexity [**DEF93**], parameterized approximation [**BFH97**], parameterized cryptosystems based on $O(n^k)$ security, etc.

We feel that the parametric complexity notions, with their implicit ultrafinitism, correspond better to the natural landscape of computational complexity, where we find ourselves overwhelmingly among hard problems, dependent on identifying and exploiting thin zones of computational viability. Many natural problem distributions are generated by processes that inhabit such zones themselves (e.g., computer code that is written in a structured manner so that it can be comprehensible to the programmer), and these distributions then inherit limited parameter ranges because of the computational parameters that implicitly govern the feasibility of the generative processes, though the relevant parameters may not be immediately obvious. [2]

8. Some Open Problems

Parameterized complexity has been primarily motivated by concrete computational problems. The structure theory of parameterized complexity is very rich, and also not very well developed. One of the main questions is: how many of the important structural theorems of classical complexity have parameterized analogs? We know that there is a good analog of Cook's Theorem [**CCDF96**], and there is a good analog of Mahaney's Theorem on sparse hard sets [**CF96**]. There is a partial analog of Ladner's Density Theorem [**DF93, DF98**]. Notably lacking are parameterized analogs of probability amplification. An analog of Toda's Theorem [**To91**] would be very interesting, as together with the results of [**DFR98a**] it would show that UNIQUE k-CLIQUE is hard for $W[t]$ for all t via randomized parametric reductions.

***FPT* or W-Hard?** The following concrete problems seem important to classify because of their significant applications.

1. The naturally parameterized DIRECTED FEEDBACK VERTEX SET problem. This can be shown to be equivalent to the naturally parameterized DIRECTED FEEDBACK ARC SET problem. For undirected graphs the problem is in *FPT*.
2. The problem of determining whether it is possible to delete at most k edges of a graph so that the resulting graph is bipartite.
3. The problem of determining whether it is possible to delete at most k vertices from a graph so that the result is an interval graph. This has important applications in DNA sequence analysis [**Sha97**].
4. The problem of determining whether a graph H is immersed in a graph G, where the parameter k is the size of H. This is in *XP* and the immersion partial ordering of graphs is a well-partial-order by the major theorems of Robertson and Seymour [**RS85, RS96**]. GRAPH TOPOLOGICAL CONTAINMENT is a similarly interesting problem, currently known only to belong to *XP*.
5. The LONGEST COMMON SUBSEQUENCE and SHORTEST COMMON SUPERSEQUENCE problems for k sequences over an alphabet of constant size (as is the case in analyzing biological sequences). If the size is not constant and the parameter is the pair (s, k) where s is the size of alphabet and k is the number of sequences, then both problems are hard for $W[t]$ for all t [**BDFHW95, Hal96**].
6. The GRAPH ISOMORPHISM problem where the parameter k is the maximum degree of the graphs. This is in *XP* by the results of Luks [**Luks82**]. This problem would seem to be a reasonable candidate for representing a parameterized degree intermediate between *FPT* and $W[1]$, for the same reasons that GRAPH ISOMORPHISM classically seems to represent a polynomial time degree intermediate between *P* and *NP*. On the other hand, parameterized complexity behaves differently. TOURNAMENT DOMINATING SET, for example, which is classically intermediate, is precisely $W[2]$-complete [**DF95c**].

[2] For a philosophically similar discussion see [**Gur89**].

Maximum degree parameterized GRAPH ISOMORPHISM could even be in *FPT* without this entailing any known surprises. On the other hand, if it could be shown to be hard for $W[1]$, then this would give good evidence that GRAPH ISOMORPHISM in general is not in P.

It is also reasonable to wonder whether the parameterized complexity framework might provide new opportunities for addressing such central mysteries as the $P = NP$ question. For example, Theorem 6.7 (perhaps only weakly) suggests looking for a parametric reduction using an oracle for *NP* from the CAT AND MOUSE GAME to GRAPH COLORING. In general, it seems fruitful to explore the possibilities for stronger connections between the parameterized and classical frameworks.

Acknowledgments. We would like to thank Mike Hallett, Todd Wareham and Ken Regan for stimulating discussions and assistance in improving the presentation. On a number of key points, the paper greatly benefited from the detailed comments and criticisms of an earlier draft by Bernard Moret and an anonymous referee, and from conversations with Chantal Korostensky of ETH.

References

[ADF95] K. Abrahamson, R. Downey and M. Fellows, "Fixed Parameter Tractability and Completeness IV: On Completeness for $W[P]$ and *PSPACE* Analogs," *Annals of Pure and Applied Logic* 73 (1995), 235–276.

[AF93] K. Abrahamson and M. Fellows, "Finite Automata, Bounded Treewidth and Wellquasiordering," In: *Graph Structure Theory*, American Mathematical Society, Contemporary Mathematics Series, vol. 147 (1993), 539–564.

[AFGPR96] E. Allender, J. Feigenbaum, J. Goldsmith, T. Pitassi and S. Rudich, "The Future of Computational Complexity Theory: Part II," *SIGACT News* 27 (1996), 3–7.

[AK94] A. Amir and D. Keselman, "Maximum Agreement Subtree in a Set of Evolutionary Trees – Metrics and Efficient Algorithms," In: *35th Annual Symposium on Foundations of Computer Science*, 758–769, 1994.

[AL97] E. Aarts and J.K. Lenstra (eds.), *Local Search in Combinatorial Optimization*, John Wiley and Sons, 1997.

[ALMSS92] S. Arora, C. Lund, R. Motwani, M. Sudan and M. Szegedy, "Proof Verification and Intractability of Approximation Algorithms," *Proceedings of the IEEE Symposium on the Foundations of Computer Science*, 13–22, 1992.

[AMOV91] G. B. Agnew, R. C. Mullin, I. M. Onyszchuk and S. A. Vanstone, "An Implementation for a Fast Public-Key Cryptosystem," *J. Cryptology* 3 (1991), 63–79.

[Ar96] S. Arora, "Polynomial Time Approximation Schemes for Euclidean TSP and Other Geometric Problems," In: *Proceedings of the 37th IEEE Symposium on Foundations of Computer Science*, 1996.

[AYZ94] N. Alon, R. Yuster and U. Zwick, "Color-Coding: A New Method for Finding Simple Paths, Cycles and Other Small Subgraphs Within Large Graphs," *Proc. Symp. Theory of Computing (STOC)*, ACM (1994), 326–335.

[Ba94] B. Baker, "Approximation Algorithms for *NP*-Complete Problems on Planar Graphs," *J.A.C.M.* 41 (1994), 153–180.

[Baz95] C. Bazgan, "Schémas d'approximation et complexité paramétrée," Rapport de stage de DEA d'Informatique à Orsay, 1995.

[BDFHW95] H. Bodlaender, R. Downey, M. Fellows, M. Hallett and H. T. Wareham, "Parameterized Complexity Analysis in Computational Biology," *Computer Applications in the Biosciences* 11 (1995), 49–57.

[BDFW95] H. Bodlaender, R. Downey, M. Fellows and H. T. Wareham, "The Parameterized Complexity of the Longest Common Subsequence Problem," *Theoretical Computer Science A* 147 (1995), 31–54.

[BF95] H. Bodlaender and M. Fellows, "On the Complexity of k-Processor Scheduling," *Operations Research Letters* 18 (1995), 93–98.

[BFH94] H. Bodlaender, M. R. Fellows and M. T. Hallett, "Beyond *NP*-completeness for Problems of Bounded Width: Hardness for the *W* Hierarchy," *Proc. ACM Symp. on Theory of Computing (STOC)* (1994), 449–458.

[BFH97] H. Bodlaender, M. Fellows, M. Hallett, "Parameterized Complexity and Parameterized Approximation for Some Problems About Trees," manuscript, 1997.

[BFR98] R. Balasubramanian, M. Fellows and V. Raman, "An Improved Fixed-Parameter Algorithm for Vertex Cover," *Information Processing Letters* 65 (1998), 163–168.

[BFRS98] D. Bryant, M. Fellows, V. Raman and U. Stege, "On the Parameterized Complexity of MAST and 3-Hitting Sets," manuscript, 1998.

[BFW92] H. Bodlaender, M. Fellows and T. Warnow, "Two Strikes Against Perfect Phylogeny," in: *Proceedings of the 19th International Colloquium on Automata, Languages and Programming*, Springer-Verlag, Lecture Notes in Computer Science vol. 623 (1992), 273–283.

[BGKRS95] R. S. Barr, B. L. Golden, J. P. Kelly, M. G. C. Resende and W. R. Stewart, "Designing and Reporting on Computational Experiments with Heuristic Methods," *J. Heuristics* 1 (1995), 9–32.

[BGS95] M. Bellare, O. Goldreich and M. Sudan, "Free Bit and Nonapproximability," *Proceedings of the 36th Annual IEEE Symposium on the Theory of Computing* (1995), 422-431.

[Bod96] H. Bodlaender, "A Linear Time Algorithm for Finding Tree Decompositions of Small Treewidth," *SIAM J. Comp.* 25 (1996), 1305–1317.

[Bry97] D. Bryant, "Building Trees, Hunting for Trees, and Comparing Trees," Ph.D. dissertation, Department of Mathematics, Univ. Canterbury, Christchurch, New Zealand, 1997.

[Bry98] D. Bryant, private communication, 1998.

[Byl94] T. Bylander, "The Computational Complexity of Propositional STRIPS Planning," *Artificial Intelligence* 69 (1994), 165–204.

[CC97] L. Cai and J. Chen. "On Fixed-Parameter Tractability and Approximability of *NP*-Hard Optimization Problems," *J. Computer and Systems Sciences* 54 (1997), 465–474.

[CCDF96] L. Cai, J. Chen, R. G. Downey and M. R. Fellows, "On the Parameterized Complexity of Short Computation and Factorization," *Arch. for Math. Logic* 36 (1997), 321–337.

[CCDF97] L. Cai, J. Chen, R. Downey and M. Fellows, "Advice Classes of Parameterized Tractability," *Annals of Pure and Applied Logic* 84 (1997), 119–138.

[CD94] K. Cattell and M. J. Dinneen, "A Characterization of Graphs with Vertex Cover up to Five," *Proceedings ORDAL'94*, Springer Verlag, Lecture Notes in Computer Science, vol. 831 (1994), 86–99.

[CDDFL98] K. Cattell, M. Dinneen, R. Downey and M. Fellows, "On Computing Graph Minor Obstruction Sets," *Theoretical Computer Science A*, to appear.

[CDM92] A. Colorni, M. Dorigo and V. Manniezzo, "An Investigation of Some Properties of an 'Ant Algorithm'," in R. Männer and B. Manderick (eds.) *Parallel Problem Solving From Nature 2*, North-Holland, Amsterdam (1992), 509–520.

[Ces96] M. Cesati, "Structural Aspects of Parameterized Complexity," Ph.D. dissertation, University of Rome, 1995.

[CF96] M. Cesati and M. Fellows, "Sparse Parameterized Problems," *Annals of Pure and Applied Logic* 62 (1996).

[CKT91] P. Cheeseman, B. Kanefsky and W. Taylor, "Where the Really Hard Problems Are," *Proc. 12th International Joint Conference on Artificial Intelligence* (1991), 331–337.

[Cl87] K. L. Clarkson, "New Applications of Random Sampling in Computational Geometry," *Discrete and Computational Geometry* 2 (1987), 195–222.

[CL95] B. Courcelle and J. Lagergren, "Equivalent Definitions of Recognizability for Sets of Graphs of Bounded Treewidth," *Mathematical Structures in Computer Science* (1995), 141–165.

[Co90] B. Courcelle, "Graph Rewriting: An Algebraic and Logical Approach," in: *Handbook of Theoretical Computer Science, vol. B*, J. van Leeuwen, ed., North Holland (1990), Chapter 5.

[CT97] M. Cesati and L. Trevisan, "On the Efficiency of Polynomial Time Approximation Schemes," *Information Processing Letters* 64 (1997), 165–171.

[CW95] M. Cesati and H. T. Wareham, "Parameterized Complexity Analysis in Robot Motion Planning," *Proceedings 25th IEEE Intl. Conf. on Systems, Man and Cybernetics.*

[DEF93] R. Downey, P. Evans and M. Fellows, "Parameterized Learning Complexity," *Proc. 6th ACM Workshop on Computational Learning Theory* (1993), 51–57.

[Deu93] G. Deuck, "New Optimization Heuristics: the Great Deluge Algorithm and the Record-to-Record-Travel," *J. Computational Physics* 104 (1993), 86–92.

[DF93] R. Downey and M. Fellows, "Fixed Parameter Tractability and Completeness III: Some Structural Aspects of the W-Hierarchy," in: K. Ambos-Spies, S. Homer and U. Schöning, editors, *Complexity Theory: Current Research*, Cambridge Univ. Press (1993), 166–191.

[DF95a] R. G. Downey and M. R. Fellows, "Fixed Parameter Tractability and Completeness I: Basic Theory," *SIAM Journal of Computing* 24 (1995), 873–921.

[DF95b] R. G. Downey and M. R. Fellows, "Fixed Parameter Tractability and Completeness II: Completeness for W[1]," *Theoretical Computer Science A* 141 (1995), 109–131.

[DF95c] R. G. Downey and M. R. Fellows, "Parametrized Computational Feasibility," in: *Feasible Mathematics II, P. Clote and J. Remmel (eds.)* Birkhauser, Boston (1995) 219–244.

[DF98] R. G. Downey and M. R. Fellows, *Parameterized Complexity*, Springer-Verlag, 1998.

[DFKHW94] R. G. Downey, M. Fellows, B. Kapron, M. Hallett, and H. T. Wareham. "The Parameterized Complexity of Some Problems in Logic and Linguistics," *Proceedings Symposium on Logical Foundations of Computer Science (LFCS)*, Springer-Verlag, Lecture Notes in Computer Science vol. 813 (1994), 89–100.

[DFR98a] R. G. Downey, M. R. Fellows and K. W. Regan, "Parameterized Circuit Complexity and the W Hierarchy," *Theoretical Computer Science A* 191 (1998), 91–115.

[DFR98b] R. G. Downey, M. Fellows and K. Regan. "Threshold Dominating Sets and an Improved Characterization of $W[2]$," *Theoretical Computer Science A*, to appear.

[DFT96] R. G. Downey, M. Fellows and U. Taylor, "The Parameterized Complexity of Relational Database Queries and an Improved Characterization of $W[1]$," in: *Combinatorics, Complexity and Logic: Proceedings of DMTCS'96*, Springer-Verlag (1997), 194–213.

[DFVW98] R. Downey, M. Fellows, A. Vardy and G. Whittle, "The Parameterized Complexity of Some Fundamental Problems in Coding Theory," *SIAM J. Computing*, to appear.

[DKL96] T. Dean, J. Kirman and S.-H. Lin, "Theory and Practice in Planning," Technical Report, Computer Science Department, Brown University, 1996.

[Due93] G. Dueck, "New Optimization Heuristics: the Great-Deluge Algorithm and the Record-to-Record Travel," *J. Computational Physics* 104 (1993), 86–92.

[Fel97] J. Felsenstein. Private communication, 1997.

[FHK95] M. Fellows, M. Hallett and D. Kirby, "The Parameterized Complexity of the Shortest Common Supersequence Problem," manuscript, 1995.

[FHW93] M. Fellows, M. Hallett and H. T. Wareham, "DNA Physical Mapping: Three Ways Difficult," in: *Algorithms – ESA'93: Proceedings of the First European Symposium on Algorithms*, Springer-Verlag, Lecture Notes in Computer Science vol. 726 (1993), 157–168.

[Fis95] S. T. Fischer, "A Note on the Complexity of Local Search Problems," *Information Processing Letters* 53 (1995), 69–75.

[FK93] M. Fellows and N. Koblitz, "Fixed-Parameter Complexity and Cryptography," *Proceedings of the 10th Intl. Symp. on Applied Algebra, Algebraic Algorithms and Error-Correcting Codes*, Springer-Verlag, Berlin, Lecture Notes in Computer Science vol. 673 (1993), 121–131.

[FKS98] M. Fellows, C. Korostensky and U. Stege, "Theory-Based Heuristics for Editing Gaps in Multiple Sequence Alignments," manuscript, 1998.

[FL87] M. Fellows and M. Langston, "Nonconstructive Proofs of Polynomial-Time Complexity," *Information Processing Letters* 26(1987/88), 157–162.

[FL88] M. Fellows and M. Langston, "Nonconstructive Tools for Proving Polynomial-Time Complexity," *Journal of the Association for Computing Machinery* 35 (1988) 727–739.

[FL89] M. R. Fellows and M. A. Langston, "An Analogue of the Myhill-Nerode Theorem and its Use in Computing Finite-Basis Characterizations," *Proceedings of the IEEE Symposium on the Foundations of Computer Science* (1989), 520–525.

[FL94] M. R. Fellows and M. A. Langston, "On Search, Decision and the Efficiency of Polynomial-Time Algorithms," *Journal of Computer and Systems Science* 49 (1994), 769–779.

[FN71] R. E. Fikes and N. J. Nilsson, "STRIPS: A New Approach to the Application of Theorem Proving to Problem Solving," *Artificial Intelligence* 2 (1971), 189–208.

[FPT95] M. Farach, T. Przytycka, and M. Thorup. "On the agreement of many trees" *Information Processing Letters 55* (1995), 297–301.

[Fr97] J. Franco, J. Goldsmith, J. Schlipf, E. Speckenmeyer and R.P. Swaminathan, "An Algorithm for the Class of Pure Implicational Formulas," to appear in *Discrete Applied Mathematics*.

[GJ79] M. Garey and D. Johnson. *Computers and Intractability: A Guide to the Theory of* NP-*completeness.* W.H. Freeman, San Francisco, 1979.

[GLR92] R. Govindan, M. Langston and S. Ramachandramurthi, "A Practical Approach to Layout Optimization," *Proceedings of the International Conference on VLSI Design*, Bombay, India, January 1992.

[GM75] G. R. Grimmet and C. H. J. McDiarmid, "On Colouring Random Graphs," *Math. Proc. Cambridge Philos. Soc.* 77 (1975), 313–324.

[GM84] S. Goldwasser and S. Micali, "Probabilistic Encryption," *J. Computer and Systems Science* 28 (1984), 270–299.

[GS94] J. Gustedt and A. Steger, "Testing Hereditary Properties Efficiently on Average," *Orders, Algorithms and Applications: Proceedings of the International Workshop ORDAL'94*, Springer-Verlag, Lecture Notes in Computer Science, vol. 831 (1994), 100–116.

[Gur89] Y. Gurevich, "The Challenger-Solver Game: Variations on the Theme of *P=?NP*," *Bulletin EATCS* 39 (1989), 112–121.

[Hal96] M. Hallett. "An Integrated Complexity Analysis of Some Problems in Computational Biology," Ph.D. dissertation, Department of Computer Science, University of Victoria, 1996.

[Hart94] J. Hartmanis, "About the Nature of Computer Science," *Bulletin EATCS* 53 (1994), 170–190.

[Has96] J. Hastad, "Fast and Efficient Testing of the Long Code," *Proc. ACM Symposium on the Theory of Computing (STOC)* (1996).

[HGS98] M. Hallett, G. Gonnet, and U. Stege, "Vertex Cover Revisited: A Hybrid Algorithm of Theory and Heuristic," manuscript 1998.

[HL92] J. Hartmanis and H. Lin, editors, *Computing the Future: A Broader Agenda for Computer Science and Engineering*, National Academy Press, 1992.

[HM91] F. Henglein and H. G. Mairson, "The Complexity of Type Inference for Higher-Order Typed Lambda Calculi." In *Proc. Symp. on Principles of Programming Languages (POPL)* (1991), 119-130.

[Hoch97] D. Hochbaum, "Various Notions of Approximations: Good, Better, Best and More," in: *Approximation Algorithms for NP-Hard Problems* (D. Hochbaum, ed.), PWS Publishing Co., Boston (1997), 346–398.

[Hoo95] J. N. Hooker, "Testing Heuristics: We Have It All Wrong," *J. Heuristics* 1 (1995), 33–42.

[IK75] O. H. Ibarra and C. E. Kim, "Fast Approximation for the Knapsack and Sum of Subset Problems," *J. Assoc. Computing Machinery* 22 (1975), 463–468.

[IW97] R. Impagliazzo and A. Wigderson, *Proc. ACM Symposium on Theory of Computing*, 1997.

[JDUGG74] D. S. Johnson, A. Demers, J. D. Ullman, M. R. Garey and R. L. Graham, "Worst-Case Performance Bounds for Simple One-Dimensional Packing Algorithms," *SIAM J. Computing* 3 (1974), 299–325.

[Jo87] D. S. Johnson, "The *NP*-Completeness Column: An Ongoing Guide (19th edition)," *Journal of Algorithms* 8 (1987), 285–303.

[JM93] D. S. Johnson and C. McGeoch (eds.), *Network Flows and Matching: First DIMACS Implementation Challenge*, American Mathematical Society, 1993.

[JMc97] D.S. Johnson and L.A. McGeoch, "The Traveling Salesman Problem: A Case Study," in: Aarts and Lenstra (eds.), *Local Search in Combinatorial Optimization*, John Wiley and Sons, 1997.

[JPY88] D.S. Johnson, C.H. Papadimitriou and M. Yannakakis, "How Easy is Local Search?" *Journal of Computer and System Sciences* 37 (1988), 79–100.

[JT96] D. S. Johnson and M. Trick (eds.), *Cliques, Coloring and Satisfiability: Second DIMACS Implementation Challenge*, American Mathematical Society, 1996.

[Karp72] R. M. Karp, "Reducibility Among Combinatorial Problems," in: *Complexity of Computer Applications*, Plenum Press, New York (1972), 85–103.

[Karp86] R. M. Karp, "Combinatorics, Complexity and Randomness," *Communications of the ACM* 29 (1986), 98–109.

[Kr90] M.W. Krentel, "On Finding and Verifying Locally Optimal Solutions," *SIAM Journal on Computing* 19 (1990), 742–751.

[KS94] S. Kirkpatrick and B. Selman, "Critical Behavior in the Satisfiability of Boolean Formulae," *Science* 264 (1994), 1297–1301.

[KST94] H. Kaplan, R. Shamir and R. E. Tarjan, "Tractability of Parameterized Completion Problems on Chordal and Interval Graphs: Minimum Fill-In and DNA Physical Mapping," in: *Proc. 35th Annual Symposium on the Foundations of Computer Science (FOCS)*, IEEE Press (1994), 780–791.

[KT92] A. Kornai and Z. Tuza, "Narrowness, Pathwidth and Their Application in Natural Language Processing," *Discrete Applied Mathematics* 36 (1992), 87–92.

[Len90] T. Lengauer, "VLSI Theory," in: *Handbook of Theoretical Computer Science vol. A*, Elsevier (1990), 835–868.

[Lev86] L. Levin, "Average Case Complete Problems," *SIAM J. Computing* 15 (1986), 285–286.

[LK73] S. Lin and B.W. Kernighan, "An Effective Heuristic Algorithm for the Traveling Salesman Problem," *Operations Research* 21 (1973), 498–516.

[LP85] O. Lichtenstein and A. Pneuli. "Checking That Finite-State Concurrents Programs Satisfy Their Linear Specification." In: *Proceedings of the 12th ACM Symposium on Principles of Programming Languages* (1985), 97–107.

[LR91] M. Langston and S. Ramachandramurthi, "Dense Layouts for Series-Parallel Graphs," *Proc. Great Lakes Symposium on VLSI* (1991), 14–17.

[Luks82] E. Luks, "Isomorphism of Graphs of Bounded Valence Can Be Tested in Polynomial Time," *J. Comput. and Systems Sci.* 25 (1982), 42–65.

[Mad72] W. Mader, "Hinreichende Bedingungen fuer die Existenz von Teilgraphen, die zu einem vollstaendigen Graphen homomorph sind," *Abt. Math, Sem. Hamburg Univ.* 37 (1972), 86–97.

[MP94] S. Mahajan and J. G. Peters, "Regularity and Locality in k-Terminal Graphs," *Discrete Applied Mathematics* 54 (1994), 229–250.

[MR95a] R. Motwani and P. Raghavan, *Randomized Algorithms*, Cambridge Univ. Press, 1995.

[MR95b] S. Mahajan and H. Ramesh, "Derandomizing Semidefinite Programming Based on Approximation Algorithms," in *Proceedings of the 36th Annual IEEE Symposium on Foundations of Computer Science* (1995), 162–169.

[MR98] M. Mahajan and V. Raman, "Parameterizing Above the Guarantee: MaxSat and Max-Cut," to appear in *J. Algorithms.*

[MSL92] D. Mitchell, B. Selman and H. Levesque, "Hard and Easy Distributions of SAT Problems," in: *Proceedings AAAI-92*, AAAI (1992), 459–465.

[NP85] J. Nešetřil and S. Poljak, "On the Complexity of the Subgraph Problem," *Comm. Math. Univ. Carol.* 26 (1985), 415–419.

[Pet81] J. L. Peterson, *Petri Net Theory and the Modeling of Systems*, Prentice-Hall, 1981.

[PGWRS96] C. Papadimitriou, O. Goldreich, A. Wigderson, A. Razborov and M. Sipser, "The Future of Computational Complexity Theory: Part I," *SIGACT News* 27 (1996), 6–12.

[PY96] C. Papadimitriou and M. Yannakakis, "On Limited Nondeterminism and the Complexity of the VC Dimension," *J. Computer and Systems Sciences* 53 (1996), 161–170.

[Rag88] P. Raghavan, "Probabilistic Construction of Deterministic Algorithms: Approximating Packing Integer Programs," *J. Computer and Systems Sciences* 37 (1988), 130–143.

[RS85] N. Robertson and P. D. Seymour, "Graph Minors: A Survey," in *Surveys in Combinatorics*, I. Anderson, ed. (Cambridge University Press: Cambridge, 1985), 153–171.

[RS96] N. Robertson and P. D. Seymour, "Graph Minors XV. Giant Steps," *J. Comb. Theory Ser. B* 68 (1996), 112–148.

[Sha97] R. Shamir. Private communication, 1997.

[SOWH96] D. L. Swofford, G. J. Olsen, P. J. Waddell and D. M. Hillis, "Phylogenetic Inference," in: D. M. Hillis, C. Moritz and B. K. Mable (eds.) *Molecular Systematics.* Sinauer Associates, Inc. (1996), 407–514.

[SS77] R. Solovay and V. Strassen, "A Fast Monte Carlo Test for Primality," *SIAM J. Computing* (1977), 84–85.

[Th97] M. Thorup, "Structured Programs Have Small Tree-Width and Good Register Allocation," *Proceedings 23rd International Workshop on Graph-Theoretic Concepts in Computer Science, WG'97*, R. Möhring (ed.), Springer Verlag, Lecture Notes in Computer Science vol. 1335 (1997), 318–332.

[To91] S. Toda, "PP is as Hard as the Polynomial Time Hierarchy," *SIAM J. Comput.* (1991), 865–877.

[Wilf85] H. Wilf, "Some Examples of Combinatorial Averaging," *Amer. Math. Monthly* 92 (1985), 250–261.

[Yan95] M. Yannakakis. "Perspectives on Database Theory," *Proceedings of the IEEE Symposium on the Foundations of Computer Science* (1995), 224–246.

[YV70] A. S. Yessenin-Volpin, "The Ultraintuitionistic Criticism and the Antitraditional Program for Foundations of Mathematics," in: *Intuitionism and Proof Theory: Proceedings of the Summer Conference at Buffalo, N.Y., 1968*, A. Kino, J. Myhill and R. E. Vesley (eds.), North-Holland, 1970.

RODNEY G. DOWNEY, SCHOOL OF MATHEMATICS AND COMPUTING SCIENCES, P.O. BOX 600, VICTORIA UNIVERSITY, WELLINGTON, NEW ZEALAND.
E-mail address: rod.downey@vuw.ac.nz

MICHAEL R. FELLOWS, DEPARTMENT OF COMPUTER SCIENCE, UNIVERSITY OF VICTORIA, VICTORIA, B.C. V8W 3P6, CANADA.
E-mail address: mfellows@csr.uvic.ca

ULRIKE STEGE, COMPUTATIONAL BIOCHEMISTRY RESEARCH GROUP, ETH ZÜRICH, CH-8092 ZÜRICH, SWITZERLAND
E-mail address: stege@inf.ethz.ch

DIMACS Series in Discrete Mathematics
and Theoretical Computer Science
Volume **49**, 1999

On the Structure of Large Homothetic Subsets

György Elekes

Dedicated to the memory of PAUL ERDŐS

ABSTRACT. We study the structure of sets of N points in the r-dimensional Euclidean space which contain many homothetic subsets of size cN. In one dimension, we describe these structures in terms of "generalized" arithmetic and/or geometric progressions. Of these, the former extend to higher dimensions, as well, while the latter turn out to be inherently one-dimensional objects. The proofs also involve certain results on point sets which determine few directions.

1. Problems and results in one dimension

DEFINITION 1.1. For two subsets $\mathcal{A}$, $\mathcal{B} \subset \mathbf{R}$, we put

$$H(\mathcal{A},\mathcal{B}) \stackrel{\text{def}}{=} \#\{\mathcal{A}' \subset \mathcal{B} \ ; \ \mathcal{A}' \approx \mathcal{A}\},$$

where "$\approx$" means "homothetic", i.e., a magnified/shrunk image. (Reflections — and, later on, rotations in higher dimensions — will not be allowed.) Also, we denote

$$h(t,n) \stackrel{\text{def}}{=} \max\{H(\mathcal{A},\mathcal{B}) \ ; \ |\mathcal{A}| = t, \ |\mathcal{B}| = n\}.$$

1.1. Upper and lower bounds.

QUESTION 1.2. What is the order of magnitude of $h(t,n)$ as a function of two variables?

THEOREM 1.3.
There is a positive constant C (independent of t and n) for which

$$h(t,n) \leq C\frac{n^2}{t}.$$

1991 *Mathematics Subject Classification.* 52C10, 51A25, 20F12.

Research partially supported by HU-NSF grants OTKA T017580 and T019367.

If $n \geq 2t$ *then*

$$h(t,n) \geq \frac{1}{16}\frac{n^2}{t}.$$

The proof can be found in Section 2.1.

REMARK 1.4. The above result gives the exact order of magnitude (apart from the multiplicative constant) for $2t \leq n$. Even for smaller values of n, two arithmetic progressions show that

$$h(t,n) \geq \begin{cases} n-t+1 & \text{if } n < 2t-1 \\ t+1 & \text{if } n = 2t-1. \end{cases}$$

Perhaps this formula also gives the exact value, i.e., equality may hold here in place of " $\geq$".

1.2. The structure problem. After the quantitative result above, it is natural to ask for a qualitative one.

QUESTION 1.5. What is the structure of those pairs $\mathcal{A}$, $\mathcal{B}$ for which $H(\mathcal{A},\mathcal{B})$ is *asymptotically best possible,* i.e., $|\mathcal{A}| = t$, $|\mathcal{B}| = n$ and $H(\mathcal{A},\mathcal{B}) \geq c_0 n^2/t$, for a fixed $c_0 > 0$?

I. The special case when both t and $\mathcal{A}$ (with $|\mathcal{A}| = t$) are fixed, is almost completely understood. It was shown in [**4**] that a quadratic order of magnitude can be attained for various fixed patterns. More specifically, this holds for all those $\mathcal{A}$ which are homothetic images of an $\mathcal{A}_0 \in \mathbf{R}$ whose elements are all algebraic. Laczkovich and Ruzsa [**7**] proved that for all other "transcendental" $\mathcal{A}$ with $|\mathcal{A}| \geq 4$, a quadratic order of magnitude is impossible (for $n \to \infty$). They also showed there that (for algebraic patterns) asymptotically extremal structures must be contained in "generalized" arithmetic progressions (to be defined later). Historically, the first result in this direction was that of Balog-Szemerédi [**1**] who proved the special case of $\mathcal{A} = \{0,1,2\}$ as a lemma to settle a conjecture of Erdős on three-term arithmetic progressions.

II. As for the case $t \to \infty$, no structure theorem is known in general. One possible reason for this is that the (asymptotically) extremal configurations for the Szemerédi–Trotter Theorem (Theorem 2 in [**11**], see below in Section 2.1) have not been characterized so far. (Perhaps this characterization may not be easy.)

III. Our first goal is to settle the "upper extreme" case: pairs of sets $\mathcal{A}$, $\mathcal{B}$, near-optimal with respect to $h(t,n)$, for which $|\mathcal{A}| \geq c_0|\mathcal{B}|$. In other words (and with a shorter notation of one less variable) we study sets with

$$N \leq |\mathcal{A}|, |\mathcal{B}| \leq CN \text{ and } H(\mathcal{A},\mathcal{B}) \geq cN, \tag{1.1}$$

fore some fixed $C \geq 1$ and $c > 0$, while $N \to \infty$. (The last inequality in (1.1) says that $\mathcal{A}$ and $\mathcal{B}$ are asymptotically optimal.)

In this case yet another type of near–extremal structures (beyond arithmetic progressions) can be found: if both $\mathcal{A}$ and $\mathcal{B}$ are geometric progressions of the same quotient and $|\mathcal{A}| = N$, $|\mathcal{B}| = CN$ (where $C > 1$) then $H(\mathcal{A},\mathcal{B}) \geq (C-1)N = C_0 N$.

It will be our first main result (more specifically, the 1–dimensional case of our Structure Theorem) which asserts that, roughly speaking,

no other near–extremal structure exists.

This is stated more precisely in the next section, after defining certain notions.

1.3. The one dimensional Structure Theorem. Following [**10**] and [**9**], we define generalized arithmetic and geometric progressions.

DEFINITION 1.6. Let d and $n_1, n_2, \dots, n_d$ be positive integers and $\Delta_1, \dots, \Delta_d$ arbitrary real numbers. We call the set

$$\mathcal{G} = \left\{ \sum_{i=1}^{d} k_i \cdot \Delta_i \ ; \ 0 \le k_i < n_i \text{ for } i = 1 \dots d \right\}$$

a *generalized arithmetic progression* of dimension d and parameters n_i and Δ_i. Similarly — instead of differences Δ_i — with quotients $q_1, q_2, \dots, q_d$ positive reals, $\mathcal{G}$ is a *generalized geometric progression* if

$$\mathcal{G} = \{\pm \prod_{i=1}^{d} q_i^{k_i} \ ; \ 0 \le k_i < n_i \text{ for } i = 1 \dots d\}.$$

The $\pm$ sign is placed in the definition for convenience; we just wanted to avoid having to play around with quotients of different signs.

We shall use the shorthands "arithmetic GP" and "geometric GP" for the above structures.

REMARK 1.7. Note that the sums or products in the above definition may not be distinct, i.e., $|\mathcal{G}| < \prod n_i$ or $|\mathcal{G}| < 2 \prod n_i$ are possible, respectively.

In what follows, $\mathcal{G}_{d,n}$ will denote an arithmetic or geometric GP of dimension *not exceeding* d and size *at most* n. For short, we shall also use expressions like "an arithmetic or geometric $\mathcal{G}_{d,n}$".
Also, we shall denote by $M(\mathcal{G})$ the element of an arithmetic or geometric GP which includes all differences/quotients at highest possible multiplicity. (If all $\Delta_i > 0$ or $q_i > 1$ then $M(\mathcal{G})$ is the maximal element of $\mathcal{G}$.)

When speaking about arithmetic and geometric GPs simultaneously, "$\oplus$" will be used as a shorthand for the natural operation: "$+$" and "$\cdot$", respectively. Thus we denote by $\mathcal{G} \oplus \mathcal{G}$ both the sumsets $\mathcal{G} + \mathcal{G}$ and the product sets $\mathcal{G} \cdot \mathcal{G}$, whichever appropriate.

The following statement is, probably, folklore. (It is really obvious if the sums or products in Definition 1.6 are all distinct. Otherwise, see e.g., [**2**], Proposition 3 for a simple proof due to I.Z. Ruzsa).

REMARK 1.8. $|\mathcal{G} \oplus \mathcal{G}| \le 2^d |\mathcal{G}| \le 2^d n$ if $\mathcal{G} = \mathcal{G}_{d,n}$ is an arithmetic or geometric GP. Moreover, $\mathcal{G} \oplus \mathcal{G}$ obviously contains $|\mathcal{G}|$ copies of $\mathcal{G}$.

The Theorem below asserts that, if a not too big set $\mathcal{B}$ contains many copies of a not too small $\mathcal{A}$, then the latter is contained in one of the above mentioned configurations. As for the structure of $\mathcal{B}$, of course, one can only describe that portion of it, whose points are covered by at least one homothetic copy of $\mathcal{A}$. (The rest can be arbitrary.) That is why the following result involves

$$\mathcal{B}_{\mathcal{A}} \stackrel{\text{def}}{=} \bigcup \{\mathcal{A}' \subset \mathcal{B} \ ; \ \mathcal{A}' \approx \mathcal{A}\}.$$

The interesting feature of the statement is that *all* elements of $\mathcal{B}_{\mathcal{A}}$ can be accounted for — even those which are contained in as few as *one* copy $\mathcal{A}'$.

THEOREM 1.9 (Structure Theorem in one dimension). *If $\mathcal{A}$ and $\mathcal{B}$ satisfy (1.1) then there exist $C^* = C^*(C,c) > 0$, $d^* = d^*(C,c) > 0$ and an arithmetic or geometric $\mathcal{G} = \mathcal{G}_{d^*,C^*N}$, together with reals $b, b_1, \dots, b_{C^*}$ and $m_i > 0$ ($i = 1, \dots, C^*$), for which*

$$\mathcal{A} \subset \mathcal{G} + b\ ;$$

$$\mathcal{B}_{\mathcal{A}} \subset \bigcup_{i=1}^{C^*} \Big(m_i(\mathcal{G} \oplus \mathcal{G}) + b_i \Big).$$

The proof can be found in Section 2.3. It uses certain results concerning plane sets which determine few directions. The latter will be introduced in Section 2.2.

2. Proof of the one dimensional Theorems

2.1. Proof of Theorem 1.3.

1. Assume that $|\mathcal{A}| = t$, $|\mathcal{B}| = n$ and $h = h(t,n) = H(\mathcal{A},\mathcal{B})$. Denote by $\mathcal{A}_1, \dots, \mathcal{A}_h$ the subsets of $\mathcal{B}$ which are homothetic to $\mathcal{A}$ and by f_i (for $i = 1, \dots, h$) those linear functions $f_i(x) = m_i x + b_i$ which map $\mathcal{A}$ to $\mathcal{A}_i$, i.e., for which $f_i(\mathcal{A}) = \mathcal{A}_i$. Then the graph of each such f_i contains at least t points of $\mathcal{A} \times \mathcal{B}$, i.e., those of the form $(a, f_i(a))$ for $a \in \mathcal{A}$. By a result of Szemerédi–Trotter (Theorem 2 in [**11**]), the number of such graphs (straight lines) satisfies
$$h \le C\frac{|\mathcal{A} \times \mathcal{B}|^2}{t^3} = C\frac{(nt)^2}{t^3} = C\frac{n^2}{t},$$
as required.
2. The lower bound comes easily by showing that an arithmetic progression of n terms contains $\lfloor \frac{n}{2t} \rfloor \lfloor \frac{n}{2} \rfloor$ (or more) homothetic copies of an arithmetic progression of t terms. This will imply
$$h(t,n) \ge \left\lfloor \frac{n}{2t} \right\rfloor \left\lfloor \frac{n}{2} \right\rfloor \ge \frac{n}{4t}\frac{n}{4} = \frac{n^2}{16t}$$
if $n \ge 2t$, since $\lfloor x \rfloor \ge x/2$ if $x \ge 1$. Now if $\mathcal{A} = \{1, 2, \dots, t\}$ and $\mathcal{B} = \{1, 2, \dots, n\}$ then the functions $f_{ij} = m_i x + b_j$ for $m_i \in \{1, \dots, \lfloor \frac{n}{2t} \rfloor\}$ and $b_j \in \{1, \dots, \lfloor \frac{n}{2} \rfloor\}$ will all map $\mathcal{A}$ to subsets of $\mathcal{B}$, since $t\lfloor \frac{n}{2t} \rfloor + \lfloor \frac{n}{2} \rfloor \le n$. Also, these functions are all distinct and so are the images of $\mathcal{A}$. ■

2.2. Sets which determine few directions. In order to prepare for the proof of Theorem 1.9, we give some results on plane sets which determine few, i.e., a linear number of directions.

For any two (finite) disjoint point sets $\mathcal{A}_1, \mathcal{A}_2 \subset \mathbf{R}^2$, we denote by $D(\mathcal{A}_1, \mathcal{A}_2)$ the number of distinct directions determined by the segments A_1A_2 for $A_i \in \mathcal{A}_i$ ($i = 1, 2$). For the sake of simplicity, we do not distinguish segments A_1A_2 and A_2A_1; thus two segments have equal directions iff they are parallel.

We shall say that $\mathcal{A}_1$ and $\mathcal{A}_2$ determine few directions if $N \le |\mathcal{A}_1|, |\mathcal{A}_2|$ and $D(\mathcal{A}_1, \mathcal{A}_2) \le CN$. Several examples of such pairs exist (e.g., appropriate subsets of vertices of a regular polygon or of a square lattice). Two of them will play special role in the sequel:

$$\bigcup_{i-1}^{C}\Big(\mathcal{G}\cdot\overline{v}+\overline{h}_i\Big) \tag{2.1}$$

for $\mathcal{G}$ an arithmetic GP and $\overline{v},\overline{h}_1,\dots,\overline{h}_C\in\mathbf{R}^2$; and

$$\{\overline{b}\}+\bigcup_{i=1}^{C}\Big(\mathcal{G}\cdot\overline{v}_i\Big) \tag{2.2}$$

for $\mathcal{G}$ a geometric GP and $\overline{b},\overline{v}_1,\dots,\overline{v}_C\in\mathbf{R}^2$. The constant C will be called the *thickness* of these structures.

Here the first configuration consists of C copies of an arithmetic GP, one on each of C lines, all in direction $\overline{v}$, while the other one contains C copies of a geometric GP, one on each of C concurrent lines (in directions $\overline{v}_1,\dots,\overline{v}_C$), all through $\overline{b}$.

We shall call them arithmetic and geometric GP-type structures, respectively.

The following result (Theorem 2 in **[3]**) will work as our main tool in proving Theorem 1.9.

Theorem A. *Assume that $\mathcal{A}_1,\mathcal{A}_2\in\mathbf{R}^2$ with $N\le|\mathcal{A}_1|,|\mathcal{A}_2|\le CN$ and*

1. *$\mathcal{A}_1$ is collinear, located on a straight line l;*
2. *$\mathcal{A}_2\cap l=\emptyset$ (otherwise $\mathcal{A}_2$ is arbitrary);*
3. *$D(\mathcal{A}_1,\mathcal{A}_2)\le CN$.*

Then $\mathcal{A}_1\cup\mathcal{A}_2$ is contained in one of the structures of (2.1) or (2.2) of thickness C^*, based upon an arithmetic or geometric $\mathcal{G}=\mathcal{G}_{d^*,C^*N}$ where $d^*=d^*(C)$ and $C^*=C^*(C)$ do not depend on N.

The following Proposition describes the directions arising in Theorem A (and, as a by-product, it also demonstrates that arithmetic and geometric GP-type structures really determine few directions.)

PROPOSITION 2.1. *Assume that the x–axis contains $\mathcal{A}_1$. If, moreover,*

1. *$\mathcal{A}_2$ lies on the line $y=1$ and $\mathcal{A}_1,\mathcal{A}_2$ are copies of an arithmetic GP $\mathcal{G}$; or*
2. *$\mathcal{A}_2$ lies on a line through the origin and $\mathcal{A}_1,\mathcal{A}_2$ are copies of a geometric GP $\mathcal{G}$, both centered at $(0,0)$,*

then all directions counted in $D(\mathcal{A}_1,\mathcal{A}_2)$ are parallel to one of the vectors

$$(x,1)\qquad x\in\mu\cdot(\mathcal{G}\oplus\mathcal{G})+\beta,$$

for some $\mu,\beta\in\mathbf{R}$.

PROOF. In case (i), pick two generic points $A_1=(M-g_i,0)\in\mathcal{A}_1$ and $A_2=(g_j,h)\in\mathcal{A}_2$, where $M=M(\mathcal{G})$ was defined after Definition 1.6 and $g_i,g_j\in\mathcal{G}$. Then the direction of the segment A_1A_2 is

$$\Big(\frac{1}{h}(g_i+g_j)-\frac{M}{h}\ ;\ 1\Big). \tag{2.3}$$

In case (ii), let $\mathcal{A}_2$ be located on the line in direction (u,v). Also let there be given $A_1=(g_i,0)\in\mathcal{A}_1$ and $A_2=(u\frac{M}{g_j},v\frac{M}{g_j})\in\mathcal{A}_2$, for some $g_i,g_j\in\mathcal{G}$. Then the direction of the segment A_1A_2 is

$$\Big(-\frac{1}{Mv}\,g_ig_j+\frac{u}{v}\ ;\ 1\Big).\ \blacksquare \tag{2.4}$$

2.3. Proof of Theorem 1.9. We show that, for the values C^* and d^* found in Theorem A, $2C^*$ and d^*+1 make a suitable choice here.
Let $\mathcal{A}, \mathcal{B}$ satisfy (1.1). We consider $\mathcal{A}$ and $\mathcal{B}_{\mathcal{A}}$ as subsets of $\mathbf{R}^2$ as follows:

$$\mathcal{A}_1 \stackrel{\text{def}}{=} \{(a,1)\ ;\ a \in \mathcal{A}\};$$

$$\mathcal{B}_1 \stackrel{\text{def}}{=} \{(b,0)\ ;\ b \in \mathcal{B}_{\mathcal{A}}\}.$$

(In other words, we preserve $\mathcal{B}_1 = \mathcal{B}_{\mathcal{A}}$ on the x-axis while placing a copy $\mathcal{A}_1$ of $\mathcal{A}$ on the line $y = 1$.)

Now we extend our workspace even further to $\mathcal{P}^2$, the projective plane. To every homothetic copy $\mathcal{A}' \subset \mathcal{B}_1$ of $\mathcal{A}$ there is a unique (finite or infinite) point of $\mathcal{P}^2$ which, if used as the center of a projection, will map $\mathcal{A}$ to $\mathcal{A}'$. Let $\mathcal{A}_2 \subset \mathcal{P}^2$ be the set of these centers. (They can be located anywhere but on the lines $y = 0$ or $y = 1$.) The two sets will have the property that connecting all points of $\mathcal{A}_1$ to those of $\mathcal{A}_2$, the resulting straight lines will only have CN or fewer distinct points of intersection with the x-axis (i.e., those in $\mathcal{B}_1$).

We apply the following projective transform $\pi : \mathcal{P}^2 \to \mathcal{P}^2$:

$$\pi(u,v,w) \stackrel{\text{def}}{=} (u,w,v),$$

where the point $(u,v,w) \in \mathcal{P}^2$ is written in projective coordinates.

This mapping exchanges the points of the x-axis with those of the line at infinity, while preserving the line $y = 1$. Also, horizontal lines are mapped to horizontal lines. Moreover,

$$D\big(\pi(\mathcal{A}_1), \pi(\mathcal{A}_2)\big) = |\pi(\mathcal{B}_1)| = |\mathcal{B}_1| = |\mathcal{B}_{\mathcal{A}}| \le CN.$$

Use Theorem A for these $\pi(\mathcal{A}_1)$ and $\pi(\mathcal{A}_2)$ and get a cover of the $\pi(\mathcal{A}_i)$ described there, based upon an arithmetic or geometric GP $\mathcal{G}$. In both cases, using Proposition 2.1, $\pi(\mathcal{B}_1)$ will be contained in the union of certain $\mu_i(\mathcal{G} \oplus \mathcal{G}) + \beta_i$, for some $\mu_i, \beta_i \in \mathbf{R}$ where $1 < i \le C^*$. Therefore so is $\mathcal{B}_1$, while $\mathcal{A}_1 = \pi(\mathcal{A}_1) \subset \mu_1\mathcal{G} + \beta_1$. Since $\mu_1 \cdot \mathcal{G}$ is contained in another arithmetic or geometric GP of dimension at most one higher and size at most twice more than that of $\mathcal{G}$, we are done. ∎

3. Problems and results in higher dimensions

The quantity $H(\mathcal{A}, \mathcal{B})$ naturally extends to higher dimensional sets, as well. However, this does not change $h(t,n)$. Indeed, on the one hand, every 1-dimensional configuration can be considered as a subset of $\mathbf{R}^k$ — thus $h(t,n)$ does not decrease. Also, on the other hand, projecting $\mathcal{A}, \mathcal{B} \subset \mathbf{R}^k$ to a straight line (distinct points mapped to distinct points) will preserve $H(\mathcal{A}, \mathcal{B})$. Therefore, $h(t,n)$ cannot increase.

The interesting questions arise when we require that the (affine) dimension of $\mathcal{A} \subset \mathbf{R}^k$ be some $r > 1$. The corresponding quantities will be denoted by $H^{(r)}(\mathcal{A}, \mathcal{B})$ and $h^{(r)}(t,n)$, respectively. (No need for indicating the dimension of the underlying space; following the line of the previous argument, it can be seen that $h^{(r)}$ does not depend on it.)

I. For t fixed,

$$c_1 n^{1+1/r} \le h^{(r)}(t,n) \le c_2 n^{1+1/r}$$

follows from some results in [**4**]. This order of magnitude can really be attained for a fixed t (while $n \to \infty$) and any fixed set $\mathcal{A}$ (of dimension r), provided that $\mathcal{A}$ can be represented by points whose coordinates are all algebraic. (This was proven in [4].) However, nothing is known about transcendental structures.

II. For the general case $t \to \infty$, suitable portions of the r-dimensional cubic lattice give

$$h^{(r)}(t,n) \geq c_r \frac{n^{1+1/r}}{t^{1/r}},$$

where c_r may depend on r but not on t or n.
If $t = o(n)$ (i.e., $t/n \to 0$), it is not known if $h^{(r)}$ can be essentially bigger than this lower bound. The reason for this difficulty is, again, related to the Szemerédi–Trotter Theorem: sufficiently tight higher dimensional versions of that result (concerning incidences of points and hyperplanes) have not been found so far.

III. In spite of the lack of these tools, the 1-dimensional bounds on "large" subsets can be extended to any dimension. First, by projecting $\mathbf{R}^k$ onto a line (a trick we already used before), for every r and $N \leq |\mathcal{A}|, |\mathcal{B}| \leq CN$ we have

$$H^{(r)}(\mathcal{A},\mathcal{B}) \leq C^* N. \tag{3.1}$$

Also, if $\mathcal{A}$ and $\mathcal{B}$ are r-dimensional cubic lattices again, then

$$H^{(r)}(\mathcal{A},\mathcal{B}) \geq c_r' N, \tag{3.2}$$

Below we describe two other r-dimensional structures which are relatives of the 1-dimensional near-extremal configurations. Also, they play a crucial role in our higher dimensional structure theorem, as well.

Arithmetic GPs consisting of numbers extend naturally to arithmetic GPs of vectors of $\mathbf{R}^r$ — we just let the differences Δ_i be such vectors. In this generality, an arithmetic GP has two parameters related to dimension: we shall call r the *vector dimension* while d the *progression dimension*. It will still hold that $|\mathcal{G}+\mathcal{G}| \leq 2^d|\mathcal{G}|$ and it also remains true that $\mathcal{G}+\mathcal{G}$ contains $|\mathcal{G}|$ homothetic copies of $\mathcal{G}$.

Geometric GPs do not really extend to $\mathbf{R}^r$. (There is no natural multiplication there.) The best we can do in order to have proper r-dimensional geometric GP-type structures is to consider structures like those in (2.2) using directions $\overline{v}_i$ which span $\mathbf{R}^r$. It will turn out that no other geometric GP-based configurations can produce many large homothetic subsets. (This phenomenon can be considered as an "implicit evidence" of the intuitive fact that geometric GPs are inherently one-dimensional objects.)

Definition 3.1. Let C be a positive integer. We call $\mathcal{J}$ a *geometric GP–type bunch*, of center $\overline{b} \in \mathbf{R}^r$, located on C straight lines and based upon a geometric $\mathcal{G} \subset \mathbf{R}$, if there exist vectors $\overline{v}_1, \ldots, \overline{v}_C \in \mathbf{R}^r$ such that

$$\mathcal{J} = \{\overline{b}\} \cup \bigcup_{i=1}^{C} \{g\overline{v}_i + \overline{b}\ ;\ g \in \mathcal{G}\}.$$

The above structures are natural relatives of those in (2.2). These were called "geometric GP–type generalized Jamison configurations" in [3].)
Also, with the same center $\overline{b}$ and vectors $\overline{v}_i$ but with $\mathcal{G}\cdot\mathcal{G}$ in place of $\mathcal{G}$ above, we shall denote the resulting structure by $\mathcal{J}\cdot\mathcal{J}$. (Thus, in this "product", the lines do not change — just the geometric GPs on them extend.)

Note that $\mathcal{J}\cdot\mathcal{J}$ contains $|\mathcal{J}|$ homothetic copies of $\mathcal{J}$. Of course, $|\mathcal{J}\cdot\mathcal{J}| \leq 2^d|\mathcal{J}|$ still holds.

THEOREM 3.2. *If $\mathcal{A}, \mathcal{B} \subset \mathbf{R}^r$ with $N \leq |\mathcal{A}|, |\mathcal{B}| \leq CN$ and $H(\mathcal{A}, \mathcal{B}) \geq cN$, then*

1. *either a shifted copy of an arithmetic $\mathcal{G} = \mathcal{G}_{d^*, C^*N}$ contains $\mathcal{A}$, while $\mathcal{B}_{\mathcal{A}}$ is contained in C^* homothetic copies of $\mathcal{G} + \mathcal{G}$;*
2. *or a geometric GP–type bunch $\mathcal{J}$ on C^* lines, based upon a geometric GP $\mathcal{G}_{d^*, C^*N}$, contains $\mathcal{A}$, while $\mathcal{B}_{\mathcal{A}}$ is contained in C^* homothetic copies of $\mathcal{J} \cdot \mathcal{J}$,*

where d^* and C^* only depend on c and C but neither on n nor on r.

(Note again the asymmetry between several dimensional arithmetic versus one dimensional geometric GPs.)

The proof follows exactly the same lines of reasoning as that of Theorem 1.9 in Section 2.3: we use copies of $\mathcal{A}$ and $\mathcal{B}_{\mathcal{A}}$ in $\mathbf{R}^{r+1}$ on the hyperplanes $x_{r+1} = 1$ and $x_{r+1} = 0$, respectively, and then we apply the transform $\pi(x_1, \ldots, x_{r-1}, x_r, x_{r+1}) = (x_1, \ldots, x_{r-1}, x_{r+1}, x_r)$. Also Proposition 2.1 will extend easily to structures derived from $\mathcal{G} + \mathcal{G}$ or $\mathcal{J} \cdot \mathcal{J}$.
The only thing to prove is the higher dimensional version of Theorem A. That is why the rest of this paper is devoted to developing that generalization — see Theorem 4.1 below.

(Another natural attempt would be to try to use Theorem 1.9 and induction on the dimension to prove Theorem 3.2. This would work well from $r = 2$ upwards; however, we cannot go that far. Already the very first step from 1 to 2 fails, since projecting to a line loses all collinearity information. That is why no such "direct" way seems easier than the above mentioned "bypass".)

4. Few directions in several dimensions

In what follows, we consider $\mathbf{R}^r$ to be a special subspace of $\mathbf{R}^k$, for $r < k$:

$$\mathbf{R}^r = \{(x_1, x_2, \ldots, x_k)\ ;\ x_{r+1} = x_{r+2} = \ldots = x_k = 0\} \subset \mathbf{R}^k.$$

The theorem to be proven here (and required at the end of the previous section) is the following.

THEOREM 4.1. *Let $\mathcal{A}_1, \mathcal{A}_2 \subset \mathbf{R}^k$ with $N \leq |\mathcal{A}_1|, |\mathcal{A}_2| \leq CN$. Assume, moreover, that $\mathcal{A}_1 \subset \mathbf{R}^r$ (for some $r < k$) while $\mathcal{A}_2 \cap \mathbf{R}^r = \emptyset$.*
If $D(\mathcal{A}_1, \mathcal{A}_2) \leq CN$ then

1. *either $\mathcal{A}_1$ is contained in a shifted copy of an arithmetic $\mathcal{G}_{d^*, C^*N} \subset \mathbf{R}^r$ while $\mathcal{A}_2$ can be covered with at most C^* copies of $\mathcal{G}$, one on each of C^* affine subspaces of dimension r, all parallel to $\mathbf{R}^r$;*
2. *or $\mathcal{A}_1 \cup \mathcal{A}_2$ is contained in a geometric GP–type bunch $\mathcal{J}$, based upon a geometric $\mathcal{G}_{d^*, C^*N} \subset \mathbf{R}$ and located on C^* straight lines with center in $\mathbf{R}^r$.*

First we just prove some coverability results; the arithmetic or geometric GP structures will come easily then.

4.1. Covers by straight lines. In the sequel we shall always consider sets which satisfy the assumptions of Theorem 4.1, i.e., $\mathcal{A}_1, \mathcal{A}_2 \subset \mathbf{R}^k$ with $N \leq |\mathcal{A}_1|, |\mathcal{A}_2| \leq CN$ and $\mathcal{A}_1 \subset \mathbf{R}^r$ while $\mathcal{A}_2 \cap \mathbf{R}^r = \emptyset$; moreover $D(\mathcal{A}_1, \mathcal{A}_2) \leq CN$.

Also, in what follows, $C_2 = C_2(C)$ denotes the quantity $C^*(C)$ found in the 2-dimensional Theorem A.

LEMMA 4.2. *For $r = 1$ and $k \geq 2$, $\mathcal{A}_1 \cup \mathcal{A}_2$ can be covered with C_2 straight lines, either all parallel to $\mathbf{R}$ or all concurrent, with common point in $\mathbf{R}$.*

PROOF. by induction on k.
The case $k = 2$ is contained in Theorem A.

For $k \geq 3$, project $\mathcal{A}_1 \cup \mathcal{A}_2$ in a direction which is sufficiently general with respect to $\mathcal{A}_1$ and $\mathcal{A}_2$. By this we mean that none of the following occur:

- two points of $\mathcal{A}_1 \cup \mathcal{A}_2$ mapped to the same image point;
- non-collinear triples of $\mathcal{A}_1 \cup \mathcal{A}_2$ mapped to collinear points;
- two non-parallel lines determined by two pairs of points of $\mathcal{A}_1 \cup \mathcal{A}_2$ mapped to parallel lines;
- three non-concurrent lines determined by three pairs of points of $\mathcal{A}_1 \cup \mathcal{A}_2$ mapped to concurrent lines.

Obviously, each forbidden case excludes "few" directions (i.e., of measure zero according to the Lebesgue measure of appropriate dimension). Indeed, one direction is excluded in the first case for each pair of points; directions within a hyperplane are excluded in the second and third cases for any three and four points, respectively; finally, the directions involved in the fourth case satisfy a quadratic equation, for every three pairs of points. Therefore, "general" directions always exist.

Now if we project our point sets in such a direction to a hyperplane and apply the induction hypothesis, we are done, since no parallel or concurrent system of lines can appear by coincidence. ∎

4.2. Covers in general.

LEMMA 4.3. *For all $r < k$, $\mathcal{A}_2$ can be covered with either*

$$\hat{C} = \hat{C}(C) \stackrel{\text{def}}{=} C_2(C) + C_2\big(C \cdot C_2(C)\big)$$

shifted copies of $\mathbf{R}^r$ or with $\hat{C}$ concurrent straight lines whose common point lies in $\mathbf{R}^r$.

(Here $C_2 = C_2(C)$ still denotes the function $C^*(C)$ found in Theorem A.)

PROOF. by induction from $(r-1, k-1)$ to (r, k).
The case $r = 1$ comes from the foregoing Lemma.

For $r \geq 2$, we proceed as in the previous proof, except for the very first value $r = 2$ which will cause some complications. Anyway, we shall project $\mathcal{A}_1 \cup \mathcal{A}_2$ in a direction $\overline{v} \in \mathbf{R}^r$ (this restriction is the reason for the above mentioned difficulties) which is general in the following sense:

- no two points of $\mathcal{A}_1 \cup \mathcal{A}_2$ are mapped to the same image point;
- non-collinear triples of $\mathcal{A}_1 \cup \mathcal{A}_2$ are not mapped to collinear points UNLESS $r = 2$ AND THE THREE POINTS LIE IN A PLANE PARALLEL TO $\mathbf{R}^2$;
- two non-parallel lines determined by two pairs of points of $\mathcal{A}_1 \cup \mathcal{A}_2$ are not mapped to parallel lines UNLESS $r = 2$ AND THE TWO LINES ARE PARALLEL TO $\mathbf{R}^2$;
- three non-concurrent lines determined by three pairs of points of $\mathcal{A}_1 \cup \mathcal{A}_2$ are not mapped to concurrent lines UNLESS $r = 2$ AND ONE OF THE LINES LIES IN A PLANE PARALLEL TO $\mathbf{R}^2$ WHILE THE OTHER TWO INTERSECT AT A POINT OF THIS PLANE.

Again, it is not difficult to show that such a $\overline{v} \in \mathbf{R}^r$ does, indeed, exist.

Denote by π the projection of $\mathbf{R}^k$ in direction $\overline{v}$. Then $\pi(\mathcal{A}_1), \pi(\mathcal{A}_2)$ are contained in a hyperplane (of dimension $k-1$) and $\pi(\mathcal{A}_1)$ also in an $(r-1)$-dimensional affine subspace.

Case I. If $r = 2$ then we can use the induction hypothesis (i.e., Lemma 4.2) to find a cover of $\pi(\mathcal{A}_1) \cup \pi(\mathcal{A}_2)$ by C_2 lines. If they are all parallel then we have also found a cover of $\mathcal{A}_1 \cup \mathcal{A}_2$ by shifted copies of $\mathbf{R}^2$. Otherwise, $\pi(\mathcal{A}_2)$ can be covered with less that C_2 concurrent lines with common point in $\pi(\mathbf{R}^2)$. Since we excluded degeneracies, also $\mathcal{A}_2$ itself possesses such a cover. (The lines intersect in $\mathbf{R}^2$.) We are left to show that $\mathcal{A}_1$, too, can be covered by $C_2\big(C \cdot C_2(C)\big)$ appropriate lines.

If $|\mathcal{A}_1|$ does not exceed this quantity, we are done. Otherwise, we select one of the lines that cover $\mathcal{A}_2$ and which contains more than N/C_2 of its points. Denote this line by l and put $\mathcal{A}_2' \stackrel{\text{def}}{=} \mathcal{A}_2 \cap l$. Now we apply Lemma 4.3 with $N' = N/C_2$ in place of N and $C \cdot C_2(C)$ in place of C to $\mathcal{A}_2'$ in place of $\mathcal{A}_1$ and $\mathcal{A}_1$ in place of $\mathcal{A}_2$. This results in a cover of $\mathcal{A}_1$ by $C_2\big(C \cdot C_2(C)\big)$ straight lines and they must all go through $\mathbf{R}^r \cap l$. (Indeed, they cannot be parallel to l; nor can they intersect out of $\mathbf{R}^r$ since otherwise they could cover at most one point of $\mathcal{A}_1$ each but $|\mathcal{A}_1| > C_2\big(C \cdot C_2(C)\big)$ was assumed.) This settles the case $r = 2$.

Case II. For $r \geq 3$ the projection results in a cover of both $\pi(\mathcal{A}_1)$ and $\pi(\mathcal{A}_2)$. As no degeneracy can occur in this case, also $\mathcal{A}_1$ and $\mathcal{A}_2$ possess an appropriate cover. This proves Lemma 4.3. ∎

4.3. Proof of Theorem 4.1. Let $\mathcal{A}_1, \mathcal{A}_2$ satisfy the assumptions. By Lemma 4.3, there is a cover of $\mathcal{A}_1 \cup \mathcal{A}_2$ either by a concurrent bunch of straight lines or by shifted copies of $\mathbf{R}^r$. We are left to find the geometric or arithmetic GP type structures.

In the first case, we again project $\mathcal{A}_1 \cup \mathcal{A}_2$ to a two dimensional subspace such a way that the images of the covering lines be all distinct. (Coinciding points do not matter here.) Then Theorem A finishes the proof, since a collinear pre-image of a geometric GP $\mathcal{G}$ is $\lambda \cdot \mathcal{G}$, for a real constant λ.

In the second case, we shall use the following result of Freiman and Ruzsa [**5, 8, 9**].

Theorem B. *For every $C > 0$ there is a $d^* = d^*(C) > 0$ and a $C^* = C^*(C) > 0$ (both independent of k and n below) with the following property.*
If

$$|A + B| \leq Cn \text{ or } |A - B| \leq Cn$$

for some $A, B \subset \mathbf{R}^k$ with $|A|, |B| \geq n$, then $A \cup B$ is contained in an arithmetic $\mathcal{G}_{d^*, C^* n}$.

This will, indeed, prove the existence of a suitable arithmetic GP. To this end, we project $\mathcal{A}_2$ to $\mathbf{R}^r$ and denote the set of the image points by B. (It does not matter if several points are mapped to one; this can only decrease the size by a constant factor $\hat{C}$.) Then

$$|B - A_1| \leq C \cdot \hat{C} \cdot N,$$

since image points of $\mathcal{A}_2$ from one shifted copy of $\mathbf{R}^r$ produce as many differences with $\mathcal{A}_1$ as many directions the original points determine. By Theorem B, the arithmetic GP is found. ∎

4.4. Substructures from weaker assumptions. We mention without proof that also a regular *sub*structure can be found under the weaker assumption that (instead of $D(\mathcal{A}_1, \mathcal{A}_2)$), just

$$D_E(\mathcal{A}_1, \mathcal{A}_2) \stackrel{\text{def}}{=} \#\{\text{directions of segments } \overline{A_1 A_2} \mid (A_1, A_2) \in E\}$$

is small, for some (not too small) $E \subset \mathcal{A}_1 \times \mathcal{A}_2$.

Indeed, the following result was also proven in [**3**] as the 2-dimensional special case.

Theorem C. *For any positive integer C there is a $c^* = c^*(C)$ with the following property.*
Let $\mathcal{A}_1, \mathcal{A}_2 \subset \mathbf{R}^2$ with $N \leq |\mathcal{A}_1|, |\mathcal{A}_2| \leq CN$, and $\mathcal{A}_1 \subset \mathbf{R}$ while $\mathcal{A}_2 \cap \mathbf{R} = \emptyset$; moreover

$$D_E(\mathcal{A}_1, \mathcal{A}_2) \leq CN$$

for some $E \subset \mathcal{A}_1 \times \mathcal{A}_2$ with $|E| \geq N^2$.
Then there is a collinear subset $\mathcal{A}_2^* \subset \mathcal{A}_2$ such that $|E \cap (\mathcal{A}_1 \times \mathcal{A}_2^*)| \geq c^* N^2$.

The same line of reasoning as used for Theorem 4.1 yields the following.

THEOREM 4.4. *For any positive integer C there is a $c^* = c^*(C)$ with the following property.*
Let $\mathcal{A}_1, \mathcal{A}_2 \subset \mathbf{R}^k$ with $N \leq |\mathcal{A}_1|, |\mathcal{A}_2| \leq CN$ and $\mathcal{A}_1 \subset \mathbf{R}^r$ (where $r < k$) while $\mathcal{A}_2 \cap \mathbf{R}^r = \emptyset$; moreover

$$D_E(\mathcal{A}_1, \mathcal{A}_2) \leq CN$$

for some $E \subset \mathcal{A}_1 \times \mathcal{A}_2$ with $|E| \geq N^2$.
Then there is a subset $\mathcal{A}_2^* \subset \mathcal{A}_2$ contained either in an affine subspace of dimension r parallel to $\mathbf{R}^r$ or on a line not *parallel to $\mathbf{R}^r$, such that $|E \cap (\mathcal{A}_1 \times \mathcal{A}_2^*)| \geq c^* N^2$.*

(Thus, again, $\mathcal{A}_2^*$ contains $(c^*/C)N = c^{**}N$ or more points.)

References

[1] Antal Balog and Endre Szemerédi. A statistical theorem of set addition. *Combinatorica*, 14 (1994), pp 263–268.

[2] György Elekes. On linear combinatorics II. *Combinatorica*, 18(1), (1998), pp 13–25.

[3] György Elekes. On linear combinatorics III. *Combinatorica*, 18(3), (1998).

[4] György Elekes and Paul Erdős. Similar configurations and pseudo grids. In *Coll. Math. Soc. János Bolyai* 63 *Intuitive Geometry,* (1991), pp 85–104.

[5] Gregory A Freiman. *Foundations of a Structural Theory of Set Addition, Translation of Mathematical Monographs vol. 37.* Amer. Math. Soc., Providence, R.I., USA, 1973.

[6] Ron Graham and Jaroslav Nešetřil, editors. *The Mathematics of Paul Erdős.* Springer, 1996.

[7] Miklós Laczkovich and Imre Z Ruzsa. *The Number of Homothetic Subsets.* In Graham and Nešetřil [**6**], 1996.

[8] Imre Z. Ruzsa. Arithmetical progressions and sum sets. Technical Report 91-51, DIMACS Center for Discrete Math. and Theoretical Comp. Sci., Rutgers University, 1991.

[9] Imre Z Ruzsa. Generalized arithmetic progressions and sum sets. *Acta Math. Sci. Hung.*, 65 (1994), pp 379–388.

[10] Endre Szemerédi. On sets of integers containing no k elements in arithmetic progression. *Acta Arithmetica* 27 (1975)

[11] Endre Szemerédi and W. T. Trotter Jr. Extremal problems in Discrete Geometry. *Combinatorica*, 3 (3–4), (1983), pp 381–392.

EÖTVÖS UNIVERSITY, BUDAPEST, MÚZEUM KRT 6–8, H–1088
E-mail address: elekes@cs.elte.hu

DIMACS Series in Discrete Mathematics
and Theoretical Computer Science
Volume **49**, 1999

The Complexity of an Inverse Shortest Paths Problem

Sándor P. Fekete, Winfried Hochstättler, Stephan Kromberg,
and Christoph Moll

ABSTRACT. In this paper we study the complexity of an Inverse Shortest Paths Problem (ISPP). We show that the problem is intractable even in very restricted cases. In particular, we prove that the ISPP is NP-complete in the planar case. Furthermore, we give a characterization of the class of graphs G_d of given distances for which the ISPP is tractable: we give polynomial algorithms for special classes of G_d and provide evidence that no polynomial time algorithms exist if the structure of G_d is only slightly more complicated in a well-defined graph-theoretical sense. Finally, we discuss the situation for directed graphs.

1. Introduction

Computing distances in a weighted graph is one of the classical problems in combinatorial optimization. The inverse problem of finding edge weights in a given graph that induce prescribed distances for certain pairs of vertices, has attracted some attention recently. This so-called *Inverse Shortest Paths Problem* (ISPP) is of importance for estimating the transition parameters in networks. Other applications are the computation of average speeds on roads, where the road network and the typical travel times between some locations are known. Similar problems arise in traffic regulation and network control when trying to find "fair" cost allocations, as well as in biology, in particular when dealing with sequence analysis for reconstructing degrees of relationships and phylogenetic trees.

In practical vehicle routing problems a major difficulty is the estimate of the travel times. In most cases these travel times are calculated using Euclidean distances between customers. This approach obviously fails when rivers have to be crossed or the trip does not follow a highway. Better estimates are obtained when using shortest path algorithms on road networks. However, the estimate of the average speed on roads is a difficult task. Experience shows that the transport manager has a very good knowledge of the travel times between established customers, but no idea of the travel times needed for small parts of these trips. Finally the average speed heavily depends on the traffic situation, i. e., it is time dependent.

This problem leads to the idea of using tachograph data or the travel times of trips in the past to estimate the average speeds. This data is collected in modern

1991 *Mathematics Subject Classification.* Primary 05C12; Secondary 68Q25.

Key words and phrases. distance in graphs, planar graphs, shortest paths, inverse shortest paths, computational complexity, NP-completeness.

board systems and could be extracted automatically. The main problem of such an approach – as we will show in this paper – is to recover the route of the truck, even assuming that the drivers always tend to minimize the travel time.

The problem of estimating weights when the path is known has been discussed by Toint and Burton [**Bu92, To92**]. They gave a quadratic programming formulation for the task to adjust the given weights of a graph such that a certain given path becomes a shortest path. See also the paper by Burton, Pulleyblank, and Toint [**BPT97**] for the case where there are upper bounds on the costs of shortest paths. Similar kinds of problems have been considered for several other combinatorial optimization problems such as Minimum Spanning Tree, Weighted Matching and Weighted Matroid Intersection [**ML97**]. In all of these problems a weight function shall be adjusted in order to make a given solution an optimum solution. In the paper by Cai and Li [**ML97**] it is proven that the very general Inverse Matroid Intersection Problem is equivalent to a certain min-cost flow problem. As a corollary one derives strongly polynomial algorithms for all the problems mentioned above.

Motivated by the above application, we consider the more general question, where the distances are given, but not the paths achieving them. Under this assumption the ISPP is closely related to the problem of finding a set of vertex-disjoint paths connecting given pairs of vertices. The complexity of the "Vertex-Disjoint Paths Problem" (VDPP) is well-studied. An overview can be found in [**Fr90**]. Polynomial-time algorithms are known only under very strong additional assumptions, indicating that polynomial cases of the ISPP can only be expected in very restricted situations.

The objective of this paper is to examine the complexity of the ISPP. We will try to follow the notation of Garey and Johnson [**Ga79**] whenever possible. The rest of this paper is organized as follows: in Section 2, we introduce basic definitions, Section 3 gives a proof that the ISPP is NP-complete for undirected graphs, Section 4 deals with planar undirected graphs. In Section 5 we identify the classes of graphs of given distances which lead to tractable instances. Finally, in Section 6 we consider directed graphs, in particular acyclic directed graphs.

2. Notation and Preliminaries

An instance of the Inverse Shortest Paths Problem (ISPP) consists of two graphs on the same set of nodes V. The first graph $G(V, E)$ consists of the edges that are used for routing the paths. The second graph $G_d(V, E_d, w_d)$ is provided with a weight function $w_d : E_d \to \mathbb{R}^+$ describing the distances between certain vertex pairs. A weight function $w : E \longrightarrow \mathbb{R}^+$ for G is called a *G_d-satisfying weight function on* G, if for all edges $(s,t) \in E_d$ the length of a shortest path from s to t in $G(V, E, w)$ is exactly $w_d(s,t)$. We also call G the *network* and G_d the *distance graph* of the instance.

Now the Inverse Shortest Paths Problem is given as follows:

Instance: A network $G(V, E)$, a distance graph $G_d(V, E_d, w_d)$.
Question: Is there a G_d-satisfying weight function w on G?

Let G and G_d be as above, and for each edge (s,t) in G_d, let $P_{(s,t)}$ be a path from s to t in G. Then we call the family $P_d = \{P_{(s,t)} | (s,t) \in E_d\}$ a *G_d-collection of paths in* G.

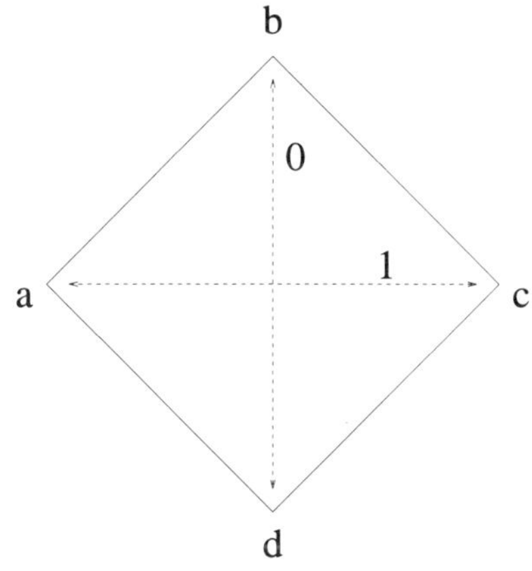

FIGURE 1. A diamond

Given an instance G, G_d for the ISPP, one might expect the set $\{w \in \mathbb{R}^E \mid w \text{ is } G_d\text{-satisfying}\}$ to be some nice geometric object. However, the following simple example shows that this set does not even have to be connected.

EXAMPLE 1. Let $V = \{a,b,c,d\}, E = \{(a,b),(b,c),(c,d),(a,d)\}$, $E_d = \{(a,c),(b,d)\}$, and $w_d(a,c) = 1, w_d(b,d) = 0$ (see Figure 1). The G_d-satisfying weight functions form two disjoint components, corresponding to the two possible paths from b to d. Each of these components consists of two rays meeting at their origins. More precisely the set of G_d-satisfying weight functions $w = (w_{(a,b)}, w_{(b,c)}, w_{(c,d)}, w_{(a,d)})$ is given by

$$\{(1,0,0,z)|z \geq 1\} \cup \{(z,0,0,1)|z \geq 1\} \mathbin{\dot{\cup}} \{(0,z,1,0)|z \geq 1\} \cup \{(0,1,z,0)|z \geq 1\}.$$

In the following section we will show how this example can be utilized to reduce the vertex-disjoint paths problem to the ISPP.

3. A Basic NP-completeness Result

The simple proof of the following theorem establishes the close relation between the ISPP and vertex-disjoint paths problems.

THEOREM 1 *The ISPP is NP-complete.*

PROOF. We give a reduction of the ISPP from the following version of the vertex-disjoint paths problem, denoted by (VDPP1).

Instance: A graph $G(V,E)$ and a "demand graph" $G_d(V,E_d)$ with maximum degree 1, $V_d = \{v \in V | d_{G_d}(v) = 1\}$.

Question: Is there a subset of edges $P \subset E$, such that two vertices in V_d are connected in $G(V,P)$, if and only if they are connected in G_d?

This problem is NP-hard; a straightforward reduction of the general vertex-disjoint paths problem (VDPP), where E_d does not necessarily have bounded degree to this slightly restricted case VDPP1 can be found in [**Pf90**].

Given an instance $G(V,E)$, $G_d(V,E_d)$ of VDPP1, we construct an instance $\hat{G}(V,\hat{E})$, $\hat{G}_d(V,\hat{E}_d,w_d)$ of the ISPP as follows (see Figure 2): Take the complete

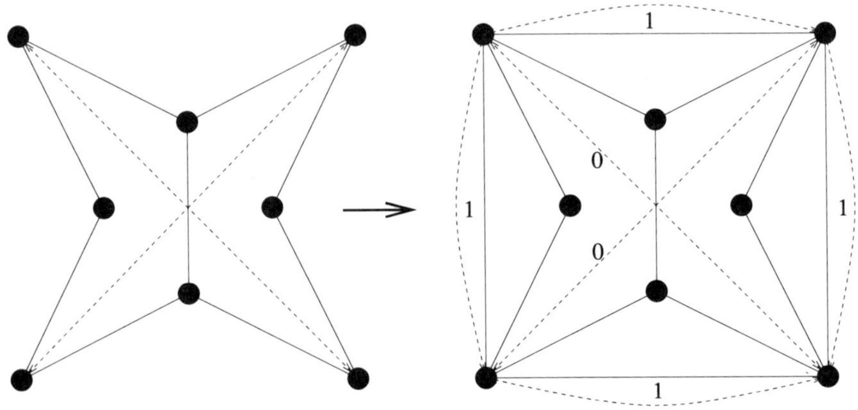

FIGURE 2. Constructing the ISPP instance for the given VDPP1 instance

graph on the "demand vertices" V_d and add the remaining isolated vertices of V to derive the weighted graph

$$\hat{G}_d(V, \hat{E}_d, w_d) \text{ with } \hat{E}_d = \binom{V_d}{2}.$$

$$\text{with weights } w_d(e) = 0 \text{ if } e \in E_d, w_d(e) = 1 \text{ else.}$$

The graph $\hat{G}$ is derived from G by adding the "non-demand edges" from the complete graph on V_d.

$$\hat{G} = \hat{G}(V, \hat{E}) \text{ with } \hat{E} = E \cup \binom{V_d}{2} \setminus E_d.$$

Given a solution P of VDPP1, we define the following weight function for $\hat{G}$:

$$w(e) = 0 \text{ if } e \in P, w(e) = 1 \text{ else.}$$

This weight function yields a distance of 0 for a pair $(s,t) \in \binom{V_d}{2}$ if and only if they are joined by an edge in E_d. On the other hand, we have an edge of weight 1 between all pairs $(s,t) \in \binom{V_d}{2} \setminus E_d$.

Conversely, let w be a solution of the ISPP instance. Consider

$$P = \{e \in \hat{E} | w(e) = 0\}.$$

All edges $(s,t) \in \hat{E} \setminus E \subset \binom{V_d}{2} \setminus E_d$ must have weight at least 1, since that distance is required for vertices s and t, therefore we have $P \subset E$.

By construction of w_d, two vertices $s, t \in V_d$ are connected by a path consisting of edges from P if and only if they have induced distance 0, so these are precisely the pairs in E_d. Hence P solves the Vertex Disjoint Paths Problem.

To complete the proof we have to argue that a solvable problem always has a short certificate. For a given weight function w, we can compute the induced distances in polynomial time using Dijkstra's algorithm (see Section 6.4 in [**Pa82**]). So it is left to show that there is a solution w with a description which is polynomial in the coding length of the ISPP instance. For a fixed family of (s,t)-paths between two vertices that are adjacent in E_d the possible solutions of the ISPP form a polyhedron. To be more precise, a weight function $w : E \rightarrow \mathbb{R}^+$ is G_d-satisfying and achieves weight $w_d(f)$ on some path P_f for each $f = (s_f, t_f) \in E_d$, if and only if

$$\begin{aligned} w(P_f) &= w_d(f) \quad \text{for all } f \in F \quad \text{and} \\ w(P) &\geq w_d(f) \quad \text{for all } f \in F \quad \text{and all } (s_f, t_f)\text{-paths } P \text{ in } G. \end{aligned}$$

By Cramer's rule, non-emptiness of the polyhedron implies the existence of a weight function of polynomial length. □

Note that the above reduction does not necessarily preserve planarity of a given instance, in particular if $|V_d| > 5$. The case of planar ISPP instances will be dealt with in the following section.

4. The Planar Case

The main motivation for considering our form of the ISPP is the computation of average travel times on roads. Since road networks can usually be considered to be planar graphs, the case of planar graphs is of particular importance.

The complexity of disjoint paths problems in the planar case has been studied extensively. The vertex disjoint paths problem remains NP-complete for instances where the graph $G + G_d := (V, E \cup E_d)$ is planar, as was shown by Middendorf and Pfeiffer [**MP88**]. It was shown by Seymour [**Se81**] that the latter problem is polynomially solvable if $G + G_d$ is planar and Eulerian. Okamura and Seymour [**OS81**], and Schrijver [**Sc88**] showed that it is also polynomial in the cases where the vertices of degree 1 in G_d lie in a bounded number of faces of $G + G_d$.

Planar Eulerian graphs do not seem to define an appropriate subclass for the ISPP: The satisfying collection of paths is usually overlapping and thus, their union does not have to define a Eulerian graph.

THEOREM 2. *The ISPP is NP-complete, even if restricted to instances G, G_d for which $G + G_d$ is planar.*

PROOF. From the argument given above ISPP-planar is in NP.
In the following we will show: ISPP-planar $\geq$ PLANAR 3SAT ([**Ga79, Li82**]).

An instance I of PLANAR 3SAT is defined as a 3SAT instance with clauses $c_1, \ldots, c_m$ and variables $x_1, \ldots, x_n$ where the following (bipartite) graph $H_I = (V_I, E_I)$ is planar:

$$\begin{aligned} V_I &= \{c_1, \ldots, c_m, x_1, \ldots, x_n\}, \\ E_I &= \{(c_j, x_i) \mid \text{variable } x_i \text{ occurs in clause } c_j\}. \end{aligned}$$

Fix a planar embedding of H_I and consider its medial graph $\hat{H}_I$ (see Figure 3). The vertices of $\hat{H}_I$ are the center points of edges of H_I; two vertices of $\hat{H}_I$ are adjacent if the corresponding edges of H_I share a vertex and lie on the same face of the embedding. Since this yields a four-regular planar graph, its dual graph $\hat{D}$ is bipartite. Furthermore, one color class of $\hat{D}$ corresponds to the faces of $\hat{H}$, the other to the vertices V_I. Therefore, the edge set $\hat{E}$ may be partitioned into

- "variable cycles", corresponding to the face boundaries of x_i faces and
- "clause cycles", corresponding to faces for the clauses c_j.

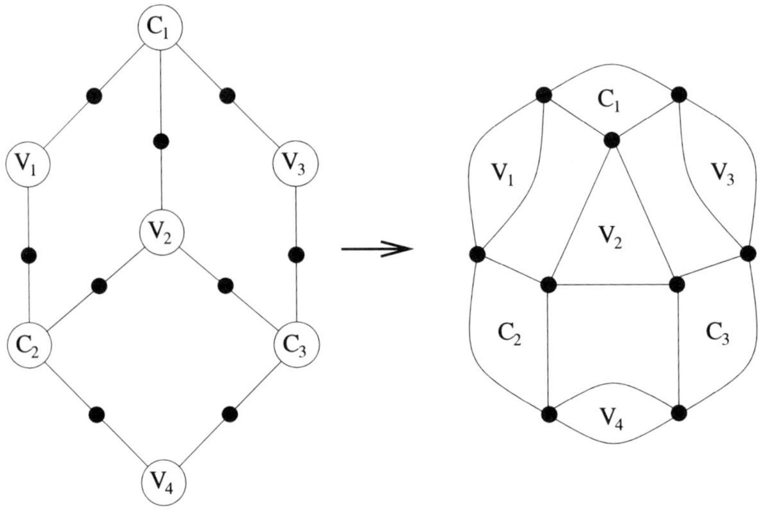

FIGURE 3. Constructing the medial graph

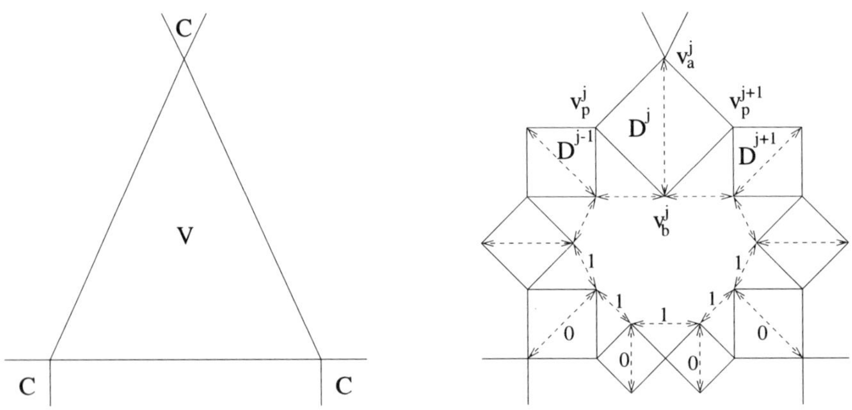

FIGURE 4. Constructing the diamond chains for variable cycles

Note that the vertices of $\hat{H}$ represent the literals of I and that each vertex is incident to exactly one variable cycle.

In the following we will first replace the variable cycles by slightly more complicated graphs and then modify the clause cycles.

As shown in Figure 4, we replace each variable cycle C_{x_i} of length $L(x_i)$ by a closed chain $K(x_i)$ of $q := 3L(x_i)$ "diamonds" (the configuration shown in Figure 1), denoted by $D_0, \ldots, D_k, \ldots, D_q = D_0$. We identify the original vertices of $\hat{H}$, and therefore the literals of I, with those diamonds D_i satisfying $i \equiv 0 \bmod 3$.

For a fixed embedding of the resulting graph, we have an exterior vertex v_a^j and an interior vertex v_b^j for each diamond D_j. The vertex connecting D_j and D_{j-1} is denoted by v_p^j.

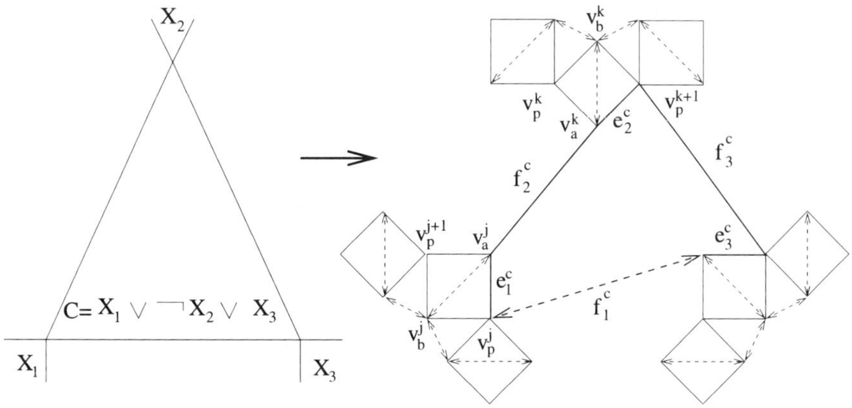

FIGURE 5. Constructing the clause paths

Using this construction, we get a "switch" gadget that forces each variable gadget into one of two feasible states. This is achieved with the help of the following distance edges (see Figure 4) for each chain $K(x_i)$ and all $j = 0, \dots, 3L(x_i) - 1$:

$$(v_b^j, v_a^j) \quad \text{with weight } w_d(v_b^j, v_a^j) = 0 \tag{1}$$

$$(v_b^j, v_b^{j+1}) \quad \text{with weight } w_d(v_b^j, v_b^{j+1}) = 1. \tag{2}$$

Now let $l_1 l_2 (l_3)$ be the literals of a clause c and assume that the corresponding cycle of length 2 or 3 is traversed in clockwise order in H and $\hat{H}$. Let D_{k_i} denote the diamond corresponding to literal l_i.

If l_i is a positive literal, let e_i^c denote the edge $(v_p^{k_i}, v_a^{k_i})$, otherwise let e_i^c be $(v_p^{k_i+1}, v_a^{k_i})$. Note that the e_i^c are edges of diamonds. Let $f_1^c, f_2^c(, f_3^c)$ denote the unique edges which turn $e_1^c, e_2^c(, e_3^c)$ into a planar hexagon $f_1^c, e_1^c, f_2^c, e_2^c, f_3^c, e_3^c$, rsp. a quadrilateral. (See Figure 5.) Add $f_2^c(, f_3^c)$ to the edges E of G and f_1^c to the edges E_d of G_d with weight

$$w_d(f_1^c) = |c| - 1.$$

This construction can be performed in polynomial time and yields an instance of the ISPP where $G + G_d$ is planar.

Now let $T : X \to \{0, 1\}$ be a satisfying variable assignment for I. We construct a weight function $w : E \to \{0, 1, 2\}$ for G as follows:

$$\begin{aligned} w(v_b^j, v_p^{j+1}) = w(v_a^j, v_p^{j+1})) &= 1 && \text{for edges in } K(x) \text{ with } T(x) = 1, \\ w(v_b^j, v_p^j)) = w((v_a^j, v_p^j)) &= 1 && \text{for edges in } K(x) \text{ with } T(x) = 0, \\ w(f_2^c) &= M - 1 && \text{if clause } c \text{ is satisfied by } M \text{ literals}, \\ w(f) &= 0 && \text{else.} \end{aligned}$$

We claim that w is G_d-satisfying. Clearly, we have distance 0 between any two nodes joined by an edge as in (1). The path $P = (v_b^j, v_p^j, v_b^{j+1})$ has length 1, and removing all edges with non-zero weight disconnects v_b^j from v_b^{j+1} (see e. g. cut C_0 in Figure 6), hence $d_w(v, w) = 1$ for all edges in (2).

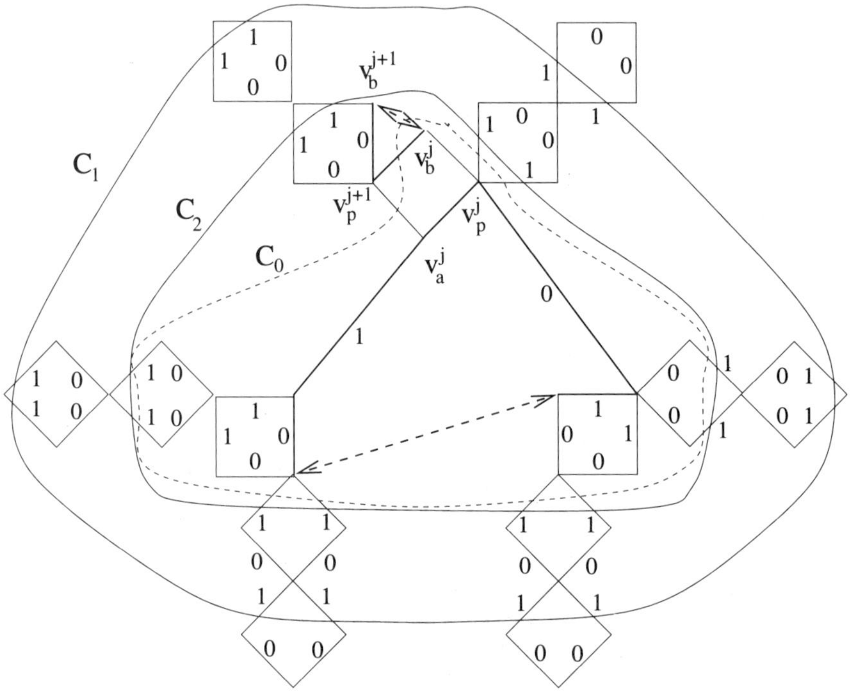

FIGURE 6. Paths and cuts

Finally, for $f_1^c \in E_d$ we compute the length of the path $e_1^c, f_2^c, e_2^c, (f_3^c, e_3^c)$ to $w(f_2^c) + |\{\text{unsatisfied literals}\}| = |c| - 1$ and we are left to show that there cannot be any shorter path. Clearly, removing the non-zero edges from G disconnects the end points of f_1^c and it is impossible to join their components by a single edge from G (see cuts C_1 and C_2 in Figure 6).

Now assume $w : E \to \mathbb{R}^+$ is a G_d-satisfying weight function for G. We claim that for any diamond chain $K(x)$ for all $j = 0, \dots L(x) - 1$ we must have either

$$\begin{aligned} w(v_b^j, v_p^j) = w(v_a^j, v_p^j) = 0 \text{ and} \\ w(v_b^j, v_p^{j+1}) \geq 1, w(v_a^j, v_p^{j+1}) \geq 1, \end{aligned} \tag{3}$$

or

$$\begin{aligned} w(v_b^j, v_p^{j+1}) = w(v_a^j, v_p^{j+1}) = 0 \text{ and} \\ w(v_b^j, v_p^j) \geq 1, w(v_a^j, v_p^j) \geq 1. \end{aligned} \tag{4}$$

Because of (1), at least one of the edges leaving v_b^0 must get weight zero. We proceed to show that $w(v_b^0, v_p^0) = 0$ implies that (3) holds. The proof that $w(v_b^0, v_p^1) = 0$ implies (4) is similar. Since $d(v_b^{3L(x)-1}, v_b^0) = 1$ by (2), we must have $w(v_b^{3L(x)-1}, v_p^0) \geq 1$ and, inductively, we conclude that $w(v_b^j, v_p^j) = 0$ and $w(v_p^j, v_p^{j+1}) \geq 1$ for all $j = 0, \dots, 3L(x) - 1$. Let $(v_b^k, u_1, \dots, u_s, v_a^k)$ denote some path of length zero from v_b^k to v_a^k. Then $(v_b^k, u_1, \dots, u_s, v_a^k, v_p^{k+1}, v_b^{k+1})$ is a path that enforces $w(v_a^k, v_p^{k+1}) \geq 1$. Observe, this implies that the length of a shortest path between any two "literals" v_a^{3i} and v_a^{3j} in a variable cycle is at least one. We

1: $M = \{s\}, w(e) = \infty$ for all $e \in E$, $D(v) = \infty$ for all $v \in V$.
2: Choose a vertex $v \in V \setminus M$, adjacent to a vertex v_1 in M, such that either $(s,v) \notin E_d$ or $w_d(s,v) = \min\{w_d(s,w)|(s,w) \in E_d, w \notin M\}$. If there is no such vertex, go to (5)
3: If $(s,v) \in E_d$, then set $D(v) = w_d(s,v)$ and $w(v_1,v) = D(v) - D(v_1)$,
else set $w((v_1,v)) = 0, D(v) = D(v_1)$
4: $M = M \cup \{v\}$ and go to (2).
5: If $M = V$, then w solves the ISPP, otherwise the neighborhood $N(M) \setminus M$ and the vertex t with $w_d(s,t) = \min\{w_d(s,w)|(s,w) \in E_d, w \notin M\}$ satisfy the condition (a) of Theorem 3.

FIGURE 7. Algorithm for $E_d \subset \{s\} \times V$

conclude that any path from v_a^0 to v_b^0 that avoids the edge (v_a^0, v_p^0) has length at least one and (3) follows.

If (3) holds, we set $T(x) = 1$, otherwise $T(x) = 0$, and claim that T is a satisfying variable assignment. To see this, consider an edge $f_1^c \in E_d$ and recall that $w_d(f_1^c) = |c| - 1$. The length of the path $e_1^c, f_2^c, e_2^c, (f_3^c, e_3^c)$ is at least the number of unsatisfied literals in c. Figure 6 shows that any other path connecting the endpoints of f_1^c must be longer. Therefore at least one literal of c must be satisfied. This concludes the proof. □

5. Special Distance Graphs

Following the line of research on disjoint paths problems, this section considers the complexity of the ISPP for special distance graphs G_d. First we identify two polynomial cases, and then we prove that the problem becomes NP-complete if we consider slightly more complicated distance graphs.

A set of edges S is called a *star*, if all edges in S are incident to one vertex s, hence $(S \subset s \times V)$. If S covers all vertices $v \in V$, then we call S a *complete star*. If the distance graph is a star then a condition which is trivially necessary for the existence of a G_d-satisfying weight function is already sufficient. For $s, t \in V$, a vertex set $W \subset V \setminus \{s,t\}$ is called *s,t-separating*, if s and t lie in different components of $G \setminus W$.

THEOREM 3. *Let (G, G_d) be an ISPP instance.*

(a) *Assume there is an edge $(s,t) \in E_d$ and a vertex set $W \subseteq V$ which is s-t-separating in G, such that the star $(s \times W) \subseteq E_d$ and for all $v \in W$: $w_d(s,v) > w_d(s,t)$. Then there is no satisfying weight function.*

(b) *If E_d is a star, then (G, G_d) has a satisfying weight function if and only if the condition in (a) does not hold. Furthermore, there exists a polynomial time algorithm to solve the problem.*

PROOF. It is straightforward to see that the first statement holds. So assume that the condition in (a) does not hold and that E_d is a star. In that case we apply the algorithm in Figure 7 to solve the ISPP.

During the course of the algorithm, all edges with finite weights form a shortest paths tree with respect to w and the subgraph of G that is induced by M. D

describes the distance to s for all vertices in M. In step **3**, the weights are computed in such a way that the induced distances D to s are equal to w_d for all v in M that are adjacent to s in G_d. Since v was chosen appropriately in step **2**, the computed weights are all positive.

If the algorithm reaches step **5** with $M \neq V$, then vertex t with $w_d(s,t) = \min\{w_d(s,w))|(s,w) \in E_d, w \notin M\}$ is not adjacent to a vertex in M (see step **2**) and the set W of all vertices not in M but adjacent to a vertex in M is an s-t-separating vertex set satisfying condition (a). Hence the algorithm solves the ISPP. Its running time is clearly polynomial, thus proving Theorem 3. □

It is not surprising that the special case of the distance edges forming a complete star is even simpler. In that situation, Algorithm 7 reduces to the construction of an arborescence in the digraph $D(V, \hat{E})$ with

$$\hat{E} = \{(v,w)|(v,w) \in E \text{ and } w_d(s,v) \leq w_d(s,w)\}.$$

Here the ISPP is solvable if and only if D contains an arborescence T with root s. If there is such an arborescence, the following weight function w solves the ISPP:

$$w(e) = \begin{cases} w_d(s,w) - w_d(s,v) & \text{for } (v,w) \in T \\ \infty & \text{else.} \end{cases}$$

This method can be generalized to the case where E_d is the union of complete stars, i. e. the case where the distance to all vertices of V is given from a subset of the vertices.

THEOREM 4. *The ISPP is polynomially solvable if the distance graph is the union of complete stars.*

PROOF. Let $E_d = \{s_1, \dots, s_k\} \times V$. Since any satisfying weight function $\tilde{w}$ must satisfy

$$\tilde{w}(v,w) \geq |w_d(v,s_i) - w_d(w,s_i)| \text{ for all } i, \tag{5}$$

there is an obvious candidate for a weight function:

$$w(v,w) := \max_{i=1,\dots,k} |w_d(s_i,v) - w_d(s_i,w)|.$$

The following computation shows that w is a solution of the ISPP if there is one. For that purpose, assume $\tilde{w}$ to be a G_d-satisfying weight function and let d_w and $d_{\tilde{w}}$ denote distance functions induced by w or $\tilde{w}$, rsp. By definition, we have $\tilde{w} \geq w$ implying that $d_{\tilde{w}} \geq d_w$. Let $(s_i, v) \in E_d$ and $s_i = x_0, x_1, \dots, x_n = v$ be a shortest s_i-v-path with respect to w. Then

$$\begin{aligned} w_d(s_i,v) = d_{\tilde{w}}(s_i,v) &\geq d_w(s_i,v) \\ &= \sum_{j=1}^{n} w(x_{j-1},x_j) \\ &\geq \sum_{j=1}^{n} (w_d(s_i,x_j) - w_d(s_i,x_{j-1})) = w_d(s_i,v). \end{aligned}$$

□

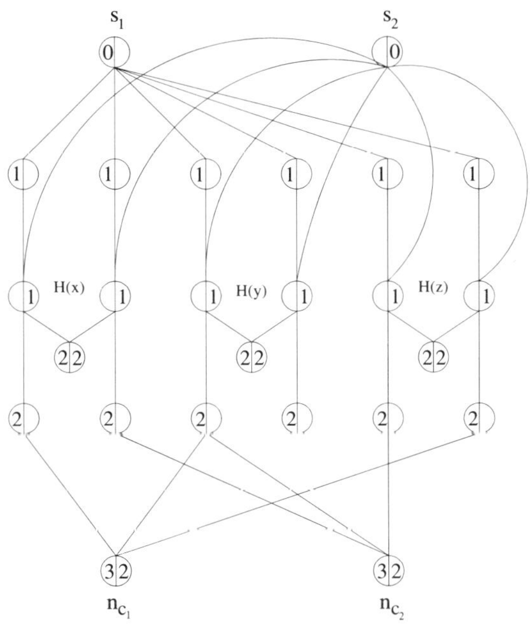

FIGURE 8. 2-star instance for $(x \vee y \vee \overline{z}) \wedge (\overline{x} \vee y \vee z)$

As in the star case, we can check the solvability of the ISPP instance by searching for arborescences in the following digraphs $D_i(V, \hat{E}_i)$:

$$\hat{E}_i = \{(v, w) | (v, w) \in E \text{ and } w_d(s_i, w) = w_d(s_i, v) + w(v, w)\}.$$

The ISPP is solvable if and only if for all D_i, there is an arborescence T_i with root s_i.

Unfortunately, these polynomial algorithms cannot be modified to solve the 2-star problem ($E_d \subset \{s_1, s_2\} \times V$). In fact, the existence of any polynomial algorithm for this problem is rather unlikely, as can be seen from the following theorem.

THEOREM 5. *The ISPP is NP-complete, even when restricted to instances where the edges of the distance graph can be covered by two stars.*

PROOF. We give a reduction from SAT [**Ga79**]. See Figure 8.

Let I be an instance of SAT. Each variable x of I is represented by a graph $H(x)$, consisting of seven vertices and six edges and looking like an H – see Figure 9. The four vertices of degree 1 are called the *right* (left) *upper* (lower) *vertices* for variable x, those of degree 3 are the *right* (left) *middle vertices*, and the remaining vertex is called the *center vertex* for x. For each clause C, we insert a vertex n_C and, if $x \in C$ is positive, we join it to the left lower vertex of $H(x)$; if $x \in C$ is negative, we join it to the right lower vertex of $H(x)$. Finally, we introduce two more vertices s_1 and s_2 and connect s_1 with all upper vertices, s_2 with all middle

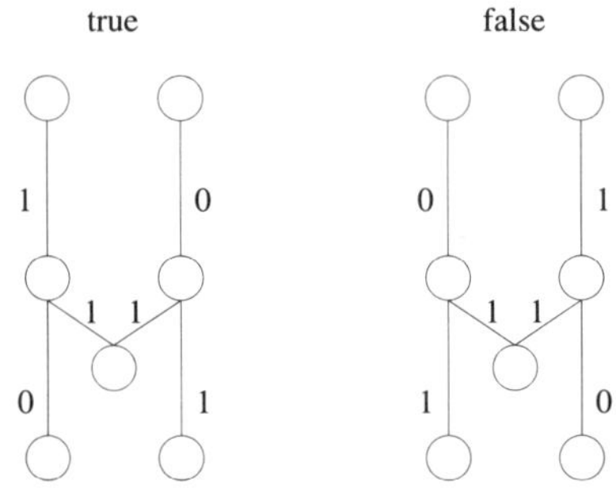

FIGURE 9. Weights of the variable components

vertices. The graph G_d is defined by the following edges and their weights:

$$\begin{array}{lcll} w_d(s_1,u) &=& 1 & \text{for all upper vertices } u, \\ w_d(s_2,m) &=& 1 & \text{for all middle vertices } m, \\ w_d(s_1,c) &=& 2 & \text{for all center vertices } c, \\ w_d(s_2,c) &=& 2 & \text{for all center vertices } c, \\ w_d(s_1,l) &=& 2 & \text{for all lower vertices } l, \\ w_d(s_1,n_C) &=& 3 & \text{for all clause vertices } n_C, \\ w_d(s_2,n_C) &=& 2 & \text{for all clause vertices } n_C. \end{array}$$

Let T be a satisfying variable assignment for the SAT instance. For all variable components, choose the edge weights as shown in Figure 9 and set all other edge weights to 1. Since T is satisfying, for each clause C there is an s_2-n_C path of length 2. It is easy to check that this weight function solves the constructed ISPP.

To see the converse, assume that w is a G_d-satisfying weight function of the constructed ISPP instance. All edges with the exception of those between upper and middle and middle and lower vertices must have at least weight 1. The distance of s_1 to the center vertices of the literals forces at least one of the right and left edge between upper and middle vertex to have weight 0. Since the upper and lower vertices have distance at least 1, the weight of any lower edge incident to an upper edge of weight 0 must be at least 1. However, for any clause, there must be an s_2-n_C-path of length 2. Hence there must be a variable with corresponding edge between lower and middle vertex having weight 0, enforcing a weight of at least 1 for the incident upper edge. Therefore, the following assignment satisfies the SAT instance: $T(x) = 1$ if and only if the edge between left middle vertex and left lower vertex has weight 0. □

6. Digraphs

It is straightforward to define the ISPP for directed graphs. In this section we discuss results for digraphs corresponding to those of the previous section. Finally we consider the case where the underlying network is an acyclic digraph.

In general, weighted digraphs have an asymmetric distance function, i. e., it does make sense to have distance edges in both directions, such that $w_d((s,t)) \neq w_d((t,s))$.

The Vertex Disjoint Paths Problem is significantly harder for digraphs than for graphs. In particular, the existence of disjoint paths in graphs can be checked in polynomial time, if there is a bounded number of edges in the demand graph [**Ro86**];

on the other hand, this problem is hard for the case of digraphs, even if there are only two edges in the demand graph [**FHW80**]. The latter implies the following theorem:

THEOREM 6. *The ISPP for digraphs is NP-complete, even restricted to instances where the distance graph has only 3 edges.*

PROOF. We give a reduction from the vertex disjoint paths problem with two independent edges:

Instance: A digraph $D(V, E)$, $s_1, s_2, t_1, t_2 \in V$.
Question: Are there directed vertex-disjoint (s_i, t_i)-paths in D?

Without loss of generality, we may assume $|\{s_1, s_2, t_1, t_2\}| = 4$. Set $E_d = \{(s_1, t_1), (s_2, t_2), (s_1, t_2)\}$, $w_d((s_i, t_i)) = 0, w_d((s_1, t_2)) = 1$ and, if necessary, add edge (s_1, t_2) to E. As in the proof of Theorem 1, it follows that this ISPP instance has a solution if and only if the VDPP instance has a solution. □

Both polynomial algorithms of the preceding section imply polynomial algorithms for the corresponding cases on digraphs. In order to transfer the results, we have to adapt the term "star" to digraphs. We call an edge set $S \subset V^2$ a *source star (sink star)* from (to) s, if all edges have tails (heads) in common. In this section, a "star" will be either a source star or a sink star. A *complete source star* is an edge set $S = \{s\} \times W \subset \{s\} \times V$, where W contains all $v \in V$ for which there is a directed (s, v)-path. A *complete sink star* is defined analogously.

COROLLARY 1. *Let $(G(V, E), G_d(V, E_d, w_d)$ be a directed ISPP instance.*

(a) *The ISPP is polynomially solvable if the edges of the distance graph G_d form a star.*
(b) *The ISPP is polynomially solvable if the edges of the distance graph are the union of complete stars.*

PROOF. The algorithms for the undirected case immediately generalize to directed graphs. For the second claim inequality (5) is replaced by either

$$w((v, w)) \geq w_d((s, w)) - w_d((s, v)) \text{ for all } (v, w) \in E \text{ or}$$

$$w((v, w)) \geq w_d((v, s)) - w_d((w, s)) \text{ for all } (v, w) \in E.$$

□

Acyclic digraphs have to be considered separately. The VDPP for bounded size of G_d on acyclic digraphs can be solved in polynomial time [**FHW80**]. In general, however, Theorem 5 implies that the problem is intractable even when restricting the distance graphs to unions of two stars. More precisely, we get:

COROLLARY 2. *The ISPP on acyclic digraphs $G(V, E)$ is NP-complete, even when restricted to instances where the distance graph consists of two stars, i. e.*

(a) *a source and a sink star*
(b) *two source stars*
(c) *two sink stars.*

PROOF. Cases (b) and (c) can be proven by orienting the graph in Figure 8, which was constructed in the proof of Theorem 5 (see Figure 10). Case (a) needs a slight modification of the construction, shown in Figure 11.

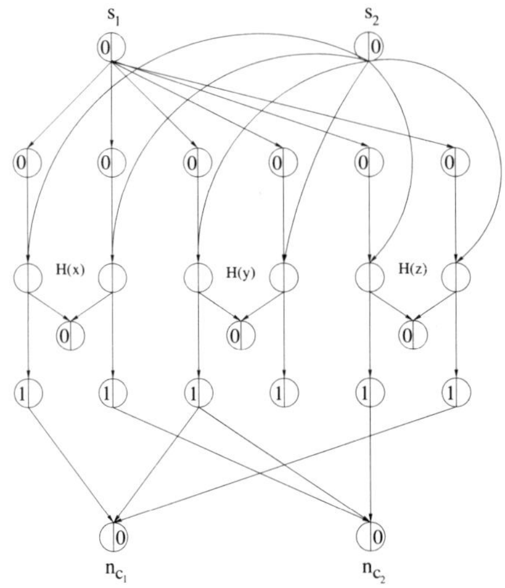

FIGURE 10. NP-completeness for two source stars

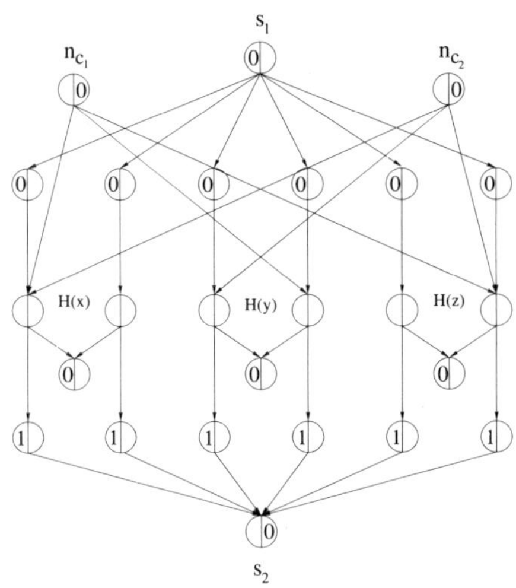

FIGURE 11. NP-completeness for a source star and a sink star

Most of the edges in the construction of Figure 8 have the purpose of excluding possible shortest paths. When considering acyclic digraphs, construction and proof become considerably easier. We leave the details to the reader. □

By identifying the vertices s_1 and s_2 in Figure 11, we obtain the following Corollary.

COROLLARY 3. *The ISPP is NP-complete, even when restricted to instances where the distance edges can be covered by a sink star and a source star with fixed center s.*

References

[Bu92] B. Burton. *On the inverse shortest path problem.* Dissertation, University of Namur, 1992.

[BPT97] B. Burton, W. R. Pulleyblank, and Ph. L. Toint. The inverse shortest path problem with upper bounds on shortest path costs, In: P. Pardalos, D. W. Hearn, and W. H. Hager, eds. *Network Optimization*, Springer Lecture Notes in Economics and Mathematical Systems, **450** (1997), 156–171.

[To92] B. Burton and Ph. L. Toint. On an instance of the inverse shortest paths problem. *Math. Prog.*, **53** (1992), 45–62.

[Da92] H. Daugs. *Betrachtung des inversen Kürzeste-Wege-Problems.* Diploma Thesis, Universität zu Köln, 1992.

[FHW80] S. Fortune, J. E. Hopcroft, and J. Wyllie. The directed subgraph homeomorphism problem. *Theor. Comp. Sci.*, **10** (1980), 111–121.

[Fr90] A. Frank. Packing paths, circuits and cuts – a survey. In: Korte, Lovász, Prömel, Schrijver (Eds.), *Paths, Flows and VLSI-Layout*, Springer Verlag, 1990, 47–100.

[Ga79] M. R. Garey and D. S. Johnson. *Computers and Intractability. A Guide to the Theory of NP-Completeness.* W. H. Freeman, New York , 1979

[Li82] D. Lichtenstein. Planar formulae and their uses. *SIAM J. Computing*, **11** (1982), 329–343

[ML97] M. Cai and Y. Li. Inverse Matroid Intersection Problems. *Math. Methods of OR*, **45** (1997), 235–243.

[MP88] M. Middendorf and F. Pfeiffer. On the Complexity of the disjoint paths problem. *Combinatorica*, **13** (1993), 91–107.

[Mo95] C. Moll. *Das inverse Kürzeste-Wege-Problem.* Dissertation, Universität zu Köln, 1995.

[OS81] H. Okamura and P. D. Seymour. Multicommodity flows in planar graphs. *J. Comb. Theory, Ser. B*, **31** (1981), 75–81.

[Pa82] C. H. Papadimitriou and K. Steiglitz. *Combinatorial Optimization: Algorithms and Complexity.* Prentice Hall, 1982.

[Pf90] F. Pfeiffer. *Zur Komplexität des Disjunkte-Wege-Problems.* Dissertation, University of Bonn, 1990.

[Ro86] N. Robertson and P. D. Seymour. Graph minors XIII: The disjoint paths problem. *J. Comb. Theory, Ser. B*, **63** (1995), 65–110.

[Sc86] A. Schrijver. *Theory of linear and integer programming.* Wiley-Interscience Publication, 1986.

[Sc88] A. Schriver. Homotopic routing methods. In: Korte, Lovász, Prömel, Schrijver, eds. *Paths, flows and VLSI-layout*, Springer Verlag, 1990, 329–371.

[Se81] P. D. Seymour. On odd cuts and planar multi-commodity-flows, *Proc. Lond. Math. Soc. III*, **42** (1981), 178–192.

CENTER FOR PARALLEL COMPUTING, UNIVERSITÄT ZU KÖLN, D–50923 KÖLN, GERMANY
E-mail address: sandor@zpr.uni-koeln.de

CENTER FOR PARALLEL COMPUTING, UNIVERSITÄT ZU KÖLN, D–50923 KÖLN, GERMANY
E-mail address: wh@zpr.uni-koeln.de

CENTER FOR PARALLEL COMPUTING, UNIVERSITÄT ZU KÖLN, D–50923 KÖLN, GERMANY
Current address: Collogia Consulting, Ubierring 11, D–50678 Köln, GERMANY
E-mail address: kromberg@collogia.de

CENTER FOR PARALLEL COMPUTING, UNIVERSITÄT ZU KÖLN, D–50923 KÖLN, GERMANY
Current address: Siemens AG, ZT SE 4, D–81730 München, GERMANY
E-mail address: christoph.moll@mchp.siemens.de

DIMACS Series in Discrete Mathematics
and Theoretical Computer Science
Volume **49**, 1999

FINDING MINIMUM WEIGHTED GENERATORS OF A PATH SYSTEM

ANDRÁS FRANK

ABSTRACT. In a recent paper the author [5] described a short algorithmic proof of a min-max theorem of E. Győri [8] on minimum generators of a system of subpaths of an underlying path. A. Frank and T. Jordán [4] generalized Győri's theorem in several ways; the underlying path was replaced by a circuit, generators were extended to weighted generators, and minimum cost generators were also treated for node-induced cost-functions. The original proof method however was not constructive and the only known algorithm so far to solve the optimization problems in question relied on the ellipsoid method. In this paper a constructive proof is described which gives rise to a purely combinatorial strongly polynomial algorithm.

1. INTRODUCTION

Let $P = (v_0, j_1, v_1, j_2, v_2, \ldots, j_n, v_n = v_0)$ be a directed circuit, that is, each directed edge j_i has tail v_{i-1} and head v_i, and the nodes $v_1, \ldots, v_n$ are distinct. Let $V := \{v_0, \ldots, v_n\}$ and $E := \{j_1, \ldots, j_n\}$ denote the node set and edge set of P, respectively. Let $E^* := \{uv : u, v \in V, u \neq v\}$ denote the set of all directed edges on the node set V. Suppose that a non-negative integer-valued weight-function $p : E \to \mathbf{Z}_+$ is given.

Let $\mathcal{P}$ be a system of distinct subpaths of P. We use the convention that the edge set of P_i will be denoted by the same letter P_i. The node set of P_i is denoted by $V(P_i)$. For a path J, let $f(J)$ and $l(J)$ denote the first and last nodes of J, respectively. Also, for a (directed) edge $e = uv$ let $f(e) := u$ and $l(e) := v$.

We say that a system $\mathcal{G}$ of subpaths of P **generates** a path J if J is the union of some members of $\mathcal{G}$. $\mathcal{G}$ **generates** $\mathcal{P}$ or $\mathcal{G}$ is a **generator of** $\mathcal{P}$ if every member of $\mathcal{P}$ is generated by $\mathcal{G}$. Let $\gamma(\mathcal{P})$ denote the minimum cardinality of a generator of $\mathcal{P}$. More generally, a family $\mathcal{G}$ of (not-necessarily distinct) subpaths of P is a p-**generator** or a **weighted generator** of $\mathcal{P}$ if, for every path $J \in \mathcal{P}$ and edge $j \in J$, $\mathcal{G}$ contains at least $p(j)$ paths containing j and included in J. Let $\gamma_p(\mathcal{F})$ denote the minimum cardinality of a p-generator of $\mathcal{P}$. Clearly, for $p \equiv 1$, a p-generator is a generator. One of our main concerns will be to find algorithmically a minimum weighted generator.

To formulate a min-max theorem for γ_p we need the following terminology. Let $J \in \mathcal{P}$ be a path and j an edge of J. We say that the pair (J, j) is a **path-edge pair.** Let $\mathcal{R}$ denote the set of all path-edge pairs (J, j) for which $j \in J \in \mathcal{P}$. A

1991 *Mathematics Subject Classification.* 05C40, 90B10, 90C35.

Key words and phrases. combinatorial algorithms, path systems.

Research supported by the Hungarian National Foundation for Scientific Research Grant, OTKA T17580.

member (J, j) of $\mathcal{R}$ is called **essential,** or an **essential pair** if there is no member J' $(\neq J)$ of $\mathcal{P}$ for which $j \in J' \subset J$. Let $\mathcal{E}$ denote the set of essential pairs of $\mathcal{R}$. We will say that an essential pair (J, j) is **sitting** at j. Weight $p(j)$ is called the **weight** of (J, j) and denoted by $p(J, j)$. (Hopefully, no confusion may arise from this kind of notation that symbol p is used for a weighting on E and for a weighting on the set of path-edge pairs, as well.)

Let $(J, j)^-$ denote the set of nodes of J preceding edge j, and let $(J, j)^+$ denote the set of nodes of J following j. That is, $(J, j)^-$ and $(J, j)^+$ form a partition of the node-set $V(J)$ of J. For any edge $e = uv \in E^*$, e **covers** (J, j) if $u \in (J, j)^-$ and $v \in (J, j)^+$. For a non-negative integer vector $z : E^* \to \mathbf{Z}_+$, let $\varrho_z(J, j) := \sum(z(e) :$ e covers $(J, j))$. The **deficiency** of a path-edge pair (J, j) is defined by $h_z(J, j) :=$ $\max(0, p(j) - \varrho_z(J, j))$. The **deficiency** of z is $h(z) := \sum(h_z(J, j) : (J, j) \in \mathcal{E})$.

If the deficiency of (J, j) is zero, z is said to **p-cover** (J, j). Otherwise (J, j) is called **deficient.** We say that z is a **p-covering** or a **weighted covering** of $\mathcal{R}$ if $\varrho_z(J, j) \geq p(j)$ holds for every path-edge pair $(J, j) \in \mathcal{R}$. The **size** of a p-covering z of $\mathcal{R}$ is $z(E^*) := \sum(z(e) : e \in E^*)$.

There is a natural correspondence between p-generators of $\mathcal{P}$ and p-coverings of $\mathcal{R}$. Namely, from a p-covering z we can obtain a p-generator by associating with each edge $uv \in E^*$ $z(e)$ copies of the subpath of P from u to v, and conversely, from a p-generator $\mathcal{G}$ we obtain a p-covering z by letting $z(uv)$ be the number of paths in $\mathcal{G}$ from u to v. By this correspondence, $z(E^*) = |\mathcal{G}|$ and hence $\gamma_p(\mathcal{P})$ is the minimum size of p-covers of the set $\mathcal{R}$ of path-edge pairs.

Two path-edge pairs $(I, i), (J, j)$ are called **independent** if $(I, i)^- \cap (J, j)^- = \emptyset$ or $(I, i)^+ \cap (J, j)^+ = \emptyset$ which is equivalent to saying that no edge in E^* can cover both of them. A set of represented paths is called **independent** if its members are pairwise independent. Let $\sigma_p(\mathcal{P})$ denote the maximum total weight of an independent subset of $\mathcal{R}$. For $p \equiv 1$, σ_p is abbreviated by σ. Because two independent path-edge pairs cannot be covered by one edge, $z(E^*) \geq \sigma_p(\mathcal{P})$ holds for every p-covering z of $\mathcal{R}$, or equivalently, every p-generator of $\mathcal{P}$ has at least $\sigma_p(\mathcal{P})$ members, and hence $\sigma_p(\mathcal{P}) \leq \gamma_p(\mathcal{P})$.

Actually, here we always have equality:

Theorem 1.1. *For a set $\mathcal{P}$ of subpaths of a directed circuit (V, E) and for a weight function $p : E \to \mathbf{Z}_+$,*

$$\sigma_p(\mathcal{P}) = \gamma_p(\mathcal{P}),$$

that is, the maximum total weight of an independent subset of $\mathcal{R}$ is equal to the minimum size of a p-covering of $\mathcal{R}$.

This result was proved by Frank and Jordán [4] in a more abstract form concerning coverings of bi-supermodular functions. The special case of Theorem 1.1 when no member of $\mathcal{P}$ contains edge j_n of P, that is, when $\mathcal{P}$ may be considered as a set of subpaths of an underlying directed path $P - j_n$ was proved by A. Lubiw [10]. When $p \equiv 1$, Lubiw's theorem specializes to a theorem of Győri [8] which has actually been the starting point of this area.

Győri's original proof for his theorem is not constructive but its ideas could be used to develop a combinatorial algorithm. That was achieved by D.S. Franzblau and D.J. Kleitman [7]. For a clear explanation and simplification of their algorithm, along with a computer code, see [9]. The algorithm of Franzblau and Kleitman was extended by A. Lubiw [10] to the weighted case. A completely different approach, relying on Dilworth's theorem, was introduced in [5] to obtain a short algorithmic proof of Győri's theorem.

The proof of Theorem 1.1 in [4] is not constructive. Although the same paper describes a polynomial time algorithm to compute the two extrema in question, that algorithm relies on the theorem itself and uses the ellipsoid method. The main purpose of the present work is to extend ideas of [5] to develop an alternative proof of Theorem 1.1 which is constructive and gives rise to a strongly polynomial (purely combinatorial) algorithm to compute $\sigma_p(\mathcal{P})$ and $\gamma_p(\mathcal{P})$. A slight modification of the algorithm will allow us to solve a min-cost version of the weighted generator problem, as well.

The main min-max theorem of [4] on covering bi-supermudular functions by a minimum number of directed edges (whose special case is Theorem 1.1) remains algorithmically unsolved (apart from an ellipsoid-method based algorithm). In particular, no combinatorial polynomial time (exact) algorithm is known for the other important special case of the general model when the objective is adding a minimum number of new edges to make a given directed graph k-connected. However, for the even more special problem when the starting digraph is $(k-1)$-connected, that is, when the connectivity of the digraph is required to be increased by one by adding a minimum number of new edges, it was possible to develop such an algorithm [6].

Throughout we use the notational convention for any given function $g : S \to \mathbf{R}$ and subset $X \subseteq S$ to denote $\sum(g(x) : x \in X)$ by $g(X)$.

2. PREPARATIONS

Let $(\mathcal{K}, \leq)$ be a partially ordered set and $p : \mathcal{K} \to \mathbf{Z}_+$ a weight function. Two elements K, L of $\mathcal{K}$ are called **comparable** if $K \leq L$ or $L \leq K$, otherwise they are **uncomparable.** By a **chain** (respectively, **antichain)** we mean a subset consisting of pairwise comparable (resp., uncomparable) elements. A family $\mathbf{C}$ of (not necessarily distinct) chains is said to **cover** $(\mathcal{K}, p)$ or that $\mathbf{C}$ is a **chain-covering** of $(\mathcal{K}, p)$ if every element $K \in \mathcal{K}$ belongs to at least $p(K)$ members of $\mathbf{C}$. The number $|\mathbf{C}|$ of chains is called the **size of the chain-covering** of $(\mathcal{K}, p)$. The **weight** $p(\mathcal{A})$ of a subset $\mathcal{A} \subseteq \mathcal{K}$ is $\sum(p(K) : K \in \mathcal{A})$. In the algorithm we are going to describe, a weighted version of Dilworth's theorem plays a fundamental role.

Theorem 2.1 (Weighted Dilworth). *In a partial order $(\mathcal{K}, \leq)$ with a weight function p, the minimum size of a chain-covering of $(\mathcal{K}, p)$ is equal to the maximum weight of an antichain. Moreover, there is a set $\{C_1, \ldots, C_k\}$ of $k \leq |\mathcal{K}|^2$ distinct chains along with positive integer coefficients $\gamma_1, \ldots, \gamma_k$ so that the family consisting of γ_i copies of every chain C_i $(i = 1, \ldots, k)$ is a minimum size chain-covering of p.* □

This form immediately follows from the original version of Dilworth's theorem (where $p \equiv 1$) by replacing each element K by $p(K)$ pairwise uncomparable elements. Ford and Fulkerson [3] described an elegant way to deduce Dilworth's theorem from that of Kőnig. An analogous reduction can be used to formulate the weighted Dilworth problem as a maximum flow problem and therefore both a minimum chain-covering of p and a maximum weight antichain can be computed in (strongly) polynomial time via a max-flow min-cut algorithm. We will refer to this algorithm as the **weighted Dilworth algorithm** (see the Appendix). Note that the second half of the theorem follows from this max-flow approach.

With the help of this (rather standard) technique even the following minimum cost version becomes algorithmically tractable. Suppose that, in addition to the weight function p, we are given two cost functions d_f and d_l on $\mathcal{K}$. The **cost** of a chain with minimal element K and maximal element L is defined by $d_f(K)+d_l(L)$. We will refer to the algorithm computing a minimum cost chain-covering of p as the **min-cost weighted Dilworth algorithm.** (See the Appendix).

We will work with a particular partial ordered set $(\mathcal{E}, \preceq)$ which is defined, as follows. For two essential pairs (I,i) and (J,j) let $(I,i) \preceq (J,j)$ if $(I,i)^- \subseteq (J,j)^-$ and $(I,i)^+ \supseteq (J,j)^+$. If $(I,i) \preceq (J,j)$ but $(I,i) \neq (J,j)$, we write $(I,i) \prec (J,j)$. Two essential pairs are **crossing** if they are neither comparable nor independent. (That is, two uncomparable essential pairs are either independent or crossing). Note that the essential pairs sitting at the same edge form a chain. A subset $\mathcal{K}$ of $\mathcal{E}$ is called **cross-free** if it contains no two crossing members.

Claim 2.2. *If a function $z : E^* \to \mathbf{Z}_+$ p-covers $\mathcal{E}$, then z p-covers $\mathcal{R}$, as well.*

Proof. It is enough to prove that if (J,j) is a path-edge pair not p-covered by z so that $|J|$ is minimum, then (J,j) is essential. If there is a path $J' \in \mathcal{P}$ with $j \in J' \subset J$, then, by the minimality of J, $\varrho_z(J',j) \geq p(j)$. Since every edge covering (J',j) covers (J,j), as well, we obtain that $\varrho_z(J,j) \geq \varrho_z(J',j) \geq p(j)$, contradicting the assumption that z does not p-cover (J,j), that is, (J,j) is indeed essential. □

The claim implies that in order to p-cover $\mathcal{R}$ it suffices to p-cover $\mathcal{E}$.

Claim 2.3. *The members of a chain $\mathcal{C}$ of $(\mathcal{E}, \preceq)$ can be covered by one edge. If $\mathbf{C}$ is a family of chains and $p'(K,k)$ denotes the number of chains in $\mathbf{C}$ containing (K,k) $((K,k) \in \mathcal{E})$, then there is an integer vector $z : E^* \to \mathbf{Z}_+$ so that $\varrho_z(K,k) \geq p'(K,k)$ for every $(K,k) \in \mathcal{E}$ and so that the size $z(E^*)$ of z is $|\mathbf{C}|$.*

Proof. If (K,k) and (J,j) are the minimal and maximal members of $\mathcal{C}$, respectively, then any edge uv with $u \in (K,k)^-$, $v \in (J,j)^+$ covers each member of the chain $\mathcal{C}$. The second half follows easily from the first. Namely, for each chain $\mathcal{C}$ in $\mathbf{C}$ choose an edge $e_{\mathcal{C}}$ covering the members of $\mathcal{C}$, define $z_{\mathcal{C}} : E^* \to \mathbf{Z}_+$ by $z_{\mathcal{C}}(e_{\mathcal{C}}) = 1$ and $z_{\mathcal{C}}(e) = 0$ otherwise, and finally define $z := \sum(z_{\mathcal{C}} : \mathcal{C} \in \mathbf{C}\ )$. This function z satisfies the requirements. □

The following claim is staightforward.

Claim 2.4. *Let (K,k) and (J,j) be two crossing essential pairs. Then both (K,j) and (J,k) are essential.* □

Lemma 2.5. *Let (K,k) and (J,j) be two crossing essential pairs. If an essential pair (X,x) crosses neither of them, then (X,x) crosses neither (K,j) nor (J,k).*

Proof. First assume that (X,x) is comparable with both (K,k) and (J,j). It is not possible that $(K,k) \preceq (X,x) \preceq (J,j)$ or $(J,j) \preceq (X,x) \preceq (K,k)$ since then (K,k) and (J,j) would be comparable and not crossing. Therefore either $(X,x) \succeq (K,k),(J,j)$ or $(X,x) \preceq (K,k),(J,j)$. By symmetry we may assume that $(X,x) \succeq (K,k),(J,j)$. But in this case $(X,x) \succeq (K,j)$ and $(X,x) \succeq (J,k)$, that is, (X,x) does not cross (K,j) and (J,k).

Second, assume that (X,x) is independent from both (K,k) and (J,j). Then $x \notin K \cup J$ and hence (X,x) is independent from both (K,j) and (J,k).

In the third case (X,x) is independent from one of (K,k) and (J,j), say from (K,k), and comparable with the other, (J,j). Then $x \notin K$ and hence (X,x) is independent of (K,j). If $(X,x) \preceq (J,j)$, then $(X,x) \preceq (J,k)$. If $(X,x) \succeq (J,j)$, then $(X,x) \succeq (J,k)$. □

3. ALGORITHMIC PROOF OF THEOREM 1.1

We have seen that $\sigma_p(\mathcal{P}) \le \gamma_p(\mathcal{P})$. In order to prove equality, it suffices to find a p-covering z of $\mathcal{R}$ and an independent subset $\mathcal{A}$ of $\mathcal{E}$ for which $z(E^*) = p(\mathcal{A})$ ($:= \sum(p(A) : A \in \mathcal{A})$). Below we describe a constructive proof of the existence of such z and $\mathcal{A}$, and indicate how this proof can be turned into a strongly polynomial time algorithm.

PHASE 1. Compute a cross-free subset $\mathcal{K}$ of $\mathcal{E}$, as follows. At the beginning, let $\mathcal{K}$ be empty. Choose a total ordering of the elements of $\mathcal{E}$ so that (K,k) precedes (J,j) whenever $p(K,k) > p(J,j)$. Consider the elements of $\mathcal{E}$ in this order and put one into $\mathcal{K}$ if it does not cross any previously chosen member of $\mathcal{K}$.

Let $(K_1,k_1),(K_2,k_2),\dots,(K_N,k_N)$ denote the members of the final $\mathcal{K}$ where the indices reflect the order in which the members have been put into $\mathcal{K}$. For each $(J,j) \in \mathcal{E} - \mathcal{K}$ there is a member of $\mathcal{K}$ crossing (J,j), and let $t(J,j)$ denote the least index i for which $(K_i,k_i) \in \mathcal{K}$ crosses (J,j).

PHASE 2. With the help of the weighted Dilworth algorithm determine a maximum weight antichain $\mathcal{A}$ of $\mathcal{K}$ and a minimum cardinality chain-covering $\mathbf{C}$ of $(\mathcal{K},p)$ (for which $|\mathbf{C}| = p(\mathcal{A})$). By Claim 2.3, there is an integer-valued p-covering $z : E^* \to \mathbf{Z}_+$ of $\mathcal{K}$ with size $|\mathbf{C}|$. If z p-covers $\mathcal{E}$ as well, then by Claim 2.2 we have found a required p-covering of $\mathcal{R}$ and the algorithm (as well as the proof of the theorem) terminates.

So suppose that there are deficient essential pairs. We are going to improve the current p-covering z of $\mathcal{K}$ step by step without changing its size so as to obtain a final p-covering of $\mathcal{K}$ which p-covers the whole $\mathcal{E}_p$, as well, (and hence $\mathcal{R}_p$, too).

Let (J,j) be a deficient essential pair which is **latest** in the sense that $t(J,j)$ is maximum. Let $(K,k) := (K_{t(J,j)}, k_{t(J,j)})$. By Claim 2.4, (J,k) and (K,j) are essential.

Claim 3.1. *Both (J,k) and (K,j) are p-covered by z.*

Proof. Suppose indirectly that (J,k) is not covered by z. (The proof for (K,j) is analogous). Since every member of $\mathcal{K}$ is p-covered, we have $(J,k) \notin \mathcal{K}$. Since (J,j) is latest, $i' := t(J,k) \le t(J,j)$. Here we cannot have equality because (K,k) and (J,k) are comparable and not crossing. Now $(K_{i'},k_{i'})$ crosses neither (K,k) (since both are in $\mathcal{K}$) nor (J,j) (by the definition of $t(J,j)$) but $(K_{i'},k_{i'})$ does cross (J,k), contradicting Lemma 2.5. □

By reversing the edges, if necessary, (and thereby interchanging $(X,x)^-$ and $(X,x)^+$ for every $(X,x) \in \mathcal{R}$), we may assume that $f(k) \in (J,j)^-$ and $l(k) \in (J,j)^-$.

Claim 3.2. *There is an edge $e_1 = u_1v_1 \in E^*$ with $u_1 \in (J,j)^- \cap (K,k)^-$, $v_1 \in (J,j)^- \cap (K,k)^+$) so that $z(e_1) > 0$. There is an edge $e_2 = u_2v_2$ with $u_2 \in (K,k)^- - (J,j)^-$, $v_2 \in (J,j)^+ \cap (K,k)^+$ so that $z(e_2) > 0$.*

Proof. We claim that $p(k) \ge p(j)$ for otherwise (J,j) would precede (K,k) in the total ordering of $\mathcal{E}$ introduced in Phase 1, that is, during the construction of $\mathcal{K}$ pair (J,j) would have been considered earlier than (K,k) and then, by the definition of $i = t(J,j)$, (J,j) would have been chosen to belong to $\mathcal{K}$.

By $p(k) \geq p(j)$ we have $p(J,k) = p(K,k) \geq p(J,j)$. Since z does not p-cover (J,j) but, by Claim 3.1, it p-covers (J,k), there must be an edge e_1 required by the Claim. Similarly, since z p-covers (K,j), an edge e_2 required by the Claim must exist. □

EXTREME CHOICE OF e_1 AND e_2. Choose edges e_1 and e_2 ensured by the previous claim so that e_1 is minimal and e_2 is maximal in the sense that the subpath of P from u_1 to v_1 is as small as possible and the subpath of P from u_2 to v_2 is as large as possible. (This particular choice of e_1 and e_2 will not be used in the proof of the theorem. It will play a role in getting a strongly polynomial upper bound for the number of steps of the algorithm.)
FLIPPING OPERATION. Let $\delta := \min(z(e_1), z(e_2))$ and let $e_1' := u_1v_2, e_2' := u_2v_1$. Revise z by letting $z'(e_1) := z(e_1)-\delta, z'(e_2) := z(e_2)-\delta, z'(e_1') := z(e_1')+\delta, z'(e_2') := z(e_2)+\delta$, and let $z'(e) := z(e)$ otherwise. Note that $z'(E^*) = z(E^*)$, that is, the size of z' and z is the same. We call this revision of z a **flipping operation** or, in short, a **flipping** at e_1 and e_2.
FIXING A LATEST DEFICIENT PAIR (J,j). As long as (J,j) is deficient, choose edges e_1 and e_2 as described in *extreme choice*, and apply a flipping operation at e_1 and e_2.
PHASE 3. As long as there are deficient pairs, choose a latest deficient pair (J,j) and fix it.

In order to conclude the proof of the theorem, it suffices to show that the deficiency of z' obtained by applying one flipping is smaller than that of z, that is, $h(z') < h(z)$. (We will prove this by assuming only the properties of e_1 and e_2 given in Claim 3.2 and not using the extreme choice of e_1, e_2.)

Claim 3.3. *$\varrho_{z'}(X,x) \geq \varrho_z(X,x)$ for every essential pair (X,x) and $\varrho_{z'}(J,j) > \varrho_z(J,j)$.*

Proof. As the second part is straightforward, we prove only the first. Obviously, if e_2 covers (X,x), then e_1' or e_2' covers (X,x). It is also trivially seen that if both e_1 and e_2 cover (X,x), then so do e_1' and e_2'. Finally, if (X,x) is covered by e_1, then (X,x) must be covered by at least one of e_1' and e_2' for otherwise we have $k \in X \subset K$ contradicting that (K,k) is essential. □

By this claim z' is also a p-covering of $\mathcal{K}$ for which $h(z') < h(z)$. Therefore, after a finite number of flippings, a vector z_0 is obtained with $h(z_0) = 0$, that is, z_0 is a p-covering of $\mathcal{E}$ and its size is equal to the total weight of an independent set $\mathcal{A}$ of essential pairs, as required for the theorem. □

By now we have finished the description of the steps of the algorithm and proved its finiteness. Our next goal is to get a polynomial bound on the number of steps. To this end first we show that Phase 3 terminates after at most n^4 flippings.

Claim 3.4. *The precedure of fixing a latest deficient pair (J,j) terminates after at most $n(n-1)$ flippings.*

Proof. When a flipping is carried out at e_1 and e_2, at least one of the z-values $z(e_1)$ and $z(e_2)$ becomes zero. By the extreme choice of e_1 and e_2, their z-values cannot increase during the subsequent flippings as long as (J,j) is deficient. Therefore after at most $|E^*| = n(n-1)$ flippings (J,j) must become p-covered. □

Claim 3.5. $|\mathcal{E}| \leq n^2$.

Proof. Let $\mathcal{P}_i$ denote the set of paths in P with first node v_i. For every edge $e \in P$ there is at most one member X of $\mathcal{P}_i$ for which $e \in X$ and (X, e) is essential. That is, there are at most n essential pairs (X, e) with $f(X) = v_i$. Hence $|\mathcal{E}| \leq n^2$. □

Combining Claims 3.4 and 3.5 we obtain that after at most n^4 flipping operations Phase 3 terminates with a p-covering z_0 of $\mathcal{R}$ for which $z_0(E^*) = p(\mathcal{A})$.

A. Benczúr, J. Förster, and Z. Király [2] proved that any cross-free set of essential pairs, and hence $\mathcal{K}$, can have at most $n \log n$ members. Since there are max-flow min-cut algorithms of complexity $O(p^3)$, where p is the number of nodes of the network, Phase 2 will terminate in $O(nm + n^2\sqrt{n \log n})$ steps. Note that if one uses only the naive upper bound $|\mathcal{K}| \leq |\mathcal{E}| \leq n^2$, then the number of steps in Phase 2 could be estimated only by $O(n^6)$.

Since $|\mathcal{E}| \leq n^2$, Phase 1 can be carried out in $O(n^4)$ steps, we can conclude that the complexity of the whole algorithm is $O(n^4)$. Benczúr et al. [2] made further simplifications for the unweighted case ($p \equiv 1$) and obtained an algorithm of complexity $O(n^{2.5} \log^{3/2} n)$.

4. MINIMUM COST WEIGHTED p-COVERINGS

Our next goal is to show how a slight modification of the algorithm above helps us compute a minimum cost p-covering of $\mathcal{R}$ for a node-induced cost function. The only change in the whole algorithm occurs in Phase 2 where finding a minimum cost chain-covering of $(\mathcal{K}, p)$ will be demanded rather than a minimum size.

Suppose we are given two cost-functions $d_f : V \to \mathbf{R}_+$ and $d_l : V \to \mathbf{R}_+$. The cost $d(e)$ of an edge $e = uv \in E^*$ is defined by $d(e) := d_f(u) + d_l(v)$. (This is why we call such a cost-function node-induced). For a function $z : E^* \to \mathbf{Z}_+$ let the cost of z be defined by $d(z) := \sum(z(uv)d(uv) : uv \in E^*)$. (When $d_f \equiv 1, d_l \equiv 0$, the cost $d(z)$ is the size of z.) Extend the domain of d_f and d_l to the set of path-edge pairs, as follows. Let $d_f(K, k) := \min(d_f(u) : u \in (K, k)^-)$ and $d_l(K, k) := \min(d_l(u) : u \in (K, k)^+)$. Note that d_f is **monotonously decreasing** and d_l is **monotonously increasing** in the sense that if $(J, j) \preceq (K, k)$, then $d_f(K, k) \leq d_f(J, j)$ and $d_l(K, k) \geq d_f(J, j)$. Define the cost of a chain $\mathcal{C}$ by $d(\mathcal{C}) := d_f(K, k) + d_l(L, l)$ where (K, k) and (L, l) are the minimal and maximal elements of $\mathcal{C}$, respectively.

Let $\mathcal{K}$ be the cross-free family determined by Phase 1. It follows from the definitions that for any edge $e \in E^*$ the members of $\mathcal{K}$ covered by e form a chain of cost at most $d(e)$, and conversely (similarly to the proof of Claim 2.4) any chain $\mathcal{C}$ of $\mathcal{K}$ can be covered by an edge e of cost $d(C)$. Therefore a minimum cost p-covering of $\mathcal{K}$ can be determined by computing a minimum cost chain-covering of $(\mathcal{K}, p)$ which, in turn, can be done in Phase 2 with the help of the min-cost weighted Dilworth algorithm (to be outlined in the Appendix). Hence we can find a minimum cost integer-valued p-covering z of $\mathcal{K}$.

Finally, apply Phase 3. The correctness of the algorithm (that is, that the final p-covering of $\mathcal{E}$ is of minimum cost) follows from the easy observation that if z' denotes the function obtained from z by a flipping operation, then $d(z') = d(z)$, that is, the cost of z and z' are the same. (This is the place where we exploit that d is node-induced and not a general cost-function.)

We remark that the above cost function can be made slightly more general by allowing a specified subset $F \subset E^*$ of edges to be used freely, that is, with cost zero. Everything above (proof and algorithm) remains unchanged if one works with the subset of represented pairs not covered by any element of F (including that $\mathcal{E}$ and $\mathcal{K}$ are also restricted in this way.)

APPENDIX: (MIN-COST) WEIGHTED DILWORTH ALGORITHM

Ford and Fulkerson [3] showed how to derive Dilworth's theorem from Kőnig's theorem on maximum matchings in bipartite graphs. Since a maximum matching can be efficiently computed, this reduction made it possible to compute in polynomial time a minimum chain decomposition of a partially ordered set along with a maximum antichain. In Section 2 we have formulated the weighted Dilworth theorem. Here we show how the reduction technique of Ford and Fulkerson extends to the weighted case. For the closely related problem of finding a *minimum* flow in a network satisfying lower bound requirements on the arc-flows, see [1].

Let $(\mathcal{K}, \leq)$ be a partially ordered set endowed with an integer-valued weight-function $p : \mathcal{K} \to \mathbf{Z}$. With $(\mathcal{K}, \leq)$ we associate a directed graph $D := (U, F)$ with edge-capacities, as follows. For each element $K \in \mathcal{K}$ let k' and k'' be two corresponding elements. Let s be a source node and t a sink node and let $U := \{s,t\} \cup \{k' : K \in \mathcal{K}\} \cup \{k'' : K \in \mathcal{K}\}$. For every element $K \in \mathcal{K}$, let sk' and $k''t$ belong to F with capacities $g(sk') := g(k''t) := p(K)$. Furthermore, let $k'l''$ belong to F if $K > L$ and define $g(k'l'') = \infty$. (Note that $k'k''$ does not belong to F.)

By a max-flow min-cut algorithm we can compute a maximum integer-valued feasible flow x and a set S for which $s \in S \subseteq U - t$ and $\delta_g(S)$ is minimum where $\delta_g(X) := \sum(g(f) : f \in F \text{ leaves } S)$. By the MFMC theorem the value of maximum flow x (that is, by definition, $\delta_x(s)$) is equal to $\delta_g(S)$.

Define another directed graph $D' = (U, F')$, as follows. For every element $K \in \mathcal{K}$, let sk', $k''t$, $k'k''$ belong to F', and let $l''k'$ belong to F' if $K > L$. Let us define $x' : F' \to \mathbf{Z}_+$, as follows. For $K \in \mathcal{K}$, let $x'(sk') := p(K) - x(sk'), x'(k''t) := p(K) - x(k''t)$. Let $x'(k'k'') := p(K)$, and in case $K > L$ let $x'(l''k') := x(k'l'')$. It is not difficult to see that x' is a non-negative integer-valued flow in D' whose value is $p(\mathcal{K}) - \delta_x(s)$.

It is well-known (and can be proved easily by induction) that in an acyclic digraph with n' nodes and m' edges, any flow can be decomposed into at most $m' - n' + 2$ path-flows. (A **path-flow** is a flow that is positive on the edges of a directed path from s to t and zero otherwise). Moreover the inner nodes of any directed path of D' from s to t form a chain of $(\mathcal{K}, \leq)$. Let $N := |\mathcal{K}|$. Since D' has $2N+2$ nodes and at most $3N + N(N-1)/2$ edges, we obtain that x' can be decomposed into at most $3N + N(N-1)/2 - (2N+2) + 2 = N(N+1)/2 \leq N^2$ path-flows. Hence there is a set of chains $\mathcal{C}_1, \ldots, \mathcal{C}_k$ $(k \leq N^2)$ of $(\mathcal{K}, \leq)$ and positive integer-valued coefficients $\gamma_1, \ldots, \gamma_k$ so that the family $\mathbf{C}$ of chains consisting of γ_i copies of $\mathcal{C}_i$ $(i = 1, .., k)$ is a chain-covering of $(\mathcal{K}, p)$ and $\sum \gamma_i = \delta_{x'}(s) = p(\mathcal{K}) - \delta_x(s) = p(\mathcal{K}) - \delta_g(S)$.

Let us now consider the minimum cut in D determined by S.

Claim A.1. *If $k'' \in S$, then $k' \in S$.*

Proof. If $k'' \in S$ and $k' \notin S$, then there must be an element $L_1 \in \mathcal{K}$ for which $l_1'' \notin S$ and $K > L_1$ for otherwise we would have $\delta_g(S + k') < \delta_g(S)$, contradicting the choice of S. Similarly, there must be an element $L_2 \in \mathcal{K}$ for which $l_2' \in S$ and $K < L_2$, for otherwise we would have $\delta_g(S - k'') < \delta_g(S)$. But now $L_2 > L_1$ and the edge $l_2'l_1''$ leaves S and hence $\delta_g(S) = \infty$, contradicting the minimality of $\delta_g(S)$. □

Let $\mathcal{A} := \{K \in \mathcal{K} : k' \in S, k'' \notin S\}$. Then $\mathcal{A}$ is an antichain of $(\mathcal{K}, \leq)$ and $p(\mathcal{A}) = p(\mathcal{K}) - \delta_g(S)$. Therefore $p(\mathcal{A})$ is equal to the flow-value of x', that is, $\mathcal{A}$ is a maximum p-weight antichain of $(\mathcal{K}, \leq)$ and $\mathbf{C}$ is a minimum chain-covering of $(\mathcal{K}, p)$.

In conclusion, both a minimum chain-covering $\mathbf{C}$ of $(\mathcal{K}, p)$ and an antichain $\mathcal{A}$ of maximum weight can be computed with the help of a max-flow min-cut algorithm.

We remark that the minimum chain-covering constructed this way contains *exactly* $p(K)$ chains containing K for every element K of $\mathcal{K}$. We call such a chain-covering a **chain-decomposition** of $(\mathcal{K}, p)$. The above construction shows that there is a one-to-one correspondence between the chain-decompositions of $(\mathcal{K}, p)$ and the integer-valued flows of the network (D, g) defined above.

Suppose now that we have two cost-functions d_f and d_l defined on $\mathcal{K}$ so that they are monotonously decreasing and increasing, respectively. Define the cost of a chain $\mathcal{C}$ by $d(\mathcal{C}) := d_f(K) + d_l(L)$ where K and L are the minimal and the maximal elements of $\mathcal{C}$, respectively. It follows from the monotonicity of d_f and d_l that the cost function d is monotonously increasing in the sense that $d(\mathcal{C}_1) \geq d(\mathcal{C}_2)$ holds for every two chains with $\mathcal{C}_1 \supseteq \mathcal{C}_2$.

Our goal is to show how to modify the algorithm above to find a mininum cost chain-decomposition of $(\mathcal{K}, p)$. Use the same digraph $D = (U, F)$ introduced above along with the capacity function g and define the cost of edges sk', $k''t$ by $c(sk') := -d_l(K)$ and $c(k''t) := -d_f(K)$ $(K \in \mathcal{K})$ while the cost of all other edges is zero. By a min-cost flow algorithm (see e.g. [1]) determine a maximum flow x of minimum cost. By the above-mentioned correspondence between flows and chain-decompositions of $(\mathcal{K}, p)$, one can easily see that the chain-decomposition $\mathbf{C}$ assigned to a minimum cost maximum flow of D will be of minimum cost.

We can conclude that the minimum cost chain-decomposition problem can solved with the help of a min-cost flow algorithm. There are strongly polynomial combinatorial algorithms for the latter (for a comprehensive account, see Ahuja et al. [1]) and hence we have found one for the minimum cost weighted generator problem.

It should be remarked however that we need the min-cost flow problem only for a special type of cost function, namely when there are non-zero costs only on the edges leaving the source-node s or entering the sink-node t. It can be shown that in such a case there is no need for a general purpose min-cost flow algorithm. Instead, a modification of a max-flow min-cut algorithm will do by taking into consideration the edges of D at s and t one by one according to the increasing order of their costs.

References

1. R.K. Ahuja, T.L. Magnanti, and J.B. Orlin, *Network Flows*, Prentice Hall, 1993.
2. A. Benczúr, J. Förster, and Z. Király, *Finding minimum generators for path systems of a cycle – implementation and analysis*, submitted (1997).
3. L.R. Ford, D.R. Fulkerson, *Flows in Networks*, Princeton Univ. Press, Princeton NJ., 1962.
4. A. Frank and T. Jordán, *Minimal edge-coverings of pairs of sets*, Combinatorial Theory **65** (1995), 73-110.
5. A. Frank, *Finding minimum generators of path systems*, J. Combinatorial Theory **Ser B.** (1997), to appear.
6. A. Frank, *Finding minimum edge-coverings of pairs of sets*, in preparation (1997).
7. D.S. Franzblau and D.J. Kleitman, *An algorithm for covering polygons with rectangles*, Information and Control **63** (1984), 164-189.
8. E. Győri, *A minimax theorem on intervals*, J. Combinatorial Theory **Ser. B 37** (1984), 1-9.
9. D. E. Knuth, *Irredundant Intervals*, ACM Journal of Experimental Algorithmics **1** (1996), article 1, 19.

10. A. Lubiw, *A weighted min-max relation for intervals*, J. Combinatorial Theory **Ser.B 53** (1991), 151-172.

Acknowledgement. Many thanks are due to A. Benczúr, J. Förster, and Z. Király for the helpful conversations on the topic.

DEPARTMENT OF OPERATIONS RESEARCH, EÖTVÖS UNIVERSITY, MÚZEUM KRT. 6-8, BUDAPEST, HUNGARY, H-1088 AND ERICSSON TRAFFIC LABORATORY, LABORC U. 1. BUDAPEST, HUNGARY H-1037.

E-mail address: frank@cs.elte.hu

DIMACS Series in Discrete Mathematics
and Theoretical Computer Science
Volume **49**, 1999

On the Distribution of Sums of Vectors in General Position

Jerrold R. Griggs and Günter Rote

ABSTRACT. An analogue of the Littlewood-Offord problem posed by the first author is to find the maximum number of subset sums equal to the same vector over all ways of selecting n vectors in $\mathbb{R}^d$ in general position. This note reviews past progress and motivation for this problem, and presents a construction that gives a respectable new lower bound, $\Omega(2^n n^{1-3d/2})$, which compares for $d \geq 2$ to the previously known upper bound $O(2^n n^{-1-d/2})$.

One version of the famous Littlewood-Offord problem [**10**] asks how to select complex numbers $a_1, \dots, a_n$, not necessarily distinct, with each $|a_i| \geq 1$, and a target open ball $T \subseteq \mathbb{C}$ of unit diameter to maximize the number of the 2^n subset sums $\sum_{i \in I} a_i$, where $I \subseteq [n]$, lying in T. Viewing $\mathbb{C}$ as $\mathbb{R}^2$, one can extend this problem to arbitrary dimension d, and ask the same thing, where now the a_i's are vectors in $\mathbb{R}^d$. By setting all a_i equal to the same vector, it is possible to have $\binom{n}{\lfloor \frac{n}{2} \rfloor}$ subset sums lying inside T. Erdős [**3**] showed this was best-possible for the reals ($d = 1$); Katona [**7**] and Kleitman [**8**] independently proved the same for the original case of complex numbers ($d = 2$); Kleitman [**9**] later found an ingenious inductive proof that $\binom{n}{\lfloor \frac{n}{2} \rfloor}$ is best-possible for general d.

Even if we restrict the target set to just a single point t, this bound is still achieved. But what if we must also spread out the vectors a_i in the sense of asking that any d of them be linearly independent? Had we only needed to hit a unit diameter ball target, the answer would have remained at $\binom{n}{\lfloor \frac{n}{2} \rfloor}$, but by shrinking the target to a single point, it will be tougher in general to get as many sums to hit the target. With a single point target, the restriction that each $|a_i| \geq 1$ no longer affects the answer, so it can be dropped. Therefore, we are now interested in the following:

GENERAL POSITION SUBSET PROBLEM. *Given positive integers n, d, how can one select vectors $a_1, \dots, a_n \in \mathbb{R}^d$ and a target $t \in \mathbb{R}^d$ to achieve the maximum number $f_d(n)$ of the 2^n subset sums $\sum_{i \in I} a_i$, where $I \subseteq [n]$, equal to t, provided that every d of the vectors a_i are linearly independent?*

1991 *Mathematics Subject Classification.* Primary 52C10, 05D05; Secondary 11B75.

The first author was supported in part by NSF grant DMS–9701211.

The second author was supported in part by the Spezialforschungbereich "Optimierung und Kontrolle".

Griggs [**5**] arrived at exactly this problem in connection with a model of database security. In the database security studies of Mirka Miller *et al.* [**12, 11, 1**], *cf.* [**4**], there is a database of numerical records, $\{x_1, \dots, x_n\}$, *e.g.*, the salaries of the n members of a department. One may request the sums $\sum_{j \in J} x_j$ of certain subsets $J \subseteq \{1, \dots, n\}$, and an answer will be given by the control mechanism, provided that no "compromise" results. In the basic model, a compromise means that the requester is able to determine, by taking an appropriate linear combination of the answered queries (sums), some individual entry x_i. The problem is to maximize the number of queries that can be answered without compromising the database. Griggs proposed an extension of this problem to prevent compromise by anyone with prior knowledge of $d-1$ records: We say a compromise results whenever one can determine some linear combination of at most d records, $\sum_{j \in J} \alpha_j x_j$, where all $\alpha_j \neq 0$ and $0 < |J| \leq d$. It turns out that the maximum number of queries that can be answered without compromise is precisely $f_d(n)$.

In one dimension, $f_1(n)$ is equivalent to the real case of the Littlewood-Offord problem, and so

$$f_1(n) = \binom{n}{\lfloor \frac{n}{2} \rfloor}.$$

Determining $f_d(n)$ for fixed dimension $d \geq 2$ remains an intriguing and fundamental open problem with connections to many fields of mathematics. As shown in [**5**], an $O(2^n n^{-\lfloor d/2 \rfloor})$ upper bound on $f_d(n)$ can be deduced by a simple sphere-packing argument applied to the equivalent database security problem.

Deeper work of Halász provides a slight improvement, an $O(2^n n^{-1-d/2})$ upper bound for $d \geq 2$. More sophisticated analytical methods, especially Fourier analysis, are apparently needed to obtain this bound. An accessible proof would be valuable, in that it might adapt nicely to other variations of the Littlewood-Offord problem, such as when varying lower bounds on different vectors a_i are imposed.

Is the Halász bound correct for general d to within a constant factor? We still cannot say, not having a suitable construction achieving the bound for $d > 2$. In [**5**] two sets of vectors for $d = 2$ are presented, each achieving the Halász bound, to within a constant factor, for general n. Until the DIMATIA-DIMACS conference on "The Future of Discrete Mathematics" in Štiřín castle in May, 1997, no reasonable lower bound for dimensions $d > 2$ had been described. In this note, we provide such a construction.

We first review and extend the method which was applied in [**5**] to obtain a lower bound from a construction. Suppose we have a set of integer vectors $a_1, \dots, a_n \in \mathbb{R}^d$ such that any d of them are linearly independent, $n \geq d$. Consider any particular coordinate of the a_i's, say coordinate j, and denote this component of a_i by $u_i \in \mathbb{R}$. Then the distribution of the jth components of the 2^n subset sums $\sum_{i \in I} a_i$, when multiplied by 2^{-n}, is the same as the probability distribution for the random variable $X = \sum_{i=1}^{n} u_i X_i$, where the X_i are i.i.d. random variables, each equal to 0 or 1 with probability 1/2. One checks routinely that X has mean $\mu_j := \sum u_i/2$ and variance $\sigma_j^2 := \sum_i u_i^2/4$. Chebyshev's inequality implies that a proportion at most $1/K^2$ out of our 2^n subset sums have jth component differing from μ_j by more than $K\sigma_j$.

Applying this for all j with $K = \sqrt{2d}$, we learn that at least half of the 2^n subset sums vectors are within $K\sigma_j$ of μ_j for all j. That is, at least 2^{n-1} of the subset sums lie within a box, with sides parallel to the coordinate planes, with side lengths $2K\sigma_j$, $1 \leq j \leq d$. Since our vectors a_i have integer coordinates, the lattice points are the only possible subset sums. The number of lattice points in this box is roughly

$$\prod_{j=1}^{d} 2K\sigma_j = (8d)^{d/2} \prod_{j=1}^{d} \sigma_j.$$

Consequently, some lattice point occurs as a subset sum for at least

$$(1/2)2^n \left((8d)^{d/2} \prod_{j=1}^{d} \sigma_j \right)^{-1}$$

different subset sums.

Here is one construction that gives the desired independence, yet keeps the product of σ_j's under control: Let $a_i = (1, i, i^2, \ldots, i^{d-1})$, $1 \leq i \leq n$. Then any d of these vectors, say $a_{i_1}, \ldots, a_{i_d}$ with $i_1 < i_2 < \cdots < i_d$, are linearly independent, since they form a Vandermonde matrix, with determinant

$$\prod_{j<k} (i_k - i_j) \neq 0.$$

Thus, our n vectors are in general position. As for the bound they give, we have

$$\sigma_j = \left(\sum_i i^{2(j-1)}/4 \right)^{1/2} = \Theta(n^{\frac{2j-1}{2}}).$$

Thus, $\prod_{j=1}^{d} \sigma_j = \Theta(n^{d^2/2})$, giving us a lower bound on $f_d(n)$ of order $\Theta(2^n n^{-d^2/2})$.

For $d = 1$ and $d = 2$, this bound is best-possible, up to a constant factor. For $d > 2$ we can modify this construction so that it still works, while the vector coordinates stay much smaller ($< 2n$). Several months of continued reflection on the problem have not led to any further improvement, so we shall describe this progress now. Perhaps it is the upper bound that needs tightening, rather than the lower bound. Here is our new result.

THEOREM. *For fixed $d \geq 2$, there exist constants $C, C' > 0$ such that the maximum number $f_d(n)$ of subset sums equal to the same value, for any set of n vectors in $\mathbb{R}^d$ in general position, satisfies*

$$C 2^n n^{1-(3/2)d} < f_d(n) < C' 2^n n^{-1-d/2}.$$

PROOF. As noted above, the upper bound follows from a result of Halász [**6**]. We present a construction for the lower bound. Choose a prime number p with $n \leq p < 2n$, which is well-known to exist. Take the a_i as in the construction above, except reduce the coordinates modulo p, so that every coordinate belongs to $\{0, \ldots, p-1\}$. The new vectors are in general position: For any d of them, each entry of the corresponding determinant is the same, mod p, as before. Thus, the new determinant is congruent mod p to $\prod_{j<k} (i_k - i_j)$, which is not zero mod p since each of its factors is a positive integer $< n \leq p$. It follows that the new determinant is not zero, since it is not divisible by p.

The first coordinates are all 1, so we have $\sigma_1^2 = n/4$. For $j > 1$, since all coordinates are at most $p-1$, we get that $\sigma_j = O((np^2)^{1/2}) = O(n^{3/2})$. Completing the analysis as before gives us the stated lower bound. □

The point sets in general position with limited coordinates which were constructed above have also been proposed in computational geometry, where they were useful to deal with problems of geometric degeneracy by perturbation methods, see Emiris, Canny, and Seidel [**2**].

Acknowledgements

We are grateful to the centers DIMATIA in Prague and DIMACS in New Jersey for organizing the exciting workshop at Štiřín last May and for supporting its participants, including the authors, who had not met previously. The collaboration described in this note was a direct result of this contact.

We thank Torsten Thiele and Emo Welzl for discussions about the problem of finding many points with small coordinates in general position.

References

[1] L. Branković, M. Miller, and J. Širáň, *Towards a practical auditing method for the prevention of statistical database compromise,* Proc. 7th Australasian Database Conf., Austral. Comp. Sci. Commun.**18 (no. 2)**(1996), 177–184.

[2] I. Z. Emiris, J. F. Canny, and R. Seidel, *Efficient perturbations for handling geometric degeneracies,* Algorithmica **19**(1997), 219–242.

[3] P. Erdős, *On a lemma of Littlewood and Offord,* Bull. Amer. Math. Soc. (2nd ser.)**51**(1945), 898–902.

[4] J. R. Griggs, *Concentrating subset sums at k points,* Bull. Inst. Combin. Applns. **20**(1997), 65–74.

[5] J. R. Griggs, *Database security and the distribution of subset sums in* $\mathbb{R}^m$, Proc. Intern. Colloq. Combin. Graph Th. 1996, Balatonlelle, Hungary, to appear.

[6] G. Halász, *Estimates for the concentration function of combinatorial number theory and probability,* Periodica Math. Hungar. **8 (3–4)**(1977), 197–211.

[7] G. O. H. Katona, *On a conjecture of Erdős and a stronger form of Sperner's theorem,* Studia Sci. Math. Hungar. **1**(1966), 59–63.

[8] D. J. Kleitman, *On a lemma of Littlewood and Offord on the distribution of certain sums,* Math. Z. **90**(1965), 251–259.

[9] D. J. Kleitman, *On a lemma of Littlewood and Offord on the distributions of linear combinations of vectors,* Advances in Math. **5**(1970), 155–157.

[10] J. Littlewood and C. Offord, *On the number of real roots of a random algebraic equation III,* Mat. Sbornik **12**(1943), 277–285.

[11] M. Miller, I. Roberts, and J. Simpson, *Application of symmetric chains to an optimization problem in the security of statistical databases,* Bull. Inst. Combin. Applns. **2**(1991), 47–58.

[12] M. Miller and J. Seberry, *Relative compromise of statistical databases,* Austral. Computer J. **21 (2)**(1989), 56–61.

Department of Mathematics, University of South Carolina, Columbia, SC 29208 USA

E-mail address: griggs@math.sc.edu

Technische Universität Graz, Institut für Mathematik, Steyrergasse 30, A-8010 Graz, AUSTRIA

E-mail address: rote@opt.math.tu-graz.ac.at

DIMACS Series in Discrete Mathematics
and Theoretical Computer Science
Volume **49**, 1999

The Generalized Matching Problem on Partial k-Trees*

A. Gupta D. Kaller S. Mahajan T. Shermer

January 13, 1999

Abstract

If the solutions of a given graph problem can be encoded by some suitable extension of Monadic Second-order (or MS) logic, then the problem can be solved over partial k-trees in polynomial time by dynamic-programming techniques. The solutions of many different graph problems can be encoded very easily in this way. For other problems, however, it seems difficult (if not impossible) for MS logic to encode the solutions for an arbitrary input graph. But in order to obtain a polynomial-time algorithm, we need only obtain an MS formula that encodes some solution for any input partial k-tree. By restricting our attention in this way, we can often encode problems over partial k-trees when they cannot be encoded over general graphs. In this paper we show that GENERALIZED MATCHING can be encoded in "Extended" MS logic by exploiting structural features of the partial k-trees. Thus it follows that the problem can be solved over partial k-trees by a polynomial-time dynamic-programming algorithm. Although this algorithm was previously known to exist, it was not thought to be predicted by the MS logical formalism.

Mathematics Subject Classification: 05C85

1 Introduction

Many NP-hard graph problems can be solved over partial k-trees in polynomial time (often even linear-time) by dynamic-programming algorithms [1, 2, 3, 4, 6, 7, 9, 10, 12, 14, 25, 27]. The partial k-trees encompass many important graph families including trees and forests ($k = 1$), series-parallel and outerplanar graphs ($k = 2$) and Halin graphs ($k = 3$) [5]. Any partial k-tree can be parsed into a tree with each node labeled by a symbol from a finite alphabet (whose size is dependent only on k); Bodlaender [8] has shown that this parsing can be done in linear time. The labeled tree (i.e., the *tree*

*Some of these results were presented at the 5th Scandinavian Workshop on Algorithm Theory [16].

decomposition) is the input for dynamic-programming algorithms that can be modeled by *tree automata.* These machines process a tree decomposition (in order from its leaves to its root), assigning each node to one of a possibly infinite number of states; the algorithm's output is a function of the state to which the root is thus assigned. Several different variations of the Monadic Second-order (or MS) logic were proposed to encode the problems amenable to this approach [2, 10, 12, 14]. A formula in MS logic can be automatically translated into a tree automaton to solve the encoded problem over partial k-trees in linear time. Often, it is very easy to encode a given graph problem in some suitable extension of MS logic. However, there are problems that can be solved in polynomial time by dynamic programming on tree decompositions, and yet these problems can not (seemingly) be encoded in any MS logical formalism [2, 7, 9, 14]. The CHROMATIC INDEX problem is a good example of this. There other are problems, however, such as GENERALIZED MATCHING, that perhaps cannot be encoded over general graphs but can be encoded over partial k-trees in the *Extended* MS (or EMS) logic of Arnborg et al. [2]. GENERALIZED MATCHING is the restriction of PARTITION INTO ISOMORPHIC SUBGRAPHS (problem GT12 in [15]) to some fixed *pattern graph.*

Problem: GENERALIZED MATCHING. For a fixed pattern graph H, a graph G is a yes-instance iff $V(G)$ can be partitioned into $\frac{|V(G)|}{|V(H)|}$ sets, each inducing a subgraph of G that is isomorphic to H.

Kirkpatrick and Hell [23] have shown this problem to be NP-complete (over general graphs) for any pattern graph on three or more vertices. Arnborg et al. [2] had claimed that an EMS formula can encode the problem over general graphs but the claim has been withdrawn [24]. On the other hand, Borie [9] conjectured that GENERALIZED MATCHING is not EMS-encodable over partial k-trees; and he developed a general technique to solve a family of apparently non-EMS-encodable problems (including GENERALIZED MATCHING) in polynomial time, by dynamic programming on tree decompositions. In this paper, we establish that EMS logic does, in fact, predict that GENERALIZED MATCHING can be solved in this way.

The problems that can be solved by tree automata have a quality of "locality" [25] or "regularity" [4] which enables a solution (for a given partial k-tree G) to be computed from sub-solutions on subgraphs of G. As a tree automaton processes a tree decomposition of G, sub-solutions are combined together as a function of the label of the node that is currently being processed. A tree automaton could keep track of *all* the possible sub-solutions that have been seen so far; but this could be an exponential quantity of information to maintain. However, if the sub-solutions can be divided into a constant number of equivalence classes (with respect to their contribution to any global solution), then the tree automaton need only remember one of a finite number of states for each node of a tree decomposition; thus the problem can be solved in linear time by a *finite-state tree automaton.* Courcelle [12] (see also [2, 10]) has shown that if a sentence in the *Counting MS* (or CMS) logic can encode a graph decision problem, then the problem can be solved over partial k-trees in linear time by a finite-state tree automaton. Here, CMS logic is the extension of MS logic with predicates to count the cardinality of sets modulo fixed integers. It has been conjectured (conversely) that a graph decision problem can be encoded by CMS logic *only* if it can be solved over partial k-trees in linear time by a finite-state tree automaton; proofs of this equivalence are known in several restricted cases [12, 13, 20, 22].

The standard MS logic can encode, for example, whether a graph is Hamiltonian, whether it is properly c-colorable (for any constant c) and whether it is connected; the "counting" predicates of CMS logic are not needed to encode any of these properties. On the other hand, the "counting" predicates are needed to encode, for example, whether a graph has an even number of vertices [12]. In other cases, CMS logic cannot encode all the graphs with a given property, but it can nevertheless encode the partial k-trees with the property; an example of this is given by graphs whose vertices can be partitioned into fixed-size independent sets [18]. There are other graph properties that provably cannot be encoded by CMS logic. To obtain negative results of this form, one can use a lemma of Courcelle [12], which shows that, for any CMS-encodable property, it is decidable whether a given "context-free" graph grammar generates a graph with the property. Alternatively, one can use a pumping lemma [17, 21], reminiscent of the pumping lemma for regular sets. This provides a proof that CMS logic cannot GENERALIZED MATCHING over partial k-trees (for certain disconnected pattern graphs).

Arnborg et al. [2] and Courcelle and Mosbah [14] discuss ways to extend MS logic, so that it can encode some problems that are not CMS-encodable. The *Extended* MS (or EMS) logic [2] allows arithmetic calculations to be performed on the cardinalities of a constant number of vertex and edge sets that satisfy particular conditions (these conditions will be explained in more detail in Section 5). A formula in EMS logic can be automatically translated into an polynomial-time algorithm for the encoded problem over partial k-trees. Since there is no finite bound on the set cardinalities, these are not *finite state* tree automata. But each cardinality is only linear in the size of the input graph; so there are only a polynomial number of different combinations of set cardinalities for the automaton to consider. Hence, a "polynomial-state" tree automaton can be used to model the algorithm obtained from an EMS formula.

The rest of this paper is organized as follows: Section 2 contains some graph-theoretic preliminaries. In Section 3 we discuss partial k-trees and tree decompositions. In Section 4 we review CMS logic, demonstrating how it can encode the Hamiltonian graphs (as an example). In Section 5 we discuss how EMS logic can encode some decision problems that CMS logic cannot. In Section 6 we consider GENERALIZED MATCHING for several special types of pattern graph: In some cases, the problem can be encoded by the standard MS logic; in other cases, the "counting" predicates of CMS logic are necessary; and for other pattern graphs, the problem cannot be encoded by CMS logic. In Section 7 we use interesting combinatorial techniques to show that the problem can always be encoded in EMS logic (for any fixed pattern graph). Finally, in Section 8, we make some concluding remarks.

2 Preliminaries

The graphs in this paper are finite, simple, loop-free and undirected. The vertex set of a graph G is denoted by $V(G)$, and its edge set is denoted by $E(G)$. We write $\{u, v\} \in E(G)$ to mean that u and v are *adjacent* vertices of G. We say that a pair of vertex-disjoint subgraphs (say H and H') of G are *adjacent* if there is some edge of G with one endpoint in $V(H)$ and the other endpoint in $V(H')$.

If G is a graph, and V' is a subset of $V(G)$, then $G_{[V']}$ is the subgraph of G induced by V'. If V' is a set (but not necessarily a subset of $V(G)$) then $G\backslash V'$ is the subgraph of G induced by $V(G) - V'$. The disjoint union of two graphs G and G' is denoted by $G \sqcup G'$: this is the graph with vertex set $V(G) \cup V(G')$ and edge set $E(G) \cup E(G')$. We write $G \cong G'$ to indicate that G and G' are isomorphic graphs.

A *c-colored* graph is one whose vertices are partitioned into c disjoint color classes (we do not require this to be a proper vertex coloring). In the following lemma, we wish to partition the vertices of a c-colored graph into equivalence classes in which no pair of *differently*-colored vertices are adjacent.

Lemma 2.1. *Let $c, r \in \mathbb{N}$, $c \geq 2$; and suppose G is a c-colored graph (on rc vertices) that has exactly r vertices of each color. If no component of G contains more than $\lfloor \frac{r}{c-1} \rfloor$ vertices, then $V(G)$ can be partitioned into $(c+1)!$ sets, each of which satisfies the following:*

- *It contains an equal number of vertices of each color; and*
- *it contains no pair of adjacent vertices with different colors.*

Proof. Suppose no component of G contains more than $\lfloor \frac{r}{c-1} \rfloor$ vertices; and let G be partitioned into $c+1$ subgraphs $G = G_0 \sqcup G_1 \sqcup \ldots \sqcup G_c$, each with r or fewer vertices. To see that this is possible, let the vertices of G be enumerated $v_1, v_2, \ldots, v_{rc}$ such that the vertices of each component are numbered consecutively. If v_{ri} belongs to the same component as v_{ri+1} (for $1 \leq i \leq c-1$), then let G_c contain that component. Let G_i $(0 \leq i \leq c-1)$ contain the remaining vertices of $v_{ri+1}, v_{ri+2}, \ldots, v_{ri+r}$.

Claim. $V(G)$ can be partitioned into r sets, each containing exactly one vertex of each color, such that at most one vertex belongs to each subgraph G_i $(0 \leq i \leq c)$.

There are $(c+1)!$ different ways to assign c colors from $G_0, G_1, \ldots, G_c$. The sets of vertices chosen in each different way can be grouped into one of the classes (to satisfy the lemma), because each pair of differently-colored vertices in such a class belong to disjoint components of G. To prove the lemma, then, we need only prove the above claim.

Without loss of generality, assume G_0 contains $r-1$ or fewer vertices. Therefore, if G' is the disjoint union of any ℓ $(1 \leq \ell \leq c)$ of the other subgraphs $G_1, G_2, \ldots, G_c$, then G' contains vertices of at least ℓ different colors. For otherwise, G' could contain no more than $(\ell-1)r$ vertices; and since there are at most $(c-\ell)r$ vertices in $V(G)-V(G_0)-V(G')$, there would be a total of only $(r-1)+(\ell-1)r+(c-\ell)r < cr$ vertices. It follows that one of the sets (required by the claim) can be selected from $G_1 \sqcup G_2 \sqcup \ldots \sqcup G_c$. Since G_0 contained at most $r-1$ vertices to begin with, the claim follows by induction on r. □

3 Partial k-Trees and Tree Decompositions

A *k-tree* is either the clique on k vertices, or a graph that can be obtained (recursively) from a k-tree G by adding a new vertex and making it adjacent to any k distinct vertices that induce a clique in G. A *partial k-tree* is a subgraph of a k-tree. For example, a graph is a partial 0-tree iff its edge set is empty; a graph is a partial 1-tree iff it is a forest. Series-parallel graphs and outerplanar graphs are subclasses of the partial 2-trees; Halin graphs form a subclass of the partial 3-trees. It is not hard to show (see, e.g., [28]) that a graph is a partial k-tree iff it admits a *width-k tree decomposition*:

Definition 3.1. A *tree decomposition* of a graph G is a pair $(T, \mathcal{X})$ where T is a tree and $\mathcal{X} = \{X_a\}_{a \in V(T)}$ is a collection of subsets of $V(G)$, indexed by the nodes of T, for which

- $\bigcup_{a \in V(T)} X_a = V(G)$, and
- each edge of G has both endpoints in some set $X_a \in \mathcal{X}$, and
- if $a, b, c \in V(T)$ such that b lies on the path between a and c, then $X_a \cap X_c \subseteq X_b$.

We refer to elements of $V(T)$ as *nodes*, so as not to confuse them with vertices of G. The set X_b is called the *bag* indexed by $b \in V(T)$. If no bag in $\mathcal{X}$ contains more than $k+1$ vertices, then $(T, \mathcal{X})$ is called a *width-k* tree decomposition. If T' is a subgraph of T, then $X_{T'}$ is the union of the bags indexed by $V(T')$: that is, $X_{T'} = \bigcup_{a \in V(T')} X_a$. We refer to the subgraph of G that is induced by $X_{T'}$ as the subgraph *underlying* T'. The following well-known lemma is often useful when working with trees [19]; we give a proof for completeness

Lemma 3.2. *Suppose G is a partial k-tree; and let $V' \subseteq V(G)$. There exists a subset X of $V(G)$, with cardinality $|X| \leq k+1$, such that $G \backslash X$ consists of two disjoint partial k-trees, neither containing more than $\lfloor \frac{2|V'|}{3} \rfloor$ vertices of V'.*

Proof. Let $n' = |V'|$; and let $(T, \mathcal{X})$ be a width-k tree decomposition of G. By deleting an edge $\{a, b\}$ from $E(T)$, we obtain two disjoint trees, say T_a and T_b, where $a \in V(T_a)$ and $b \in V(T_b)$. Suppose that this edge becomes directed from a to b if $|X_{T_b} \cap V'| > |X_{T_a} \cap V'|$. Since T has more nodes than edges, there exists a node (say c) away from which no edge may become directed in this manner. It follows that no component of $G \backslash X_c$ contains more than $\lfloor \frac{n'}{2} \rfloor$ vertices of V'. Since V' contains only n' vertices in total, we can easily group the components of $G \backslash X_c$ into two disjoint partial k-trees, neither containing more than $\lfloor \frac{2n'}{3} \rfloor$ vertices of V'. Therefore, $X = X_c$ is the subset required by the lemma. □

4 Monadic Second-Order Logic

A graph $G = (V, E)$ can be interpreted as a logical structure over the universe $V \cup E$, with a predicate $Edge(e, v)$ that holds whenever $v \in V$ is an endpoint of $e \in E$. Many different graph classes can then be encoded by logical sentences in Monadic Second-order (or MS) logic [2, 10, 12] using the following symbols: individual variables (to represent vertices or edges); set variables (to represent sets of vertices or edges); the equality (=) and membership ($\in$) symbols; existential ($\exists$) and universal ($\forall$) quantifiers; the logical operators $\wedge$ ("and"), $\vee$ ("or"), $\neg$ ("not"), $\Rightarrow$ ("implies") $\Leftrightarrow$ ("if and only if"); and the *Edge* predicate. MS logic allows quantification over both individual and set variables. An MS sentence—in which all variables (except V and E) are quantified—is evaluated on a graph $G = (V, E)$. We write $G \models \Phi$ to indicate that an MS sentence Φ is true for a graph G. This sentence defines a certain class (say Π) of graphs: that is, $G \models \Phi$ iff $G \in \Pi$. The class of Hamiltonian graphs, for example, can be defined as follows:

Example 4.1. A graph $G = (V, E)$ has a Hamiltonian circuit iff the following MS sentence is satisfied:

$$(\exists E')(\text{ "}E' \subseteq E\text{" } \wedge \text{ "}(V, E') \text{ is connected" } \wedge \text{ "}(V, E') \text{ is 2-regular" })$$

Here, the condition "$E' \subseteq E$" can be encoded as $(\forall x)(x \in E' \Rightarrow x \in E)$. The connectivity condition can be encoded by saying that if X and Y are non-empty sets partitioning V, then some edge in E' has one endpoint in X and the other endpoint in Y:

$$(\forall X,Y)(\ \ (\ (\forall v)((v\in V)\Leftrightarrow((v\in X)\vee(v\in Y)))\wedge(\exists x)(x\in X)\wedge(\exists y)(y\in Y)\ )\\ \Rightarrow(\exists x,y,e)((x\in X)\wedge(y\in Y)\wedge(e\in E')\wedge Edge(e,x)\wedge Edge(e,y))\quad)$$

Finally, the 2-regularity condition can be encoded as follows:

$$(\forall v\in V)(\exists e,e'\in E')(\ \ \neg(e=e')\wedge Edge(e,v)\wedge Edge(e',v)\wedge\\ \neg(\exists e''\in E')(\neg(e''=e)\wedge\neg(e''=e')\wedge Edge(e'',v))\ \ )$$

For the rest of this paper, we will use high-level expressions (such as those of Example 4.1) without providing a detailed translation into the low-level symbols of MS logic. Refer to Arnborg et al. [2], Borie et al. [10] or Courcelle [11, 12] for a further discussion of encoding such expressions.

Counting MS (or CMS) logic is the extension of MS logic with unary predicates $\mathbf{card}_{\ell,c}$ for nonnegative integer constants ℓ, c (with $\ell < c$). If S is a set, then $\mathbf{card}_{\ell,c}(S)$ is true iff S has cardinality ℓ (mod c). Courcelle [12] has shown that CMS logic is more powerful than the standard MS logic; for example, no MS sentence can define the class of graphs with an even number of vertices, but a CMS sentence can. Any CMS-definable class of partial k-trees can be recognized in linear time by a finite-state tree automaton [12].

Lemma 4.2. *If Φ is a CMS statement, then (for each $k \in \mathbf{N}$) there is a finite-state tree automaton that decides in linear time whether $G \models \Phi$, for any input partial k-tree G.*

The converse of Lemma 4.2 remains an open question: i.e., if a class of partial k-trees can be recognized in this way (by a finite-state tree automaton), then is that class necessarily CMS-definable? This converse is known to hold for partial 1-trees [12], partial 2-trees [13], partial 3-trees and k-connected partial k-trees [22], and partial k-paths [20].

5 Extended Monadic Second-Order Logic

By providing tree automata with counters (to keep track of set cardinalities), it is possible to recognize some non-CMS-definable classes of partial k-trees in linear time (and many others in polynomial time). These "extended tree automata" are not finite-state machines, because there is no bound on the value of a counter. Arnborg et al. [2] and Courcelle and Mosbah [14] discuss ways to extend MS logic to define graph classes that can be recognized in this way. Arnborg et al. [2] introduced the *Extended MS* (or EMS) logic for this purpose; an EMS sentence consists of:

- a constant number (say c) of *common* set variables $X_1, X_2, \ldots, X_c$;

- a CMS formula Φ, such that "$(\exists X_1, X_2, \ldots, X_c)\ \Phi$" is a CMS sentence; and

- an *evaluation relation* Ψ, which is a logical predicate that enforces equality (and inequality) conditions on polynomial functions of the cardinalities $|X_1|, |X_2|, \ldots, |X_c|$.

Such a formula is satisfied by a graph $G = (V, E)$ iff it is possible to choose $X_1, X_2, \ldots, X_c \subseteq V \cup E$ such that Ψ is true and $G \models (\exists X_1, X_2, \ldots, X_c)\ \Phi$.

Example 5.1. EMS logic can encode whether a graph $G = (V, E)$ can be partitioned into *noninduced* subgraphs isomorphic to a fixed graph H. Suppose H has c maximal components $H_1, H_2, \ldots, H_c$; and let h_i $(1 \leq i \leq c)$ be the number of vertices in H_i. An EMS sentence for this problem has common set variables $V_1, V_2, \ldots, V_c$, as well as set variables $E_1, E_2, \ldots, E_c$ quantified within the CMS formula Φ. Now, Φ can ensure that $V_1, V_2, \ldots, V_c$ is a partition of $V(G)$, and (for $1 \leq i \leq c$) that $E_i \subseteq E(G)$, and H_i is isomorphic to each maximal component of the subgraph (V_i, E_i). The evaluation relation Ψ can then stipulate that the value of $\frac{|V_i|}{h_i}$ is equal over $1 \leq i \leq c$. Therefore, this problem can be solved in polynomial time over partial k-trees.

The GENERALIZED MATCHING problem asks if a graph G can be partitioned into *induced* copies of a fixed pattern graph H: This is more difficult than Example 5.1 because we need to ensure the copies of H can be matched into groups of c components that are pairwise nonadjacent (with respect to the graph G). In Section 7, we will show how an EMS sentence can encode the existence of such a matching for any partial k-tree G.

6 Some Special Cases of GENERALIZED MATCHING

A graph G is a yes-instance of GENERALIZED MATCHING if its vertices can be partitioned into sets that each induce a subgraph isomorphic to some fixed *pattern graph* H. Because this partition contains an unbounded number $\frac{|V(G)|}{|V(H)|}$ of sets, an MS sentence cannot explicitly represent each of them. We will show in Section 7, however, that these sets can be represented by a constant number of equivalence classes—provided G is a partial k-tree. This enables EMS logic to encode the problem over partial k-trees for any fixed pattern graph. In this section, we consider several specific types of pattern graphs: For some pattern graphs, GENERALIZED MATCHING can be encoded by the standard MS logic (without using the "counting" predicates of CMS logic); for other pattern graphs, the problem cannot be encoded by the standard MS logic, but it can be encoded by CMS logic; and for other pattern graphs, the problem is not CMS-encodable.

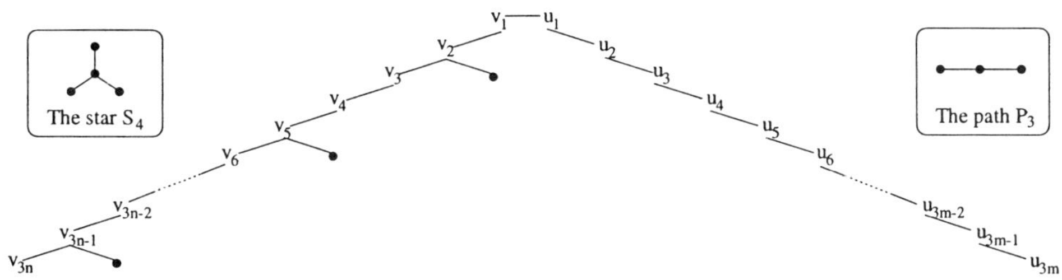

Figure 1: The tree $G_{n,m}$

Theorem 6.1. *For any connected pattern graph,* GENERALIZED MATCHING *is MS-encodable over general graphs, and hence it can be solved in linear time over partial k-trees by a finite-state tree automaton.*

Proof. A graph $G = (V, E)$ is a yes-instance of this problem iff there exists an edge subset $E' \subseteq E$ such that each component of (V, E') is isomorphic to the pattern graph. This isomorphism test can be hard-coded into an MS sentence for any fixed pattern graph. Therefore, the theorem follows from Lemma 4.2. □

For some disconnected pattern graphs, GENERALIZED MATCHING is not CMS-encodable over partial k-trees. This negative result can be obtained with a pumping lemma [21, 17]. Consider, for example, the pattern graph $S_4 \sqcup P_3$, where P_3 is the path on three vertices, and S_4 is the star on four vertices. For $n, m \in \mathbf{N}$, let $G_{n,m}$ be the tree (shown in Figure 1) on $4n+3m$ vertices, containing a path on $3n+3m$ vertices $v_{3n}, v_{3n-1}, \ldots, v_1, u_1, u_2, \ldots, u_{3m}$ (in that order), such that for $i \equiv 2 \pmod 3$, the vertex v_i is adjacent to one other pendant vertex (not including v_{3n}). So $G_{n,m}$ is a yes-instance iff $m = n \geq 2$. But if the yes-instances can be recognized by a finite-state tree automaton, then for any n that is sufficiently large, the graph $G_{n,n}$ can be "pumped" into a graph $G_{n,m}$ (for $m \neq n$) that will also be accepted by the tree automaton (a contradiction).

It is not always the case, however, that a tree automaton cannot solve GENERALIZED MATCHING when the pattern graph is disconnected. For example, if the pattern graph consists of a fixed number of isolated vertices, then the partial k-trees that are yes-instances can be defined by a CMS sentence [18] (where the "counting" predicates are necessary). This result can easily be generalized to any pattern graph consisting of pairwise isomorphic components.

7 Encoding GENERALIZED MATCHING in EMS Logic

In this section, we show how GENERALIZED MATCHING can be encoded by an EMS sentence over partial k-trees. Let H be a fixed pattern graph with c components $H_1, H_2, \ldots, H_c$; let G be a partial k-tree on n vertices that can be partitioned into r induced copies of H; and let $\mathcal{P}$ be a partition of $V(G)$ that satisfies the following:

$$\exists \mathcal{P} = \{V_1, V_2, \ldots, V_r\} \text{ such that: } \quad \begin{array}{l} \mathcal{P} \text{ is a partition of } V(G), \text{ and} \\ G_{[V_i]} \cong H, \text{ for } 1 \le i \le r \end{array} \tag{7.1}$$

Definition 7.2. A $\mathcal{P}$-*component* is a component of the subgraph $G_{[\hat{V}]}$ for any $\hat{V} \in \mathcal{P}$. We assume that the $\mathcal{P}$-components are assigned to c color classes, such that each $G_{[\hat{V}]}$ (for $\hat{V} \in \mathcal{P}$) consists of c $\mathcal{P}$-components with distinct colors, and each copy of H_i $(1 \le i \le c)$ is assigned to the same color class. Here, any pair of isomorphic components ($H_i \cong H_j$ where $i \ne j$) are not considered to be equivalent.

Thus $V(G)$ is partitioned into rc $\mathcal{P}$-components, r of each color. We will delete, from G, the vertices of a constant number of the sets in $\mathcal{P}$, resulting in a subgraph G' whose components each contain fewer than $\lfloor \frac{r}{c-1} \rfloor$ $\mathcal{P}$-components. Then we can use Lemma 2.1 to group the $\mathcal{P}$-components into a constant number of equivalence classes in which no pair of differently-colored components are adjacent. The $\mathcal{P}$-components in each such equivalence class can be matched arbitrarily to construct subgraphs isomorphic to H. Hence, each equivalence class can be represented by one vertex set variable and one edge set variable of an EMS sentence.

Proposition 7.3. *Suppose G' is the subgraph of G induced by the vertices in some collection of (say n') $\mathcal{P}$-components. There exists a set $X \subseteq V(G)$, the union of $k+1$ or fewer sets in $\mathcal{P}$, such that $G' \backslash X$ consists of two disjoint partial k-trees, neither containing more than $\lfloor \frac{2n'}{3} \rfloor$ $\mathcal{P}$-components. We will say that the deletion of X is a* balanced cut *to G'.*

Proof. Let V' be a subset of $V(G')$ that contains exactly one vertex from each of the n' $\mathcal{P}$-components. By Lemma 3.2, there exists a set $Y \subseteq V(G')$ with cardinality $|Y| \le k+1$, such that $G' \backslash Y$ consists of two disjoint partial k-trees, neither containing more than $\lfloor \frac{2}{3} n' \rfloor$ vertices of V'. We choose the set X (to satisfy the proposition) as the union of the vertex sets of the ($k+1$ or fewer) sets in $\mathcal{P}$ that each contain one or more vertices of Y. □

Lemma 7.4. *For $\ell \in \mathbf{N}$, there exists a subset $X_\ell \subseteq V(G)$ for which*

C1 *X_ℓ is the union of $(k+1)(2^\ell - 1)$ or fewer of the sets in $\mathcal{P}$, and*

C2 *$G_\ell = G\backslash X_\ell$ consists of 2^ℓ disjoint partial k-trees, none of which contain more than $\lfloor (\frac{2}{3})^\ell rc \rfloor$ $\mathcal{P}$-components.*

Proof. Let $X_0 = \emptyset$; so $G\backslash X_0 = G$, and the lemma is satisfied for $\ell = 0$. For $\ell \geq 1$, let G_ℓ be obtained (inductively) by making a balanced cut (Prop. 7.3) to each of the $2^{\ell-1}$ disjoint partial k-trees comprising $G_{\ell-1}$. For each such cut, we delete the vertices of $k+1$ or fewer sets in $\mathcal{P}$. So $G_\ell = G_{\ell-1}\backslash X$ where X is the union of $(k+1)2^{\ell-1}$ or fewer sets in $\mathcal{P}$. Inductively, $G_\ell = G_{\ell-1}\backslash X = G\backslash X_\ell$ where $X_\ell = X_{\ell-1} \cup X$ is the union of $(k+1)(2^\ell - 1)$ or fewer sets in $\mathcal{P}$ (hence **C1**). Furthermore, G_ℓ consists of $2 \cdot 2^{\ell-1} = 2^\ell$ disjoint partial k-trees, none with more than $\lfloor \frac{2n'}{3} \rfloor$ $\mathcal{P}$-components, where $n' = \lfloor (\frac{2}{3})^{\ell-1} rc \rfloor$ (hence **C2**). □

Lemma 7.5. *For $c \geq 2$, there is a constant $t_{c,k}$ (dependent only on c and k) for which the following is true: There exists a subset $X \subseteq V(G)$, the union of $d \leq t_{c,k}$ of the sets in $\mathcal{P}$, for which $G\backslash X$ has no component with more than $\frac{r-d}{c-1}$ $\mathcal{P}$-components.*

Proof. Let ℓ be a constant greater than $\lceil 2\log_{\frac{3}{2}} c \rceil$; so $c^2(\frac{2}{3})^\ell < 1$. Now, the set X_ℓ (of Lemma 7.4) is the union of some number $d \leq (2^\ell - 1)(k+1)$ of the sets in $\mathcal{P}$; and no component of $G\backslash X_\ell$ contains more than $\lfloor (\frac{2}{3})^\ell rc \rfloor$ $\mathcal{P}$-components. Since $c^2(\frac{2}{3})^\ell < 1$, there exists a constant $t'_{c,k}$ (dependent only on c and k) for which the following holds whenever $r \geq t'_{c,k}$.

$$\begin{aligned} rc^2(\tfrac{2}{3})^\ell + 2^\ell(k+1) &\leq r \\ rc(\tfrac{2}{3})^\ell &\leq \tfrac{r-2^\ell(k+1)}{c} \leq \tfrac{r-d}{c-1} \end{aligned}$$

Hence, if $r \geq t'_{c,k}$, the statement of the lemma is satisfied by $X = X_\ell$; otherwise, it is satisfied by $X = V(G)$. We need only choose the constant $t_{c,k} = \max\{t'_{c,k}, (2^\ell - 1)(k+1)\}$. □

Lemma 7.5 breaks the partial k-tree G into small enough pieces for the $\mathcal{P}$-components to be easily matched into induced copies of H. Figure 2 gives a high-level description of an EMS sentence that encodes the existence such a matching. Without loss of generality, assume $c' \leq c$ such that $H_1, H_2, \ldots, H_{c'}$ are pairwise non-isomorphic, and each of $H_{c'+1}, H_{c'+2}, \ldots, H_c$ is isomorphic to one of the first c' components. Suppose c_j maximal components of H are isomorphic to H_j (for $1 \leq j \leq c'$). The EMS sentence provides a constant number c^* of vertex set variables V_i and edge set variables E_i to represent disjoint subgraphs (V_i, E_i): These are to be collections of $\mathcal{P}$-components, and each collection is associated with c "counting sets" $W_{i,j,h}$, for $1 \leq j \leq c'$, $1 \leq h \leq c_j$. For each $\mathcal{P}$-component of (V_i, E_i) that is isomorphic to H_j, exactly one of its vertices is to be contained in the union $\bigcup_{h=1}^{c_j} W_{i,j,h}$.

Common Variables

$$W_{i,j,h} \subseteq V(G) \quad \left.\right\} \quad \begin{array}{l}\text{for } 1 \le i \le c^*,\ 1 \le j \le c',\ 1 \le h \le c_j \\ \quad \text{where } c^*, c', c_1, c_2, \ldots, c_{c'} \text{ are constants (see text)}\end{array}$$

MS Formula (Φ)

- $\{V_1, V_2, \ldots, V_{c^*}\}$ is a partition of $V(G)$.
- $E_1, E_2, \ldots, E_{c^*} \subseteq E(G)$.
- For $1 \le i \le c^*$: each component of (V_i, E_i) is isomorphic to $H_1, H_2 \ldots$ or H_c.
- For $1 \le i \le c^*$ and $1 \le j \le c'$: $\bigcup_{h=1}^{c_j} W_{i,j,h}$ consists of exactly one vertex of each component of (V_i, E_i) that is isomorphic to H_j.
- For $1 \le i \le c^*$: if C and C' are components of (V_i, E_i) that are represented in distinct "counting sets", then there is no edge of G between $V(C)$ and $V(C')$.

Evaluation Relation (Ψ)

- For $1 \le i \le c^*$: $|W_{i,j,h}|$ has the same value over $1 \le j \le c',\ 1 \le h \le c_j$.

Figure 2: EMS sentence for GENERALIZED MATCHING over partial k-trees, for a pattern graph $H = H_1 \sqcup H_2 \sqcup \ldots \sqcup H_c$

Theorem 7.6. *A partial k-tree G can be partitioned into subgraphs isomorphic to a fixed graph H if and only if the EMS sentence of Figure 2 is satisfied.*

Proof. Suppose the EMS sentence is true for a partial k-tree G. So for $1 \le i \le c^*$, each subgraph (V_i, E_i) contains components of c different colors (i.e., $H_1, H_2, \ldots$ and H_c) such that no pair of differently-colored components are adjacent (in G). The "counting sets" $W_{i,j,h}$ ensure that each (V_i, E_i) contains the same number of components of each color. Therefore, the components in each (V_i, E_i) can be matched arbitrarily to form a partition $\mathcal{P}$ satisfying formula (7.1).

Conversely, suppose there exists a partition $\mathcal{P}$ to satisfy formula (7.1). If $c = 1$, the EMS sentence is satisfied by $V_1 = V(G)$, $E_1 = E(G)$, and all other $V_i, E_i = \emptyset$. Otherwise $c \ge 2$, so (by Lemma 7.5) there exists a set X that is the union of $d \le t_{c,k}$ sets in $\mathcal{P}$; $G \backslash X$ contains $(r-d)c$ $\mathcal{P}$-components; and no component of $G \backslash X$ contains more than $\frac{r-d}{c-1}$ $\mathcal{P}$-components. By contracting each $\mathcal{P}$-component of $G \backslash X$ into a single vertex, we obtain a minor of G with a total of $(r-d)c$ vertices, consisting of $(r-d)$ vertices of each color. Hence, it follows from Lemma 2.1 that the $\mathcal{P}$-components of $G \backslash X$ can be partitioned into $(c+1)!$ equivalence classes in which any c differently-colored $\mathcal{P}$-components are pairwise non-adjacent. Thus the $\mathcal{P}$-components of $G\backslash X$ can be represented by $(c+1)!$ of the subgraphs (V_i, E_i) in the EMS sentence. A distinct subgraph (V_i, E_i) can then be used for any subset of X that belongs to $\mathcal{P}$. We need only choose the constant $c^* = (c+1)! + t_{c,k}$, where $t_{c,k}$ is the constant of Lemma 7.5. □

8 Conclusion

We have shown how a formula in EMS logic can encode GENERALIZED MATCHING over partial k-trees. From such a formula, one can automatically obtain a polynomial-time dynamic-programming algorithm that solves the problem by processing a width-k tree decomposition of the input graph. Borie [9] has already shown that GENERALIZED MATCHING can be solved in this way over partial k-trees; but he conjectured that EMS logic does not predict the existence of such an algorithm. Borie described a general technique to obtain polynomial-time dynamic-programming algorithms for a family of problems (including GENERALIZED MATCHING) that do not seem to be EMS-encodable. We have shown, however, that these problems are sometimes EMS-encodable—but not in a straightforward way.

Borie [9] also lists several examples of optimization problems (including CONNECTIVITY and SUBGRAPH CONNECTIVITY) that have polynomial-time algorithms on partial k-trees; and he suggests that these algorithms are predicted by neither EMS logic nor by his technique. However, EMS logic can encode those two optimization problems over partial k-trees. The CONNECTIVITY problem asks for the maximum c such that the graph is c-connected. Since the connectivity of a partial k-tree G cannot be greater than k, an EMS formula need only maximize the size of a set $C \subseteq V(G)$ such that the following is satisfied for every i, $1 \leq i \leq k+1$:

$$(\text{“}G \text{ is not } i\text{-connected”} \Rightarrow \text{“}C \text{ contains fewer than } i \text{ vertices”})$$

An EMS formula can encode the i-connectivity condition by saying that each pair of vertices are the endpoints of i internally-vertex-disjoint paths. Furthermore, it is not difficult to modify this example to encode SUBGRAPH CONNECTIVITY, which asks for the maximum c such that some subgraph of G is c-connected.

References

[1] S. Arnborg. Efficient algorithms for combinatorial problems on graphs with bounded decomposability. *BIT*, 25:2–33, 1985.

[2] S. Arnborg, J. Lagergren, and D. Seese. Easy problems for tree decomposable graphs. *J. Algorithms*, 12:308–340, 1991.

[3] S. Arnborg and A. Proskurowski. Linear time algorithms for NP-hard problems restricted to partial k-trees. *Disc. Appl. Math.*, 23:11–24, 1989.

[4] M.W. Bern, E.L. Lawler, and A.L. Wong. Linear-time computation of optimal subgraphs of decomposable graphs. *J. Algorithms*, 8:216–235, 1987.

[5] H.L. Bodlaender. Classes of graphs with bounded treewidth. Technical Report RUU-CS-86-22, Department of CS, Utrecht University, 1986.

[6] H.L. Bodlaender. Dynamic programming on graphs with bounded treewidth. In *Lecture Notes in Computer Science (Proc.* 15th *ICALP)*, volume 317, pages 105–119. Springer-Verlag, 1988.

[7] H.L. Bodlaender. Polynomial algorithms for graph isomorphism and chromatic index on partial k-trees. *J. Algorithms*, 11:631–643, 1990.

[8] H.L. Bodlaender. A linear time algorithm for finding tree-decompositions of small treewidth. *SIAM J. Comput.*, 25:1305–1317, 1996.

[9] R.B. Borie. Generation of polynomial-time algorithms for some optimization problems on tree-decomposable graphs. *Algorithmica*, 14:123–137, 1995.

[10] R.B. Borie, R.G. Parker, and C.A. Tovey. Automatic generation of linear-time algorithms from predicate calculus descriptions of problems on recursively constructed graph families. *Algorithmica*, 7:555–581, 1992.

[11] B. Courcelle. Graph rewriting: an algebraic and logic approach. In J. van Leeuwen, editor, *Handbook of Theoretical Computer Science*, volume B, pages 193–242. Elsevier, Amsterdam, 1990.

[12] B. Courcelle. The monadic second-order logic of graphs. I. Recognizable sets of finite graphs. *Information and Computation*, 85:12–75, 1990.

[13] B. Courcelle. The monadic second-order logic of graphs. V. On closing the gap between definability and recognizability. *Theoret. Comput. Sci.*, 80:153–202, 1991.

[14] B. Courcelle and M. Mosbah. Monadic second-order evaluations on tree-decomposable graphs. *Theoretical Computer Science*, 109:49–82, 1993.

[15] M.R. Garey and D.S. Johnson. *Computers and Intractability: A Guide to the Theory of NP-Completeness.* W.H. Freeman and Company, New York, 1979.

[16] A. Gupta, D. Kaller, S. Mahajan, and T. Shermer. Vertex partitioning problems on partial k-trees. In *Lecture Notes in Computer Science (Proc. 5th SWAT)*, volume 1097, pages 161–172. Springer-Verlag, 1996.

[17] A. Gupta, D. Kaller, S. Mahajan, and T. Shermer. A pumping lemma for tree automata, manuscript.

[18] A. Gupta, D. Kaller, and T. Shermer. Linear-time algorithms for partial k-tree complements. *Algorithmica*, to appear.

[19] C. Jordon. Sur les assemblages de lignes. *Journal Reine Angew. Math.* 70: 185–190, 1869.

[20] V. Kabanets. Recognizability equals definability for partial k-paths. In *Lecture Notes in Computer Science (Proc. 24th ICALP)*. Springer-Verlag, 1997.

[21] D. Kaller. *Monadic Second-Order Logic and Linear-Time Algorithms for Graphs of Bounded Treewidth.* PhD thesis, Simon Fraser University, Burnaby, B.C., Canada, 1996.

[22] D. Kaller. Definability equals recognizability of partial 3-trees and k-connected partial k-trees. *Algorithmica*, to appear.

[23] D.G. Kirkpatrick and P. Hell. On the complexity of a generalized matching problem. In *Proc. 10th STOC*, pages 240–245, 1978.

[24] J. Lagergren, October 1994. Personal communication.

[25] S. Mahajan and J.G. Peters. Regularity and locality in k-terminal graphs. *Disc. Appl. Math.*, 54:229–250, 1994.

[26] N. Robertson and P.D. Seymour. Graph minors. II. Algorithmic aspects of tree-width. *J. Algorithms*, 7:309–322, 1986.

[27] K. Takamizawa, T. Nishizeki, and N. Saito. Linear-time computability of combinatorial problems on series-parallel graphs. *JACM*, 29:623–641, 1982.

[28] J. van Leeuwen. Graph algorithms. In J. van Leeuwen, editor, *Handbook of Theoretical Computer Science*, volume A, pages 527–631. Elsevier, Amsterdam, 1990.

Author Affiliations:

Arvind Gupta, School of Computing Science, Simon Fraser University, Burnaby, BC, Canada, V5A 1S6. email: arvind@cs.sfu.ca. Research supported by the Natural Sciences and Engineering Research Council of Canada.

Damon Kaller, Department of Computer Science, Langara College, 100 West 49th Ave, Vancouver, BC, Canada, V5Y 2Z6. email: kaller@cs.sfu.ca.

Sanjeef Mahajan, Physical Design Department, LSI Logic, 1501 McCarthy Blvd, Milpitas, CA, USA. email: msanjeev@lsil.com.

Thomas Shermer, School of Computing Science, Simon Fraser University, Burnaby, BC, Canada, V5A 1S6. email: shermer@cs.sfu.ca. Research supported by the Natural Sciences and Engineering Research Council of Canada.

DIMACS Series in Discrete Mathematics
and Theoretical Computer Science
Volume **49**, 1999

Bases of Cocycle Lattices and Submatrices of a Hadamard Matrix

Winfried Hochstättler and Martin Loebl

ABSTRACT. We study the lattice $lat(M)$ of cocycles of a binary matroid M. By an isomorphism we show that such lattices are equivalent to lattices generated by the columns of proper submatrices of Sylvester matrices of full row length. As an application we show that the cocycle lattice of a recursively defined class of matroids, including all binary matroids of rank four, always has a basis consisting of cocycles.

1. Introduction

Our aim is to propose and study the following conjecture.

CONJECTURE 1. There exists a set of cocycles of each binary matroid B, which forms a basis of the lattice generated by the cocircuits of B.

Let $\mathcal{A}$ be a set of rational vectors of the same dimension d. The *lattice* $lat(\mathcal{A})$ generated by $\mathcal{A}$ is the set

$$\{\sum_{a\in\mathcal{A}} \lambda_a a \mid \lambda_a \in \mathbb{Z}\,\forall a \in \mathcal{A}\}$$

of all integer linear combinations of vectors of $\mathcal{A}$. A set A of linearly independent vectors (over the rational numbers) such that $lat(A) = lat(\mathcal{A})$ is called *a basis* of $lat(\mathcal{A})$. Every lattice generated by rational vectors has a basis (see [**9**] also for other facts on lattices).

Let $M = (E, \mathcal{C}^*)$ be a binary matroid given by the set of its cocircuits. The lattice generated by the characteristic vectors of cocircuits

$$lat(M) := lat\left(\{\chi^{C^*} \in \{0,1\}^E \mid \exists C^* \in \mathcal{C}^* \left(\chi_e^{C^*} = 1 \Leftrightarrow e \in C^*\right)\}\right)$$

will be called *cocircuit lattice* of M. Since any cocycle of M is the disjoint union of cocircuits of M we will also speak of $lat(M)$ as the *lattice of cocycles*. It is immediate that $lat(M)$ is isomorphic to the lattice $lat(\tilde{M})$ where $\tilde{M}$ is the simplification of M. Therefore, we will assume throughout the paper that M is simple, and consider the set E as an arbitrary but fixed representation of M, i.e. $E \subseteq GF(2)^{r(M)} \setminus \{0\}$.

1991 *Mathematics Subject Classification.* Primary 05B40; Secondary 11H31.

Key words and phrases. Geometry of Numbers, Binary Matroids, Lattice Basis, Hadamard matrices.

Part of this work was done while the authors were visiting the Department of Combinatorics & Optimization at the University of Waterloo.

We assume some familiarity with binary matroids. Our notation and terminology is fairly standard as in [**8**].

Let us survey some results related to our considerations.

Following Goddyn [**3**] we say that a simple binary matroid M has the *lattice of cocircuits property* if $2\mathbb{Z}^{|E|} \subseteq lat(M)$. It was pointed out by M. Laurent that not all binary matroids have this property, the Fano-plane F_7 being an example. Cunningham (see [**1**]) proved that if a binary matroid has no F_7 minor then it has the lattice of cocircuits property.

Previous investigations of cocircuit lattices were lead mostly from an algorithmic point of view: is there a polynomial-time algorithm to decide whether a vector belongs to the cocircuit lattice of a given binary matroid?

If $\mathcal{L}$ is a lattice then its *dual* $\mathcal{L}^*$ is the set of all vectors

$$\{v \in \mathbb{R}^d \mid v^T x \text{ is integer for all } x \in \mathcal{L}\}.$$

Lovász and Seress (see [**5**]) studied sets of generators of cocircuit lattices and their duals. Their first paper introduced us into the subject. The basic idea is as follows. Let A be a $GF(2)$ basis of cocycles of a binary matroid B of dimension r. Let A^k denote the set of all intersections of at most k elements of A. Then there exists $k \leq r$ such that $A \cup 2A^2 \cup ... \cup 2^k A^{k+1}$ generates the cocircuit lattice of B. Let $k(B)$ be the smallest k with that property. Then $k(B)$ is the smallest integer such that $2^{k(B)}w$ is integer for all vectors w of the dual of the cocircuit lattice of B. Also, $k(B)$ is the smallest integer such that $A^{k(B)+1}$ generate all vectors modulo 2. This provides a polynomial set of generators of the cocircuit lattice of a binary matroid B with $k(B)$ bounded.

Making a connection to (punctured) Reed-Muller codes, Lovász and Seress proved (see [**5, 6, 7**]):

1. If a binary matroid B has no F^s minor then $k(B) < s$.
2. Let integer k be fixed. Then given $B \subset F^d$ with $|B| = n$ and a vector $w \in 2^{-k}\mathbf{Z}^n$, it can be decided whether w belongs to the dual of the cocircuit lattice of B in time polynomial in d and n.
3. Moreover there is a polynomial time algorithm to decide whether $k(B) \geq k$.

Finally let us mention that it is proved by Galluccio and Loebl (see [**2**]) that the binary matroids with no F_7 minor satisfy Conjecture 1. Hochstättler observed that the duals of projective spaces, which have the lattice of cocircuits property and contain an F_7 minor satisfy 1 as well. These results are further generalized by Hochstättler, Laurent and Loebl in [**4**].

The paper is organized as follows. In Section 2 we derive a duality result for bases of columns of submatrices of certain Hadamard matrices of full row length. In Section 3 we show that Conjecture 1 can not be proven in a "naive way", namely by an extension of a basis of the cocycle lattice of a restriction minor. In Section 4 we prove Conjecture 1 for a recursively defined class of binary matroids. In Section 5 we show that this class contains all rank four binary matroids and name an example of a binary rank five matroid not in this class.

2. Submatrices of the Sylvester matrix and cocycles of binary matroids

We start this section recalling some facts about lattices and matrices.

DEFINITION 1. Let $m \geq r$ and let A be an integer $(r \times m)-$matrix of full rank. A subset B of columns of A is an *integer basis*, if each column of A is an integer linear combination of B, i.e. B is a basis of the lattice generated by the columns of A.

The *index* $ind(A)$ *of* A is the greatest common divisor of the determinants of all $(r \times r)$-submatrices of A.

The following is well known (see e.g. [**9**]).

PROPOSITION 1. *Let* $m \geq r$. *An* $(r \times m)$*-integer-matrix* A *of full rank has an integer basis of columns if and only if* A *has an* $(r \times r)$*-submatrix* B *such that* $|\det(B)| = ind(A)$.

Proof. By Cramer's rule we have

$$x = \sum \lambda_i b_i \Leftrightarrow \lambda_i = \frac{\det(b_1, \ldots, b_{i-1}, x, b_{i+1}, \ldots, b_r)}{\det(B)}.$$

□

The following observation about orthocomplementary integral matrices will become useful for us.

THEOREM 1. *Let* A_1 *be an* $(r \times m)$*-integer-matrix let* A_2 *be an* $((m-r) \times m)$*-integer-matrix such that* A_1, A_2 *are orthocomplementary, i.e. the concatenated matrix* $\begin{pmatrix} A_1 \\ A_2 \end{pmatrix}$ *has rank* m *and* $A_1 A_2^\top = 0$. *Let* (B, N) *be a partition of the columns such that* $|B| = r$. *Then* B *is an integer basis of* A_1 *if and only if* N *is an integer basis of* A_2.

Proof. Let $(A_1)_B$ be an integer basis of A_1. Then we have

$$\begin{aligned} A_1 A_2^\top = 0 \quad &\Leftrightarrow \quad (A_1)_B (A_2)_B^\top = -(A_1)_N (A_2)_N^\top \\ &\Leftrightarrow \quad (A_2)_B^\top = -(A_1)_B^{-1} (A_1)_N (A_2)_N^\top. \end{aligned}$$

By assumption $-(A_1)_B^{-1}(A_1)_N$ is an integer matrix. Therefore each column from $(A_2)_B^\top$ is an integer linear combination of columns of $(A_2)_N^\top$. □

We now turn our attention to a certain matrix (see eg. [**10**]) which has a well known relationship to binary projective spaces.

DEFINITION 2. A *Hadamard matrix of order* n is an $(n \times n)$-matrix H with entries in $\{+1, -1\}$ satisfying

$$HH^\top = nI_n.$$

Let $n = 2^m$. Then the Sylvester Hadamard matrix H_n is recursively defined as follows:

$$H_2 = \begin{pmatrix} +1 & -1 \\ -1 & -1 \end{pmatrix} \quad H_{2^m} = \begin{pmatrix} -H_{2^{m-1}} & +H_{2^{m-1}} \\ +H_{2^{m-1}} & +H_{2^{m-1}} \end{pmatrix}$$

We consider the rows and columns of H_{2^m} indexed by the elements of $GF(2)^m$ in lexicographic decreasing order. Then the following is immediate from the definition.

PROPOSITION 2. $(H_{2^m})_{x,y} = 1$ *if and only if* $x^\top y \equiv 1 \ mod\ 2$. □

Next, we examine some properties of submatrices of this Hadamard matrix. Let $E \subseteq GF(2)^m$ denote a subset of the binary space. By $H(E)$ we denote the matrix consisting of the rows of H_{2^m} indexed by elements from E. Now we relate this matrix to binary matroids. Let $M = (E, \mathcal{C}^*)$ be a simple binary matroid. As mentioned above we assume that $E \subseteq GF(2) \setminus \{0\}$ is a fixed representation of M. Consider the matrix $H(E)$ consisting of the rows indexed by elements from M and let $\tilde{H}(E)$ be the 0-1-matrix that arises from $H(E)$ by the elementwise operation $x \mapsto \frac{x+1}{2}$, if we, additionally, delete the last (zero-)column.

PROPOSITION 3. *A vector $\chi^{C^*} \in \{0,1\}^{|E|}$ is the characteristic vector of a cocycle of M if and only if it is a column of $\tilde{H}(E)$.*

Proof. A vector χ^{C^*} is the characteristic vector of a cocycle C^* of M if and only if it is in the row space of the matrix E representing M, i.e. there is a coefficient vector $y \in GF(2)^{r(M)}$ such that for all $x \in E$ we have $x^T y \equiv 1 \bmod 2 \Leftrightarrow x \in C^*$. The claim therefore follows from Proposition 2. □

This cocycle matrix now is orthocomplementary to a piece of the Hadamard matrix. An easy computation verifies the following.

PROPOSITION 4. *Let $\overline{E} := GF(2)^{r(M)} \setminus (E \cup \{0\})$ and $H'(\overline{E})$ the matrix derived from $H(\overline{E})$ by deleting the last column (of all -1s). Then $H'(\overline{E})$ and $\tilde{H}(E)$ satisfy the assumptions from Theorem 1.* □

We summarize some of the results of this section in the following theorem.

THEOREM 2. *Let $M = (E, \mathcal{C}^*)$ be a binary matroid. Then the following statements are pairwise equivalent*

i). *$lat(M)$ has an integer basis of cocycles and thus $\tilde{H}(E)$ has an integer basis of columns,*
ii). *$H'(\overline{E})$ has an integer basis of columns,*
iii). *$H(\overline{E})$ has an integer basis of columns not containing the all -1 column,*
iv). *$H(E \cup \{0\})$ has an integer basis of columns containing the all -1 vector,*
v). *the columns of $\tilde{H}(E \cup \{0\})$ and the all -1 vector have an integer basis containing the all -1 vector.*

Proof. The first equivalence is clear from Proposition 4 and Theorem 1. The second claim implies the third since the first $2^d - 1$ columns sum up to the all -1 column. By Theorem 1 we have that $H(E \cup \{0\})$ has an integer basis of columns containing the all -1 vector if and only if $H(\overline{E})$ has an integer basis of columns not containing the all -1 vector, which is exactly the third statement.

We are left to verify the equivalence to the last statement. So assume we have a basis B as in v.) Note that if x is a column of $H(E \cup \{0\})$ different from the -1 column then it can be written as $x = 2\chi^C + J$ where J is the all -1 vector and χ^C is a column of $\tilde{H}(E \cup \{0\})$. Thus we get

$$x = 2\chi^C + J = \mu J + 2 \sum_{C_b \in B \setminus J} \lambda_b \chi^{C_b} + J = \tilde{\mu} J + \sum_{C_b \in B \setminus J} \lambda_b (2\chi^{C_b} + J)$$

for some integers $\mu, \tilde{\mu}$.

Vice versa we get

$$\begin{aligned}\chi^C &= \frac{1}{2}(x - J) \\ &= \frac{\kappa - 1}{2} J + \frac{1}{2}\left(\sum_{b \in B \setminus J} \lambda_b x_b - J\right) \\ &= \sum_{b \in B \setminus J} \lambda_B \left(\frac{1}{2}(x_B - J)\right) + \frac{\tilde{\kappa}}{2} J,\end{aligned}$$

where $\frac{\tilde{\kappa}}{2}$ must be integral since all other summands in the equation are. □

We want to end this section with a remark on the vector of all 1s. If it is not a cocycle of M then the lattice generated by the cocycles and the vector of all -1s is the cocycle lattice of a matroid M' which we get by first coextending M by an element which makes each odd circuit even and then deleting it.

3. On solution adjustment

In this section we will show that an integer basis of the cocycle lattice in general can not be constructed in an augmentation process from a restriction minor.

By Theorem 2 the existence of an integer basis of the cocycle lattice of each matroid $M = (E, \mathcal{C}^*)$ where $E \subseteq GF(2)^{r(M)} \setminus \{0\}$ is equivalent to the existence of a basis of columns of submatrices of H_{2^m}. One might try to prove this existence by induction on the number of rows.

Therefore, we may restrict our attention to the case where some row is added to – or by Theorem 1 – deleted from a submatrix of H_n. Now consider the following matrix

$$A = \begin{pmatrix} 1 & 1 & 1 & 1 & -1 & -1 & 1 & 1 & 1 & -1 \\ 1 & 1 & -1 & 1 & 1 & 1 & -1 & -1 & -1 & 1 \\ 1 & 1 & 1 & -1 & -1 & -1 & 1 & 1 & -1 & -1 \\ 1 & 1 & 1 & -1 & 1 & -1 & -1 & -1 & 1 & 1 \\ 1 & 1 & -1 & 1 & -1 & -1 & 1 & -1 & -1 & 1 \\ 1 & 1 & 1 & -1 & 1 & -1 & -1 & 1 & -1 & -1 \\ 1 & -1 & 1 & 1 & -1 & 1 & -1 & -1 & -1 & -1 \\ -1 & 1 & -1 & -1 & -1 & 1 & -1 & -1 & 1 & -1 \\ -1 & -1 & 1 & 1 & 1 & -1 & -1 & -1 & -1 & -1 \\ -1 & -1 & -1 & -1 & -1 & -1 & -1 & -1 & 1 & -1 \end{pmatrix}$$

and note that $\det(A) = -512$. This is smallest possible for a (10×10)-matrix since one easily observes the following.

LEMMA 1. *Let M be a $d \times d$ (1,-1)-matrix. Then $2^{d-1} | \det(M)$.*

Proof. Adding the first row to the other $d - 1$ rows does not change the determinant. But it makes all entries not in the first row divisible by 2 and the claim follows by multilinearity of the determinant. □

Therefore, if A is a submatrix of a (10×2^m) submatrix of H_{2^m} for some m then it must be a basis by Proposition 1. The following Lemma shows that any $(1,-1)$-matrix can be embedded into some H_{2^m}.

LEMMA 2. *Let $M \in \{1,-1\}^{m\times n}$ be a $(1,-1)$-matrix. There is some d such that M is a submatrix of H_{2^d}.*

Proof. By Proposition 2 it suffices to give binary vectors R_i for each row i and C_j for each column j such that

$$m_{ij} = 1 \Leftrightarrow R_i^T C_j \equiv 1 \bmod 2.$$

Let $\tilde{M}$ denote the matrix which arises from M by replacing -1s by 0s. Let $\tilde{M}_j$ denote column j of $\tilde{M}$. For each column j we define

$$C_j = (\underbrace{0,\dots,0}_{1,\dots,m(j-1)},\ \tilde{M}_j,\ \underbrace{0,\dots,0}_{mj+1,\dots,mn}).$$

For each row i we set

$$R_i = \underbrace{(e_i,\dots,e_i)}_{n \text{ times}}$$

where e_i denotes the i-th unit vector of length m. □

Therefore, A is the basis of some matrix $H(M)$.

Assume now, for a contradiction, that an integer basis of the cocycle lattice of a binary matroid M could always be found by augmenting an integer basis of $M \setminus e$ by a single cocycle. By Theorem 1 this is equivalent to assuming that a basis of $H(\overline{M} \cup \{e\})$ could always be reduced to a basis of $H(\overline{M})$ by deleting a single column. In particular if we delete the last row of A, then the 10 reduced columns should contain a basis of the lattice which they generate. But, deleting the last row of A the first nine columns are linearly independent and multiplying with their inverse yields

$$\begin{pmatrix} 1&0&0&0&0&0&0&0&0&5/6\\ 0&1&0&0&0&0&0&0&0&-1/2\\ 0&0&1&0&0&0&0&0&0&-2/3\\ 0&0&0&1&0&0&0&0&0&-1/3\\ 0&0&0&0&1&0&0&0&0&5/6\\ 0&0&0&0&0&1&0&0&0&0\\ 0&0&0&0&0&0&1&0&0&5/6\\ 0&0&0&0&0&0&0&1&0&-2/3\\ 0&0&0&0&0&0&0&0&1&1/3 \end{pmatrix}.$$

From this it is clear that the reduced matrix does not have a basis of columns. This contradiction implies that

THEOREM 3. *There is a binary matroid $M = (E, \mathcal{C}^*)$ and an $e \in E$ such that the cocycle lattice $M \setminus e$ has an integer basis B and for any integer basis B' of the cocycle lattice of $M : B \not\subseteq B'$.*

4. Recursive monotone binary matroids

The recursive structure of the Sylvester matrix can be used to construct integer bases of the cocycle lattice of matroids with a recursive structure.

DEFINITION 3. A binary matroid $M = (E, \mathcal{C}^*)$ is *d-monotone*, if $r(M) \leq d$ and there is a hyperplane h of $GF(2)^d$ and a point $x \in (GF(2)^d \setminus h)$ such that one of the following maps is well-defined:

$$\begin{array}{ccc} E \setminus (h \cup \{x\}) & \rightarrow & E \cap h \\ a & \mapsto & a \oplus x, \end{array} \qquad \begin{array}{ccc} E \cap h & \rightarrow & E \setminus h \\ a & \mapsto & a \oplus x, \end{array}$$

where $\oplus$ denotes the (binary) sum in $GF(2)^d$. Note, that the definition of being d-monotone is independent of the representation chosen.

Call a binary matroid $M = (E, \mathcal{C}^*)$ *recursive d-monotone*, if $r(M) \leq d$ and, additionally, $d = 1$ or M is d-monotone with respect to some x and h, and the matroids defined by $E \cap h$ and $E \setminus (h \cup \{x\})$ are recursive $(d-1)$-monotone.

Note, that the empty set is recursive d-monotone for arbitrary d.

THEOREM 4. *Let $M = (E, \mathcal{C}^*)$ be recursive d-monotone. Then the matroid defined by $\bar{E}$ is also recursive d-monotone.*

Proof. We proceed by induction on d. By definition there is a hyperplane h of $GF(2)^d$ and a point $x \in (GF(2)^d \setminus h)$ such that one of the following maps is well-defined:

$$\begin{array}{ccc} E \setminus (h \cup \{x\}) & \rightarrow & E \cap h \\ a & \mapsto & a \oplus x, \end{array} \qquad \begin{array}{ccc} E \cap h & \rightarrow & E \setminus h \\ a & \mapsto & a \oplus x. \end{array}$$

Thus, also one of the maps

$$\begin{array}{ccc} \bar{E} \cap h & \rightarrow & \bar{E} \setminus h \\ a & \mapsto & a \oplus x. \end{array} \qquad \begin{array}{ccc} \bar{E} \setminus (h \cup \{x\}) & \rightarrow & \bar{E} \cap h \\ a & \mapsto & a \oplus x \end{array}$$

is well defined.

We may assume that the representation of M has been chosen such that $x = (1, 0, \ldots, 0)$ and $h = \{y \in GF(2)^d \mid y_1 = 0\}$. By definition the sets

$$E^h := \{v \in GF(2)^{d-1} \mid (1, v) \in E\} \; E_h := \{v \in GF(2)^{d-1} \mid (0, v) \in E\}$$

define matroids which are recursive $(d-1)$-monotone. By induction their complements, which are isomorphic to the domains of the above map are recursive $(d-1)$-monotone. □

We are now ready to apply the results of Section 2 to monotone matroids.

THEOREM 5. *Let $M = (E, \mathcal{C}^*)$ be a simple binary recursive d-monotone matroid. Then its cocycle lattice has a basis of cocycles.*

Proof. We will prove this by induction on d, considering two of four possible cases, the other cases being similar.

First, consider the case where the map

$$\begin{array}{ccc} E \cap h & \rightarrow & (E \setminus (h \cup \{x\})) \\ a & \mapsto & a \oplus x, \end{array}$$

is well defined and $x \notin E$.

By Theorem 2 it suffices to show that the matrix $H(\bar{E})$ has an integer basis of columns not containing the all -1 vector. Again we assume that the representation of M has been chosen such that $x = (1, 0, \dots, 0)$ and $h = \{y \in GF(2)^d \mid y_1 = 0\}$. Thus $H_d(\bar{E})$ has the following structure

$$H_d(\bar{E}) = \left(\begin{array}{ccc|c|ccc|c} & & & 1 & & & & -1 \\ & -A' & & 1 & & A' & & -1 \\ & & & 1 & & & & -1 \\ \hline 1 & 1 & 1 & 1 & -1 & -1 & -1 & -1 \\ \hline & & & -1 & & & & -1 \\ & A & & -1 & & A & & -1 \\ & & & -1 & & & & -1 \end{array}\right) \begin{array}{c} \\ \bar{E} \setminus (h \cup \{x\}) \\ \\ x \\ \\ h \cap \bar{E} \\ \\ \end{array} .$$

With the notation defined above we have $(A' | -1) = H_{d-1}(\bar{E}^h)$ and $(A, -1) = H_{d-1}(\bar{E}_h)$. Since the above map is well defined, A' is a submatrix of A. Therefore, by elementary row operations we can transform $H(\bar{E})$ into the matrix

$$H_2 = \left(\begin{array}{ccc|c|ccc|c} & & & 0 & & & & -1 \\ & 0 & & 0 & & A' & & -1 \\ & & & 0 & & & & -1 \\ \hline -1 & -1 & -1 & -1 & 1 & 1 & 1 & 1 \\ \hline & & & -1 & & & & -1 \\ & A & & -1 & & A & & -1 \\ & & & -1 & & & & -1 \end{array}\right).$$

By induction the matrix $H_{d-1}(\bar{E}^h)$ has an integer basis of columns not containing the all -1 vector. Furthermore, using induction and Theorems 2 and 4 we conclude that $H_{d-1}(\bar{E}_h \cup \{0\})$ has an integer basis of columns containing the all -1 vector. Therefore by elementary row operations and permutation of columns we can transform H_2 into

$$H_3 = \left(\begin{array}{cc|c|cc|c} 0 & 0 & 0 & I & B & v \\ \hline z^T & 0 & 1 & -z^T & 0 & -1 \\ \hline C & I & 0 & C & I & 0 \end{array}\right).$$

and finally

$$H_4 = \left(\begin{array}{cc|c|cc|c} 0 & 0 & 0 & I & B & v \\ \hline z^T & 0 & 1 & 0 & w^T & k \\ \hline C & I & 0 & 0 & D & u \end{array}\right).$$

with integer matrices B, C, D, integer vectors u, v, w, z and some integer k. Thus, the claim follows.

The second case that we consider, is when the map

$$\begin{array}{ccc} E \cap h & \to & (E \setminus (h \cup \{x\})) \\ a & \mapsto & a \oplus x, \end{array}$$

is well defined and $x \in E$.

Here, we show that the matrix $H_d(E \cup \{0\})$ has an integer basis of columns containing the all -1 vector. Again, we assume that the representation of M has been chosen such that $x = (1, 0, \ldots, 0)$ and $h = \{y \in GF(2)^d \mid y_1 = 0\}$. Then $H_d(E \cup \{0\})$ has the following structure

$$H_d(E\cup\{0\}) = \left(\begin{array}{ccc|c|ccc|c} & & & 1 & & & & -1 \\ & -A' & & 1 & & A' & & -1 \\ & & & 1 & & & & -1 \\ \hline 1 & 1 & 1 & 1 & -1 & -1 & -1 & -1 \\ \hline & & & -1 & & & & -1 \\ & A & & -1 & & A & & -1 \\ & & & -1 & & & & -1 \\ \hline -1 & -1 & -1 & -1 & -1 & -1 & -1 & -1 \end{array}\right) \begin{array}{c} \\ E\setminus(h\cup\{x\}) \\ \\ x \\ \\ E\cap h \\ \\ 0 \end{array}$$

and we proceed similar to the first case. □

5. The rank four case

In this section we will prove that all binary matroids of rank at most four are recursive 4-monotone, implying that their lattice has a basis of cocycles. Furthermore, we give an example of a rank five binary matroid which is not monotone.

It is immediate that subsets of the Fano-plane are always recursive 3-monotone. The rank four case will be treated by case checking.

LEMMA 3. *Let M be a binary matroid of rank four. Then M is 4-monotone.*

Proof. By Theorem 4 it suffices to consider matroids with at most seven elements. We have the following cases:

Case 1: All three-point-lines in M intersect in a point $a \in M$. If there is at most one point which is not on any three-point-line, then M is monotone with respect to a and any hyperplane h not containing a. Therefore, we may assume that there are at least two points $b, c \in M$ such that $b \oplus a, c \oplus a \notin M$. For any such pair of points we must have $(b \oplus a) \oplus (c \oplus a) = b \oplus c \notin M$. If $rank(\{e \oplus a \notin M \mid e \in M\}) \leq 2$ then M is monotone with respect to any hyperplane containing this line and a. Therefore, we may consider the $GF(2)$-span of three elements $b \oplus a, c \oplus a, d \oplus a$ from $\{e \oplus a \notin M \mid e \in M\}$ which are linearly independent. We are done if this hyperplane does not meet M. So assume it does and hence $a \oplus b \oplus c \oplus d \in M$ since by the above the binary sum of two of the generators cannot be in M. Since $b, c, d \in M$ we derive $d \oplus e \oplus a, c \oplus e \oplus a, c \oplus d \oplus a \notin M$. Therefore, $c \oplus d, d \oplus e$ and a span a hyperplane h which intersects M only in a. M now is monotone with respect to $b \oplus c \oplus d$ and h.

Case 2: There are three pairwise intersecting three-point-lines with empty total intersection. They span a hyperplane h satisfying $|h \cap E| = 6$. If $E \setminus h \neq \emptyset$ let x be that point, otherwise choose x arbitrarily and M is monotone with resp. to x and h.

Case 3: We are disjoint from Cases 1 and 2. So there are two three-point-lines $x, y, x \oplus y, a, b, a \oplus b \in E$ such that a, b, x, y are linearly independent. If there is no seventh point we are done, thus, by symmetry we may assume that $a \oplus x \in E$. Then M is monotone with respect to b and the hyperplane spanned by a, x and y.

□

Unfortunately, this does not extend to rank five matroids. Examining randomly generated examples by computer we found that the following matroid on 12 points is not monotone.

$$\begin{pmatrix} 1 & 0 & 0 & 0 & 0 & 1 & 1 & 1 & 0 & 0 & 1 & 1 \\ 0 & 1 & 0 & 0 & 0 & 1 & 1 & 0 & 0 & 1 & 1 & 0 \\ 0 & 0 & 1 & 0 & 0 & 0 & 0 & 1 & 1 & 0 & 1 & 0 \\ 0 & 0 & 0 & 1 & 0 & 1 & 1 & 0 & 1 & 1 & 0 & 0 \\ 0 & 0 & 0 & 0 & 1 & 0 & 1 & 0 & 0 & 1 & 0 & 1 \end{pmatrix}.$$

It is not clear to us whether the class of recursive monotone matroids is dense or sparse in the set of all binary matroids.

Acknowledgement. We would like to thank to Chris Godsil, Monique Laurent and Andras Sebö for helpful discussions.

References

[1] Cunningham, W. Chords and disjoint paths in matroids. *Discrete Mathematics 19* (1977), 7–15.

[2] Gallucio, A., and Loebl, M. Cycles of binary matroids without an F_7^*-minor. Tech. Rep. 431, IASI, 1996.

[3] Goddyn, L. A. Cones, lattices and Hilbert bases of circuits and perfect matchings. In *Graph Structure Theory: Proc. Joint Summer Conf. Graph Minors* (1993), no. 147 in Contemporary Mathematics, Amer. Math. Soc., pp. 419–440.

[4] Hochstättler, W., Laurent, M., and Loebl, M. Cycle bases for lattices of matroids with no Fano dual minor and their one-element extensions. *submitted.*

[5] Lovász, L., and Seress, A. The cocycle lattice of binary matroids. *European Journal of Combinatorics 14* (1993), 241–250.

[6] Lovász, L., and Seress, A. The cocycle lattice of binary matroids ii. *preprint* (1995).

[7] MacWilliams, F., and Sloane, N. *The Theory of Error-correcting Codes.* North-Holland, Amsterdam, 1977.

[8] Oxley, J. *Matroid Theory.* Oxford University Press, 1992.

[9] Schrijver, A. *Theory of Integer and Linear Programming.* Wiley, Chichester, 1986.

[10] Wallis, W., Street, A., and Wallis, J. *Combinatorics: Room Squares, Sum-Free Sets, Hadamard Matrices.* Springer, Berlin etc., 1972.

Center for Parallel Computing, Universität zu Köln, D–50923 Köln, GERMANY
E-mail address: `wh@zpr.uni-koeln.de`

Applied Mathematics (KAM), Charles University, 118 00 Praha 1, CZECH REPUBLIC
E-mail address: `loebl@kam.ms.mff.cuni.cz`

DIMACS Series in Discrete Mathematics
and Theoretical Computer Science
Volume **49**, 1999

On the Maximum Lengths of Davenport–Schinzel Sequences

Martin Klazar

Abstract

The quantity $N_5(n)$ is the maximum length of a finite sequence over n symbols which has no two identical consecutive elements and no 5-term alternating subsequence. Improving the constant factor in the previous bounds of Hart and Sharir, and of Sharir and Agarwal, we prove that

$$N_5(n) < 2n\alpha(n) + O(n\alpha(n)^{1/2}),$$

where $\alpha(n)$ is the inverse to the Ackermann function. Quantities $N_s(n)$ can be generalized and any finite sequence, not just an alternating one, can be assigned extremal function. We present a sequence with no 5-term alternating subsequence and with an extremal function $\gg n2^{\alpha(n)}$.

1 Introduction

Sequences are finite strings of symbols taken from a fixed infinite alphabet. For u a sequence, $|u|$ and $\|u\|$ denote its length and the number of its distinct symbols. Always $|u| \geq \|u\|$, in the case of equality u has no repeated symbol and it is called a *chain*. We say that $u = x_1x_2 \dots x_l$ is *sparse* if $x_i \neq x_{i+1}$ for each $i = 1, 2, \dots, l-1$. We say that u is *alternating* if $u = ababa \dots$ and $a \neq b$. The maximum length s of an alternating subsequence $x_{i_1}x_{i_2} \dots x_{i_s}, 1 \leq i_1 < i_2 < \dots < i_s \leq l$, in u is denoted $\mathrm{al}(u)$.

Sparse sequences u with bounded $\mathrm{al}(u)$ arise naturally in computational and combinatorial geometry. Davenport and Schinzel introduced them in 1965 [3] in connection with a geometric problem from control theory. They were interested in determining the quantities, s is fixed and $n \to \infty$,

$$N_s(n) = \max\{|u| : \; u \text{ is sparse \& } \mathrm{al}(u) < s \text{ \& } \|u\| \leq n\}. \qquad (1)$$

1991 *Mathematics subject Classification.* Primary 05D99, 11B83.

Supported by AvH Stiftung fellowship and by grants GAČR 0194/1996 and GAUK 194/1996.

It is trivial that $N_1(n) = 0, N_2(n) = 1$, and $N_3(n) = n$. It is easy to prove [3] that $N_4(n) = 2n-1$. For $s > 4$ things get complicated. We mention only few important results and suggest as further reading [10], [8], and the article of P. Valtr in this volume.

In 1986 Hart and Sharir [4] found the rough asymptotics of the fifth function:

$$n\alpha(n) \ll N_5(n) \ll n\alpha(n). \tag{2}$$

We remind that $f(n) \ll g(n)$ is an abbreviation for $f(n) < cg(n)$ where $n > n_0$ and $c > 0$ is a constant. The function $\alpha(n)$, the inverse to the Ackermann function, is integral, nondecreasing, and unbounded. Its growth to infinity is enormously slow. Agarwal, Sharir and Shor [2] gave later a similar bound to the sixth function:

$$n2^{\alpha(n)} \ll N_6(n) \ll n2^{\alpha(n)}. \tag{3}$$

They proved [2] strong (but not tight in the $\ll$ sense) upper and lower bounds to any $N_s(n), s > 6$.

In this paper we are concerned in the constant in the upper bound in (2). In Section 2 we prove the following estimate.

Theorem 1.1

$$N_5(n) < 2n\alpha(n) + O(n\alpha(n)^{1/2}). \tag{4}$$

Our constant 2 improves the constants 52 in [4], 68 in [10], and 4 in [7] (unpublished). The proof is selfcontained and all details are given. Those who are curious about the constant in the lower bound go to (17). In Section 3 we comment on the proof and pose a problem. Then we formulate a conjecture about growth rates of a generalization of $N_s(n)$ and support it by a consequence of the lower bound construction in (3).

2 The upper bound for $N_5(n)$

The proof of (4) follows. We use the techniques developed by Hart and Sharir [4], and by Sharir and Agarwal [10]. After the proof we will comment on lemmas and on our improvements.

We begin with the standard definition of $\alpha(n)$ and of related functions. All the functions $F_k(n)$ and $\alpha_k(n)$, $k = 1, 2, \ldots, \omega$, are mappings from $\{1, 2, \ldots\}$ to itself. First $F_1(n) = 2n$. For $k > 1$

$$F_k(n) = F_{k-1}(F_{k-1}(\ldots F_{k-1}(1)\ldots)) \ (n \text{ applications of } F_{k-1}). \tag{5}$$

For example,

$$F_2(n) = 2^n \text{ and } F_3(n) = 2^{2^{\cdot^{\cdot^{2}}}} \Big\} n.$$

For every $k, n \geq 1$ we have $F_k(n) \leq F_{k+1}(n)$ and $F_k(n) < F_k(n+1)$. The *Ackermann function* $F_\omega(n)$ is defined diagonally as $F_\omega(n) = F_n(n)$. The inverse functions are, $k = 1, 2, \ldots, \omega$,

$$\alpha_k(n) = \min\{m : \ F_k(m) \geq n\}. \tag{6}$$

Clearly, $\alpha_k(n) \geq \alpha_{k+1}(n)$ and $\alpha_k(n) \leq \alpha_k(n+1)$. The subscript of $\alpha_\omega(n)$ is usually omitted, $\alpha(n) = \alpha_\omega(n)$. For example,

$$\alpha_1(n) = \lceil n/2 \rceil \text{ and } \alpha_2(n) = \lceil \log_2 n \rceil \text{ (for } n > 1;\ \alpha_2(1) = 1).$$

Lemma 2.1 *For every $n \geq 3$ and $k \geq 2$ we have*

$$\alpha_k(\alpha_{k-1}(n)) = \alpha_k(n) - 1. \tag{7}$$

Proof. It is easy to check, using (6) and (5), that if $F_k(m) < n \leq F_k(m+1)$ then both sides of (7) are equal to m. □

Lemma 2.2 *For every $n \geq 1$ it holds*

$$\alpha_{\alpha(n)+1}(n) \leq 4. \tag{8}$$

Proof. First we show that for $k \geq 1$ and $n \geq 3$ it holds $F_{k+1}(n) \geq F_k(n+1)$. Indeed, $F_{k+1}(n) = F_k(F_{k+1}(n-1)) \geq F_k(F_2(n-1)) = F_k(2^{n-1}) \geq F_k(n+1)$. Applying repeatedly this inequality we obtain $F_{k+1}(3) \geq F_k(4) \geq \cdots \geq F_1(k+3) > k$. Thus,

$$F_{k+1}(4) = F_k(F_{k+1}(3)) > F_k(k).$$

Setting $k = \alpha(n)$ we obtain (8):

$$F_{\alpha(n)+1}(4) > F_{\alpha(n)}(\alpha(n)) = F_\omega(\alpha(n)) \geq n.$$

□

We introduce an important function $\psi(m,n)$. First few more definitions. The set of all symbols appearing in a sequence u is $S(u)$. If $u = x_1x_2\ldots x_l$ and x_i is such that $x_j \neq x_i$ for all $j < i$, x_i is said to be the first appearance (of the symbol x_i) in u. Last appearances are defined analogously. The subsequences of first and last appearances in u are denoted $F(u)$ and $L(u)$, respectively. Thus, $|F(u)| = |L(u)| = \|u\| = |S(u)|$. The *normal order* $(S(u), \prec)$ is the linear ordering of $S(u)$ by the natural order of $F(u)$, i.e. $x \prec y$ iff the first appearance of x in u precedes that of y.

Recall that a chain is a sequence with no repeated symbol. We say, for a positive integer m, that a sequence u *m-decomposes* if one can split u into m possibly empty chains $u = u_1u_2\ldots u_m$ such that each $u_i \backslash F(u)$ is decreasing (going from left to right) with respect to the normal order $(S(u), \prec)$. The function $\psi(m,n)$ is defined as

$$\psi(m,n) = \max\{|u| : u\ m\text{-decomposes } \&\ \mathrm{al}(u) < 5\ \&\ \|u\| \leq n\}.$$

We set $\psi(0,n) = \psi(m,0) = 0$. Note that $\psi(m,n)$ is nondecreasing in both variables.

Lemma 2.3 *Let $m, n, m_1, m_2, \ldots, m_j$ be positive integers, $j \geq 2$, such that $m = m_1 + m_2 + \cdots + m_j$. Then there exist nonnegative integers $n_0, n_1, \ldots, n_j$ such that $n = n_0 + n_1 + \cdots + n_j$ and*

$$\psi(m,n) \leq \sum_{i=1}^{j} \psi(m_i, n_i) + 2m + 2n_0 + \psi(j-1, n_0). \tag{9}$$

Proof. Suppose u is a sequence that m-decomposes, uses at most n symbols (in fact, it must be $\|u\| = n$), has no 5-term alternating subsequence, and has the maximum length $|u| = \psi(m,n)$. Let $u = u_1u_2 \dots u_m$ be its m-decomposition. For given j positive integers $m_1, \dots, m_j$ that sum up to m the first m_1 chains are concatenated to form the sequence v_1, the next m_2 chains are concatenated to form the sequence v_2 and so on. We obtain the splitting of u in j sequences $u = v_1v_2 \dots v_j$. Each v_i is partitioned into four subsequences (not necessarily contiguous blocks of v_i)

$$v_i = r_i \cup s_i \cup t_i \cup w_i$$

as follows. Subsequence r_i consists of all appearances of the symbols $x \in S(u)$ that appear only in v_i. We put $n_i = \|r_i\|$. Subsequence s_i consists of the appearances of the symbols that appear in v_i and before v_i but not after v_i. The remaining terms of v_i, i.e. the appearances of symbols appearing in v_i and after v_i and possibly before v_i, form the subsequence z_i. Then $t_i = z_i \backslash L(z_i)$ and $w_i = L(z_i)$. Let $n_0 = \|u \backslash r_1r_2 \dots r_j\|$.

Obviously, $n_0 + n_1 + \dots + n_j = n$. We estimate the contribution of each of the four subsequence types to the length of u. The intersections of r_i with the m_i chains forming v_i produce the m_i-decomposition of r_i, whence $|r_i| \leq \psi(m_i, n_i)$. Altogether,

$$|r_1r_2 \dots r_j| \leq \sum_{i=1}^{j} \psi(m_i, n_i). \tag{10}$$

To estimate the contribution of s_i's we observe first that $|(s_i \backslash F(s_i)) \cap u_k| \leq 1$ for each i and each chain u_k. Suppose to the contrary that $a \prec b$ are two symbols which appear in some $(s_i \backslash F(s_i)) \cap u_k$. Since a and b appear also before v_i and $a \prec b$, there is an ab subsequence before v_i. The first a in s_i appears before u_k. By the definition of m-decomposition, in $s_i \cap u_k$ we have a subsequence ba. We have a contradiction — the forbidden subsequence $ababa$. Thus, $|(s_i \backslash F(s_i)) \cap u_k| \leq 1$ and $|s_i \backslash F(s_i)| \leq m_i$. It follows from the definition of s_i that $S(s_i) \cap S(s_k) = \emptyset$ for $i \neq k$. Thus, $|F(s_1)F(s_2) \dots F(s_j)| \leq n_0$. Together

$$|s_1s_2 \dots s_j| \leq m + n_0. \tag{11}$$

As to t_i's, $|(t_i \backslash F(u)) \cap u_k| \leq 1$ for each i and each chain u_k. Suppose to the contrary that two symbols $a \neq b$ appear in $(t_i \backslash F(u)) \cap u_k$ in the order, say, ba. There is an a before u_k, namely the first a in u. By the definition of t_i, there is also a b in t_i after u_k (namely, the last b in t_i) and an a after v_i. These appearances form the forbidden subsequence $ababa$. Again, $|(t_i \backslash F(u)) \cap u_k| \leq 1$ and $|t_i \backslash F(u)| \leq m_i$. Since $|F(u) \cap t_1t_2 \dots t_j| \leq n_0$, we have again

$$|t_1t_2 \dots t_j| \leq m + n_0. \tag{12}$$

To estimate the last contribution we show that

$$w = w_1w_2 \dots w_j = w_1w_2 \dots w_{j-1}$$

is a $(j-1)$-decomposition of w. Clearly, $z_j = w_j = \emptyset$. Each w_i is a chain and we need to show only that $w_i \backslash F(w)$ decreases in the normal order $(S(w), \prec)$. Suppose not, then two distinct symbols a and b appear before some w_i in this order ab and in w_i also in the same order. By the definition of w_i, a appears (in u) also after v_i and we obtain again $ababa$. Hence, we have a $(j-1)$-decomposition and can estimate $|w|$ by ψ:

$$|w_1 w_2 \dots w_j| \le \psi(j-1, n_0). \tag{13}$$

Summing up (10), (11), (12), and (13), we obtain (9). □

Lemma 2.4 *For integers $k \ge 2$ and $m, n \ge 1$,*

$$\psi(m,n) \le 2k(\alpha_k(m)m + n). \tag{14}$$

Proof. We proceed by induction on k and for k fixed we use induction on m. The latter is started easily because by the trivial inequality $\psi(m,n) \le mn$ (14) is certainly true for $m \le 2k$. The induction on k starts with $k = 2$. We need to prove that

$$\psi(m,n) \le 4m\lceil \log_2 m \rceil + 4n. \tag{15}$$

Let $m \ge 2$ and let $m = m_1 + m_2$ where $m_1 = \lceil m/2 \rceil$ and $m_2 = \lfloor m/2 \rfloor$. By (9), there are n_0, n_1, and n_2 such that $n = n_0 + n_1 + n_2$ and

$$\begin{aligned} \psi(m,n) &\le \psi(m_1,n_1) + \psi(m_2,n_2) + 2m + 2n_0 + \psi(1,n_0) \\ &= \psi(m_1,n_1) + \psi(m_2,n_2) + 2m + 3n_0. \end{aligned}$$

We estimate $\psi(m_i, n_i)$ by the inductive assumption for m,

$$\psi(m,n) \le 4m_1\lceil \log_2 m_1 \rceil + 4m_2\lceil \log_2 m_2 \rceil + 4n_1 + 4n_2 + 2m + 3n_0.$$

Since $4n_1 + 4n_2 + 3n_0 \le 4n$, it suffices to show

$$m_1\lceil \log_2 m_1 \rceil + m_2\lceil \log_2 m_2 \rceil \le m(\lceil \log_2 m \rceil - 1).$$

The last inequality is immediate to check, thus (15) holds.

For $k > 2$ and $m \ge 3$ we apply (9) with the partition $m = m_1 + m_2 + \dots + m_j$, where $j = \lceil m/\alpha_{k-1}(m) \rceil > 1$, $m_1 = \dots = m_{j-1} = \alpha_{k-1}(m)$, and $1 \le m_j \le \alpha_{k-1}(m)$. By (9), there are n_i, $i = 0, 1, \dots, j$, that sum up to n and

$$\psi(m,n) \le \sum_{i=1}^{j} \psi(m_i, n_i) + 2(m + n_0) + \psi(j-1, n_0).$$

Each $\psi(m_i, n_i)$ is estimated by (14) (induction on m) for the current k, $\psi(j-1, n_0)$ is estimated by (14) for $k-1$. By the definition of j,

$$(j-1)\alpha_{k-1}(j-1) \le (j-1)\alpha_{k-1}(m) \le m.$$

By (7),

$$\alpha_k(m_i) \le \alpha_k(\alpha_{k-1}(m)) = \alpha_k(m) - 1.$$

Thus,

$$\begin{aligned}\psi(m,n) &\leq \sum_{i=1}^{j} 2k(m_i\alpha_k(m_i)+n_i)+2(m+n_0)\\ &\quad +2(k-1)((j-1)\alpha_{k-1}(j-1)+n_0)\\ &\leq 2km(\alpha_k(m)-1)+2k(n-n_0)\\ &\quad +2(m+n_0)+2(k-1)(m+n_0)\\ &= 2k(m\alpha_k(m)+n).\end{aligned}$$

□

Lemma 2.5 *For all positive integers $l \geq 2$ and n,*

$$N_5(n) \leq \psi(\lceil 2n/l\rceil, n) + 2l(l-1)\lceil 2n/l\rceil. \tag{16}$$

Proof. Let u be a sparse sequence with $\mathrm{al}(u) < 5$, $|u| = N_5(n)$, and $\|u\| \leq n$ (thus, $\|u\| = n$). *Bad elements* are the elements in $F(u) \cup L(u)$. *Repetition* $I(a), a \in S(u)$, is any subinterval in u that begins and ends with a and has no a inside. Note that the interior of each $I(a)$ is nonempty because u is sparse.

Consider the splitting $u = u_1u_2\dots u_j$ in which each u_i starts with a bad element and contains, for $1 \leq i \leq j-1$, exactly l bad elements. The last block u_j may contain fewer bad elements. Hence, $j \leq \lceil 2n/l\rceil$. We claim that there are at most $(2l-1)(l-1)$ repetitions in each u_i.

Suppose, for the contrary, that u_i contains $(2l-1)(l-1)+1$ repetitions. There cannot be l repetitions with mutually disjoint interiors, otherwise we would have a repetition $I(a)$ in u_i having inside no bad element. But this forces the forbidden subsequence $babab$. Hence, for each symbol a there are at most $l-1$ repetitions $I(a)$ of a in u_i. It follows that in u_i there are l repetitions $I(a_1), I(a_2), \dots, I(a_l)$ where $a_1, a_2, \dots, a_l$ are l distinct symbols that are in addition distinct to those at most l symbols appearing in u_i as bad elements. Two of these repetitions, say $I(a_1)$ and $I(a_2)$, must intersect. Say a_1 appears inside $I(a_2)$. This again forces the forbidden subsequence $a_1a_2a_1a_2a_1$ because a_1 appears before and after u_i. Again a contradiction.

Therefore, $|u_i| - \|u_i\| \leq (2l-1)(l-1)$. Deleting all terms from u_i except $F(u_i)$ we delete at most $(2l-1)(l-1)$ elements and turn u_i into a chain. We obtain the splitting into j chains

$$v = F(u_1)F(u_2)\dots F(u_j),$$

where $|v| \geq |u| - (2l-1)(l-1)j$.

Finally, we delete $L(v)$. We have the splitting into j chains

$$w = w_1w_2\dots w_j,$$

where $w_i = F(u_i)\backslash L(v)$ and $|w| \geq |u| - (2l-1)(l-1)j - n$. We show it is a j-decomposition of w. If not then $a \prec b$ are two elements from $(S(w), \prec)$ that appear in some $w_i\backslash F(w)$ in the order ab. We have ab before w_i (the elements in

$F(w)$), ab in w_i and an a after w_i (the element in $L(v)$). Thus u contains $ababa$, a contradiction. The splitting of w is a j-decomposition and (16) follows:

$$|u| \leq |w| + (2l-1)(l-1)j + n \leq \psi(\lceil 2n/l \rceil, n) + 2l(l-1)\lceil 2n/l \rceil.$$

□

From (14), setting $k = \alpha(m) + 1$, we obtain, using (8),

$$\psi(m, n) \leq 8m\alpha(m) + 8m + 2n\alpha(m) + 2n.$$

Using this bound in (16) with $l = \lfloor \alpha(n)^{1/2} \rfloor$ we obtain

$$\begin{aligned} N_5(n) &\leq \psi(\lfloor 2n/l \rfloor, n) + 2l(l-1)\lfloor 2n/l \rfloor \\ &\leq 8\alpha(\lfloor 2n/l \rfloor)\lfloor 2n/l \rfloor + 2\alpha(\lfloor 2n/l \rfloor)n \\ &\quad +8\lfloor 2n/l \rfloor + 2n + 2l(l-1)\lfloor 2n/l \rfloor \\ &\leq 2n\alpha(n) + O(n\alpha(n)^{1/2}). \end{aligned}$$

This finishes the proof of (4).

3 Concluding comments and remarks

Lemma 2.1 is standard. Lemma 2.2 was proved in Appendix 1 in [2], see also [10]. Function $\psi(m, n)$ and Lemma 2.3 form the heart of the proof. The coefficient at n_0 in (9) is the crucial one because it produces the same constant factor in (4). The coefficient at m is irrelevant. Our $\psi(m, n)$ is a combination of the versions in [4] and [10]. From [4] we took the idea of ordered chains. Our proof of Lemma 2.3 is inspired by the ingenious proof in [4]. However, the normal order $(S(u), \prec)$ is not essential and one can obtain 2 at n_0 working only with unordered chains in the spirit of [10] (in [10] there is 4 at n_0). For unordered chains one can use in the proof of Lemma 2.3 the partition of v_i

$$v_i = \overline{r_i} \cup \overline{s_i} \cup \overline{t_i} \cup \overline{w_i},$$

where

$$\overline{r_i} = r_i,\ \overline{s_i} = s_i \backslash F(s_i),\ \overline{t_i} = t_i,\ \text{and}\ \overline{w_i} = w_i \cup F(s_i).$$

A little technical complication for the proof of Lemma 2.4 is that then $j - 1$ in (9) increases to j. We leave it as an exercise for the interested reader to fill in the details. Lemma 2.4 is similar to the corresponding lemmas in [4] and [10]. The main improvement is Lemma 2.5 ([7]); [4] and [10] use the instance with $l = 1$.

As to the constant factor in the lower bound in (2), in 1988 Wiernik and Sharir [11] proved that

$$N_5(n) \geq \tfrac{1}{2}n\alpha(n) - 2n. \qquad (17)$$

See also pp. 21–29 in [10]. Estimates (4) and (17) suggest the following problem.

Problem 3.1 *Does the limit*

$$\lim_{n \to \infty} \frac{N_5(n)}{n\alpha(n)}$$

exist?

If it exists then it lies in the interval $[1/2, 2]$. An easier problem might be to narrow this interval.

In [1] the following generalization of $N_s(n)$ was proposed. Two sequences $v = a_1a_2 \dots a_k$ and $w = b_1b_2 \dots b_k$ of the same length are *equivalent* if, for each i and j, $a_i = a_j$ iff $b_i = b_j$. A sequence v is *contained* in other sequence u if u has a subsequence equivalent to v. We denote this relation as $v \prec u$. Alternating sequence *abab*... of length s is denoted al_s. Note that $\mathrm{al}(u) < s$ expresses in the new notation as $\mathrm{al}_s \not\prec u$. We say that u is *k-sparse* if each interval in u of length $\le k$ is a chain. We have extended [1] the definition (1) to any sequence v:

$$\mathrm{Ex}(v, n) = \max\{|u| :\ u \text{ is } \|v\|\text{-sparse } \& \ v \not\prec u \ \& \ \|u\| \le n\}.$$

Note that $N_s(n) = \mathrm{Ex}(\mathrm{al}_s, n)$.

The next two bounds are the basic facts abouth the growth rates of $\mathrm{Ex}(v, n)$.

$$\forall c \, \exists s \ N_s(n) = \mathrm{Ex}(\mathrm{al}_s, n) \gg n2^{\alpha(n)^c} \quad \text{and} \tag{18}$$

$$\forall v \, \exists c \ \mathrm{Ex}(v, n) \ll n2^{\alpha(n)^c}. \tag{19}$$

(18) was proved in [2] and (19) in [5] (both results are actually stronger). Since $u \prec v$ implies easily $\mathrm{Ex}(u, n) \ll \mathrm{Ex}(v, n)$ (see [1]; this is *not* true with $\le$ in place of $\ll$), it follows from (18) that the containment $\mathrm{al}_s \prec v$ for big s makes $\mathrm{Ex}(v, n)$ grow "fast". Perhaps $\mathrm{Ex}(v, n)$ can grow "fast" even if $v \not\succ \mathrm{al}_5 = ababa$.

Problem 3.2 *We conjecture that*

$$\forall c \, \exists v \ ababa \not\prec v \ \& \ \mathrm{Ex}(v, n) \gg n2^{\alpha(n)^c}. \tag{20}$$

In [9] (for details see [6]) a sequence v was presented, namely $v = abcbadadbcd$, with $ababa \not\prec v$ and $\mathrm{Ex}(v, n) \gg n\alpha(n)$. To support the conjecture even more we show now that (20) is true for $c = 1$.

We make use of the construction of Agarwal, Sharir and Shor [2] proving the lower bound in (3). We describe it as on pp. 53–54 in [10]. A *fan*, more precisely an *m-fan*, is any sequence of length $2m - 1$ equivalent to the sequence $1\ 2\ \dots\ (m - 1)\ m\ (m - 1) \dots\ 2\ 1$. We define by double induction a two-dimensional array $(S(k, m))_{k,m=1}^{\infty}$ of sequences. $S(k, m)$ is sparse and is a concatenation of several m-fans (their number will be uniquely determined by induction). One symbol will appear typically in more fans of $S(k, m)$.

$S(1, m)$ consists of just one m-fan. $S(k, 1), k > 1$, equals to $S(k - 1, 2^{k-1})$, where each 2^{k-1}-fan is regarded in $S(k, 1)$ as $2^k - 1$ 1-fans. The sequence $S(k, m)$ for $k, m > 1$ is obtained from $T = S(k, m - 1)$ and $U = S(k - 1, M)$, where M is the number of $(m - 1)$-fans in T. Suppose U has p M-fans. Create $2p$ copies of T (with disjoint sets of symbols which are also disjoint to the set of symbols of U) $T_1, \dots, T_{2p}$ and merge them with U as follows. First double the middle element in each fan in each T_i and in each fan of U. Then separate the twins in the middle of the k-th expanded $(m - 1)$-fan of T_{2i-1} by the k-th element of the first half of the i-th expanded M-fan of U (this way an m-fan is obtained). The k-th element (counted from the left) of the second half does the same job in T_{2i}. Denote the modified copies as T_i^m. Set $S(k, m) = T_1^m T_2^m \dots T_{2p}^m$.

It can be shown (in Lemma 3.1 we prove a more general statement) that $\mathrm{al}_6 \not\prec S(k,m)$ for all k and m. One can construct — for details see pp. 52–56 in [10] — an infinite sequence of sequences

$$(u_1, u_2, \ldots) \tag{21}$$

with the following properties. Each u_i equals to some $S(k,m)$ (thus u_i is sparse and $\mathrm{al}_6 \not\prec u_i$), $\|u_i\| = n_i < n_{i+1} = \|u_{i+1}\|$, and $|u_i| \gg n_i 2^{\alpha(n_i)}$.

For a sequence u an oriented graph $D(u) = (V, E)$ is defined by $V = S(u)$ (the symbols of u) and $a \to b$ iff $abba$ is a subsequence of u. For example, $D(\mathrm{al}_6)$ is $a \leftrightarrow b$. We remind that an oriented graph is *strongly connected* if each two distinct vertices x_1 and x_2 can be joined by a directed path going from x_1 to x_2.

Lemma 3.1 *Suppose u is a sparse sequence, $\|u\| > 1$, and $D(u)$ is strongly connected. Then $u \not\prec S(k,m)$ for all k and m.*

Proof. By double induction on k and m. Obviously, $u \not\prec S(1,m)$. By induction, $u \not\prec S(k,1) = S(k-1, 2^{k-1})$. It remains to show that $u \not\prec S(k,m)$ provided $u \not\prec T = S(k, m-1)$ and $u \not\prec U = S(k-1, M)$. Suppose v is a subsequence of $S(k,m)$ equivalent to u. It follows easily from the construction that if $x \in S(v)$ comes from a copy of T (with expanded fans) and $x \to y$ in $D(v) = D(u)$, then y must come from the same copy of T. Because $D(v)$ is strongly connected, the whole v comes from a copy of T with expanded fans or from U with expanded fans. Because u is sparse, u is contained already in T or in U. □

Lemma 3.2 *For u from the previous lemma*

$$\mathrm{Ex}(u,n) \gg n2^{\alpha(n)}.$$

Proof. Consider the sequences (21). We have $|u_i| \gg n_i 2^{\alpha(n_i)}$ and, by the previous lemma, $u \not\prec u_i$. There are two small troubles. The first is that u_i is sparse but may not be $\|u\|$-sparse. Taking from u_i an appropriate subsequence we can keep a constant fraction of length and achieve $\|u\|$-simplicity (we use that $\mathrm{al}_6 \not\prec u_i$). We leave this to the reader as an exercise; see [1] for this technique. Second, we need the lower bound $|u_i| \gg n_i 2^{\alpha(n_i)}$ for all n and not only for infinitely many. This is achieved by the same interpolation as in [10]. □

Now consider the sequence

$$u^* = abcbadadbecfcfedef,$$

$S(u^*) = \{a,b,c,d,e,f\}$. It does not contain $ababa$ but at the same time it satisfies the hypothesis of Lemma 3.1 since it is sparse and $D(u^*)$ contains the oriented Hamiltonian cycle $abdfec$. Thus, by Lemma 3.2, u^* witnesses (20) for $c = 1$.

One cannot stregthen the conjecture (20) by replacing $ababa$ with $abab$. It follows from the results in [9] that

$$abab \not\prec v \Rightarrow \mathrm{Ex}(v,n) \ll n.$$

References

[1] R. Adamec, M. Klazar and P. Valtr, *Generalized Davenport–Schinzel sequences with linear upper bound,* Discrete Math. **108** (1992), 219–229.

[2] P.K. Agarwal, M. Sharir and P. Shor, *Sharp upper and lower bounds on the length of general Davenport–Schinzel sequences,* J. Combinatorial Theory Ser. A **52** (1989), 228–274.

[3] H. Davenport and A. Schinzel, *A combinatorial problem connected with differential equations,* Amer. J. Math. **87** (1965), 684–694.

[4] S. Hart and M. Sharir, *Nonlinearity of Davenport–Schinzel sequences and of generalized path compression schemes,* Combinatorica **6** (1986), 151–177.

[5] M. Klazar, *A general upper bound in Extremal theory of sequences,* Comment. Math. Univ. Carolin. **33** (1992), 737–746.

[6] M. Klazar, *Two results on a partial ordering of finite sequences,* Comment. Math. Univ. Carolin. **34** (1993), 697–705.

[7] M. Klazar, *Combinatorial aspects of Davenport–Schinzel sequences,* Doctoral thesis, Charles University, Prague, 1995.

[8] M. Klazar, *Combinatorial aspects of Davenport–Schinzel sequences,* Discrete Math. **165/166** (1997), 431–445.

[9] M. Klazar and P. Valtr, *Generalized Davenport–Schinzel sequences,* Combinatorica **14** (1994), 463–476.

[10] M. Sharir and P.K. Agarwal, *Davenport–Schinzel sequences and their geometric applications,* Cambridge University Press, Cambridge, 1995.

[11] A. Wiernik and M. Sharir, *Planar realization of nonlinear Davenport–Schinzel sequences by segments,* Discrete Comput. Geom. **3** (1988), 15–47.

Forschungsinstitut für Diskrete Mathematik, Lennéstr. 2, 53113 Bonn, Germany

Current address : Department of Applied Mathematics, Malostranské nám. 25, 118 00 Praha, Czech Republic

Email address: klazar@kam.ms.mff.cuni.cz

DIMACS Series in Discrete Mathematics
and Theoretical Computer Science
Volume **49**, 1999

On the minimum number of edges giving maximum oriented chromatic number

Alexandr V. Kostochka, Tomasz Łuczak,
Gábor Simonyi and Eric Sopena

ABSTRACT. We show that the minimum number of edges in a graph on n vertices with oriented chromatic number n is $(1 + o(1))n \log_2 n$.

In 1995, in a conversation with the French member of the set of the authors of this note, Pál Erdős asked about the minimal number of edges a graph on n vertices with oriented chromatic number n can have. During the conference on the Future of Discrete Mathematics in the cosy but fruitful atmosphere of the Štiřín Castle we found an elementary answer to this question which we present below.[1]

We consider first an extremal question closely related to Erdős's problem. A colouring of the vertices of a graph G whose edges are initially coloured with k colours is *admissible*, if every colour class spans an independent set and each pair of colour classes is joined by edges of one colour only. After Alon and Marshall [1], we define the *k-chromatic number $\chi_k(G)$ of G* as the minimal number ℓ such that for any colouring of the edges of G with k colours there exists an admissible ℓ-colouring of the vertices of G. Finally, by $f_k(n)$ we denote the minimum number of edges in a graph G on n vertices for which $\chi_k(G) = n$. We shall show that for any fixed k the function $f_k(n)$ grows roughly as $n \log_k n$.

Theorem 1. *For a fixed $k \geq 2$ and n large enough*

$$n(\log_k n - 4 \log_k \log_k n - 5) \leq f_k(n) \leq \lceil \log_k n \rceil (n - \lceil \log_k n \rceil) .$$

1991 *Mathematics Subject Classification.* 05C15, 05C20, 05C35.

The first author was partially supported by the grant 96-01-01614 of the Russian Foundation for Fundamental Research and by the Cooperative Grant Award RM1-181 of the US Civilian Research and Development Foundation. The second author acknowledges support of KBN grant 2 P03A 023 09. The research of the third author was partially supported by the Hungarian National Foundation for Scientific Research (OTKA) Grant Nos. F023442 and T016386 while the forth author was partially supported by the Barrande Grant no. 97137.

[1]After the final version of this note had been sent to the publisher we were informed that a very similar result had been proved independently by Z. Füredi, P. Horak, C.M. Pareek and X. Zhu in the paper *Minimal oriented graphs of diameter 2*, which is to appear in *Graphs and Combinatorics.*

The argument we use in the proof of Theorem 1 in the case when $k = 2$ can be easily employed to answer Erdős's original question. Recall (see [2–4]) that the oriented chromatic number of a graph G is defined as the smallest number ℓ, such that for every orientation $\vec{G}$ of G there exists a tournament $\vec{T}(\vec{G})$ on ℓ vertices such that $\vec{G}$ can be homomorphically embedded into $\vec{T}(\vec{G})$.

Theorem 2. *Let $g(n)$ be the smallest number for which there exists a graph G with n vertices, $g(n)$ edges, and oriented chromatic number n. Then, for large enough n, we have*

$$n(\log_2 n - 4\log_2 \log_2 n - 5) \le g(n) \le \lceil \log_2 n \rceil (n - \lceil \log_2 n \rceil)\,.$$

Proof of Theorem 1. In order to see the upper bound for $f_k(n)$ observe that the complete bipartite graph on n vertices, with bipartition (W', W''), where $|W'| = \lceil \log_k n \rceil$ and $|W''| = n - \lceil \log_k n \rceil$, has k-chromatic number n. Indeed, label elements of W' by $1, \ldots, \lceil \log_k n \rceil$ and elements of W'' by $0, 1, \ldots, n - \lceil \log_k n \rceil - 1$. Now colour the edge $\{i', i''\}$, where $i' \in W'$, $i'' \in W''$, with the jth colour, $j = 0, 1, \ldots, k-1$, if j appears at the i'th position in the expansion of i'' in the k-ary system. Note that for each pair of vertices $x', y' \in W'$ there exists $z'' \in W''$ such that the edges $\{x', z''\}$ and $\{y', z''\}$ are coloured with different colours; it is enough to take z'' which has different digits at positions x' and y'. Similarly, for every $x'', y'' \in W''$ one can find a position z' at which the digits of the k-ary expansions of x'' and y'' differ; then the colours of the edges $\{x'', z'\}$ and $\{y'', z'\}$ must differ as well. Hence, every admissible colouring of the vertices of such a coloured graph must use different colours for different vertices, i.e. the k-chromatic number of the graph equals n.

The proof of the lower bound for $f_k(n)$ is slightly less immediate. Let G be a graph with n vertices and

$$e(G) \le n(\log_k n - 4\log_k \log_k n - 5)$$

edges, which are coloured with k colours. We need to show that for each such colouring there exists an admissible colouring of the vertices of G which uses only $n-1$ colours.

Our argument will be based on the following observation (see Tuza [5]).

Claim. *If the edges of the complete graph K_n are covered by a family of k-partite graphs $G_1, \ldots, G_m$, of $r_1, \ldots, r_m$ vertices, respectively, then $\sum_i r_i \ge n \log_k n$.* □

Let us partition the vertices of G into two classes W' and W'', where W' consists of all vertices of G of degree at least $\log_k^2 n$. Then

$$|W'| \le \frac{2e(G)}{\log_k^2 n} < \frac{2n}{\log_k n}\,,$$

and so $|W''| \ge n(1 - 2/\log_k n)$. Now, for every vertex $w' \in W'$ we define a complete k-partite graph $G_{w'}$, choosing as the ith set of the k-partition of $G_{w'}$ the set of all vertices of W'' which are connected to w' by edges of the ith colour. Thus, the total number of vertices of $G_{w'}$ is the same as the number of neighbours of w' in W''.

Furthermore, let H'' be the graph with vertex set W'', in which two vertices are adjacent if they lie within distance two in the subgraph $G[W'']$ induced by W'' in G. Note that every graph F contains a k-partite subgraph F' such that the maximum degree of the graph obtained from F by removing the edges of F' is at most $\lfloor \Delta(F)/k \rfloor$. Thus, since the maximum degree $\Delta(H'')$ of H'' is bounded from above by $\Delta^2(G[W'']) \le \log_k^4 n$, H'' can be decomposed into

$$m'' \le 1 + \log_k (\log_k n)^4 = 4 \log_k \log_k n + 1$$

k-partite subgraphs $H_1, \ldots, H_{m''}$. Observe also that the sum of the orders of all graphs $\{G_{w'}\}_{w' \in W'}$ and $\{H_i\}_{i=1}^{m''}$ can be bounded from above by

$$\begin{aligned}
\sum_{w' \in W'} |G_{w'}| + \sum_{i=1}^{m''} |H_i| \\
&\le e(G) + |W''|(4 \log_k \log_k n + 1) \\
&\le |W''|\Big(\frac{\log_k n - 4 \log_k \log_k n - 5}{1 - 2/\log_k n} + 4 \log_k \log_k n + 1\Big) \\
&\le |W''|\Big\{\big(\log_k n - 4 \log_k \log_k n - 5\big)\Big[1 + \frac{2}{\log_k n} + O\Big(\frac{1}{\log_k^2 n}\Big)\Big] \\
&\qquad\qquad + 4 \log_k \log_k n + 1\Big\} \\
&< |W''|\big(\log_k n - 1\big) < |W''| \log_k |W''| \,.
\end{aligned}$$

Hence, the Claim implies that there exists a pair of vertices $\{x'', y''\}$ which appears as an edge in none of the graphs $\{G_{w'}\}_{w' \in W'}$ and $\{H_i\}_{i=1}^{m''}$, i.e. vertices x'' and y'' are not adjacent, they have no common neighbours in W'', and for each of their common neighbours $z' \in W'$ both edges $\{x'', z'\}$ and $\{y'', z'\}$ are coloured with the same colour. Thus, the colouring in which x'' and y'' are coloured with the same colour, while all other vertices of G receive different colours, is admissible. Consequently, $\chi_k(G) \le n - 1$ and the assertion follows. □

Proof of Theorem 2. Note that a graph G on n vertices has oriented chromatic number n if and only if in some orientation $\vec{G}$ of G each pair of non-adjacent vertices of G is connected by a directed path of length two. Now to estimate $g(n)$ from above take the 2-coloured complete bipartite graph with bipartition (W', W'') used in the proof of Theorem 1 to get an upper bound for $f_2(n)$, the edges coloured with the first colour direct from W' to W'', for the edges of the second colour choose the other direction.

Similarly, to get a lower bound for $g(n)$ divide the set of vertices of a graph G into two classes W' and W'', where W' consists of all vertices of degree at least $\log_2^2 n$. Then one can mimic the proof of Theorem 1 and show that in order to orient all edges in such a way that all non-adjacent vertices of W'' are joined by a directed path of length two, G must have at least $n(\log_2 n - 4 \log_2 \log_2 n - 5)$ edges. □

We conclude with a few words on the expected behaviour of $f_k(n)$ and $g(n)$. It is tempting to conjecture that the upper bounds given in Theorems 1 and 2 are close to the truth and $f_k(n) = n\log_k n + O(n)$ and $g(n) = n\log_2 n + O(n)$, where the term $O(n)$ depends on the arithmetic properties of n. On the other hand, we should mention that the elementary bipartite construction we used to get the upper bound for $f_k(n)$ is far from being best possible. For instance, one can modify it slightly by deleting one vertex from the smaller set of the bipartition at the same time adding a perfect matching to the larger of the sets; the graph obtained in such a way has k-chromatic number $n-1$ and, typically, less than $\lceil\log_k(n-1)\rceil(n-1-\lceil\log_k(n-1)\rceil)$ edges. Furthermore, if we have a family of graphs $G_1, G_2, \ldots, G_r$, such that G_i has n_i vertices and k-chromatic number n_i, then the graph $1 + \sum_{i=1}^r G_i$ obtained by taking disjoint copies of $G_1, G_2, \ldots, G_r$ and adding to it one more vertex of degree $\sum_{i=1}^r n_i$ has the largest possible k-chromatic number as well. Hence, one can take small "extremal" graphs and build out of them graphs with a large k-chromatic number, which also improve the upper bound for $f_k(n)$ given in Theorem 1. The same observation applies for the oriented chromatic number; the Reader can easily provide examples of sparse small extremal graphs for this problem with n vertices and less than $\lceil\log_2 n\rceil(n - \lceil\log_2 n\rceil)$ edges, and describe a recursive construction for obtaining such an extremal graph out of two smaller ones. However, to find an exact guess for the structure of the extremal graphs does not seem to be easy for us.

References

1. N. Alon and T. H. Marshall, *Homomorphisms of edge-coloured graphs and Coxeter groups*, J. Algebraic Combinatorics, to appear.
2. A. V. Kostochka, E. Sopena and X. Zhu, *Acyclic and oriented chromatic numbers of graphs*, J. Graph Theory **24** (1997), 331–340.
3. A. Raspaud and E. Sopena, *Good and semi-strong colorings of oriented planar graphs*, Information Processing Letters **51** (1994), 171–174.
4. E. Sopena, *The chromatic number of oriented graphs*, J. Graph Theory **25** (1997), 191–205.
5. Zs. Tuza, *Intersection properties and extremal problems for set systems*, In: "*Irregularities of partitions*" (G. Halász, V. T. Sós, Eds.), Springer-Verlag, 1989, 141–151.

Alexandr V. Kostochka
Novosibirsk State University
Novosibirsk, Russia 630090
sasha@math.nsc.ru

Tomasz Łuczak
Adam Mickiewicz University
60-769 Poznań, Poland
tomasz@math.amu.edu.pl

Gábor Simonyi
Mathematical Institute of the
Hungarian Academy of Sciences
P.O.B.127
Budapest H-1364, Hungary
simonyi@math-inst.hu

Eric Sopena
LaBRI
Université Bordeaux I
33405 Talence Cedex, France
Eric.Sopena@labri.u-bordeaux.fr

DIMACS Series in Discrete Mathematics
and Theoretical Computer Science
Volume **49**, 1999

New trends in the theory of graph colorings: Choosability and list coloring

Jan Kratochvíl, Zsolt Tuza, and Margit Voigt

ABSTRACT. We survey recent developments and open problems on graph colorings where the color of each vertex has to be chosen from a restricted set of admissible colors.

1. Preliminaries

Graph colorings belong to classical graph theoretical problems that are important both for their practical applications and richness of theoretical results. E.g., the Four Color Conjecture has stimulated research in discrete mathematics for more than hundred years, and the recent computer-aided proofs of extensions of the Four Color Theorem by Robertson et al. highlight the development of graph theory in the latest years. Also, graph coloring viewed as graph homomorphism is an important interdisciplinary meeting point of algebra and graph theory.

A real abundance of results in the last 10 years appeared in the area of colorings with local constraints, including list colorings, precoloring extension, and choosability. The concept of list coloring was introduced in the seventies by Vizing [**53**] and independently by Erdős, Rubin and Taylor [**21**], and though it seemed to have been forgotten for more than 10 years, the beginning of nineties suddenly brought this subject into the center of interest. This might have been triggered by the solution of several problems from [**21**] concerning choosability of planar graphs. First Alon and Tarsi [**9**] proved that bipartite planar graphs are 3-choosable, then M. Voigt showed that not all planar graphs are 4-choosable [**54**], and finally C. Thomassen proved that every planar graph is 5-choosable [**46**]. Another milestone was Galvin's proof of the Dinitz conjecture on Latin squares, showing more generally that the list chromatic index of a bipartite multigraph equals the maximum degree of the multigraph in question [**23**]. (More generally, allowing that monochromatic subgraphs have maximum degree at most d, for an arbitrarily fixed natural number d, lists of $\lceil \Delta/d \rceil$ colors suffice for the edges in every graph of maximum degree Δ; see [**27**] and [**31**].)

1991 *Mathematics Subject Classification.* Primary 05C15; Secondary 05C20, 05C65, 05C80.

The first author acknowledges partial support of Czech research grants GAČR 201/1996/0194 and GAUK 194/1996.

Research of the second author supported in part by the Hungarian Scientific Research Fund, grant OTKA T–016416.

A 60-page survey of list coloring and precoloring extension problems appeared in 1997 [**48**]. Many new results have appeared during the little over a year since the publication of this survey, and collecting them and reporting on the latest developments is the aim of this article.

The key concept discussed in this survey paper is the following. Given a graph G and an assignment $\mathcal{L} = \{L(v) \mid v \in V(G)\}$ of lists of admissible colors for its vertices, we say that G is *$\mathcal{L}$-list colorable* if the vertices of G can be properly colored (i.e., adjacent vertices receive distinct colors) so that each vertex v is colored with a color from $L(v)$. If $\mathcal{L}$ is obvious from the context, we simply say that G is *list colorable*. If all list of $\mathcal{L}$ have the same size k, $\mathcal{L}$ is called a k-assignment. The minimum integer k such that G is $\mathcal{L}$-list colorable for every k-assignment $\mathcal{L}$, is called the *list chromatic number* of G (or the *choice number* of G) and is denoted by $\chi_\ell(G)$. The graph G is called *k-choosable* if $\chi_\ell(G) \leq k$. If edges are colored instead of vertices (and lists of admissible colors are assigned to edges), we talk about *list chromatic index* of G, denoted by $\chi'_\ell(G)$.

One of the most important longstanding open problems in this area is the conjecture of Vizing (re-raised later independently by several other authors as well) stating that $\chi'_\ell(G) = \chi'(G)$ for every multigraph G. As mentioned above, the affirmative solution for bipartite multigraphs by Galvin was an important and elegant step forward. However, the conjecture in its full generality is still far from being solved. Instead of a solution, several stronger conjectures have been born; see e.g. Section 3.3 of [**48**].

A more general concept of *list set colorings* stems from set colorings of graphs. In this case vertices of a given graph should be assigned sets of colors of prescribed size, say q, so that adjacent vertices are assigned disjoint sets. A graph G is called *$\mathcal{L}$-list q-set colorable* if G has a proper q-set coloring f with $f(u) \subseteq L(u)$ for every u. The graph G is *(p,q)-choosable* if it is $\mathcal{L}$-list q-set colorable for every p-assignment $\mathcal{L}$.

A rather innocent-looking problem of [**21**] asks if every (p,q)-choosable graph is also (pm,qm)-choosable (for every positive integer m). Now it seems that this problem is rather tough and it is still open. An affirmative solution would imply that choosability is a submultiplicative function on the union of graphs. More precisely, the conjecture implies that

$$\chi_\ell(G_1 \cup G_2) \leq \chi_\ell(G_1) \cdot \chi_\ell(G_2) \qquad (?)$$

Even some very simple particular cases of this conjecture are open, e.g. doubling the vertices of a bipartite graph (cf. [**48**, p. 175]).

These were just three longstanding open problems in the area of list colorings and choosability, that we consider of basic importance. In the rest of the paper we will map the recent list coloring and choosability results and new open questions and problems.

2. Random graphs and related structures

A standard way of describing the typical behavior of a graph invariant is to investigate its value on the random graph $\mathbf{G}_{n,p}$, the graph with n vertices where each vertex pair uv is chosen to be an edge with probability p, independently of the choices of other edges. For constant p, the asymptotics of $\chi_\ell(\mathbf{G}_{n,p})$ was determined by Kahn (1992) and by Tuza and Voigt (1994) (see [**5**, p. 18] and [**48**, pp. 172–173]).

These results have been extended for edge probability $p = p(n)$ in several recent works. Alon, Krivelevich and Sudakov [**7**] prove that the quotient $\chi_\ell(\mathbf{G}_{n,p})/\chi(\mathbf{G}_{n,p})$ lies between two positive constants in the entire range $2/n < p \le 1/2$, with probability $1 - o(1)$ as $n \to \infty$. Moreover, in the interval $n^{\epsilon-1/4} \le p(n) \le 3/4$ (with any constant ϵ, $0 < \epsilon < 1/4$), $\chi_\ell(\mathbf{G}_{n,p}) = (1 + o(1))\chi(\mathbf{G}_{n,p})$ holds almost surely, as shown by Krivelevich [**40**]. It is conjectured in [**40**] that this asymptotic equality holds for every function $p(n)$. For constant p, this is indeed the case, by the results cited in the first paragraph of this subsection.

It is interesting to note that a deterministic version of the inequality "$\chi_\ell(G) < c\,\chi(G)$" can also be proved. Alon et al. [**7**] and Van Vu [**59**] have found such types of results, assuming that the vertex degrees are about pn (or not much less than that), and each edge is contained in approximately (or not much more than) $p^2 n$ triangles. Almost regular triangle-free graphs occur as important particular cases.

Krivelevich [**6**] has announced a very stong asymptotic result on the choice number of *random bipartite* graphs, too. Denote by $\mathbf{B}_{n,p}$ the random subgraph of the complete bipartite graph $K_{n,n}$ where the edges are chosen independently at random, with probability p each. Krivelevich proves that for all $c/n \le p \le 9/10$, the choice number of $\mathbf{B}_{n,p}$ almost surely equals asymptotically $\log_2(pn)$.

3. Hall-type conditions

In this section we consider problems where the lists associated with the vertices may have different lengths. Observe first that a complete graph admits a list coloring if and only if the color sets of its lists have distinct representatives. Thus, by Hall's Marriage Theorem, the necessary and sufficient condition for the colorability of K_n is that the union of any i lists should contain at least i colors, for every $1 \le i \le n$.

Consider next an arbitrary graph $G = (V, E)$ with lists $L(v)$ on its vertices $v \in V$. A *necessary* condition, as a generalization of the trivial part of Hall's theorem, can be formulated in the following way. Let $Y \subseteq V(G)$ be any subset of the vertex set. For each color $i \in C(Y) := \bigcup_{y \in Y} L(y)$, we consider the subgraph $G[Y; i]$ induced by those vertices $y \in Y$ for which $i \in L(y)$. Denote by $\alpha(Y; i)$ the independence number of $G[Y; i]$.

Hall Condition: For every $Y \subseteq V(G)$,

$$\sum_{i \in C(Y)} \alpha(Y; i) \ \ge\ |Y|\,.$$

Hall number: The *Hall number*, $h(G)$, of the graph G is the smallest positive integer k with the following property. If a list assignment $\mathcal{L}$ on G satisfies the Hall Condition, and $L(v) \ge k$ holds for every $v \in V(G)$, then G is $\mathcal{L}$-list colorable.

It has to be noted that "$L(v) \ge k$" in the definition is essentially different from assuming that $\mathcal{L}$ is a k-assignment. Indeed, every 1-assignment satisfying the Hall Condition is a proper coloring of G, while every graph containing a non-complete block has Hall number ≥ 2. (For instance, in C_4 the lists $\{1\}$, $\{1,2\}$, $\{2,3\}$, $\{1,3\}$ assigned in this order along the cycle satisfy the Hall Condition but admit no list coloring.)

Contrary to many graph parameters, the Hall number has a rather irregular behavior. For every t, there exists a graph G and a vertex $v \in V(G)$ such that

$h(G) - h(G - v) > t$, even in the restricted class of line graphs (a problem raised for line graphs in [**30**] and solved in [**18**]). For unrestricted graphs of order n, the exact maximum is known, too; it follows from the results of [**49**] that the extremum is attained by $K_n - e$, and also that the largest value of $h(G - e) - h(G)$ occurs for $G = K_n$).

It is an open problem to determine the Hall number of line graphs of complete graphs. So far the best estimates of [**18**] are, depending on the parity of the number of vertices, $2t - 2 \leq h(L(K_{2t})) \leq 2t + 1$ and $2t \leq h(L(K_{2t+1})) \leq 2t + 1$, for every $t \geq 2$. Also, for the total graph $T(K_n)$ (where the vertices represent the vertices and edges of K_n, and edges of $T(K_n)$ correspond to incidences and adjacencies in K_n), it is observed in [**18**] that $h(T(K_n)) \geq h(L(K_{n+1}))$ and $\chi_\ell(L(K_n)) = \chi_\ell(T(K_n))$.

Let us mention that another graph invariant related to the Hall Condition (but not so much to colorings) is the *Hall-Condition number*, denoted $s(G)$. It is defined as the smallest positive integer k such that every k-assignment — or, equivalently, every list assignment with $|L(v)| \geq k$ for all $v \in V$ — satisfies the Hall Condition. Hilton and Johnson proved in [**28**] that $s(G)$ is in fact equal to the maximum value of $\lceil |V(H)|/\alpha(H) \rceil$, where H runs over all induced subgraphs of G. Similarly to the chromatic index, the Hall-Condition number of the line graph $L(G)$ of G is either $\Delta(G)$ (the maximum degree) or $\Delta(G) + 1$, and the latter case can be characterized in terms of overfull subgraphs, as proved by Eslahchi and Hilton [**22**].

Finally, the fractional versions of $h(G)$ and $s(G)$ are studied in [**29**], in connection with generalized list colorings where each vertex has to get a given number of colors from its list and no color may be selected for both vertices of any edge.

4. Coloring extension

This section deals with the question whether it is possible to extend a precoloring of a subset of the vertex set to a proper coloring or list coloring of the whole graph. (The precolored vertices may be viewed as vertices with lists of length 1.)

Coloring extension is a powerful technique used in many proofs in graph coloring theory. One of the interesting recent theorems applying this method will be presented in Section 4.2.

4.1. Distance constraints. Interesting results for the coloring version are obtained recently by Albertson [**1**], Albertson and Moore [**4**] and Albertson and Hutchinson [**3**]. Let us mention some of their results concerning distance constraints.

Let $W \subseteq V$ denote a subset of the vertex set of G. Assume that G is k-colorable and the distance between any two vertices of W is at least 4. Then any $(k+1)$-coloring of W can be extended to a $(k+1)$-coloring of $V(G)$ [**1**]. Under a weaker assumption, namely if the distance between two vertices of W is at least 3, then any $(k+1)$-coloring of W can be extended to a $\lceil \frac{3k+1}{2} \rceil$-coloring of $V(G)$ [**4**].

The situation changes if we consider a k-choosable graph instead of a k-colorable graph. If G is k-choosable and the distance between any two vertices of W is at least 3, then every $(k+1)$-coloring of W can be extended to a $(k+1)$-coloring of $V(G)$ [**1**]. Furthermore, for every $(k+1)$-assignment any coloring of W can beextended to a list coloring of G.

Let us consider now a planar graph G with a specified set $W \subset V(G)$, and assume that we would like to find a list coloring for every 5-assignment $\mathcal{L}$ extending any feasible (list) precoloring of W. It turns out that a distance of at least 4 between any two vertices of W is not sufficient to allow always such an extension. Tuza and Voigt [**50**] constructed a planar graph where any two vertices of W have distance ≥ 4, the vertices of W have lists of 3 colors, the vertices of $V(G) \setminus W$ have lists of 5 colors, and the graph is not list colorable for this list assignment.

On the other hand, Thomassen's proof [**46**] on planar graph 5-choosability shows that some precolorings (two adjacent precolored vertices) and short lists (on the same face) allow a list coloring extension.

Thus, even for the special case of planar graphs it remains an open question which precolored vertex subsets W allow list coloring extensions.

4.2. Dirac's Map-Color Theorem for choosability. An extension of Thomassen's 5-choosabilty result for planar graphs has been found by Böhme, Mohar, and Stiebitz [**16**]. Assume G is a plane graph, C the set of vertices on the outer face, and $P = (v_1, \ldots, v_k)$ a path along the outer face. Furthermore, there is a list assignment $\mathcal{L}$ with $|L(v)| \geq 5$ if $v \in V \setminus C$, $|L(v)| \geq 4$ if $v \in V(P) \setminus \{v_1, v_k\}$, $|L(v)| \geq 2$ if $v \in \{v_1, v_k\}$ and $|L(v)| \geq 3$ if $v \in C \setminus V(P)$. Then G is $\mathcal{L}$-list colorable.

It turns out that this extension together with a generalization of Dirac's ideas is useful to prove the choosability version of Dirac's Map-Color Theorem [**16**]:

The choice number $\chi_\ell(G)$ of every graph G embedded on a surface of Euler genus $\epsilon \geq 1$ and $\epsilon \neq 3$ is at most the Heawood number $H(\epsilon) = \lfloor (\sqrt{24\epsilon + 1} + 7)/2 \rfloor$, and equality holds if and only if G contains the complete graph $K_{H(\epsilon)}$.

4.3. Precoloring extension on planar bipartite graphs. A subproblem of Precoloring Extension, introduced in [**15**], is the 1-PREXT problem. An instance of 1-PREXT is a graph $G = (V, E)$, a subset $W \subset V$ of *precolored* vertices, and a positive integer k called *color bound*. The question is whether G admits a proper coloring with at most k colors, such that no two vertices of W get the same color.

It was conjectured by Woeginger in 1992 (see [**48**, Conjecture 4.2]) that on *planar bipartite graphs* 1-PREXT is polynomially solvable for *color bound* 3. Recently Mohar, Tuza and Woeginger [**44**] proved that in fact there exists a linear-time algorithm for this problem. One key observation is that if a hexagon is cyclically precolored with 123123, and its interior is dissected into quadrangles, then the precoloring is not extendable. It is shown in [**44**] that if G is planar and bipartite, and $k = 3$, then every non-extendable precoloring of G forces this obstruction as a subconfiguration, and it can be found in linear time.

5. Set choosability

One of the generalizations is described in terms of three parameters p, q, r. Let $G = (V, E)$ be a graph. A p-assignment $\mathcal{L} = \{L(v) \mid v \in V\}$ is called a (p, r)-*assignment* if

$$|L(u) \cap L(v)| \leq p - r \text{ for all } uv \in E,$$

i.e., if adjacent vertices are assigned lists that are "far apart" (the size of their intersection is bounded). The graph G is said to be (p, q, r)-*choosable* if it admits an $\mathcal{L}$-list q-set coloring for every (p, r)-assignment $\mathcal{L}$. Note that it may happen that for some parameters p, q, r, a (p, q, r)-choosable graph is not (p, q)-choosable. To avoid trivialities, we assume $2q < p + r$ and $r < p$.

This concept was introduced and studied in [**38**] and [**39**]. In [**38**], mostly complexity issues were considered. The effect of this separation constraints on planar graphs was noted in [**39**], where it is shown that every planar graph is

$(4,1,3)$-choosable. (In other words, though not every planar graph is 4-choosable, under the further restriction that lists of admissible colors assigned to adjacent vertices are almost disjoint — have at most one color in common — list size 4 is enough to guarantee the existence of a list coloring.) Let us note here that both the (4,1,2)-choosability and (3,1,2)-choosability of planar graphs are still open. (These problems ask if every planar graph allows a list coloring for every 4-assignment where the lists of any two adjacent vertices share at most 2 colors, or for every 3-assignment with adjacent vertices being assigned almost disjoint lists.)

Another issue studied in [**39**] was the relation to the maximum degree of G, mainly in the case of $q = 1$. For a graph G and a nonnegative integer c, $\zeta(G,c)$ denotes the minimum integer k such that G is $(k,1,k-c)$-choosable. Here c represents the maximum allowed size of the intersection of lists assigned to adjacent vertices. It was shown in [**39**] that $\zeta(G,c)$ is bounded from above by a function of the maximum degree, namely $\zeta(G,c) \leq \lceil\sqrt{\mathrm{e}c\,(2\Delta-1)}\rceil$, where $\mathrm{e} = 2.718\ldots$ is the Euler constant and Δ is the maximum degree of G. It was also shown therein that this bound is asymptotically tight, apart from a multiplicative constant. The bad guys in this case are the complete graphs, as for any prime $t > \frac{c}{2}$ and $n = \lfloor\frac{c}{2}\rfloor t^2+1$, the complete graph K_n is not $(\lfloor\frac{c}{2}\rfloor t+1, 1, \lfloor\frac{c}{2}\rfloor t+1-c)$-choosable. It turns out that the lower and upper bounds can be extended to arbitrary q, but they do not match anymore as q gets large.

Let us define $\zeta(G,q,c)$ to be the minimum integer p such that G is $(p,q,p-c)$-choosable. Then we have the following theorem:

THEOREM 5.1. *Let $\Delta(G)$ be the maximum degree of a vertex in a graph G. Then $\zeta(G,q,c) \leq \lceil q\sqrt{\mathrm{e}c(2\Delta(G)-1)}\rceil$, where $\mathrm{e} = 2.718\ldots$ is the Euler constant.*

PROOF. Suppose lists of size $k \geq q\sqrt{\mathrm{e}c(2\Delta(G)-1)}$ are assigned to the vertices of G in such a way that $|L(u)\cap L(v)| \leq c$ whenever $uv \in E(G)$. Using the well known Lovász Local Lemma ([**19**]), we will show that a list q-set coloring always exists. Consider a random q-set coloring f of G (not necessarily proper), i.e., every vertex u is colored with a set $f(u)$ of q colors from $L(u)$, independently with probability $\binom{k}{q}^{-1}$. For every edge $e = uv \in E(G)$, we define the event $B(e) = \langle f(u)\cap f(v) \neq \emptyset\rangle$. We have

$$Prob(B(e)) = \frac{\sum_{i=1}^{m}(-1)^{i-1}\binom{m}{i}\binom{k-i}{q-i}^2}{\binom{k}{q}^2} \leq \frac{m\binom{k-1}{q-1}^2}{\binom{k}{q}^2} \leq c\frac{q^2}{k^2}$$

for every edge $e = uv \in E(G)$ and $m = |L(u)\cap L(v)|$.

The line graph of G, $L(G) = (E(G), \{ee' : e\cap e' \neq \emptyset\})$ serves as the dependency graph, because for every edge e, the event $B(e)$ is totally independent of the events $B(e')$ for edges e' which do not share vertices with e. The maximum degree in the line graph is at most $2\Delta(G)-2$. Due to the choice of k,

$$(2\Delta(G)-1)\cdot\frac{cq^2}{k^2} < \frac{1}{\mathrm{e}}$$

and therefore the Lovász Local Lemma ([**19**]) implies

$$Prob(G \text{ is } \mathcal{L}\text{-list } q\text{-set colored}) = Prob\Big(\bigwedge_{e\in E(G)} \overline{B(e)}\Big) > 0.$$

It follows that there exists a proper q-set list coloring. □

Again, as in [**39**] for $q = 1$, we can prove a lower bound for complete graphs:

THEOREM 5.2. *For every integer q and $c \geq 2$, there are infinitely many values of n such that $\zeta(K_n, q, c) \geq \frac{\sqrt{2}}{4}\sqrt{qc(n-1)}$.*

PROOF. Let $t > c$ be a prime. We consider the point set $M = \{0, 1, \ldots, t\lfloor \frac{c}{2} \rfloor - 1\} \times GF(t)$, where $GF(t)$ denotes the Galois field of order t. For any quadratic function f on $GF(t)$, we set

$$M_f = \{(x, f(x \bmod t)) : 0 \leq x \leq t\lfloor \frac{c}{2} \rfloor - 1\},$$

i.e., M_f consists of the graph of the function f and its $\lfloor \frac{c}{2} \rfloor - 1$ translates. Set

$$n = \lceil \frac{ct^2}{2q} \rceil + 1.$$

Let $\mathcal{F}$ be a set of n distinct quadratic functions over $GF(t)$. Since the total number of quadratic functions is $t^3 > ct^2 \geq n$, we can find such $\mathcal{F}$.

Given a complete graph with n vertices, we associate its vertices with the functions in $\mathcal{F}$, and view the sets $M_f, f \in \mathcal{F}$ as lists assigned to the vertices. The size of each list is $t\lfloor \frac{c}{2} \rfloor$. Any two distinct quadratic functions on $GF(t)$ coincide in at most two values, and hence $|M_f \cap M_g| \leq 2\lfloor \frac{c}{2} \rfloor \leq c$. Thus, our lists form a $(t\lfloor \frac{c}{2} \rfloor, t\lfloor \frac{c}{2} \rfloor - c)$-assignment. Since the total number $t^2\lfloor \frac{c}{2} \rfloor$ of colors is less than nq, this assignment does not allow a proper q-set coloring. Therefore, $\zeta(K_n, q, c) > t\lfloor \frac{c}{2} \rfloor \geq t \cdot \frac{c-1}{2}$. Since $n - 1 \leq \frac{ct^2}{2q}$, we have $\zeta(K_n, q, c) > \sqrt{\frac{t^2(c-1)^2}{4}} \geq \sqrt{\frac{2qc(n-1)}{c^2t^2}\frac{t^2(c-1)^2}{4}} = \frac{c-1}{c} \cdot \sqrt{\frac{qc(n-1)}{2}} \geq \frac{1}{2\sqrt{2}}\sqrt{qc(n-1)}$. □

Let us note that for $c = 1$ a construction based on finite affine planes yields a lower bound of order $\sqrt{qn}$ as well, similarly as in case $q = 1$ in [**39**]. Note further that one can show, similarly as in [**39**], that in fact $\zeta(K_n, q, c) = \Omega(\sqrt{qcn})$ for n large enough. It follows that for every *fixed* q, up to a multiplicative constant the right asymptotics for $\zeta(G, q, c)$ is $\sqrt{c\Delta}$. However, this multiplicative constant depends on q. This leaves the following open problem:

PROBLEM 5.1. Are there constants ε, C_1, C_2 such that

$$C_1 q^{\varepsilon}\sqrt{nc} \leq \zeta(K_n, q, c) \leq C_2 q^{\varepsilon}\sqrt{nc}$$

for every q, n, and c?

Or, even more generally:

PROBLEM 5.2. Is there a constant $\tilde{\varepsilon} < 1$ such that

$$0 < \limsup_{G,q,c} \frac{\zeta(G, q, c)}{q^{\tilde{\varepsilon}}\sqrt{\Delta(G)c}} < \infty?$$

Our results show that $0.5 \leq \tilde{\varepsilon} \leq 1$ if such $\tilde{\varepsilon}$ exists. Perhaps the probabilistic method does not give the right order of magnitude in terms of q (though it does in terms of c and Δ), or maybe for general q the structure of the critical examples is richer than just complete graphs.

6. Partial colorings

It is a natural question to ask how many vertices of a graph can be colored properly from lists with fewer than $\chi_\ell(G)$ elements.

Let $\mathcal{L}_t$ be a list assignment such that $|L_t(v)| = t$ for all $v \in V(G)$ and $t \leq \chi_\ell(G)$. A *partial list coloring* of G is a coloring c of the vertices of a subset $V' \subseteq V$ such that $c(v) \in L_t(v)$ for all $v \in V'$, and $c(v) \neq c(w)$ for all $v, w \in V'$ and $vw \in E$. Albertson, Grossman and Haas [**2**] introduced the following parameter:

$$\lambda_t := \min_{\mathcal{L}_t}\{max.\ no.\ of\ vertices\ of\ G\ colorable\ from\ \mathcal{L}_t\}.$$

They conjectured that this parameter has a similar behavior as the analogous paramater for ordinary colorings. More formally, let n be the number of vertices of G. It is conjectured in [**2**] that $\lambda_t(G) \geq \frac{t}{\chi_\ell(G)} n$ for all $0 \leq t \leq \chi_\ell(G)$.

Obviously, the conjectured inequality is true for $t \leq 1$ and $t = \chi_\ell(G)$. A first non-trivial result was also given by Albertson, Grossman and Haas. They proved in [**2**] that $\lambda_t \geq \left(1 - \left(\frac{\chi-1}{\chi}\right)^t\right) n$ for all $t \leq \chi_\ell(G)$ and that this is asymptotically best possible. In particular, it follows that the conjecture is true for $t = 2$ if $\chi(G) < \chi_\ell(G)$.

Furthermore, it is easy to see by induction that $\left(1 - \left(\frac{\chi-1}{\chi}\right)^t\right) \geq \frac{t}{\chi+t-1}$. Thus, we have the following partial affirmative answer to the above conjecture:

$$\lambda_t(G) \geq \frac{t}{\chi_\ell(G)} n \text{ for all } 0 \leq t \leq \chi_\ell(G) - \chi + 1.$$

This also includes all $t < \chi_\ell(G)$ for the case $\chi = 2$, and thus the conjecture holds true for bipartite graphs.

By applying methods similar to those in [**2**], Chappell [**17**] proved the lower bound $\lambda_t \geq \frac{6}{7} \frac{t}{\chi_\ell(G)} n$ for all $t \leq \chi_\ell(G)$.

For both mentioned results it is possible to find the coloring of this number of vertices in polynomial time if a proper ordinary coloring or a special list coloring of the entire graph is given. The algorithms can be obtained applying the method of conditional probabilities, [**8**] using modified proofs of the results.

Let $\mathcal{L}_t$ be a list assignment with $|L_t(v)| = t \leq \chi_\ell(G)$ for all $v \in V$, and $V = V_1 \cup \cdots \cup V_\chi$ a partition of the vertex set generated by a proper coloring of G. Then there is a polynomial algorithm to color at least $\left(1 - \left(\frac{\chi-1}{\chi}\right)^t\right) n$ vertices of G properly from these lists ([**56**]).

For bipartite graphs it follows that a partial list coloring of at least $\frac{2^t-1}{2^t} n$ vertices can be found in polynomial time, for any $t \leq \chi_\ell(G) - 1$ and for every $\mathcal{L}_t$ list assignment.

In the sequel, let $\mathcal{L}_t$ be a given list assignment with $|L_t(v)| = t$ for all $v \in V$. Define a corresponding list assignment $\mathcal{L}'$ in the following way. Assume $s = \chi_\ell(G) - t$ and, without loss of generality, $1, \ldots, s \notin \bigcup_{v \in V} L_t(v)$. Set $L'(v) = L_t(v) \cup \{1, \ldots, s\}$, obtaining lists of $\chi_\ell(G)$ elements for all vertices. Assuming that an $\mathcal{L}'$-list coloring of G is given, there exists a polynomial algorithm to find a partial list coloring of at least $\frac{6}{7} \frac{t}{\chi_\ell(G)} n$ vertices from the lists $L_t(v)$ ([**56**]).

7. Oriented colorings

Perhaps the most recent variations on the theme are the oriented choice number and the dichoice number introduced by Sali and Simonyi [**45**]. Though that paper appears in this volume, let us give the essential definitions to discuss some more results on the oriented choice number. For the concept and first results on the dichoice number we refer to [**45**].

A directed graph D is called *k-choosable* if for every k-assignment $\mathcal{L}$ it is possible to find a color $c(v) \in L(v)$ for all $v \in V(D)$ such that adjacent vertices have different colors and, moreover, if there is a directed edge vw with $c(v) = a$ and $c(w) = b$ then there is no directed edge $v'w'$ with $c(v') = b$ and $c(w') = a$.

The *oriented choice number* $\chi_\ell^{\rightarrow}$ of an undirected graph G is the smallest k for which there exists an orientation of the edges of G resulting in a k-choosable directed graph D. We denote by $OCH(k)$ the class of graphs G with $\chi_\ell^{\rightarrow}(G) \leq k$.

The behavior of $\chi_\ell^{\rightarrow}$ differs significantly from that of χ_ℓ. For instance, as proved in [**45**], for every natural number k there exists an integer $m = m(k)$ such that $\chi_\ell^{\rightarrow}(mK_2) > k$. Thus, the oriented choice number is not bounded above by any function of the choice number. The asymptotic for $m(k)$ is not known, however, neither is the exact value of $m(2)$.

The value of $m(2)$ is bounded above by another number m_2 introduced in [**45**]. We say that a directed triangle D covers a tournament T if $V(D) = \{v_1, v_2, v_3\} \subseteq V(T)$ and v_1, v_2, v_3 induce D in T. Denote by m_2 the smallest possible cardinality of a family $Q(n)$ of transitive triangles such that every tournament on the vertex set $\{1, \ldots, n\}$ is covered by at least one member of $Q(n)$.

For every $n \geq 4$ it is shown in [**45**] that 12 transitive triangles are always sufficient to cover every tournament on $\{1, \ldots, n\}$. Recently, Tuza and Voigt [**51**] proved that the upper bound of 12 is best possible, i.e. a set of 11 transitive triangles cannot cover all tournaments on the vertices $\{1, \ldots, n\}$, for any $n \geq 4$. Thus, $m_2 = 12$.

Another interesting open question is how the oriented choice number depends on restrictions on the total number of colors in the union of lists. For $2 \leq k \leq r$, denote by $OCH(k, r)$ the class of graphs for which every k-assignment *from a set of r colors* admits an oriented list coloring. Clearly, $G \in OCH(k, k)$ if and only if G is k-colorable, and $OCH(k, r+1) \subsetneq OCH(k, r)$ for every k and r. Moreover, $OCH(k) = \bigcap_{r \geq k} OCH(k, r)$.

It is observed in [**45**] that $OCH(2, 4)$ is a *proper* subclass of $OCH(2, 3)$. It is not known, however, whether or not the class of graphs with oriented choice number 2 is identical to $C(2, 4)$.

So far it is an open problem to characterize the members of $OCH(2)$ and also of $OCH(2, 4)$ (and $OCH(2, 3)$, too). Tuza and Voigt [**51**] have given a structural characterization for those $G \in OCH(2)$ which are connected or contain at least one cycle. The necessary and sufficient condition is that G be a subgraph of one of two infinite graphs, either the infinite star with precisely three edges subdivided into paths of length two, or the one obtained from the complete bipartite graph $K_{2,3}$ in such a way that precisely one vertex of each vertex class is adjacent to infinitely many pendant vertices. It follows, in particular, that if $OCH(2) \neq OCH(2, 4)$, then every $G \in OCH(4, 2) \setminus OCH(2)$ is a disconnected forest.

8. Hypergraph colorings

One of the areas about which still very little is known is the list coloring of hypergraphs. The methods of J. Kahn, developed for the edge choosability problem of (multi)graphs (see e.g. [**32, 33**]) can be applied for hypergraphs as well. To some extent, consequences can be derived in this way for vertex list coloring, too, by vertex–edge duality. But the structural theory of hypergraph list-colrings still waits for development.

In this section we mention four types of problems where new results have been found.

Unless otherwise stated, $\mathcal{H} = (X, \mathcal{F})$ will denote a hypergraph with (finite) vertex set X and edge set $\mathcal{F}$; i.e., the members of $\mathcal{F}$ — the edges of $\mathcal{H}$ — are subsets of X.

8.1. The size of critical hypergraphs. For every topic it is interesting to investigate the properties of critical structures. Many results are known for color critical graphs, also some bounds on the minimum number of edges in such a graph.

The first investigations in this direction for list colorings of graphs and hypergraphs were done in [**36**] and [**35**].

Assume a list $L(x)$ of allowed colors is given for every vertex $x \in X$. Then the hypergraph $\mathcal{H}$ is called $\mathcal{L}$*-list colorable* if there is a color $c(x) \in L(x)$ for all $x \in X$ such that no edge becomes monochromatic. We say $\mathcal{H}$ is $\mathcal{L}$*-critical* for a given list assignment $\mathcal{L}$ if $\mathcal{H}$ is not $\mathcal{L}$-list colorable but every proper subgraph of $\mathcal{H}$ is $\mathcal{L}$-list colorable.

Now assume that $\mathcal{H}$ is a hypergraph on n vertices and without ordinary edges (edges of cardinality 2), and that $\mathcal{H}$ is $\mathcal{L}$-critical where $\mathcal{L}$ is a list assignment for $\mathcal{H}$ satisfying $|L(x)| = k$ for every $x \in X$. Kostochka and Stiebitz [**35**] proved by probabilistic means that such a hypergraph has at least $(k - 3k^{2/3})\,n$ edges.

As a particular case, the same lower bound follows for $(k+1)$-color-critical hypergraphs (i.e., $L(x) = \{1, \ldots, k+1\}$ for all $x \in X$) on n vertices and without ordinary edges. Recently, it was proved in [**34**] that, for every $k \geq 3$, $r \geq 3$, $g \geq 3$, and infinitely many n, there are $(k+1)$-color-critical r-uniform hypergraphs on n vertices having girth g and less than kn hyperedges.

8.2. Choice ratio and fractional colorings. The choice ratio of a graph (that is, the smallest possible ratio a/b for which the graph in question is (a,b)-choosable) was first investigated in the M.Sc. Thesis of Gutner [**25**] and in the paper by Gutner and Tarsi [**26**]. They proved that this parameter is at most the chromatic number of a graph. Alon, Tuza and Voigt [**10**] showed the equality of choice ratio and fractional chromatic number for graphs.

Considering the corresponding concepts for hypergraphs it is possible to generalize this result.

First, let us introduce independent sets and the fractional chromatic number of a hypergraph.

For $\mathcal{H} = (X, \mathcal{F})$, define

$$\mathcal{A} := \{A \subseteq X \mid \text{there is no } F \in \mathcal{F} \text{ with } F \subseteq A\},$$

the collection of independent sets in $\mathcal{H}$.

A *fractional coloring* of $\mathcal{H}$ is a mapping $\varphi : \mathcal{A} \to [0,1]$ such that

$$\sum_{\substack{A\in\mathcal{A}\\ x\in A}} \varphi(A) \geq 1 \qquad \forall x \in X\,. \tag{$*$}$$

The *fractional chromatic number*, denoted $\chi^*(\mathcal{H})$, is the solution of the linear program with constraints $(*)$ and objective function

$$\min \sum_{A\in\mathcal{A}} \varphi(A)\,.$$

Obviously, the fractional chromatic number of a hypergraph is at most the chromatic number, the minumum number of colors needed for coloring the vertices in such a way that no edge becomes monochromatic.

Furthermore, call a hypergraph $\mathcal{H} = (X,\mathcal{F})$ *(a,b)-choosable* if, for every list assignment $\mathcal{L}$ with $|L(x)| = a$ for all $x \in X$, it is possible to choose subsets $C(x) \subseteq L(x)$ with $|C(x)| = b$ such that

$$\bigcap_{x\in F} C(x) = \emptyset$$

holds for all $F \in \mathcal{F}$.

It is easy to see that, in general, for $b = 1$ the number a of colors in each list has to be at least the chromatic number of the hypergraph in question. However, there is quite a different situation for the ratio a/b.

Mihók, Tuza and Voigt [**41**] proved that the choice ratio $chr(\mathcal{H}) := \inf\{\frac{a}{b} \mid \mathcal{H}$ is (a,b)-choosable$\}$ for any hypergraph $\mathcal{H}$ equals its fractional chromatic number $\chi^*(\mathcal{H})$.

Moreover, analogously to the case of graphs, it is shown in [**41**] that for every hypergraph $\mathcal{H}$ there is a pair a, b such that $\mathcal{H}$ is (a,b)-choosable where $a/b = \chi^*(\mathcal{H})$.

These results have immediate consequences for generalized graph colorings where, instead of being monochromatic, the subgraph induced by each color has to satisfy a given structural property $\mathcal{P}$. The only (quite natural) requirement is that $\mathcal{P}$ should be "induced hereditary," i.e., if a graph satisfies $\mathcal{P}$, then so do all of its induced subgraphs as well. The machinery works for oriented (and directed) graphs, too; for instance, the above equality holds for the condition that no color class in a loopless digraph contains a directed cycle.

8.3. Clique hypergraphs. Given a graph $G = (V, E)$, *clique hypergraph* $\mathcal{C}(G)$ is defined on the same vertex set V, and a subset $E \subseteq V$ is an edge of $\mathcal{C}(G)$ if and only if E induces a complete subgraph maximal under inclusion and $|E| > 1$. That is, isolated vertices are not considered to be cliques in the present context.

The study of $\mathcal{C}(G)$ was proposed by professor Gallai a decade ago (unpublished), and the *transversal number* (minimum number of vertices meeting all cliques) was investigated in a sequence of papers [**47, 13, 20, 11, 12**], still leaving several basic problems open.

The *chromatic number* of the clique hypergraph was considered recently by Bacsó et al. [**14**], mostly for perfect graphs. More generally, Mohar and Škrekovski [**43**] study not only colorability but also choosability, especially for graphs embedded on surfaces. They prove that if G is planar, and also if G is embeddable in the projective plane, then $\mathcal{C}(G)$ is 4-choosable. The result is tight in the sense that 3-choosability is not valid in general. Indeed, if G is triangle-free, then $\mathcal{C}(G) = G$, therefore the choice number of G and $\mathcal{C}(G)$ is the same. Voigt [**55**] has constructed planar graphs of girth 4 which are not 3-choosable.

In contrast, as a strengthening of Grötzsch's theorem [**24**], $\mathcal{C}(G)$ is 3-colorable for every planar graph G [**43**]. From the algorithmic point of view, the chromatic number of clique hypergraphs of planar graphs can be determined — i.e., 2-colorability and 3-colorability can be distinguished — in polynomial time ([**37**]). On the other hand, the 2-colorability of $\mathcal{C}(G)$ for general instance graphs G is NP-complete to decide.

8.4. Mixed hypergraphs: List coloring without lists. We mention this new — and very general — area of hypergraph coloring because of its relevance in connection with the list colorability of graphs.

A *mixed hypergraph* $\mathcal{H}$ is a triple $(X, \mathcal{C}, \mathcal{D})$ where X is the *vertex set*, and both $\mathcal{C}$ and $\mathcal{D}$ are set systems over X, and for convenience we assume that neither $\emptyset$ nor singletons appear in $\mathcal{C} \cup \mathcal{D}$. The sets $C \in \mathcal{C}$ and $D \in \mathcal{D}$ are termed C-*edges* and D-*edges*, respectively, and this distinction becomes relevant when colorings are considered.

A *coloring* of a mixed hypergraph $\mathcal{H}$ is a vertex coloring that satisfies the following two requirements:

- each C-edge contains two vertices having a **c**ommon color, and
- each D-edge contains two vertices of **d**ifferent colors.

If $\mathcal{H}$ admits at least one coloring with these properties, then we say that $\mathcal{H}$ is *colorable*, and otherwise $\mathcal{H}$ is called *uncolorable*. Observe that, contrary to the usual concept of hypergraph coloring, a large family of mixed hypergraphs turns out to be uncolorable. The simplest uncolorable example is given by $|X| = 2$ and $\mathcal{C} = \mathcal{D} = \{X\}$. For colorable mixed hypergraphs it is of great interest to determine the smallest and largest possible numbers of colors in its feasible colorings (*lower* and *upper chromatic number*, respectively).

The mixed hypergraphs were introduced by Voloshin some half decade ago [**57, 58**], and recently several papers have been written on this new subject. The problem of uncolorability has been studied for mixed hypergraphs derived from Steiner systems (where each block is both a C-edge and a D-edge) by Milazzo and Tuza [**42**], and for general mixed hypergraphs by Tuza and Voloshin [**52**]. In the latter paper it is shown that the list colorability of graphs is equivalent to the uncolorability of some particular hypergraphs.

Suppose the graph $G = (V, E)$ and lists $L(v)$ for its vertices $v \in V$ are given. Denote by $c_1, \ldots, c_m$ the colors occurring in $\bigcup_{v \in V} L(v)$. We take an additional set $Y = \{y_1, \ldots, y_m\}$ of new vertices representing the colors. The mixed hypergraph $\mathcal{H}$ will be constructed on the vertex set $X = v \cup Y$. The vertex subset V induces a D-hypergraph isomorphic to G, and Y induces a complete 2-uniform D-hypergraph (i.e., a complete D-graph). The C-edges of $\mathcal{H}$ represent the lists of G in the following way. For each $v \in V$ we take a C-edge consisting of v together with those y_i for which $c_i \in L(v)$ holds.

One can see that this $\mathcal{H}$ is colorable if and only if G admits a list coloring. What is more, the list colorings of G are in one-to-one correspondence with the colorings of $\mathcal{H}$.

One interesting aspect of this transformation is that, contrary to the traditional types of coloring reductions where some parameter (e.g., the number of colors) appears as a constraint, in the present setting the list colorability is interpreted by a purely structural property.

Note added in proof. Quite recently, Zs. Tuza and M. Voigt have proved that the smallest number of edges in a matching with oriented choice number 3 is precisely 12.

References

[1] M. O. Albertson: You can't paint yourself into a corner, *J. Combinatorial Theory Ser. B* **78** (1998), 189–194

[2] M. O. Albertson, S. Grossman, R. Haas, Partial list colorings, manuscript, 1998.

[3] M. O. Albertson, J. P. Hutchinson: Extending colorings of locally planar graphs, manuscript, 1998.

[4] M. O. Albertson, E. H. Moore: Extending graph colorings, submitted to *J. Combinatorial Theory Ser. B*.

[5] N. Alon: Restricted colorings of graphs, in: *Surveys in Combinatorics* (K. Walker, ed.), Proc. 14th British Combinatorial Conference, London Math. Soc. Lecture Notes Series **187**, Cambridge University Press, 1993, 1–33.

[6] N. Alon, M. Krivelevich: The choice number of random bipartite graphs, *Annals of Combinatorics*, to appear.

[7] N. Alon, M. Krivelevich, B. Sudakov: List coloring of random and pseudo-random graphs, manuscript, 1998.

[8] N. Alon, J. H. Spencer: *The probabilistic method*, John Wiley & Sons, 1992.

[9] N. Alon, M. Tarsi: Colorings and orientations of graphs, *Combinatorica* **12** (1992), 125-134.

[10] N. Alon, Zs. Tuza, M. Voigt: Choosability and fractional chromatic number, *Discrete Mathematics* **165/166** (1997), 31–38.

[11] T. Andreae: On the clique-transversal number of chordal graphs, *Math. Semin. Univ. Hamburg*, Heft 50, 1996.

[12] T. Andreae, C. Flotow: On covering all cliques of a chordal graphs, *Discrete Mathematics* **149** (1996), 299–302.

[13] T. Andreae, M. Schughart, Zs. Tuza: Clique-transversal sets of line graphs and complemetns of line graphs, *Discrete Mathematics* **88** (1991), 11–20.

[14] G. Bacsó, S. Gravier, A. Gyárfás, M. Preissmann, A. Sebő: About coloring the hypergraph of maximal cliques of a graph, manuscript, June 1998.

[15] M. Biró, M. Hujter, Zs. Tuza: Precoloring Extension. I. Interval graphs, *Discrete Mathematics* **100** (1992), 267–279.

[16] T. Böhme, B. Mohar, M. Stiebitz: Dirac's Map Color Theorem for choosability, Preprint Series, University of Ljubljana, 30 (1998), 616.

[17] G. G. Chappell: A lower bound for partial list colorings, manuscript, 1998.

[18] M. M. Cropper, A. J. W. Hilton: Hall parameters of complete and complete bipartite graphs, manuscript, 1998.

[19] P. Erdős, L. Lovász: Problems and results on 3-chromatic hypergraphs and related questions, in: *Infinite and Finite Sets* (A. Hajnal, L. Lovász and V. Sós, eds.), Colloq. Math. Soc. J. Bolyai **10**, North-Holland, Amsterdam, 1974, 609–627.

[20] P. Erdős, T. Gallai, Zs. Tuza: Covering the cliques of a graph with vertices, *Discrete Mathematics* **108** (1992), 279–289

[21] P. Erdős, A. L. Rubin, H. Taylor: Choosability in graphs, in: *Proc. West-Coast Conference on Combinatorics, Graph Theory and Computing*, Arcata, California. *Congressus Numerantium* **XXVI** (1979), 125–157.

[22] Ch. Eslahchi, A. J. W. Hilton: The Hall-Condition index of a graph and the overfull conjecture, manuscript, 1998.

[23] F. Galvin: The list chomatic index of bipartite multigraphs, *J. Combinatorial Theory Ser. B* **63** (1995), 153–158.

[24] H. Grötzsch: Ein Dreifarbensatz für dreikreisfreieNetze auf der Kugel, *Wiss. Z. Martin Luther Univ. Halle-Wittenberg, Math. Naturwiss. Reihe* **8** (1959), 109–120.

[25] S. Gutner: Choice Numbers of Graphs, M.Sc. Thesis, Tel Aviv University, 1992.

[26] S. Gutner, M. Tarsi: Some results on $(a : b)$-choosability, to appear.

[27] A. J. W. Hilton: Some improper list colouring theorems, *Congressus Numerantium* **113** (1996), 171–178.

[28] A. J. W. Hilton, P. D. Johnson: Extending Hall's theorem, in: *Topics in Combinatorics and Graph Theory – Essays in Honour of Gerhard Ringel*, Physica-Verlag, Heidelberg, 1990, 360–371.

[29] A. J. W. Hilton, P. D. Johnson Jr., D. A. Leonard: Hall's condition for multicoloring, *Congressus Numerantium* **128** (1997), 195–203.

[30] A. J. W. Hilton, P. D. Johnson Jr., E. B. Wantland: The Hall number of a simple graph, *Congressus Numerantium* **121** (1996), 161–182.

[31] A. J. W. Hilton, T. Slivnik, D. S. G. Stirling: A vertex-splitting lema, de Werra's theorem and improper list colourings, to appear in *J. Combinatorial Theory Ser. B*.

[32] J. Kahn: Asymptotically good list-colorings, *J. Combinatorial Theory Ser. A* **73** (1996), 1–59.

[33] J. Kahn: Asymptotics of the chromatic index for multigraphs, *J. Combinatorial Theory Ser. B* **68** (1996), 233–254.

[34] A. V. Kostochka, J. Nešetřil: Properties of the Descartes' construction of triangle-free graphs with high chromatic number, to appear in *Combinatorics, Probability and Computing*.

[35] A. V. Kostochka, M. Stiebitz: On the number of edges in colour-critical graphs and hypergraphs, Mathematics Preprint Series, University of Nottingham, 98-02, submitted to *Combinatorica*.

[36] A. V. Kostochka, M. Stiebitz, B. Wirth: The colour theorem of Brooks and Gallai extended, *Discrete Mathematics* **162** (1996), 299–303.

[37] J. Kratochvíl, Zs. Tuza: On the complexity of bicoloring clique hypergraphs, manuscript in preparation, September 1998.

[38] J. Kratochvíl, Zs. Tuza, M. Voigt: Choosing subsets from color sets, *Discrete Mathematics*, to appear.

[39] J. Kratochvíl, Zs. Tuza, M. Voigt: Brooks-type theorems for choosability with separation, *J. Graph Theory* **27** (1998), 43–49.

[40] M. Krivelevich: The choice number of dense random graphs, manuscript, 1998.

[41] P. Mihók, Zs. Tuza, M. Voigt: Fractional $\mathcal{P}$-colorings and $\mathcal{P}$-choice-ratio, to appear in *Tatra Mountains Math. Publications*.

[42] L. Milazzo, Zs. Tuza: Upper chromatic number of Steiner triple and quadruple systems, *Discrete Mathematics* **174** (1997), 247–259.

[43] B. Mohar, R. Škrekovski: Grötzsch theorems for the hypergraph of maximal cliques, manuscript, September 1998.

[44] B. Mohar, Zs. Tuza, G. Woeginger: manuscript in preparation, September 1998.

[45] A. Sali, G. Simonyi: Oriented list colouring of undirected graphs, manuscript, January 1998.

[46] C. Thomassen: Every planar graph is 5-choosable, *J. Combin. Theory Ser. B* **62** (1994), 180–181.

[47] Zs. Tuza: Covering all cliques of a graph, *Discrete Mathematics* **86** (1990), 117–126.

[48] Zs. Tuza: Graph colorings with local constraints — A survey, *Discussiones Mathematicae, Graph Theory* **17** (1997), 161–228.

[49] Zs. Tuza: Choice-perfect graphs and Hall numbers, manuscript, 1997.

[50] Zs. Tuza, M. Voigt: Planar graphs with lists of 3 and 5 colors, Technische Universität Ilmenau, Preprint 8/98.

[51] Zs. Tuza, M. Voigt: Oriented list colorings, manuscript in preparation, September 1998.

[52] Zs. Tuza, V. Voloshin: Uncolorable mixed hypergraphs, to appear in *Discrete Applied Mathematics*.

[53] V. G. Vizing: Coloring the vertices of a graph in prescribed colors, *Metody Diskret. Anal. v Teorii Kodov i Schem* **29** (1976), 3–10. (In Russian)

[54] M. Voigt: List colorings of planar graphs, *Discrete Mathematics* **120** (1993), 215–219.

[55] M. Voigt: A not 3-choosable planar graph without 3-cycles, *Discrete Mathematics* **146** (1995), 325–328.

[56] M.Voigt, Algorithmic aspects of partial list colorings, manuscript, 1998.

[57] V. Voloshin: The mixed hypergraphs, *Computer Science Journal of Moldova* **1** (1993), 45–52.

[58] V. Voloshin: On the upper chromatic number of a hypergraph, *Australasian Journal of Combinatorics* **11** (1995), 25–45.

[59] V. H. Vu: On some simple degree conditions which guarantee the upper bound on the chromatic (choice) number of random graphs, manuscript, 1998.

DEPARTMENT OF APPLIED MATHEMATICS, CHARLES UNIVERSITY, MALOSTRANSKÉ NÁM. 25, 118 00 PRAGUE, CZECH REPUBLIC

E-mail address: `honza@kam.ms.mff.cuni.cz`

COMPUTER AND AUTOMATION INSTITUTE, HUNGARIAN ACADEMY OF SCIENCES, H–1111 BUDAPEST, KENDE U. 13–17, HUNGARY

E-mail address: `tuza@sztaki.hu`

INSTITUT FÜR MATHEMATIK, TECHNISCHE UNIVERSITÄT ILMENAU, D–98684 ILMENAU, GERMANY

E-mail address: `voigt@mathematik.tu-ilmenau.de`

DIMACS Series in Discrete Mathematics
and Theoretical Computer Science
Volume **49**, 1999

Topological Minors in Graphs of Minimum Degree n

W. Mader

ABSTRACT. We survey recent results on subdivisions of the complete graphs K_n in graphs G of given minimum degree. Especially, we study the dependence on the girth of G. We investigate also corresponding problems for digraphs.

1. Introduction

The question about minors or topological minors K_n arose first in connection with the Conjectures of Hadwiger and Hajós. (For the topic of this paper, confer also the survey article [40] of C. Thomassen and the recent graduate text [4] of R. Diestel.) H. Hadwiger conjectured in [12] that every graph G of chromatic number at least n has a minor K_n, whereas G. Hajós conjectured in the forties that it must even contain a topological minor K_n. In this, a *graph H is a minor or contraction of the graph G*, if we can assign every vertex $v \in V(H)$ a connected subgraph G_v of G such that $V(G_v) \cap V(G_u) = \emptyset$ for all vertices $v \neq u$ from H and for all *adjacent* vertices $v \neq u$ in H, there is an edge between G_v and G_u in G. A *subdivision of a graph H* arises from H replacing some edges of H with openly disjoint paths of any length, and is denoted by $\dot{H}$. If the minimum degree $\delta(H)$ of H is at least 3, then the vertices of degree at least 3 in $\dot{H}$ are called the *branch vertices* of $\dot{H}$, and the set of branch vertices of $\dot{H}$ is denoted by $B(\dot{H})$. If G contains a subgraph S isomorphic to any $\dot{H}$, we say, *G contains a subdivision of H or a topological minor H*. Whereas Hadwiger's Conjecture is open for all $n \geq 7$, Hajós' Conjecture is disproved for all $n \geq 7$ by P.A. Catlin in [3].

The first general result on Hadwiger's Conjecture was obtained by K. Wagner in [47], proving that every graph of chromatic number at least 2^n has a minor K_{n+1}. This was generalized by G.A. Dirac [9] and H.A. Jung [14], independently, to the result that a graph of sufficiently large chromatic number even contains a subdivision of K_n. After that, D. Grünbaum asked in [11], pp. 215/216, if perhaps every

1991 Mathematics Subject Classification. Primary 05C35; Secondary 05C10.

finite graph of sufficiently large connectivity contains a $\dot{K}_n$. This seems very difficult to prove using connectivity results as Menger's Theorem (as for infinite, highly connected graphs, such a result does not hold [20]), but if one weakens high connectivity to high average degree, the proof becomes astonishingly easy. The following theorem was the starting point for almost all further results mentioned in this paper.

(1.1) THEOREM [19]. *Every finite graph* G *with* $\|G\| \geq n2^{\binom{n}{2}-1}|G|$ *contains a* $\dot{K}_{n+1}$.

Herein, $|G|$ and $\|G\|$ denote $|V(G)|$ and $|E(G)|$, respectively. For the results of our paper, it is essential that the *graphs are finite without multiple edges and without loops*, which we assume throughout. Furthermore, $V(G) \neq \emptyset$ is assumed. In a similar way, a *digraph has no multiple edges of the same direction*, but may have edges in both the directions between two vertices. If multiple edges are admitted, we speak of a *multigraph*. Let $\mathbb{N}$ denote the set of positive integers and define $\mathbb{N}_n := \{1, \ldots, n\}$ for $n \in \mathbb{N}$.

The essential point of Theorem (1.1) is the *existence* of a real valued function $d(n) := \inf\{t \geq -1 : \|G\| \geq t|G| \to \dot{K}_n \subseteq G\}$ for $n \in \mathbb{N}$, not the special value given there. As in [19] for the corresponding "contraction function", it is easily seen that for $n \geq 3, d(n)$ is the minimum of the above numbers, i.e. $\|G\| \geq d(n)|G|$ implies the existence of a $\dot{K}_n \subseteq G$ for $n \geq 3$. Obviously, $d(n) = n - 2$ for $n \leq 3$, and for $n = 4$ this was proved in [6] and [46]. In general, one can show $d(n+1) \geq d(n) + 1$ as in [19]. So $d(5) \geq 3$, but equality was proved only recently, as we shall see below. Since a graph G with $\|G\| \geq c|G| > 0$ contains a subgraph H with minimum degree $\delta(H) > c$ (a least $G' \subseteq G$ with $\|G'\| \geq c|G'|$ does) and since every graph G with $\delta(G) \geq 2m$ has $\|G\| \geq m|G|$, it is not important, if we consider in "growth problems" the edge density $\frac{\|G\|}{|G|}$, the average degree $\frac{2\|G\|}{|G|}$ or the minimum degree $\delta(G)$. For, if we define for $n \in \mathbb{N}$, $\delta(n) := \min\{m \in \mathbb{N} \cup \{0\} : \delta(G) \geq m \to \dot{K}_n \subseteq G\}$, then $d(n) + 1 \leq \delta(n) \leq \lceil 2d(n) \rceil$ holds for all $n \geq 3$, where $\lceil r \rceil$ denotes the least integer not less than r.

The upper bound $d(n+1) \leq n2^{\binom{n}{2}-1}$ from Theorem (1.1) was improved in [22] to $d(n) \leq 3 \cdot 2^{n-3} - n$ for $n \geq 4$. On the other side, the complete bipartite graph $K_{\binom{n+1}{2},\binom{n+1}{2}-1}$ does not contain a $\dot{K}_{2n}$, which implies $d(m) > \frac{m^2}{16}$ for all $m \geq 3$. A (non published) "random example" of T. Luczak shows that

$$\underline{\lim}\frac{d(m)}{m^2} \geq \frac{1}{2}\,\frac{9}{64}.$$

An old conjecture of P. Erdös and A. Hajnal from [10] stated $d(n) = O(n^2)$. This was proved a few years ago by B. Bollobás/ A. Thomason and J. Komlós/ E. Szemerédi at the same time.

(1.2) THEOREM [1], [17]. *There is a constant* c *such that* $d(n) \leq cn^2$ *holds for all* n.

The better constant was proved in [17]: *For every* $\varepsilon > 0$, *there is an* n_0 *such that* $d(n) \leq \frac{n^2}{4-\varepsilon}$ *holds for all* $n \geq n_0$. So, for large $n, d(n)$ is known within a factor $\frac{1}{4}$.

REMARK. In [1] and [17], the papers [10] and [19] are quoted as sources for the conjecture proved in (1.2), but the paper of Erdös and Hajnal is there given with the wrong date 1969 of publication. I was very surprised when I noticed recently that their conjecture even preceded Theorem (1.1) and that in the same paper is found as an open problem, if there is a constant c such that every graph G with $\|G\| > c|G|$ contains a $\dot{K}_5$. Concerning the other source [19] given for the conjecture: I do not believe that I have stated this conjecture in any of my papers.

Whereas the assymptotic behaviour of $d(n)$ is hence well known, the situation becomes much worse, if we turn to exact values of $d(n)$.

2. Maximum size of a graph without $\dot{K}_n$

Let us start from a tree T. We add $n-1$ vertices $v_1, \ldots, v_{n-1}$ and edges from v_i to all other vertices for every $i \in \mathbb{N}_{n-1}$. So we get a graph G on $|T|+n-1$ vertices with $\|G\| = n|G| - \binom{n+1}{2}$. Of course, G has no minor K_{n+2}, hence no $\dot{K}_{n+2}$. For $n = 1$, this is obviously the maximum number of edges a graph without $\dot{K}_3$ can have. It was shown in [6] and [46] that every graph G with $\|G\| > 2|G| - 3 \geq 0$ contains a $\dot{K}_4$. For $n = 3$, it turned out to be very difficult to prove the corresponding formula, but it was conjectured by G.A. Dirac in [8] that it is true.

(2.1) CONJECTURE [8], (THEOREM [29]). *Every graph G with $\|G\| > 3|G| - 6$ contains a $\dot{K}_5$.*

It is easily deducible from a well known result of K. Wagner in [45] that the statement in (2.1) is true, if we replace "$\dot{K}_5$" with "minor K_5". The first approach to Dirac's Conjecture was made by C. Thomassen in [36], where he showed that every graph G with $\|G\| \geq 4|G| - 10 > 6$ contains a $\dot{K}_5$, what he improved in [42] (see also [43]) to $\|G\| \geq \frac{7}{2}|G| - 7 > 0$. More exactly, he proved in [42] that every graph G with $\|G\| \geq \frac{7}{2}|G| - 7 > 0$ contains a $\dot{K}_5$, where a prescribed vertex is no branch vertex, and with this condition the result is best possible. Dirac's Conjecture is proved in [29], as we will see below.
For a proof of Dirac's Conjecture, it might be advantageous to have a good estimate for the maximum number of edges in a graph without a $\dot{K}_5^-$, where K_n^- arises from K_n by deleting one edge. So, C. Thomassen proved

(2.2) THEOREM [36]. *Every graph G with $\|G\| > \frac{5}{2}|G| - 4 \geq 0$ contains a $\dot{K}_5^-$, and this is best possible for all $|G| \geq 2$.*

The factor $\frac{5}{2}$ seems too large for use in a proof of Dirac's Conjecture. So we impose the further condition that the girth $\tau(G)$ of G is at least 5, and this really decreases the factor.

(2.3) THEOREM [28]. *If G is a graph with $\|G\| \geq 2|G| - 5, |G| \geq 6$, and $\tau(G) \geq 5$, then G contains a $\dot{K}_5^-$ or is the Petersen graph.*

For the proof of Dirac's Conjecture we need also the following result of A.E. Kézdy and P.J. Mc Guinness.

(2.4) THEOREM [16]. *A counterexample to Dirac's Conjecture of least order is 5-connected and does not contain a subgraph* K_4^-.

Using this result, Dirac's Conjecture follows immediately from

(2.5) THEOREM [29]. *Every 5-connected graph* G *with* $\|G\| \geq 3|G| - 6$ *contains a* K_4^- *or a* $\dot{K}_5$.

We can give here only an idea of the PROOF. But before, we will introduce some notation. For $A \subseteq V(G)$ and $H \subseteq G, G(A)$ and $G(H)$ denote the subgraph of G spanned by A and $V(H)$, respectively. For $x \in V(G)$ and $H \subseteq G, N(x)$ is the set of neighbours of x in G and $N(H) := \bigcup_{x \in V(H)} N(x) - V(H)$. The (vertex-) connectivity number of G is $\kappa(G)$ and for a connected subgraph $H \subseteq G$, G/H denotes the *graph* arising from G by contracting H to a vertex (from $V(H)$). An x,y–path is a path with endvertices x and y. For vertices $x \neq y$ in G, $\kappa(x,y;G)$ and $\lambda(x,y;G)$ denote the maximum number of pairwise openly disjoint and edge-disjoint x,y–paths in G, respectively. For $z \in V(G)$ and $X \subseteq V(G-z)$, a *z,X–fan of order n* consists of z,x_i–paths W_i for $i = 1,\ldots,n$ with $V(W_i) \cap V(W_j) = \{z\}$ for $i \neq j$ and with $V(W_i) \cap X = \{x_i\}$ for $i \in \mathbb{N}_n$. Then we say, the z,X–fan *ends in* $x_1,\ldots,x_n$.

Let us suppose now that there is a 5-connected graph G *with* $\|G\| \geq 3|G| - 6$ *containing neither a* K_4^- *nor a* $\dot{K}_5$.

We consider all connected subgraphs $H \subseteq G$ which have the property that the contraction G/H has $|G/H| \geq 5$ and yet satisfies $\|G/H\| \geq 3|G/H| - 6$, and choose H_0 maximal in the set of all these H. Denote $G_0 := G/H_0$ and, say, H_0 is contracted to a vertex z_0.

Then $|G_0| \geq 6$, since $K_4^- \not\subseteq G_0 - z_0 \subseteq G$. This is important, since we know now that contracting any edge e incident to z_0 in G_0, we have $\|G_0/e\| < 3|G_0/e| - 6$. So, by standard methods, we get $\delta(H_1) \geq 3$ for $H_1 := G_0(N(z_0)) = G(N(H_0))$ and $\kappa(G_0) \geq 4$.

Now it is easy to deduce $\tau(H_1) \geq 4$. For assume, there is a triangle $T \subseteq H_1$. Then there is a $z \in V(H_0)$ such that there is a $z, V(T)$-fan of order 3 in $G(H_0 \cup T)$, since H_0 is connected. Since $\kappa(G) \geq 5$, we can apply a generalization of Menger's Theorem due to H. Perfect [33] and get the existence of a vertex $s \in V(H_1) - V(T)$ such that a $z, V(T) \cup \{s\}$-fan $\mathcal{F}$ of order 4 exists in $G(H_0 \cup T \cup \{s\})$. Since $\kappa(G_0 - z_0) \geq 3$, we have an $s, V(T)$-fan $\mathcal{F}'$ of order 3 in $G - V(H_0)$. But then $T, \mathcal{F}', \mathcal{F}$ form a $\dot{K}_5$ in G with $B(\dot{K}_5) = V(T) \cup \{z,s\}$, a contradiction to our assumption.

In a similar, but more complicated way, one can show $\tau(H_1) \geq 5$. If we can apply Theorem (2.3) to H_1, then we can complete a $\dot{K}_5^- \subseteq H_1$ by a path through H_0 to a $\dot{K}_5 \subseteq G$. But we know only $\|H_1\| \geq \frac{3}{2}|H_1|$. So we try to apply Theorem (2.3) to an appropriate subgraph of H_1 to finish the proof. Since such a subgraph cannot be located directly, we use instead that $\|H'\| \leq 3|H'| - 9$ for every $H' \subseteq G_0(H_1 \cup \{z_0\})$ containing z_0 with $|H'| \geq 7$. (It is easily seen that the Petersen graph is no subgraph of H_1.) Then the considerations become rather specified and connot be given here ■

Since Theorem (2.5) refers to graphs G with $\|G\| \geq 3|G| - 6$, this suggests the trial to determine the extremal graphs of the problem. Of course, the maximal planar graphs are extremal graphs, and it can be shown that these are the bricks to build all.

(2.6) THEOREM [31]. *The graphs G with $\|G\| = 3|G| - 6 > 0$ not containing a $\dot{K}_5$ are the graphs which we get from any disjoint maximal planar graphs of order at least 3 by pasting them together successively along triangles.*

(2.7) EXAMPLE. Let us consider the extremal graph G we have constructed at the beginning of this section for $n = 3$, using a tree T and vertices v_1, v_2. Let t be an endvertex of T and t' its neighbour in T. If we assume that $G - t$ has the form described in Theorem (2.6), then G also has, since we get G by pasting $G(\{t, t', v_1, v_2\}) \cong K_4$ along the triangle $G(t', v_1, v_2)$ on $G - t$.

For the PROOF of Theorem (2.6), we consider again a counterexample G of least order. It is rather laborious to prove that also in this case G is 5-connected. Let $e = [x, y] \in E(G)$ and consider the contraction $G' := G/e$. Then $\|G'\| \geq 3|G'| - 6$, since otherwise there are distinct vertices $z_1, z_2.z_3$ in $N(x) \cap N(y)$ and a circuit through z_1, z_2, z_3 in the 3-connected graph $G - \{x, y\}$, hence a $\dot{K}_5 \subseteq G$ with $B(\dot{K}_5) = \{x, y, z_1, z_2, z_3\}$. From the minimality of the counterexample G, one can then deduce $\dot{K}_5 \subseteq G'$. A check of the proof in [16] of Theorem (2.4) reveals that we can conclude, that G does not contain a $\dot{K}_4^-$, which contradicts Theorem (2.5) ■

There is a more general conjecture which would immediately imply Theorem (2.5).

(2.8) CONJECTURE (P.D. Seymour). *Every 5-connected, non-planar graph contains a $\dot{K}_5$.*

This conjecture remains undecided.
It was proved in [20] that also for $n = 4, 5$, every graph G with $\|G\| > n|G| - \binom{n+1}{2}$ and $|G| \geq n$ has a minor K_{n+2} and that this is not true for $n = 6$. I do not know an example of such a graph G for $n = 4, 5$, which does not even contain a $\dot{K}_{n+2}$.

3. Topological minors in graphs of large girth

Let us return now to Theorem (2.3). If we replace the edge number there with $\|G\| \geq 2|G| - 6$, then further exceptions occur. So, in this sense, Theorem (2.3) is best possible. But I should conjecture that only a finite number of further exceptions occur. Perhaps Theorem (2.3) reflects only a sharper result:

(3.1) CONJECTURE. *There is an n_0 such that every graph G with $\|G\| \geq 2|G| - 4, \tau(G) \geq 5$, and $|G| \geq n_0$ contains a $\dot{K}_5$.*

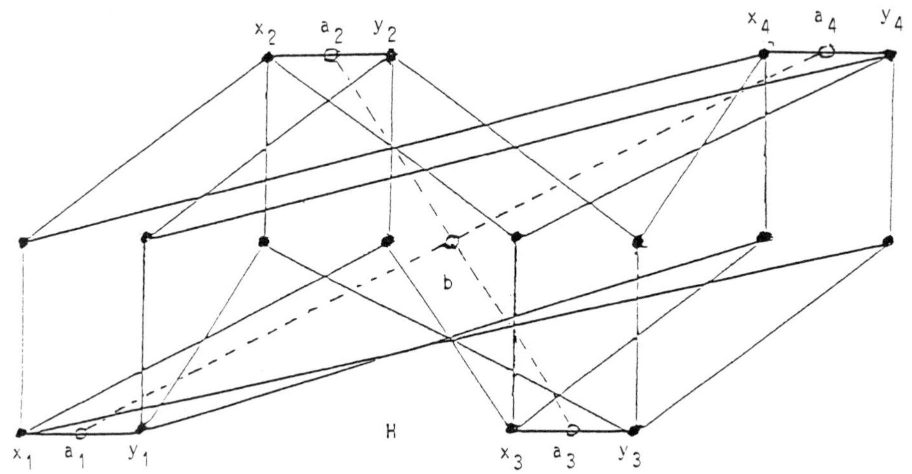

Figure 1

But we cannot take $n_0 = 4$ in (3.1), as the graph P shows which we get from the graph H in Figure 1 disregarding the blank vertices and dashed lines, i.e. $P := (H - \{b, a_1, a_2, a_3, a_4\}) \cup \{[x_i, y_i] : i \in \mathbb{N}_4\}$. For $\|P\| = 2|P| - 4, \tau(P) = 5$, and $|P| = 16$, but P does not contain a $\dot{K}_5$. Since I have not found another exception, it might be possible that P plays for (3.1) a similar role as the Petersen graph in (2.3).

If Conjecture (3.1) is true, then every graph G with $\delta(G) \geq 4$ and $\tau(G) \geq 5$ contains a $\dot{K}_5$. This is open, but we can show that there is a t_0 such that every graph G with $\delta(G) \geq 4$ and $\tau(G) \geq t_0$ has a $\dot{K}_5$.

(3.2) Theorem [30]. *For every graph H of maximum degree $\Delta(H) \geq 3$, there is a $t(H)$ such that every graph G with $\delta(G) \geq \Delta(H)$ and $\tau(G) \geq t(H)$ contains an $\dot{H}$.*

Hence we have a $\dot{K}_{n+1}$ in a graph G with $\delta(G) \geq n$, if $\tau(G)$ is large enough, whereas, by an example before Theorem (1.2), we need at least $\delta(G) > \frac{n^2}{8}$, if we have no condition on the girth. So, large girth has a similar effect for the existence of subdivisions as high minimum degree has. This was noticed for contractions by C. Thomassen in [38]. One of his results there was

(3.3) Theorem [38]. *Every graph G with $\delta(G) \geq 3$ and $\tau(G) \geq 4\delta_c(n) - 3$ contains a minor K_n.*

(Thereby, $\delta_c(n)$ is the function corresponding to $\delta(n)$ for minors, i.e. $\delta_c(n) := \min\{m \in \mathbb{N} \cup \{0\} : \delta(G) \geq m \rightarrow K_n \text{ contraction } of G\}$.)

Of course, this result (with worse bound for the sufficient girth) follows from Theorem (3.2)

Before we sketch a proof of Theorem (3.2), an example.

(3.4) EXAMPLE. Let us consider the construction of n-regular graphs of girth $t \geq 5$ given by H. Sachs in [34] (and [35], Chapter VI). By induction on n we show that *all these graphs contain a* $\dot{K}_{n+1}$. We may suppose $n \geq 3$. By the recursive construction, there are an $(n-1)$-regular graph G of girth t and a $|G|$-regular graph H of girth $\lfloor \frac{t+1}{2} \rfloor \geq 3$. For every $x \in V(H)$, we take a $G_x \cong G$ such that $V(G_x) \cap V(G_y) = \emptyset$ for $x \neq y$. The graph S may arise from $\bigcup_{x \in V(H)} G_x$ by addition of exactly one edge between G_x and G_y for every $[x, y] \in E(H)$ such that S becomes n-regular.

We use now a result from [24] which says that *the graph H contains vertices $x_1 \neq x_2$ with the following property: If we choose the same number of distinct edges $e_1^i, \ldots, e_k^i$ incident to x_i in $H - [x_1, x_2]$ for $i = 1, 2$, then there are openly disjoint x_1, x_2-paths $P_1, \ldots, P_k$ in H with $E(P_j) \cap \{e_1^i, \ldots, e_k^i\} \neq \emptyset$ for $j = 1 \ldots, k$ and $i = 1, 2$.*

First we assume that G is connected. By induction we know that there is a $\dot{K}_n \subseteq G_{x_1}$ with $B(\dot{K}_n) = \{b_1, \ldots, b_n\}$. There is exactly one edge $[b_i, a_i] \in E(S) - E(G_{x_1})$ for $i = 1, \ldots, n$. If $[x_1, x_2] \notin E(H)$, choose any $K_{1,n-1} \subseteq G_{x_2}$ with vertices $c_1, \ldots, c_n$, say, and the edges $[c_i, d_i] \in E(S) - E(G_{x_2})$. Then we can apply the above mentioned result for the corresponding edges incident to x_1 and x_2 in H and find easily a $\dot{K}_{n+1} \subseteq S$ extending $\dot{K}_n \subseteq G_{x_1}$, since G is connected. If $[x_1, x_2] \in E(H)$ with the corresponding edge $[b, c]$ in S, say, $b \in V(G_{x_1})$ and $c \in V(G_{x_2})$, we choose $\{c_1, \ldots, c_n\}$ containing c or not containing c depending on $b \in \{b_1, \ldots, b_n\}$ or $b \notin \{b_1, \ldots, b_n\}$. Then we apply again the above result.

If G is unconnected, H' arises from S by contracting every component of G_x for every $x \in V(H)$. Then we can replace H with the graph H' in the above considerations. (The above mentioned result from [24] holds for every finite graph, not only for regular ones.) ■

The main tool in the proof of Theorem (3.2) is a result proved independently by D.G. Larman and P. Mani in [18] and H.A. Jung in [15]. But first a definition.

(3.5) DEFINITION. *A graph G is n-linked for an $n \in \mathbb{N}$, if $|G| \geq 2n$ and for every sequence $a_1, \ldots, a_{2n}$ of $2n$ distinct vertices, there are disjoint a_i, a_{i+n}-paths in G for $i \in \mathbb{N}_n$.*

Using Menger's Theorem, it is easily seen $\kappa(G) \geq 2n - 1$ for all n-linked graphs G (for instance, cf. [2]), but $\kappa(G) \geq 2n$ does not follow. On the other side, the graphs G_n which we get from a K_{3n-1} by the deletion of n disjoint edges $[a_i, a_{i+n}]$ for $i \in \mathbb{N}_n$, show that a $3n-3$ connected graph is not necessarily n-linked. It is not at all clear, if there is an integer $f(n)$ such that every graph G with $\kappa(G) \geq f(n)$ is n-linked. This is not even true in general, if we admit infinite graphs: *In* [20], *infinite graphs of arbitrarily large (finite) connectivity are constructed, which are not 2-linked.* But this changes for finite graphs.

(3.6) THEOREM [15], [18]. *For every $n \in \mathbb{N}$, there is a least integer $f(n)$ such that every $f(n)$-connected, finite graph is n-linked.*

Of course, $f(1) = 1$, and $f(2) = 6$ was proved in [15]. No further values of f are known, but recently the good upper bound $f(n) \leq 22n$ was given in [2].

The idea of the PROOF of Theorem (3.2) is easily described. If we choose $\tau(G)$ large enough, then large balls around every vertex are trees. So one can partition $V(G)$ into (vertex) disjoint tress $T_x (x \in X)$ of "small" diameter such that the contraction of T_x to a vertex $\overline{x}$ gives a graph G' (without multiple edges!) with $\delta(G') \geq 4(f(\|H\|) + |H|)$, since $\delta(G) \geq n := \Delta(H) \geq 3$. For the notation, we have chosen here an $x \in V(T_x)$, and for all $v \in V(T_x)$, we set $\overline{v} := \overline{x}$. By a result in [23], G' has a subgraph G'' with $\kappa(G'') \geq f(\|H\|) + |H|$. Since the diameter of the T_x is "small", we may assume $\tau(G')$ yet "large". So we can find a set Y of $|H|$ vertices in G'' with distance $d_{G'}(x, y) \geq 3$ in G' for all $x \neq y$ from Y. Assume T_{x_y} is contracted to y for $y \in Y$. Since $G(T_{x_y} \cup N_G(T_{x_y}))$ is a tree and $\delta(G) \geq n$, we can find an $x_y, N_G(T_{x_y})$-fan of order n in G ending in $x_y^1, \ldots, x_y^n$, say. Then $\overline{x_y^1}, \ldots, \overline{x_y^n}$ are distinct and $\{\overline{x_y^1}, \ldots, \overline{x_y^n}\} \cap \{\overline{x_z^1}, \ldots, \overline{x_z^n}\} = \emptyset$ for $y \neq z$ from Y, since $d_{G'}(y, z) \geq 3$. If all these vertices $\overline{x_y^1}, \ldots, \overline{x_y^n}$ for $y \in Y$ belong to G'', then we can apply Theorem (3.6) to $G'' - Y$ and get all paths we need for an $\dot{H} \subseteq G$ with $B(\dot{H}) = \{x_y : y \in Y\}$. But $\{\overline{x_y^i} : y \in Y \text{ and } i \in \mathbb{N}_n\} \subseteq V(G'')$ is not guaranteed by application of the general result from [23]. So we have to use another procedure where we can better observe what we delete to get a highly connected graph from G' ■

It is an interesting problem to determine the least girth t_n sufficient for a graph G with $\delta(G) \geq n$ to contain a $\dot{K}_{n+1}$. In the proof of Theorem (3.2) is given only a very rough upper bound $t_n \leq cn^2$, but I should believe that t_n grows linearly in n.

We have seen above $3n - 2 \leq f(n) \leq 22n$. We will now study what happens for the function f itself, if we confine its definition to graphs of girth at least t. Please notice that there are graphs of high connectivity and large girth, since every graph G with $\delta(G) \geq 4n$ has an n-connected subgraph by [23]. At first, I thought that the straightforward strongest necessary conditions become always sufficient, if we admit only graphs of large girth. But this is not quite true.

(3.7) THEOREM [30]. *Every $2n$-connected graph G of girth large enough is n-linked. For all $n \geq 2$, this statement is not true for $2n - 1$ instead of $2n$.*

For the next result we need a stronger property of an n-linked graph than the defining one.

(3.8) LEMMA. *For every n (not necessarily distinct) pairs (a_i, a_{i+n}) of vertices $a_i \neq a_{i+n}$ for $i \in \mathbb{N}_n$ in an n-linked graph G, there are different a_i, a_{i+n}-paths P_i for $i \in \mathbb{N}_n$, such that $a \in V(P_i) \cap V(P_j)$ for $i \neq j$ implies that a is an endvertex of P_i and of P_j.*

PROOF. Set $A := \{a_i : i \in \mathbb{N}_{2n}\}$. For $|A| = 2n$, the assertion holds per definition. Inducing on $|A|$, we assume $|A| < 2n$. Then there is an a occurring at least twice in $a_1, \ldots, a_{2n}$, say, (a, b_1) and (a, b_2). Since $\kappa(G) \geq 2n - 1$, hence $\delta(G) \geq 2n - 1$, there is an $a' \in N(a) - A$. Then replace (a, b_2) with (a', b_2) and the result follows by induction, since the a', b_2-path can be enlarged to an a, b_2-path using the edge $[a, a'] \in E(G)$ ■

By Lemma (3.8), for every choice of n distinct vertices $b_1, \ldots, b_n$ in an $f(\binom{n}{2})$-connected graph G, there is a $\dot{K}_n \subseteq G$ with branch vertices $b_1, \ldots, b_n$. We say, G contains $\dot{K}_n$ with prescribed branch vertices. So we get from Theorem (3.7) and Lemma (3.8).

(3.9) COROLLARY [30]. *Every $2 \cdot \binom{n}{2}$-connected graph of sufficiently large girth contains $\dot{K}_n$ with prescribed branch vertices.*

It is easily seen that for all $n \geq 2$, there are graphs of arbitrarily large girth with connectivity number $\binom{n}{2} - 1$, which do not contain $\dot{K}_n$ with prescribed branch vertices.

(3.10) CONJECTURE [30]. *Every $\binom{n}{2}$-connected graph of sufficiently large girth contains $\dot{K}_n$ with prescribed branch vertices.*

Theorem (3.2) implies that every graph G with $\|G\| \geq (n-1)|G| > 0$ and sufficiently large girth contains a $\dot{K}_{n+1}$, since a minimal $H \subseteq G$ yet satisfying the above inequalities has $\delta(H) \geq n$. But this is not best possible for $n \geq 3$. For $n = 3$, it was shown in [28], that *every graph G with $\|G\| \geq \frac{3}{2}|G| - 2, |G| \geq 3$, and $\tau(G) \geq 5$ contains a $\dot{K}_4$*. This is best possible for every vertex number, as the graphs show, which consist of $m \geq 2$ openly disjoint a, b-paths, all of length 3 but at most one, which has length 2. - In general, I have not even a really plausible conjecture. Perhaps the following holds.

(3.11) PROBLEM [30]. *Does every graph G with $\|G\| > \frac{n}{2}(|G| - (n-1))$ and $6 \leq n + 2 \leq \tau(G) < \infty$ contain a $\dot{K}_{n+1}$?*

For $n = 4$, I know only one graph H with $\|H\| = 2|H| - 6$ and $6 \leq \tau(H) < \infty$ not containing a $\dot{K}_5$, but there are infinitely many such graphs G with $\|G\| = 2|G| - 7$ and $\tau(G) = 6$. The just mentioned graph H with $\|H\| = 2|H| - 6$ is displayed in Figure 1. This graph H has $|H| = 21$, $\|H\| = 2|H| - 6$, and $\tau(H) = 6$, but does not contain a $\dot{K}_5$. For assume $\dot{K}_5 \subseteq H$. If $b \notin B(\dot{K}_5)$, then $B(\dot{K}_5) \subset \{x_i, y_i : i \in \mathbb{N}_4\}$, but there is a vertex z of degree 3 with $N(z) \subseteq B(\dot{K}_5)$. So we must have $b \in B(\dot{K}_5)$. But then $|B(\dot{K}_5) \cap \{x_i, y_i\}| = 1$ for $i \in \mathbb{N}_4$. Then it is again easy to check that there is a z with $d(z) = 3$ and $N(z) \subseteq B(\dot{K}_5) - \{b\}$.

What happens now, when the girth becomes "very large"? It is surprising that then even an average degree less than n implies the existence of a $\dot{K}_{n+1}$.

(3.12) THEOREM [32]. *For every integer $n \geq 3$ and every $\varepsilon > 0$, there is a $t(n, \varepsilon) \in \mathbb{N}$ such that every graph G with $\|G\| \geq (\frac{n-1}{2} + \varepsilon)|G| > 0$ and $\tau(G) \geq t(n, \varepsilon)$ contains a $\dot{K}_{n+1}$.*

Of course, $\frac{n-1}{2}$ can never be reached, since there are $(n-1)$-regular graphs of arbitrarily large girth (for instance, see [35] Chap. 6, [44] Chap. 8.8 or [23]). The question, which average degree in a graph of given girth guarantees the existence of a $\dot{K}_{n+1}$, is completely answered for $n = 3$ in [32]: *Every graph G with $t := \tau(G) < \infty$ and $||G|| > \lfloor\frac{t-1}{2}\rfloor^{-1}(\lfloor\frac{t+1}{2}\rfloor|G|-t)$ contains a $\dot{K}_4$, and this inequality is best possible.*

4. Subdivisions in digraphs

A *subdivision $\dot{D}$ of a directed multigraph D*, of course, arises from D by replacing some edges by openly disjoint (continuously directed) paths. We emphasize once again that*all digraphs are finite and do not have multiple edges of the same diretion. Paths, circuits etc. in digraphs are always continuously directed and connectivity, edge-connectivity etc. mean strong connectivity, strong edge-connectivity etc.* Let $\delta^+(D)$ and $\delta^-(D)$ denote the minimum outdegree and minimum indegree, respectively. For a graph G, let the digraph $\overleftrightarrow{G}$ arise from G by replacing every edge of G by a pair of oppositely directed edges. Hence $\overleftrightarrow{K}_n$ is the complete digraph on n vertices. For instance, we get exactly all circuits as $\dot{\overleftrightarrow{K}}_2$. Let us first ask, for which directed multigraphs D we get $\dot{D}$ with prescribed branch vertices in every digraph F of sufficiently high connectivity. (Notice that the branch vertices of $\dot{D}$ are prescribed by an injective function $f : V(D) \to V(F)$, and not only as a set.)

It was proved by C. Thomassen in [41] that no high connectivity insures the existence of a circuit through two given vertices in a digraph.

(4.1) THEOREM [41]. *For every $n \in \mathbb{N}$, there is an n-connected digraph which does not contain $\dot{\overleftrightarrow{K}}_2$ with prescribed branch vertices.*

This implies also that for a path D of length 2 and for a digraph D consisting of two disjoint edges, no high connectivity insures a $\dot{D}$ with prescribed branch vertices. So the only directed multigraphs without isolated vertices which may have subdivisions with prescribed branch vertices in all digraphs of sufficiently high connectivity are got from a star by multiplying some edges and directing all edges to or all edges from the "centre". But these directed multigraphs really have subdivisions with prescribed branch vertices by a well known generalization of Menger's theorem due to G.A. Dirac [7](for instance, see [13] Chap. IV, §3). So no positive results on subdivisions with prescribed branch vertices beyond Menger type theorems are possible for digraphs.

Let us now turn to the average degree. The digraphs $\vec{K}_{n,n}$, where all edges of $K_{n,n}$ are oriented "in the same direction", show that no high average degree can insure the existence of an interesting subdivion. For $||\vec{K}_{n,n}|| = \frac{n}{2}|\vec{K}_{n,n}|$, and no proper subdivision exists in $\vec{K}_{n,n}$. Whereas for undirected graphs we could slip from large average degree to large minimum degree, and vice versa, this changes for digraphs. So it might be possible that there are interesting results on the existence of certain configurations in a digraph D depending on the minimum outdegree $\delta^+(D)$, say. A first positive result was proved by C. Thomassen.

(4.2) THEOREM [37]. *If $\delta^+(D)$ is large enough, then there are n vertex disjoint $\overset{\leftrightarrow}{\dot{K}}_2$ in the digraph D.*

But an example from [25] puts a damper on too great expectation. Modifying a counterexample of C. Thomassen [39] to a conjecture of Y.O. Hamidoune, in [25] *digraphs D were constructed with $\delta^+(D)$ and $\delta^-(D)$ arbitrarily large such that $\kappa(x,y;D) \leq 1$ or $\kappa(y,x;D) \leq 1$ holds for every pair $x \neq y$ from $V(D)$*. So, in particular, large minimum outdegree does never force the existence of a $\overset{\leftrightarrow}{\dot{K}}_3$. This fact follows also from the result of C. Thomassen [39] that a large minimum outdegree does not insure the existence of a circuit of even length.

Perhaps we have a better chance for positive results asking for paths in one direction only. For undirected graphs $G, \delta(G) = n$ implies the existence of vertices $x \neq y$ with $\kappa(x,y;G) \geq n$, as shown in [21]. A direct analogue is not true for digraphs. It is shown by examples in [25] that *for every $m \in \mathbb{N}$, there are a digraph D with $\delta^+(D) = 12m$, but $\max_{x\neq y} \kappa(x,y;D) = 11m$, and a digraph D_m with $\max_{x\neq y} \kappa(x,y;D_m) \leq \min\{\delta^+(D_m), \delta^-(D_m)\} - m$.*

It looks better for the local edge connectivity $\lambda(x,y;D)$.

(4.3) THEOREM [25]. *Every digraph D with $|D| > 1$ has vertices $x \neq y$ with $\lambda(x,y;D) \geq \delta^+(D) - 1$.*

(4.4) CONJECTURE [25]. *There are $x \neq y$ in D with $\lambda(x,y;D) \geq \delta^+(D)$.*

The above examples show that for the local (vertex) connectivity $\kappa(x,y;D)$, a result as Theorem (4.3) does not hold, but it is proved in [26] that the maximal local connectivity grows to infinity, when the minimum outdegree does.

(4.5) THEOREM [26]. *For every $n \in \mathbb{N}$, there is a least integer $g^+(n)$ such that every digraph D with $\delta^+(D) \geq g^+(n)$ has vertices $x \neq y$ with $\kappa(x,y;D) \geq n$.*

The above examples show $g^+(11m+1) > 12m$ for all $m \in \mathbb{N}$. In [26] $g^+(n) = n$ for $n \leq 3$ and $g^+(n) > n$ for $n \geq 9$ are proved. For $n = 4$, only $4 \leq g^+(4) \leq 40$ is known.

Perhaps, Theorem (4.5) can be generalized to the following result, which would be essentially the only possible counterpart of Theorem (1.1) for digraphs.

(4.6) CONJECTURE [25]. *For every $n \in \mathbb{N}$, there is a least non-negative integer $\delta^+(n)$ such that every digraph D with $\delta^+(D) \geq \delta^+(n)$ contains a subdivision of the transitive (=acyclic) tournament of order n.*

Obviously, $\delta^+(n) = n - 1$ for $n \leq 3$. In [27], $\delta^+(4) = 3$ was proved, which generalizes to digraphs the well known result of G.A. Dirac [5] that every graph G with $\delta(G) = 3$ contains a $\dot{K}_4$. If $\delta^+(n)$ exists, then obviously $\delta^+(n) \geq \delta(n)$, since $\delta^+(\overset{\leftrightarrow}{G}) = \delta(G)$ for every graph G. Proving Dirac's Conjecture in section 2, we have shown $\delta(5) = 6$. Also $\delta^+(5) = 6$ could be true, but even the existence of $\delta^+(5)$ is not known.

References

1. B. Bollobás and A. Thomason, *Proof of a conjecture of Mader, Erdös and Hajnal on topological complete subgraphs*, to appear in European Journal of Combinatorics.
2. B. Bollobás and A. Thomason, *Highly linked graphs*, Combinatorica 16 (1996), 313–320.
3. P.A. Catlin, *Hajós' graph-coloring conjecture: variations and counterexamples*, J. Combinatorial Theory (B) 26 (1979), 268–274.
4. R. Diestel, *Graph Theory*, Graduate Texts in Mathematics 173, Springer–Verlag, Berlin – New York 1997.
5. G.A. Dirac, *A property of 4-chromatic graphs and some remarks on critical graphs*, J. London Math. Soc. 27 (1952), 85–92.
6. G.A. Dirac, *In abstrakten Graphen vorhandene vollständige 4–Graphen und ihre Unterteilungen*, Math. Nachr. 22 (1960), 61–85.
7. G.A. Dirac, *Généralistions du théorème de Menger*, C.R. Acad.Sci. (Paris) 250 (1960), 4252–4253.
8. G.A. Dirac, *Homomorphism Theorems for graphs*, Math. Ann. 153 (1964), 69–80.
9. G.A. Dirac, *Chromatic number and topological complete subgraphs*, Canad. Math. Bull. 8 (1965), 711–715.
10. P. Erdös and A. Hajnal, *On complete topological subgraphs of certain graphs*, Ann. Univ. Sci. Budapest, Sect. Math. 7 (1964), 143–149.
11. B. Grünbaum, *Convex Polytopes*, Interscience Publishers, London, New York, Sydney 1967.
12. H. Hadwiger, *Über eine Klassifikation der Streckenkomplexe*, Vierteljahrsschrift der Naturf. Gesellschaft in Zürich 88 (1943) 133–142.
13. R. Halin, *Graphentheorie I*, Wissenschaftliche Buchgesellschaft, Darmstadt 1980.
14. H.A. Jung, *Anwendung einer Methode von K. Wagner bei Färbungsproblemen für Graphen*, Math. Ann. 161 (1965), 325–326.
15. H.A. Jung, *Eine Verallgemeinerung des n–fachen Zusammenhangs für Graphen*, Math. Ann. 187 (1970), 95–103.
16. A.E. Kézdy an P.J. McGuinness, *Do $3n-5$ edges force a subdivision of K_5?*, J. Graph Theory 15 (1991), 389–406.
17. J. Komlós and E. Szemerédi, *Topological cliques in graphs II*, Combinatorics, Probability and Computing 5 (1996), 79–90.
18. D.G. Larman and P. Mani, *On the existence of certain configurations within graphs and the 1–skeletons of polytopes*, Proc. London Math. Soc. 20 (1970), 144–160.
19. W. Mader, *Homomorphieeigenschaften und mittlere Kantendichte von Graphen*, Math. Ann. 174 (1967), 265–268.
20. W. Mader, *Homomorphiesätze für Graphen*, Math. Ann. 178 (1968), 154–168.
21. W. Mader, *Existenz gewisser Konfigurationen in n–gesättigten Graphen und in Graphen genügend großer Kantendichte*, Math. Ann. 194 (1971), 295–312.
22. W. Mader, *Hinreichende Bedingungen für die Existenz von Teilgraphen, die zu einem vollständigen Graphen homöomorph sind*, Math. Nachr. 53 (1972), 145–150.
23. W. Mader, *Existenz n–fach zusammenhängender Teilgraphen in Graphen genügend großer Kantendichte*, Abh. Math. Sem. Univ. Hamburg 37 (1972), 86–97.
24. W. Mader, *Ecken mit starken Zusammenhangseigenschaften in endlichen Graphen*, Math. Ann. 216 (1975), 123-126.
25. W. Mader, *Degree and local connectivity in digraphs*, Combinatorica 5 (1985), 161–165.
26. W. Mader, *Existence of vertices of local connectivity k in digraphs of large outdegree* , Combinatorica 15 (1995), 533–539.
27. W. Mader, *On topological tournaments of order 4 in digraphs of outdegree 3*, J. Graph Theory 21 (1996), 371–376.

28. W. Mader, *An extremal problem for subdivisions of* K_5^-, to appear in Journal of Graph Theory.

29. W. Mader, $3n - 5$ *edges do force a subdivision of* K_5, to appear in Combinatorica.

30. W. Mader, *Topological subgraphs in graphs of large girth*, to appear in Combinatorica.

31. W. Mader, *Graphs without a subdivision of* K_5 *of maximum size*, in preparation.

32. W. Mader, *Subdivisions of* K_{n+2} *in graphs of average degree* $n + \varepsilon$ *and large girth*, in preparation.

33. H. Perfect, *Applications of Menger's Graph Theorem*, J. Math. Analysis and Applications 22 (1968), 96–111.

34. H. Sachs, *Regular graphs with given girth and restricted circuits*, J. London Math. Soc. 38 (1963), 423–429.

35. H. Sachs, *Einführung in die Theorie der endlichen Graphen*, Carl Hanser Verlag, München 1971.

36. C. Thomassen, *Some homeomorphism properties of graphs*, Math. Nachr. 64 (1974), 119–133.

37. C. Thomassen, *Disjoint cycles in digraphs*, Combinatorica 3 (1983), 393–396.

38. C. Thomassen, *Girth in graphs*, J. Combinatorial Theory (B) 35 (1983), 129–141.

39. C. Thomassen, *Even cycles in directed graphs*, Europ. J. Combinatorics 6 (1985), 85–89.

40. C. Thomassen, *Paths, circuits and subdivisions*, in *Selected Topics in Graph Theory 3*, (Eds. L.W. Beineke and R.J. Wilson), Academic Press, London – New York 1988, 97–131.

41. C. Thomassen, *Highly connected non–2–linked digraphs*, Combinatorica 11 (1991), 393–395.

42. C. Thomassen, K_5*–subdivisions in graphs*, Combinatorics, Probability and Computing 5 (1996), 179–189.

43. C. Thomassen, *Dirac's conjecture on* K_5*–subdivisions*, Discrete Mathematics 165/166 (1997), 607–608.

44. W. T. Tutte, *Connectivity in Graphs*, University of Toronto Press 1966.

45. K. Wagner, *Über eine Eigenschaft der ebenen Komplexe*, Math. Ann. 114 (1937), 570–590.

46. K. Wagner, *Bemerkungen zu Hadwigers Vermutung*, Math. Ann. 141 (1960), 433–451.

47. K. Wagner, *Beweis einer Abschwächung der Hadwiger–Vermutung*, Math. Ann. 153 (1964), 139–141.

UNIVERSITY OF HANNOVER, INSTITUTE FOR MATHEMATICS, D 30167 HANNOVER, GERMANY

e–mail address: mader@math.uni–hannover.de

DIMACS Series in Discrete Mathematics
and Theoretical Computer Science
Volume **49**, 1999

Reducible properties and uniquely partitionable graphs

Peter Mihók

ABSTRACT. Let $\mathcal{P}$ be an additive and hereditary property of graphs. A vertex $(n, \mathcal{P})$-partition of a graph G is a partition $\{V_1, V_2, \dots, V_n\}$ of its vertex set $V(G)$ into n classes such that each V_i induces a subgraph $G[V_i]$ with property $\mathcal{P}$. A graph G is said to be uniquely $(n, \mathcal{P})$-partitionable, $n \geq 2$, if G has exactly one $(n, \mathcal{P})$-partition. In this paper we present a survey of recent results on the existence of uniquely $(n, \mathcal{P})$-partitionable graphs. We show that such graphs exist if and only if the additive hereditary property $\mathcal{P}$ is irreducible. Some other criteria of reducibility are presented.

1. Introduction and notation

A graph G is said to be uniquely n-colorable if any two proper n-colorings of G induce the same partition of the vertex set $V(G)$. The concept of uniquely n-colorable graphs was introduced in [**hahe69**]. Open problems and results on uniquely colorable planar graphs can be found in [**jeto95**], Chapter 2.16.

There are many papers dealing with the different generalizations of this concept (see e.g. [**bebr96, both77, brfr97b, brco87, fr93, mi71**]). In this paper we present a survey of recent results on the existence of uniquely partitionable graphs with respect to additive and hereditary properties of graphs.

We consider finite undirected simple graphs and in general we use the standard notation (see e.g. [**grgr95, jeto95**]). Let us denote by $\mathcal{I}$ the class of all simple graphs. A property of graphs $\mathcal{P}$ is any nonempty proper isomorphism closed subclass of $\mathcal{I}$. Let $\mathcal{P}_1, \mathcal{P}_2, \dots, \mathcal{P}_n$ be properties of graphs, a vertex $(\mathcal{P}_1, \mathcal{P}_2, \dots, \mathcal{P}_n)$-partition of a graph G is a partition $\{V_1, V_2, \dots, V_n\}$ of $V(G)$ such that each partition class V_i induces a subgraph $G[V_i]$ of property $\mathcal{P}_i, i = 1, 2, ..., n$. A graph G is said to be uniquely $(\mathcal{P}_1, \mathcal{P}_2, \dots, \mathcal{P}_n)$-partitionable if G has exactly one vertex $(\mathcal{P}_1, \mathcal{P}_2, \dots, \mathcal{P}_n)$-partition i.e. if any two vertex $(\mathcal{P}_1, \mathcal{P}_2, \dots, \mathcal{P}_n)$-partitions of G form the same (unordered) partition of $V(G)$. Let us denote by $\mathcal{P}_1 \circ \mathcal{P}_2 \circ \dots \circ \mathcal{P}_n$ the set of all vertex $(\mathcal{P}_1, \mathcal{P}_2, \dots, \mathcal{P}_n)$-partitionable graphs and by $\boldsymbol{U}(\mathcal{P}_1 \circ \mathcal{P}_2 \circ \dots \circ \mathcal{P}_n)$ the set of uniquely $(\mathcal{P}_1, \mathcal{P}_2, \dots, \mathcal{P}_n)$-partitionable graphs. In the case $\mathcal{P}_1 = \mathcal{P}_2 =$

1991 *Mathematics Subject Classification.* Primary 05C15, 05C35; Secondary 05C75.

Research supported in part by Slovak VEGA Grant 1/4377/97. The author is very grateful to J. Kratochvíl for his help in preparing the final version of the paper.

$\dots = \mathcal{P}_n = \mathcal{P}$ we write $\mathcal{P}_1 \circ \mathcal{P}_2 \circ \dots \circ \mathcal{P}_n = \mathcal{P}^n$, we say that $G \in \mathcal{P}^n$ is $(n, \mathcal{P})$-partitionable and the graphs belonging to $\boldsymbol{U}(\mathcal{P}^n)$ are called (see [**both77, mi85**]) uniquely $(n, \mathcal{P})$-partitionable.

A property of graphs is said to be hereditary if it is closed under taking subgraphs (see [**bobr97, bomi91**]) and additive if it is closed under disjoint union of graphs. (Hereditary properties are called also "monotone", see e.g. [**both97**]). It is easy to see that for $\mathcal{P}_1, \mathcal{P}_2, \dots, \mathcal{P}_n$ additive and hereditary properties of graphs the property $\mathcal{P}_1 \circ \mathcal{P}_2 \circ \dots \circ \mathcal{P}_n$ is additive and hereditary, too. Let us denote the set of all additive and hereditary properties by $\mathbb{L}^a$. The structure of the complete algebraic distributive lattice $(\mathbb{L}^a, \subseteq)$ has been studied in [**bobr97, bomi91**].

For any hereditary property $\mathcal{P}$ there exists a number $c(\mathcal{P})$, called *the completeness of* $\mathcal{P}$ defined in the following way: $c(\mathcal{P}) = \max\{k : K_{k+1} \in \mathcal{P}\}$. It is quite easy to see that the unique hereditary additive property of completeness equal to zero, is the property to be an *edgeless graph*, denoted by $\mathcal{O}$.

An additive and hereditary property $\mathcal{P}$ is said to be *reducible* if there exist $\mathcal{P}_1, \mathcal{P}_2 \in \mathbb{L}^a$ such that $\mathcal{P} = \mathcal{P}_1 \circ \mathcal{P}_2$. Otherwise $\mathcal{P}$ is called *irreducible*. For example, $\mathcal{O}$ (the smallest property of $\mathbb{L}^a$) is irreducible and the smallest reducible property in $\mathbb{L}^a$ is the class $\mathcal{O}^2$ of all bipartite graphs.

In this paper we present a short survey on recent results on the existence of uniquely partitionable graphs. In general the set $\boldsymbol{U}(\mathcal{P}_1 \circ \mathcal{P}_2 \circ \dots \circ \mathcal{P}_n)$ may be empty. In [**mi85**] it was mentioned that $\boldsymbol{U}(\mathcal{P}^n)$ is empty whenever the additive hereditary property $\mathcal{P}$ is reducible. Using the results of the recent paper [**mise97b**] we shall show that for any $n \geq 2$, uniquely $(n, \mathcal{P})$-partitionable graphs exist if and only if the property $\mathcal{P} \in \mathbb{L}^a$ is irreducible.

In Section 2 we summarize some characterizations of reducible properties, in Section 3 results on uniquely partitionable graphs and in the concluding Section 4 some open problems are presented.

2. Reducible properties

Reducible properties of graphs are investigated in the relation to generalized colorings. Their structure plays an important role in the lattice $\mathbb{L}^a$ (see e.g. [**jeto95, bobr97**]). Let us summarize some recent results on the characterization of reducible properties. In [**krsc97**] it is proved that for any additive hereditary property $\mathcal{P} \neq \mathcal{O}$ it is NP-hard to decide if a given graph G belongs to $\mathcal{O} \circ \mathcal{P}$. This leads to the conjecture that the same is true for arbitrary reducible property $\mathcal{R} \neq \mathcal{O}^2$.

The structure of minimal forbidden subgraphs for any reducible property $\mathcal{R} \neq \mathcal{O}^2$ is very complicated. There are many partial results to support the conjecture that, for any reducible property $\mathcal{P} \in \mathbb{L}^a$, the set $\boldsymbol{F}(\mathcal{P})$ of minimal forbidden subgraphs for $\mathcal{P}$ is infinite (see e.g. [**mi85, bomi91**]). Some other results on the structure of $\boldsymbol{F}(\mathcal{P})$ (for example the generalization of the famous Gallai's theorem) are presented in [**bomi91**].

To present our main results we shall substantially use the concept of $\mathcal{P}$-maximal graphs. Any hereditary property $\mathcal{P}$ can be uniquely determined by the set $\boldsymbol{M}(\mathcal{P})$ of *$\mathcal{P}$-maximal graphs* defined by the set $\boldsymbol{M}(n, \mathcal{P})$ of $\mathcal{P}$-maximal graphs of given order n in the following way:

$$\boldsymbol{M}(n, \mathcal{P}) = \{G \in \mathcal{P} : |V(G)| = n \text{ and } G + e \notin \mathcal{P} \text{ for each } e \in E(\overline{G})\}$$

and $\boldsymbol{M}(\mathcal{P}) = \bigcup_{n=1}^{\infty} \boldsymbol{M}(n, \mathcal{P})$.

It is easy to see that $\boldsymbol{M}(n,\mathcal{P}) = \{K_n\}$ for $n = 1, 2, \ldots, c(\mathcal{P})+1$ and it is useful to define the set $\boldsymbol{M}^*(\mathcal{P}) = \bigcup_{n=c(\mathcal{P})+1}^{\infty} \boldsymbol{M}(n,\mathcal{P})$. We will use also the related notion of $\mathcal{P}$-strict graphs. If G is a graph with property $\mathcal{P}$, such that the graph $G + K_1$ does not belong to $\mathcal{P}$, then G is called *$\mathcal{P}$-strict*. It is easy to see that all $\mathcal{P}$-maximal graphs in $\boldsymbol{M}^*(\mathcal{P})$ are $\mathcal{P}$-strict. The structure of $\mathcal{P}$-maximal graphs for $\mathcal{P} \in \mathbf{L}^a$ has been investigated e.g. in [**bobr97, brfr97a, krmi97, se97**].

The *join* $G_1 + G_2 + \cdots + G_n$ of n graphs $G_1, G_2, \ldots, G_n$ is the graph consisting of the disjoint union of G_i's and all the edges between $V(G_i)$ and $V(G_j)$ for any $1 \leq i < j \leq n$. If a graph G is a join we say that it is *decomposable.* A graph that is not decomposable is called *indecomposable.* It is easy to see that a graph G is decomposable if and only if its complement $\overline{G}$ is not connected. Then G is the join of the complements of the components of $\overline{G}$. Thus every decomposable graph G can be expressed in a unique way as the join of indecomposable graphs. The number of those indecomposable graphs is called *the decomposability number* of G and denoted by $dc(G)$.

The next Lemma provides a simple characterization of $\mathcal{P}_1 \circ \mathcal{P}_2 \circ \cdots \circ \mathcal{P}_n$-maximal graphs:

LEMMA 2.1. *Let $\mathcal{P}_1, \mathcal{P}_2, \ldots, \mathcal{P}_n$ be hereditary properties of graphs. Then a graph G belongs to $\boldsymbol{M}(\mathcal{P}_1 \circ \mathcal{P}_2 \circ \cdots \circ \mathcal{P}_n)$, $n \geq 2$, if and only if for every $(\mathcal{P}_1 \circ \mathcal{P}_2 \circ \cdots \circ \mathcal{P}_n)$-partition $\{V_1, V_2, \ldots, V_n\}$ of $V(G)$ the following holds: $G[V_i] \in \boldsymbol{M}(\mathcal{P}_i)$ for $i = 1, 2, \ldots, n$ and $G = G[V_1] + G[V_2] + \cdots + G[V_n]$.*

Thus, if $\mathcal{R} \in \mathbf{L}^a$ is a reducible property, then the set $\boldsymbol{M}^*(\mathcal{R})$ contains only decomposable graphs. It is proved in [**mise97b**] that for any additive reducible property the converse assertion is also valid:

THEOREM 2.2. [**mise97b**] *A hereditary additive property $\mathcal{R}$ is reducible if and only if all the $\mathcal{R}$-maximal graphs with at least two vertices are decomposable.*

Let us recall the main idea of the proof, the details are presented in [**mise97b**]:

A set $\mathcal{G} \subseteq \boldsymbol{M}^*(\mathcal{P})$ is a *generating set of* $\mathcal{P}$ if every graph from $\mathcal{P}$ is a subgraph of some graph from $\mathcal{G}$. The members of $\mathcal{G}$ are called *generators* of $\mathcal{P}$. If $\mathcal{P}$ is a hereditary property and $\mathcal{G}$ is a generating set of $\mathcal{P}$ then the *decomposability number of a generating set* $\mathcal{G}$ (denoted by $dc(\mathcal{G})$) is defined as the minimum of the decomposability numbers of the graphs belonging to $\mathcal{G}$. It is easy to prove that any generating set $\mathcal{G}$ of an additive hereditary property $\mathcal{P}$ has the same decomposability number $dc(\mathcal{G}) = dc(\boldsymbol{M}^*(\mathcal{P}))$, called the decomposability number of $\mathcal{P}$ and denoted by $dc(\mathcal{P})$. Moreover it is proved that there exists a generating set $\mathcal{G}^* \subset \boldsymbol{M}^*(\mathcal{P})$ of $\mathcal{P}$ which contains only the graphs with decomposability number equal to $dc(\mathcal{P})$.

The proof of Theorem 2.2 is based on the construction of the corresponding irreducible factors of the given reducible property $\mathcal{P}$. It is proved that if $\mathcal{P} = \mathcal{P}_1 \circ \mathcal{P}_2 \circ \ldots \circ \mathcal{P}_n$ is a factorization of $\mathcal{P}$ into irreducible factors, then $dc(\mathcal{P}) = n$. Since the generators of $\mathcal{G}^*$ are uniquely decomposable into n indecomposable graphs, we can obtain the affirmative answer to the Problem 17.9. of [**jeto95**]:

THEOREM 2.3. [**mise97b**] *Any additive and hereditary property is in $\mathbf{L}^a$ uniquely factorizable into irreducible factors.*

The proof of Theorem 2.3 implies also an affirmative answer to the problem posted in [**sztu97**] (see Problem 4, page 144):

THEOREM 2.4. **[mise97b]** *Let $\mathcal{P}_1, \mathcal{P}_2, \ldots, \mathcal{P}_n$ be hereditary properties of graphs such that the property $\mathcal{P}_1 \circ \mathcal{P}_2 \circ \cdots \circ \mathcal{P}_n$, $n \geq 2$, is additive and hereditary. Then also the properties $\mathcal{P}_i$, $1 \leq i \leq n$, are additive.*

3. Uniquely partitionable graphs

If we follow the proof of Theorem 2.2, we can see that in fact any reducible additive and hereditary property $\mathcal{R}$ is generated by uniquely partitionable graphs with respect to its irreducible factors. More precisely:
Let $\mathcal{G}^*$ be a generating set of a reducible property $\mathcal{P} = \mathcal{P}_1 \circ \mathcal{P}_2 \circ \cdots \circ \mathcal{P}_n$, $n \geq 2$ consisting of $\mathcal{P}$-maximal graphs of decomposability number equal to $dc(\mathcal{P}) = n$. (The existence of such generating set is proved in **[mise97b]**). Since the generators of $\mathcal{G}^*$ are uniquely decomposable into n indecomposable graphs, by Lemma 2.1 these generators have exactly one vertex $(\mathcal{P}_1, \mathcal{P}_2, \ldots, \mathcal{P}_n)$-partition. This implies also an affirmative answer to the Conjecture 47 of **[bobr97]**:

THEOREM 3.1. *Let $\mathcal{R} = \mathcal{P}_1, \mathcal{P}_2, \ldots, \mathcal{P}_n$, $n \geq 2$ be a factorization of a reducible property $\mathcal{R} \in \mathbf{L}^a$ into irreducible factors.*
Then $\boldsymbol{U}(\mathcal{P}_1 \circ \mathcal{P}_2 \circ \cdots \circ \mathcal{P}_n) \neq \emptyset$.
In particular, $\boldsymbol{U}(\mathcal{P} \circ \mathcal{P} \circ \cdots \circ \mathcal{P}) \neq \emptyset$ if and only if $\mathcal{P}$ is irreducible.

In the case that $\mathcal{P} = \mathcal{P}_1 \circ \mathcal{P}_2 \circ \cdots \circ \mathcal{P}_n$ and the properties $\mathcal{P}_i$ are not necessarily irreducible the existence of uniquely $(\mathcal{P}_1, \mathcal{P}_2, \ldots, \mathcal{P}_n)$-partitionable graphs is characterized in **[bu97]**. It is shown that the natural necessary "divisibility condition" that any pair of properties $\mathcal{P}_i, \mathcal{P}_j$ has "no proper common divisor" is also sufficient. We are omitting the technical details.

Let us summarize the previous results in the form of criteria of irreducibility:

THEOREM 3.2. *Let $\mathcal{P}$ be an additive and hereditary property of graphs. Then the following statements are equivalent:*

1. *the property $\mathcal{P}$ is irreducible*
2. *for every $n \geq 2$ the property $\mathcal{P}^n$ can be generated by a set of uniquely $(n, \mathcal{P})$-partitionable graphs*
3. *there exist uniquely $(n, \mathcal{P})$-partitionable graphs for every $n \geq 2$*
4. *the property $\mathcal{P}$ can be generated by a set of indecomposable graphs*
5. *there exists an indecomposable $\mathcal{P}$-maximal graph*

As an immediate consequence we have:

THEOREM 3.3. *Let $\mathcal{P}$ be an additive and hereditary property of graphs. Then the following statements are equivalent:*

1. *the property $\mathcal{P}$ is reducible*
2. *there are no uniquely $(n, \mathcal{P})$-partitionable graphs for $n \geq 2$*
3. *every $\mathcal{P}$-maximal graph of order at least two is decomposable*

Some results on uniquely $\mathcal{P}^n$-partitionable graphs obtained in **[mi85]** have been generalized in **[bufr97]** for uniquely $(\mathcal{P}_1, \mathcal{P}_2, \ldots, \mathcal{P}_n)$-partitionable graphs. Let us recall their basic properties in the following Theorem.

THEOREM 3.4. **[bufr97]** *Let $\mathcal{P}_1, \mathcal{P}_2, \ldots, \mathcal{P}_n$ be additive hereditary properties of graphs, $n \geq 2$. If G is a uniquely $(\mathcal{P}_1, \mathcal{P}_2, \ldots, \mathcal{P}_n)$-partitionable graph and $\{V_1, V_2, \ldots, V_n\}$ is the unique $(\mathcal{P}_1, \mathcal{P}_2, \ldots, \mathcal{P}_n)$-partition of $V(G)$, then*

1. *the subgraphs $G[V_i]$ are $\mathcal{P}_i$-strict, $i = 1, 2, \ldots, n$;*

2. *if* $\{i_1,\ldots,i_k\} \subseteq \{1,\ldots,n\}$, *then* $V_{i_1}\cup\cdots\cup V_{i_k}$ *induces a uniquely* $(\mathcal{P}_{i_1},\ldots,\mathcal{P}_{i_n})$-*partitionable subgraph of* G,
3. $|V(G)| \geq \sum_{i=1}^{n}(c(\mathcal{P}_i)+2)-1$,
4. *the graph* $G = G[V_1]+\cdots+G[V_n]$ *is uniquely* $(\mathcal{P}_1,\mathcal{P}_2,\ldots,\mathcal{P}_n)$-*partitionable.*

In the first paper on uniquely colorable graphs [**hahe69**] it has been proved that every k-colorable graph is an induced subgraph of a uniquely k-colorable graph. Also the generalization of this fact holds:

THEOREM 3.5. [**bu97**] *If* $H \in \mathcal{P}_1\circ\mathcal{P}_2\circ\ldots\circ\mathcal{P}_n$ *and* $\mathbf{U}(\mathcal{P}_1,\mathcal{P}_2,\cdots,\mathcal{P}_n) \neq \emptyset$, *then* H *is an induced subgraph of some uniquely* $(\mathcal{P}_1,\mathcal{P}_2,\ldots,\mathcal{P}_n)$-*partitionable graph* G.

4. Generalizations and open problems

Let us remark, that in all presented results the requirement of additivity is substantial and it cannot be omitted. To show it, let us recall the example presented in [**mise97b**]:

Let the hereditary (nonadditive!!) properties $\mathcal{Q}_i$, $i = 0,1,\ldots,4$ be generated by the following sets of graphs:

$$\mathcal{G}_j = \{\{C_k : k \geq 5 \text{ and } k \not\equiv j (\text{mod } 3)\}, j = 0,1,2.$$

$$\mathcal{G}_3 = \{C_k : k \geq 5\}$$

and

$$\mathcal{G}_4 = \{C_k + C_l :\ 5 \leq k < l \text{ and } k \not\equiv l (\text{mod } 3)\}.$$

Then the properties $\mathcal{Q}_i$, $i = 0,1,\ldots,4$ are irreducible, however

$$\mathcal{Q}_0\circ\mathcal{Q}_1\circ\mathcal{Q}_2 = \mathcal{Q}_3\circ\mathcal{Q}_4$$

and there are no uniquely $(\mathcal{Q}_4,\mathcal{Q}_4)$-partitionable graphs.

A property $\mathcal{P}$ is said to be *induced hereditary* if it is closed under taking induced subgraphs. Obviously any hereditary property is induced hereditary, but there are many important induced hereditary properties which are not hereditary, e.g. *to be a perfect graph, claw-free graphs* etc.

All presented results remain open for additive and induced hereditary properties of graphs.

A comprehensive list of references to the papers dealing with the existence of uniquely $(n,\mathcal{P})$-partitionable graphs with respect to induced hereditary properties may be found in the recent paper [**acbr98**]. In this paper it is proved that such graphs exist whenever the induced hereditary property $\mathcal{P}$ has exactly one minimal forbidden induced subgraph. This result and some other unpublished partial results lead to the following conjecture:

CONJECTURE 4.1. Let $\mathcal{P}$ be any additive and induced hereditary property of graphs.

Then $\boldsymbol{U}(\mathcal{P}^n) \neq \emptyset$ if and only if $\mathcal{P}$ is irreducible.

References

[acbr98] D. Achlioptas, J.I. Brown, D.G. Corneil, M.S.O. Molloy, *The existence of uniquely* $-G$ *colourable graphs*, Discrete Math. **179** (1998), 1–11.

[bebr96] G. Benadé, I. Broere, B. Jonck and M. Frick, *Uniquely* $(m,k)^\tau$ *- colourable graphs and* $k-\tau$*-saturated graphs*, Discrete Math. **162** (1996), 13–22.

[both77] B. Bollobás and A. G. Thomason, *Uniquely partitionable graphs*, J. London Math. Soc. (2) **16** (1977), 403–410.

[both97] B. Bollobás and A. G. Thomason, *Hereditary and monotone properties of graphs*, in: R. L. Graham and J. Nešetřil, eds., The mathematics of Paul Erdős, II, Algorithms and Combinatorics vol. 14 (Springer-Verlag, 1997), pp. 70–78.

[bobr97] M. Borowiecki, I. Broere, M. Frick, P. Mihók and G. Semanišin, *Survey of hereditary properties of graphs*, Discussiones Mathematicae - Graph Theory **17** (1997), 5–50.

[bomi91] M. Borowiecki and P. Mihók, *Hereditary properties of graphs*, in: V. R. Kulli, ed., Advances in Graph Theory, Vishwa International Publication, Gulbarga, 1991, pp. 42–69.

[brfr97a] I. Broere, M. Frick and P. Mihók, *The order of uniquely partitionable graphs*, Discussiones Mathematicae - Graph Theory **17** (1997), 115–126.

[brfr97b] I. Broere, M. Frick and G. Semanišin, *Maximal graphs with respect to hereditary properties*, Discussiones Mathematicae - Graph Theory **17** (1997), 51–66.

[brco87] J. I. Brown and D. G. Corneil, *On generalized graph colourings*, J. Graph Theory **11** (1987), 86–99.

[bu97] I. Broere, J. Bucko, *Divisibility in properties of graphs*, submitted to the Tatra Mountains, Proceedings of the Workshop: Cycles and Colorings, Stará Lesná 1997.

[bufr97] J. Bucko, M. Frick, P. Mihók and R. Vasky, *Uniquely partitionable graphs*, Discussiones Mathematicae - Graph Theory **17** (1997), 103–114.

[fr93] M. Frick, *A survey of (m,k)-colourings*, in: J. Gimbel c.a, ed., Quo Vadis, Graph Theory? Annals of Discrete Mathematics vol. 55 (North-Holland, Amsterdam, 1993), 45–58.

[grgr95] R. L. Graham, M. Grötschel and L. Lovász, Handbook of combinatorics (Elsevier Science B.V., Amsterdam, 1995).

[hahe69] F. Harary, S. T. Hedetniemi and R. W. Robinson, *Uniquely colourable graphs*, J. Comb. Theory **6** (1969), 264–270.

[jeto95] T. R. Jensen and B. Toft, Graph colouring problems (Wiley–Interscience Publications, New York, 1995).

[krmi97] J. Kratochvíl, P. Mihók and G. Semanišin, *Graphs maximal with respect to hom-properties*, Discussiones Mathematicae - Graph Theory **17** (1997), 77–88.

[krsc97] J. Kratochvíl, I. Schiermeyer, *On the computational complexity of $(\mathcal{O},\mathcal{P})$-partition problems*, Discussiones Mathematicae - Graph Theory **17** (1997), 253–258.

[mi85] P. Mihók, *Additive hereditary properties and uniquely partitionable graphs*, in: M. Borowiecki and Z. Skupien, eds., Graphs, hypergraphs and matroids (Zielona Góra, 1985) 49–58.

[mise97b] P. Mihók, G. Semanišin and R. Vasky, *Additive and hereditary properties are uniquely factorizable into irreducible factors*, Report No. IM - 5 - 97, December 1997, Zielona Góra, submitted to J. Graph Theory.

[mi71] J. Mitchem, *Uniquely k-arborable graphs*, Israel Journal of Mathematics **10** (1971), 17–25.

[se97] G. Semanišin, *On some variations of extremal graph problems*, Discussiones Mathematicae - Graph Theory **17** (1997), 67–76.

[sztu97] J. Szigeti and Zs. Tuza, *Generalized colorings and avoidable orientations*, Discussiones Mathematicae - Graph Theory **17** (1997), 137–146.

Mathematical Institute of Slovak Academy of Sciences, Grešákova 6, 040 01 Košice, Slovakia

E-mail address: mihok@kosice.upjs.sk

DIMACS Series in Discrete Mathematics
and Theoretical Computer Science
Volume **49**, 1999

Induced Monochromatic Subconfigurations

Jaroslav Nešetřil, József Solymosi, and Pavel Valtr

ABSTRACT. We consider induced monochromatic configurations in edge-colored planar point configurations. Contrary to the positive results for partitions of points ([**NV94, NV98**]) we prove two strong negative results which indicate that the Erdős–Szekeres Theorem (Happy-End Theorem) is the only positive instance of the problem.

1. Introduction

We say that a set of points in the plane is *in general position* if no three points lie on a line. A finite set of points in general position in the plane is called a *configuration.* The *orientation* of a triple (a, b, c) of non-collinear points in the plane is *positive* if the counterclockwise angle from $b-a$ to $c-a$ lies in the interval $(0, \pi)$,

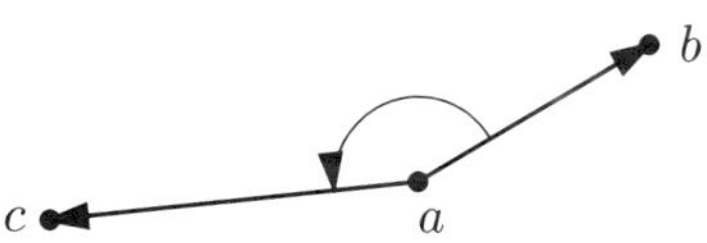

FIGURE 1. A positively oriented triple (a, b, c) (the triple (a, c, b) has negative orientation).

otherwise the orientation is said to be *negative,* see Fig. 1. Two configurations are said to be *equivalent* if there exists a bijection between them preserving orientations of non-collinear triples of points. A configuration A *contains* a configuration B (or: B is a *subconfiguration* of A), if $B \subseteq A$. The convex hull of a configuration A is denoted by $\operatorname{conv} A$. For a point p, its x-coordinate is denoted by $x(p)$. We say that a subconfiguration B of A is *induced (in A)*, if $A \cap \operatorname{conv} B = B$. A configuration is in *convex position* if its points are vertices of a convex polygon. Induced subconfiguration of A which is in convex position is also called a *hole (in A)*. Inspired by the celebrated Erdős–Szekeres Theorem and by other results in Ramsey Theory (see [**GRS80, NR90**]) the following result has been (among others) shown in [**NV94**]:

1991 *Mathematics Subject Classification.* Primary 05D10; Secondary 05C55.

Research of J.S. and P.V. was supported by DIMACS Center, research of J.N. and P.V. was supported by Czech Republic Grant GAČR 201/1996/0194 and by Charles University grants No. 193/1996 and 194/1996.

THEOREM 1. *For any configuration X and any positive integer k there exists a configuration Y such that for any coloring of points of Y by k colors there is a monochromatic induced subconfiguration of Y which is equivalent to X.*

In this paper we show that an analogous statement for colorings of pairs of points is false in a very strong sense. In fact we show that up to two open cases all positive instances are guaranteed by a combination of the Erdős–Szekeres and the Ramsey Theorems. More precisely this can be formulated as follows:

Let X_1, X_2, Y be configurations. We write $Y \longrightarrow (X_1, X_2)$ to denote the validity of the following statement: For any coloring of pairs of points of Y by two colors (say red and blue) Y contains an induced subconfiguration which either is equivalent to X_1 and with all pairs of its points colored by red, or is equivalent to X_2 and with all pairs of its points colored by blue. The non-validity of the statement will be denoted by $Y \not\longrightarrow (X_1, X_2)$. In this paper we show the following two results:

THEOREM 2. *Let $K(3,1)$ be the non-convex 4-point configuration (see Fig. 2). Then for any configuration Y holds $Y \not\longrightarrow (K(3,1), K(3,1))$.*

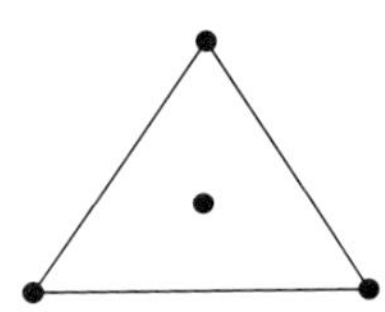

$K(3,1)$

FIGURE 2. The configuration $K(3,1)$.

THEOREM 3. *Let T be the 3-point configuration and let K be a configuration such that at least two points of K lie inside the convex hull of K. Then for any configuration Y holds $Y \not\longrightarrow (T, K)$.*

Theorem 2 will be proved in Section 2 and Theorem 3 will be proved in Section 3. Before doing so we add a few remarks:

REMARK. It follows from the Ramsey Theorem that for any two configurations C_1, C_2 in convex position there is a configuration Y with $Y \longrightarrow (C_1, C_2)$. (Any large set in convex position will do.) On the other hand, any non-convex configuration contains a subconfiguration equivalent to $K(3,1)$. Therefore, Theorems 2 and 3 nearly characterize all pairs (K_1, K_2) of configurations for which there exists at least one configuration Y satisfying $Y \longrightarrow (K_1, K_2)$. The only remaining case is when K_1 is in convex position and K_2 is in convex position after removal of one point only.

REMARK. This nearly complete solution is another indication that preserving the holes is a very restrictive condition for plane configurations. This should be compared with [**NV98**] where it has been proved that there are configurations X for which there is no Y such that $Y \longrightarrow (X, X)$ holds even for subconfigurations (which are not necessarily induced). This one can be restated (in a slightly stronger form) as follows [**NV98**]: There exists a configuration X and a 2-coloring of all the pairs in the plane such that no subconfiguration equivalent to X is monochromatic.

The configuration X in [**NV98**] is finite but very large and in fact the proof given in [**NV98**] doesn't produce any bound on its size. On the other hand for many configurations X the existence of a configuration Y with $Y \longrightarrow (X, X)$ (for non-induced case) is guaranteed by [**NV94**] and our companion paper [**NSV99**].

2. Proof of Theorem 2

Let Y be a configuration. Choose a point p outside $\mathrm{conv}(Y)$ such that $Y \cup \{p\}$ is in general position. We color pairs of points of Y by red and blue so that a pair $\{a, b\} \in \binom{Y}{2}$ is colored red if the interior of the triangle abp contains an even number of points of Y, otherwise it is colored blue. We claim that no induced subconfiguration equivalent to $K(3,1)$ has all pairs of points colored by the same color.

Let K be an induced subconfiguration of Y equivalent to $K(3,1)$. We denote the points of K by a_1, a_2, a_3, a_4 in the order in which they are seen from p (see Fig. 3). It is easy to check that one of the 4-point configurations $\{p, a_1, a_2, a_3\}$,

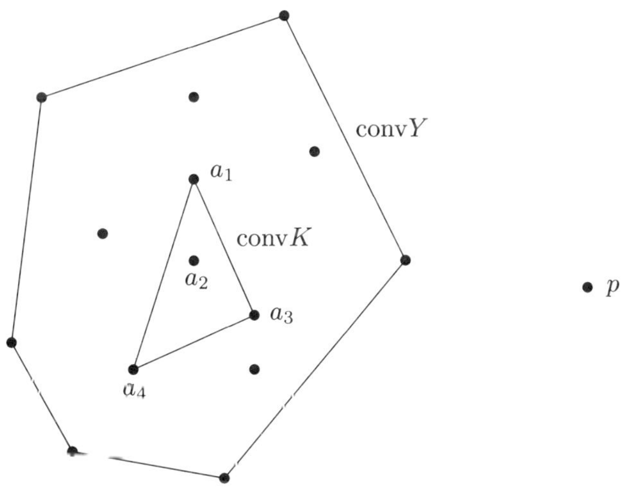

FIGURE 3. The points a_1 through a_4.

$\{p, a_2, a_3, a_4\}$ is in convex position and the other one is not. However, if $\{p, a_i, a_{i+1}, a_{i+2}\}$ is in convex position, then the interior of at least one of the triangles pa_ia_{i+1}, pa_ia_{i+2}, $pa_{i+1}a_{i+2}$ contains an even number of points of Y. Similarly, if $\{p, a_i, a_{i+1}, a_{i+2}\}$ is not in convex position, then the interior of at least one of the triangles pa_ia_{i+1}, pa_ia_{i+2}, $pa_{i+1}a_{i+2}$ contains an odd number of points of Y. It follows that the pairs $\{a_1, a_2\}$, $\{a_1, a_3\}$, $\{a_2, a_3\}$, $\{a_2, a_4\}$, $\{a_3, a_4\}$ cannot be colored all by the same color.

Theorem 2 is proved.

3. Proof of Theorem 3

Our proof of Theorem 3 relies on the following lemma:

LEMMA 4. *Let K be a configuration such that the convex hull of K contains at least two points of K in the interior. Then K contains an induced subconfiguration equivalent to one of the three configurations $K(3,2)$, $K(4,2)$, $K(3,3)$ shown in Fig. 4.*

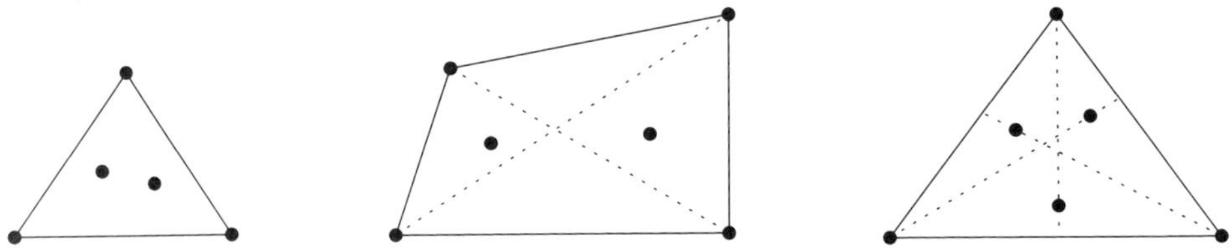

$K(3,2)$ $K(4,2)$ $K(3,3)$

FIGURE 4. The configurations $K(3,2)$, $K(4,2)$, $K(3,3)$.

PROOF. First, observe that K contains an induced configuration K_0 such that $\operatorname{conv} K_0$ has 3 or 4 vertices and contains at least two points of K in the interior. To see this, take a line through any pair of points of K inside $\operatorname{conv} K$. It crosses the boundary of $\operatorname{conv} K$ in two different sides (segments) s_1 and s_2. Then we can take $K_0 = K \cap \operatorname{conv}(s_1 \cup s_2)$.

Now, consider the configuration K_0. If K_0 is equivalent to $K(4,2)$, then the lemma holds. Otherwise, if we take the convex hull of some three vertices of $\operatorname{conv} K_0$, we obtain an induced subconfiguration K_1 such that $\operatorname{conv} K_1$ is a triangle with at least two points of K inside it.

If there are more than three points inside $\operatorname{conv} K_1$, then we can find three points in K_1 with fewer (but at least two) points of K in their convex hull as follows. Denote the vertices of $\operatorname{conv} K_1$ by a, b, c, and let d be a point inside abc nearest to the line ab. The triangle dab contains no point inside. Thus, the interior of one of the triangles dac, dbc contains at least two points. Repeating this argument, we obtain an induced subconfiguration K_2 such that $\operatorname{conv} K_2$ is a triangle with two or three points of K in the interior. If the interior of $\operatorname{conv} K_2$ contains two points of K, then K_2 is equivalent to $K(3,2)$. So suppose that it contains three points of K.

Denote the vertices of $\operatorname{conv} K_2$ by a, b, c, and let d be a point of K lying inside $\operatorname{conv} K_2$. If the remaining two points of K inside $\operatorname{conv} K_2$ lie in the same of the triangles abd, acd, bcd, then there is an induced configuration equivalent to $K(3,2)$. Thus, we may assume that the interior of abd contains a single point u of K and the interior of acd contains a single point v of K (see Fig. 5). If K_2 is equivalent to $K(3,3)$, then the lemma holds. Otherwise, either the line cu separates the points b and d, or the line bv separates the points c and d. In the first case $K_2 \setminus \{b\}$ is equivalent to $K(3,2)$, in the second case $K_2 \setminus \{c\}$ is equivalent to $K(3,2)$. □

We now prove Theorem 3. Let K be a configuration such that at least two points of K lie inside $\operatorname{conv} K$. Let Y be an arbitrary configuration. In the following, we define a 2-coloring of pairs of points of Y showing that $Y \not\longrightarrow (T, K)$.

We may assume that no pair of points of Y lies on a vertical line. For a pair of points $a, b, x(a) < x(b)$, we define S_{ab} as the set of points lying under the segment ab:

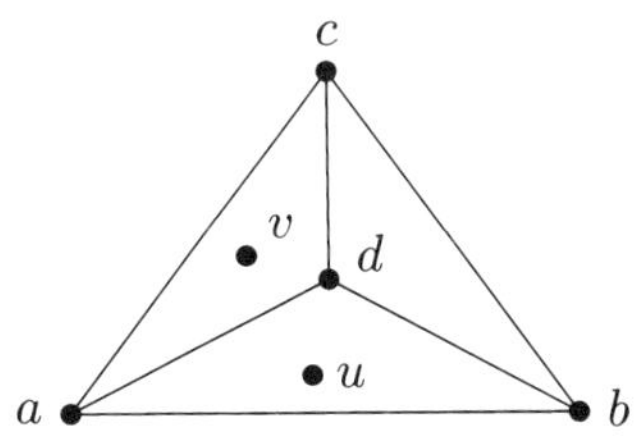

FIGURE 5. The points d, u, v.

$$S_{ab} =: \{p \in \mathbb{R}^2 : x(a) < x(p) < x(b) \text{ and } p \text{ lies below the line } ab\}.$$

We color a pair $\{a, b\}$ of points of Y red, if $|Y \cap S_{ab}| \equiv 1 (\mathrm{mod}\ 3)$ (i.e., the number of points of Y in the strip S_{ab} is $1 (\mathrm{mod}\ 3)$). Otherwise we color it blue. We call a configuration C *red*, if all pairs of points of C are colored red. Analogously we define *blue* configurations. It is easy to see that Y contains no empty red configuration equivalent to T.

To show Theorem 3, it therefore suffices to show that Y contains no empty blue configuration equivalent to K. According to Lemma 4, Theorem 3 follows from the following lemma.

LEMMA 5. *Let Y be a configuration with pairs of points colored as described above. Let $K_0 \subseteq Y$ be an empty blue subconfiguration of Y. Then*

(i) *K_0 is not equivalent to $K(3,2)$,*
(ii) *K_0 is not equivalent to $K(4,2)$, and*
(iii) *K_0 is not equivalent to $K(3,3)$.*

We prove Lemma 5 in the following four sections.

4. Proof of Lemma 5(i)

Let K_0 be an empty subconfiguration of Y equivalent to $K(3,2)$. We will show that K_0 is not blue.

Let $a_1, a_2, \ldots, a_5$ be the points of K_0 listed according to increasing x-coordinate. We denote by S_1, S_2, S_3, S_4 the four strips (from left to right) determined by the lower envelope of conv K_0 and by the 5 vertical halflines emanating from the points a_i downward (see schematic Fig. 6). For $i = 1, 2, 3, 4$, we define $c_i \in \{0, 1, 2\}$ by $c_i \equiv |(Y \setminus K_0) \cap S_i| (\mathrm{mod}\ 3)$. Further, we put $C = c_1 + c_2 + c_3 + c_4$.

LEMMA 6. *If K_0 is blue then*

(i) *$c_i = 0$ or $c_i = 2$, for every $i = 1, 2, 3, 4$, and*
(ii) *either $C = 2$ or $C = 6$.*

PROOF. (i) If $c_i = 1$ for some i, then $\{a_i, a_{i+1}\}$ is red.

(ii) By (i), C is even.

If $C = 0$ (i.e., $c_1 = c_2 = c_3 = c_4 = 0$), choose i such that a_{i+1} lies below the line $a_i a_{i+2}$. Then the pair $\{a_i, a_{i+2}\}$ is colored red.

If $C = 8$ (i.e., $c_1 = c_2 = c_3 = c_4 = 2$), choose i such that a_{i+1} lies above the line $a_i a_{i+2}$. Then the pair $\{a_i, a_{i+2}\}$ is colored red.

Finally, if $C = 4$, then $\{a_1, a_5\}$ is colored red. □

We now suppose that K_0 is blue, and distinguish several cases (see Fig. 7 for the first four cases).

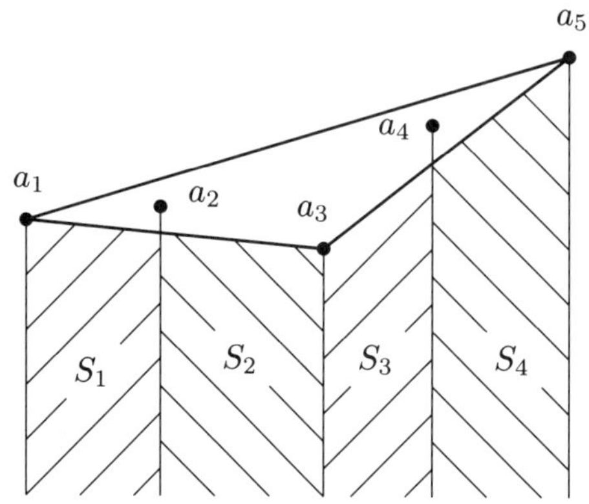

FIGURE 6. The strips S_i.

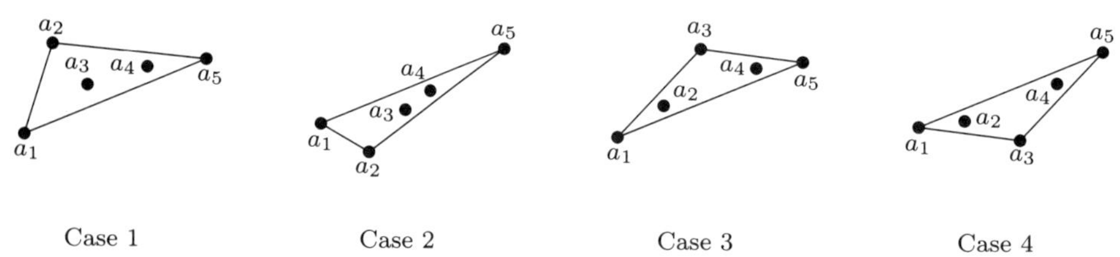

FIGURE 7. Cases 1–4.

Case 1: conv $K_0 = \{a_1, a_2, a_5\}$ *and* a_2 *lies above the line* a_1a_5. If $C = 2$, then either $c_1 = 0$ and $\{a_2, a_5\}$ is red, or $c_1 = 2$ and (at least) one of the pairs $\{a_2, a_4\}$, $\{a_3, a_5\}$ is red.

If $C = 6$, then either $c_1 = c_2 = 2$ and $\{a_1, a_3\}$ is red, or $c_1 + c_2 = c_3 = c_4 = 2$ and one of the pairs $\{a_1, a_4\}$, $\{a_3, a_5\}$ is red.

Case 2: conv $K_0 = \{a_1, a_2, a_5\}$ *and* a_2 *lies below the line* a_1a_5. If $C = 2$, then either $c_1 + c_2 = 0$ and $\{a_1, a_3\}$ is red, or $c_1 + c_2 = 2$ and (at least) one of the pairs $\{a_1, a_4\}$, $\{a_3, a_5\}$ is red.

If $C = 6$, then either $c_1 = 2$ and $\{a_2, a_5\}$ is red, or $c_1 = 0$ and one of the pairs $\{a_2, a_4\}$, $\{a_3, a_5\}$ is red.

Case 3: conv $K_0 = \{a_1, a_3, a_5\}$ *and* a_3 *lies above the line* a_1a_5. If $C = 2$, then either $c_1 + c_2 = 0$ and $\{a_1, a_3\}$ is red, or $c_3 + c_4 = 0$ and $\{a_3, a_5\}$ is red.

If $C = 6$, then either $c_2 + c_3 = 4$ and $\{a_2, a_4\}$ is red, or $c_1 = c_2 + c_3 = c_4 = 2$ and one of the pairs $\{a_1, a_4\}$, $\{a_2, a_5\}$ is red.

Case 4: conv $K_0 = \{a_1, a_3, a_5\}$ *and* a_3 *lies below the line* a_1a_5. If $C = 2$, then either $c_2 + c_3 = 0$ and $\{a_2, a_4\}$ is red, or $c_2 + c_3 = 2$ and one of the pairs $\{a_1, a_4\}$, $\{a_2, a_5\}$ is red.

If $C = 6$, then either $c_1 + c_2 = 4$ and $\{a_1, a_3\}$ is red, or $c_3 + c_4 = 4$ and $\{a_3, a_5\}$ is red.

Case 5: conv $K_0 = \{a_1, a_4, a_5\}$. This case is always symmetric to Case 1 or 2.

Lemma 5(i) is proved.

5. Configuration $K(3,1)$

In this section we prove a lemma which we apply in the proof of Lemma 5(ii),(iii). Let $K(3,1)$ be the non-convex 4-point configuration (see Fig. 2), and let K_1 be an empty blue subconfiguration of Y equivalent to $K(3,1)$. Let a_1, a_2, a_3, a_4 be the points of K_1 listed according to increasing x-coordinate. Similarly as in the previous section, let S_1, S_2, S_3 be the three strips bounded by the lower envelope of $\operatorname{conv} K_1$ and by the four vertical halflines emanating from the points a_i downward. Define $c_i \in \{0,1,2\}$ by $c_i \equiv |(Y \setminus K_1) \cap S_i| (\mathrm{mod}\ 3)$ and put $C = c_1 + c_2 + c_3$.

LEMMA 7. *Let K_1 be a configuration as above. Then*
(i) *if a_2, a_3 lie below the line a_1a_4 then $C = 4$, and*
(ii) *if a_2, a_3 lie above the line a_1a_4 then $C = 2$.*

PROOF. Certainly, each c_i is 0 or 2.
If $C = 0$ then either $\{a_1, a_3\}$ or $\{a_2, a_4\}$ is red.
If $C = 6$ then also either $\{a_1, a_3\}$ or $\{a_2, a_4\}$ is red.
In case (i), if $C = 2$ then $\{a_1, a_4\}$ is red.
In case (ii), if $C = 4$ then $\{a_1, a_4\}$ is red.
This completes the proof. □

6. Proof of Lemma 5(ii)

Let K_0 be an empty blue subconfiguration of Y equivalent to $K(4,2)$. Let a, b, c, d be the vertices of $\operatorname{conv} K_0$ listed in the counterclockwise order so that a is the leftmost of them. We say that a point p *lies under* a segment qr if $x(q) < x(p) < x(r)$ and p lies strictly under the line qr. We distinguish three cases.

Case 1: d is the rightmost point of K_0. By Lemma 7(i) for the configuration $Y \cap \operatorname{conv}\{a,b,c\}$, $0 (\mathrm{mod}\ 3)$ points of Y lie under the segment ac. By Lemma 7(i) for the configuration $Y \cap \operatorname{conv}\{a,c,d\}$, $0 (\mathrm{mod}\ 3)$ points of Y lie under the segment ad. Since there is one point of Y inside the triangle acd, it follows that $1 (\mathrm{mod}\ 3)$ points of Y lie under the segment cd. Thus $\{c, d\}$ is red.

Case 2: c is the rightmost point of K_0. By Lemma 7(i) for $Y \cap \operatorname{conv}\{a,b,c\}$, $0 (\mathrm{mod}\ 3)$ points of Y lie under the segment ac. By Lemma 7(ii) for $Y \cap \operatorname{conv}\{a,c,d\}$, $2 (\mathrm{mod}\ 3)$ points of Y lie under the segment ac – a contradiction.

Case 3: b is the rightmost point of K_0. By Lemma 7(ii) for $Y \cap \operatorname{conv}\{a,c,d\}$, $2 (\mathrm{mod}\ 3)$ points of Y lie under the segment ac. By Lemma 7(ii) for $Y \cap \operatorname{conv}\{a,b,c\}$, $2 (\mathrm{mod}\ 3)$ points of Y lie under the segment ab. Since there is one point of Y inside the triangle abc, it follows that $1 (\mathrm{mod}\ 3)$ points of Y lie under the segment bc. Thus $\{b, c\}$ is red. Lemma 5(ii) is proved.

7. Proof of Lemma 5(iii)

Let K_0 be an empty blue subconfiguration of Y equivalent to $K(3,3)$. Let a, b, c be the vertices of $\operatorname{conv} K_0$ denoted so that $x(a) < x(c) < x(b)$, and let a', b', c' be the remaining three points of K_0 denoted uniquely so that the line cc' separates the points a, b' from the points a', b (see Fig. 8). Without loss of generality, we may assume that $x(c') < x(c)$. We define strips $S_1, S_2, \ldots, S_5$ and numbers $c_1, c_2, \ldots, c_5$ and $C = \sum c_i$ similarly as in Sections 4 and 6. We distinguish four cases.

Case 1: c lies below the line ab and $x(b') < x(c')$. By Lemma 7(i) for $\{b, b', c, a'\}$, $C - c_1 \equiv 1 (\mathrm{mod}\ 3)$.

Similarly, by Lemma 7(i) for $\{b, c, c', a'\}$, $C - c_1 - c_2 \equiv 1 (\mathrm{mod}\ 3)$.

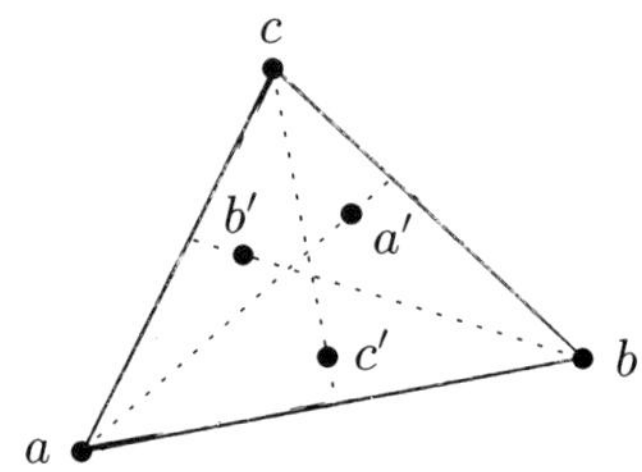

FIGURE 8. The points a', b', c'.

Thus, $c_2 = 0$. Consequently, $c_1 = 2$ (otherwise $c_1 = 0$ and $\{a, c'\}$ is red). Thus, $C \equiv 0 (\mathrm{mod}\ 3)$ and $\{a, b\}$ is red.

Case 2: c lies below the line ab and $x(c') < x(b')$. By Lemma 7(i) for $\{b, c, c', a'\}$,

$$C - c_1 + 1 \equiv 1 (\mathrm{mod}\ 3).$$

If $c_1 = 0$, then $C \equiv 0 (\mathrm{mod}\ 3)$ and $\{a, b\}$ is red.

If $c_1 = 2$, then $c_2 = 0$ (otherwise $\{a, b'\}$ is red) and $C - c_1 - c_2 \equiv 0 (\mathrm{mod}\ 3)$, but $C - c_1 - c_2 \equiv 1 (\mathrm{mod}\ 3)$ by Lemma 7(i) for $\{b, b', c, a'\}$ – a contradiction.

Case 3: c lies above the line ab and $x(b') < x(c')$. By Lemma 7(i) for $\{b, b', c, a'\}$, $C - c_1 + 1 \equiv 2 (\mathrm{mod}\ 3)$.

By Lemma 7(i) for $\{b, c, c', a'\}$, $C - c_1 - c_2 \equiv 2 (\mathrm{mod}\ 3)$.

Thus, $c_2 = 2$. Consequently, $c_1 = 0$ since otherwise $\{a, c'\}$ is red. It follows that $C \equiv 1 (\mathrm{mod}\ 3)$ and $\{a, b\}$ is red.

Case 4: c lies above the line ab and $x(c') < x(b')$. By Lemma 7(i) for $\{b, b', c, a'\}$,

$$C - c_1 - c_2 \equiv 2 (\mathrm{mod}\ 3).$$

By Lemma 7(i) for $\{b, c, c', a'\}$,

$$C - c_1 \equiv 2 (\mathrm{mod}\ 3).$$

Thus, $c_2 = 0$. Consequently, $c_1 = 2$ (otherwise $c_1 = 1$ or $\{a, b'\}$ is red). Thus, $C \equiv 1 (\mathrm{mod}\ 3)$ and $\{a, b\}$ is red.

Lemma 5(iii) is proved.

References

[GRS80] R.L. Graham, B. Rothschild, and J. Spencer, *Ramsey Theory*, J. Wiley & Sons, New York (1980).

[NR90] J. Nešetřil and V. Rödl (eds.), *Mathematics of Ramsey theory*, Springer-Verlag (1990).

[NV94] J. Nešetřil and P. Valtr, A Ramsey-type theorem in the plane, *Combinatorics, Probability and Computing* 3 (1994), 127-135.

[NV98] J. Nešetřil and P. Valtr, A Ramsey property of order types, *J. of Comb. Th. (Ser. A)* 81 (1998), 88-107.

[NSV99] J. Nešetřil, J. Solymosi, and P. Valtr, A Ramsey property of planar graphs, manuscript.

J. Nešetřil, Department of Applied Mathematics, Charles University, Malostranské nám. 25, 118 00 Praha 1, Czech Republic
E-mail address: `nesetril@kam.ms.mff.cuni.cz`

J. Solymosi, Computer and Automation Research Institute, Hungarian Academy of Sciences, H-1111 Budapest, Kende u. 13-17, Hungary
E-mail address: `solymosi@cs.elte.hu`

P. Valtr, Department of Applied Mathematics, Charles University, Malostranské nám. 25, 118 00 Praha 1, Czech Republic
E-mail address: `valtr@kam.ms.mff.cuni.cz`

DIMACS Series in Discrete Mathematics
and Theoretical Computer Science
Volume **49**, 1999

Density

Jaroslav Nešetřil and Claude Tardif

Abstract. We survey recent research of authors related to the study of homomorphisms between finite graphs from the point of view of the quasiorder defined by the existence (of homomorphism) between graphs. We demonstrate the surprising richness of this perhaps simplest case of graph comparision by proving the density theorem for both directed and undirected graphs in three different ways: Nešetřil-Perles proof using direct products (which works for undirected graphs) and two recently found proofs due to authors: a proof using arrow construction and a proof using a connection of density and homomorphism dualities. The later notion may be seen as a homomorphism approach to the class $NP \cap coNP$ and its connection to density problems may be seen as surprising. This analysis involves analysis of several other categorical notions most notably amalgamation and graph exponentiation.

1. Introduction

We consider homomorphisms between both undirected and directed graphs: given two graphs $G = (V, E)$ and $G' = (V', E')$, a *homomorphism* f of G to G' is any mapping $f : V \to V'$ which satisfies

$$(\star) \qquad [f(x), f(y)] \in E' \text{ for all } [x, y] \in E.$$

If there exists a homomorphism of G to G' then we say G is *homomorphic* to G', and write $G \to G'$. Otherwise we write $G \not\to G'$.

If $G \to G'$ and $G' \to G$ then we say G and G' are *hom-equivalent*.

The graph homomorphisms were investigated in various context: graph coloring, [**1, 2, 6, 7, 10, 11, 12, 16**], graph products [**8, 21**], automata theory and formal languages [**16, 27**], various algebraical context (as a generalization of iso morphism; categories). Various of these questions can be conveniently expressed by considering the quasiorders (and partial orders) induced by the existence of homomorphism.

Here we consider the problem motivated by the existence of homomorphism between graphs. One of the motivations of this paper is the following result due to Welzl [**27**]:

1991 *Mathematics Subject Classification.* Primary 05C99.

This paper was partially supported by GAČR 201/1996/0194 and GAUK 194/1996 grants.

THEOREM 1.1. *(density) Given two finite graphs G_1 and G_2 satisfying $G_1 \to G_2 \not\to G_1$, with G_2 not bipartite, there exists a graph G which satisfies $G_1 \to G \to G_2$ and $G_2 \not\to G \not\to G_1$.*

In the other words the quasiordering of the class of all (finite) graphs induced by the existence of a homomorphism is dense in the usual order-theoretic sense with the unique exceptional "gap-pair" $K_1 \to K_2 \not\to K_1$.

The original proof [**27**] is a complicated ad hoc argument. More recently M. Perles and J. Nešetřil independently found an alternative argument (see e.g. [**19**], [**20**] for the history of this) which simplified the proof and allowed to prove the density theorem for infinite graphs and some other classes of directed graphs [**19**]. We review this approach in Section 2. This proof is based on the direct product of graphs and it provided the motivation for other two proofs. So we decided to include it for a more clear and complete picture and we even give two variants of this proof. Let us also remark that recently there was a progress also in other directions and Tardif [**26**] solved a problem of Welzl (see [**28**]) related to density of vertex transitive graphs. However until recently for directed graphs the problem remained open. The strongest result in this direction is the following [**25**]

THEOREM 1.2. *The pairs $(P_0, P_1), (P_1, P_2), (P_{k+1}, P_k), k = 3, 4, ...$ are the only gaps for oriented paths (the graphs P_k are depicted in Fig. 1).*

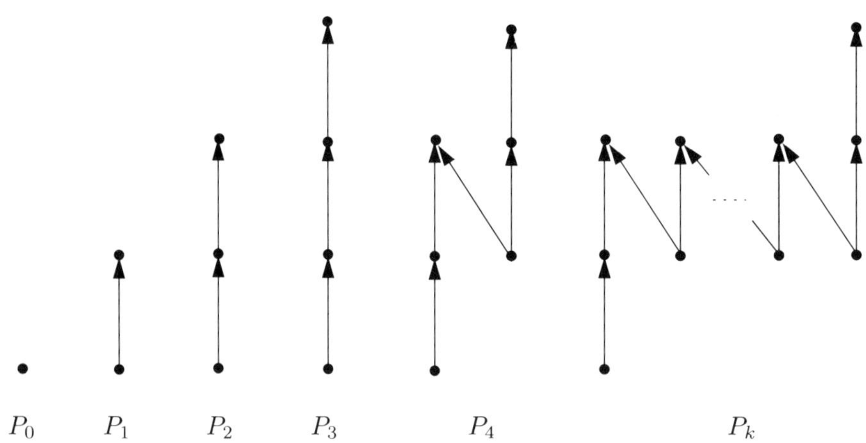

FIGURE 1. The paths P_k.

Explicitly: Let P and Q be oriented paths satisfying $P \to Q \not\to P$. Let the pair (P, Q) be none of the above pairs. Then there exists a path R which satisfies both $P \to R \to Q$ and $Q \not\to R \not\to P$.

One can see easily the the above pairs form gaps even in the class of all oriented graphs. Thus for oriented graphs we have infinitely many gaps and their full description seemed to be not easily reachable even for the class of all oriented trees ([**25**]). The problem has been finally solved in [**23**] by relating it to a seemingly unrelated result of [**13**] about homomorphism dualities [**17**],[**21**]. This together with all the necessary notions will be introduced in Section 4. Before this we introduce in Section 3 one more proof of the density of both directed and undirected graphs. This proof is based on the iterated graph amalgamation known as *arrow* *construction.* This proof is perhaps the shortest and conceptually easiest proof of the density for both undirected and directed graphs. Moreover in the more general cases (when the homomorphism dualities are not characterized) this proof is an indispensable part of the proof.

2. Nešetřil - Perles Proof

FIRST PROOF OF THEOREM 1.1. Let G_1 and G_2 be given undirected graphs, let $f : G_1 \longrightarrow G_2$ be a homomorphism, and suppose there is no homomorphism $G_2 \longrightarrow G_1$. As this pair is not equivalent to the jump (K_1, K_2), at least one component of the graph G_2 has chromatic number greater than 2. Also, at least one component of G_2 fails to be G_1-colorable, and this component may be assumed to be non-bipartite; let it contain an odd cycle of length k. Now choose a graph H with the following properties: H contains no odd cycle of length k or less, and the chromatic number of H is greater than $n^{n'}$, where n and n' denote the number of vertices of the graphs G_1 and G_2 respectively Such a graph exists by the celebrated theorem of Erdős [**4**]. Now let $G = G_1 \cup (H \times G_2)$. Here $\times$ denotes the direct product of two graphs and $\cup$ means the disjoint union. We shall prove that G has the desired properties. Obviously $G_1 \longrightarrow G$ and $G \longrightarrow G_2$ follows as the second projection of $H \times G_2$ is a homomorphism into G_2. On the other hand there is no homomorphism from G_2 into G, as homomorphisms preserve odd cycles and they cannot increase the length of the shortest of them. Thus it suffices to prove that there is no homomorphism $G \longrightarrow G_1$. Let us suppose for the contradiction that there is a homomorphism $f : H \times G_2 \longrightarrow G_1$. Thus for any vertex x of H we have an induced mapping $f_x : V(G_2) \longrightarrow V(G_1)$ defined by $f_x(y) = f(x, y)$. (This mapping need not be a homomorphism.) As there are at most $n^{n'}$ such mappings there are vertices x and x' forming an edge of H such that the mappings f_x and f'_x are identically equal, say to g. However in this case g is a homomorphism of G_2 into G_1, contrary to our assumption. $\square$

Given two graphs $G = (V, E)$ and $H = (V', E')$ one can define G *power of* H, denoted by H^G, as the following graph (W, F): $W = \{f : V \to V'\}$ and a pair (f, g) forms an edge if $(f(x), g(y)) \in E'$ for every edge $(x, y) \subset E$. (We define the G power of H by the same formula for both undirected or directed graphs.)

This construction was isolated in the graph theoretic concept in [**15**], see also [**3**], [**24**]. The following is the crucial property which we use: For every graph K holds $K \to H^G$ iff $K \times G \longrightarrow H$. (This is easy to see: Given $f : K \to H^G$ define $g : K \times G \longrightarrow H$ by $g(x, y) = f(x)(y)$. Conversely, given g we may define f by the same formula. One can easily check that f is a homomorphism $K \to H^G$ iff g is a homomorphism $K \times G \longrightarrow H$.) Thus in the above proof we have $H \times G_2 \longrightarrow G$ iff $H \longrightarrow {G_1}^{G_2}$. Thus in the above proof we may assume that the chromatic number of H is greater than the chromatic number of ${G_1}^{G_2}$. As $G_2 \not\to G_1$ there are no loops in ${G_1}^{G_2}$ and it is also clear that the chromatic number of ${G_1}^{G_2}$ is at most the number of vertices of ${G_1}^{G_2}$. (However we do not optimize at this point.)

3. Density via Amalgamation

Besides giving a new proof of undirected density theorem we prove the following result (which generalizes and solves a longstanding problem in this area). This proof appeared in [**23**].

THEOREM 3.1. *Let G, H be directed graphs which are cores. Let H be connected and assume that H fails to an orientation of a tree. Further assume that $G \to H \not\to G$ holds. Then there exists a directed graph H with $G \to K \to H$ and $H \not\to K \not\to G$.*

The proof is based on the following construction which goes back to [**5**] and [**9**], see also [**18**]:

Let $G = (V, E)$ be a directed graph, $I = (V', E')$ a graph with two distinguished vertices $a, b \in V'$. Denote by $G \star (I, a, b)$ the graph (W, F) defined as follows:

$W = E \times V' |_{\sim}$

where the equivalence $\sim$ is generated by the following rules:

$((x, y), a) \sim ((x, y'), a)$, $((x, y), b) \sim ((x', y), b)$, $((x, y), b) \sim ((y, z), a)$.

The equivalence class containing $((x, y), z)$ will be denoted by $[((x, y), z)]$. We put

$F = \{([((x, y), z)], [((x, y), z')]; (x, y) \in E, (z, z') \in E'\}$.

Clearly $G \star (I, a, b)$ arises by replacing every arrow of G by a copy of I.This construction is known as *arrow construction* (in some of the earlier papers also called šíp construction; šíp = arrow in czech). Note, that if (I, a, b) is an undirected graph (considered as a symmetric relation), then $G \star (I, a, b)$ is also an undirected graph.

We shall make use of the following obvious (but key) property of the arrow construction:

LEMMA 3.1. *Let G and H be directed graphs with $\chi(G) > |V(H)|$ and let every homomorphism $f : I \to H$ satisfies $f(a) \neq f(b)$. Then $G \star (I, a, b) \not\to H$.*

SECOND PROOF OF THEOREM 1.1. Let G, H be undirected graphs, H non bipartite, with $G \to H \not\to G$. Clearly we may assume that G and H are cores. Let $e = \{a, a'\} \in E(H)$ belong to a circuit in H. Put $I = H - e + \{a', b\}$ where $b \notin V(H)$. (Thus I arises from H by deleting the edge e, adding a new vertex $b \notin V(H)$ together with the edge $\{a', b\}$.)

It is clear that $I \to H$ (identifying vertices a and b) but any homomorphism $f : I \to G$ satisfies $f(a) \neq f(b)$ (for otherwise we get a contradiction with $H \not\to G$). Now let F be any graph satisfying $\chi(F) > |V(G)|$ and let F' be any orientation of F. Consider the arrow construction $F' \star (I, a, b)$ and define the graph K by $K = (F' \star (I, a, b)) \cup G$.

We prove that K has properties claimed by Theorem 1.1: Clearly $G \to K$. We also have $K \to H$ as the mapping f defined by $f([e, x]) = x$ for $x \in V(H)$ and $e \in E(F')$ and $f([e, b] = a$ is a homomorphism $K \to H$ (we preserve the above notation concerning the arrow construction $F' \star (I, a, b)$.) Further, by the above Lemma 3.1 $K \not\to G$ (as $\chi(F) > |V(G)|$). Thus it remains to be shown that $H \not\to K$. Suppose the contrary: Let $g : H \to F \star (I, a, b)$ be a homomorphism. Then $f \circ g : H \to H$ where f is the above defined homomorphism $F \star (I, a, b) \to H$. As H is a core $f \circ g$ is a homomorphism. Put $h = (f \circ g)^{-1}$. Then $f \circ g \circ h(x) = x$ for every $x \in V(H)$. Put $g \circ h(a) = [(e, a)]$ with $e = (u, v)$. Then the image $g \circ h(G)$ of G is a connected subgraph of $F \star (I, a, b)$ which is (by the injectivity of the mapping $f \star g \star h$) contained in the set of all $[(e', x)]$ where e' is incident with u and $x \in V(I)$ (this set is the "star" induced by those edges of F' which are incident with the vertex u). But then the edge $\{[g \circ h(a)], [g \circ h(a')]\}$ is a cut edge in the graph $g \circ h(G)$ which is the final contradiction as a, a' was contained in a cycle of H. □

PROOF OF THEOREM 3.1. Let G, H satisfy the assumption of the theorem. Let H be a core and let $(a, a') \in E(H)$ belong to a cycle in H. Put $I = H - (a, a') + (b, a')$ where $b \notin V(H)$ (i.e. we first delete arc (a, a') and then add a new vertex b together with the arc (b, a'). Let F be oriented graph with $\chi(F) > |V(G)|$ and consider the arrow construction $F \star (I, a, b)$. Put $K = G \cup (F \star (I, a, b))$. Then we have:

- $G \to K$ (by the inclusion map);
- $K \to H$ (by the same mapping as in the above proof);
- $K \not\to G$ (by the chromatic number assumption);
- $H \not\to K$ (as above in the above proof for undirected graph).

Thus the graph K has the desired properties. □

This proof can be generalized to finite models of general relational systems and thus it cunstitutes the essential part of [**24**]

4. Proof via homomorphism dualities

As we have seen Density Theorems 1.1, 3.1 do not hold without exceptional cases. To formalize this let us say that a pair (G, H) of graphs is a *gap* if $G \longrightarrow H$ and $H \not\longrightarrow G$ and every graph K with $G \longrightarrow K \longrightarrow H$ is homomorphically equivalent to either G or H. Thus any gap is a gap in the quasiorder induced by the existence of homomorphism (i.e. by $\longrightarrow$). In the above density theorems we want to characterize all gaps. It is easy to see that (K_1, K_2) is a gap for undirected graphs and thus the Density Theorem 1.1 can be formulated as follows:

THEOREM 4.1. *(K_1, K_2) and (K_0, K_1) are the only gaps for the class of indirected graphs.*

Also for directed graphs Theorem 3.1 covers all non-gaps pairs of oriented graphs. However this (perhaps a bit surprising) fact is harder to prove. We shall outline the proof, for full details see [**23**].

Let F, H be graphs. We say that (F, H) is a duality pair if for every graph G the following statement holds:

$G \longrightarrow H$ iff $F \not\longrightarrow G$.

Obviously in this case $F \not\longrightarrow H$ and thus we can also define a duality pair as follows: For any graph G we have either a homomorphism from G to H or a homomorphism from F to G while these two possibilities are mutually exclusive. This statement is also called *homomorphism duality theorem.*

Any duality pair is an example of a good characterization (in the sense of Edmonds) and in fact this was the original motivation of this notion [**21**], [**17**].

The duality pairs (and thus homomorphism duality theorems) were characterized in [**21**] for the class of undirected graphs and in [**14**] and [**24**] for the class of directed graphs:

THEOREM 4.2. *Up to a homomorphism equivalence (K_2, K_1) and (K_1, K_0) are the only duality pairs for the class of undirected graphs.*

THEOREM 4.3. *Up to a homomorphism equivalence the only homomorphism dualities for directed graphs are of the following form (T, H_T) where T is an oriented tree and for each tree T the graph H_T is uniquelly determined.*

Nešetřil and Tardif found in [**23**] and [**24**] the following (perhaps surprising) connection of duality pairs. This provided the key to the characterization of gaps for classes of directed (and undirected) graphs. We say that a gap (G, H) is *connected* if H is a connected graph. Observe that if (F, H) is a duality pair then F is necessarily connected (for if F is a core and F_1 and F_2 are distinct components of F then $F \not\to F_i$, and thus $F_i \to H$ for $i = 1, 2$. Thus $F \to H$, which a contradiction).

THEOREM 4.4. *There is a one-to-one correspondence between duality pairs and gaps for the class of directed (and also undirected) graphs. Explicitly, given a duality pair (F, H) then $(F \times H, F)$ is a gap. Conversely, given a connected gap (G, H) then (H, G^H) is a duality pair.*

PROOF. First, suppose that $G \to H \not\to G$ is a gap. We prove that for any graph K holds $H \not\to K$ iff $K \to G^H$. Thus let a graph K be such that $H \not\to K$ and suppose for the contrary that $K \not\to G^H$. Then obviously $G \to G \cup (G \times H) \to H$. If $H \to G \cup (K \times H)$ then (by the connectivity and the assumption $H \not\to G$) $H \to K \times H$ and thus $H \to K$,contrary to our assumpiton. If $G \cup (K \times H) \to G$ then $K \times H \to G$ and $K \to G^H$ again contrary to our assumption. Thus the class $H \not\to$ is a subclass of the class $\to G^H$. In order to prove the reverse inclusion let K be a graph satisfying $K \to G^H$ and $H \to K$. This implies $H \to H \times G^H \to G$ (as $H \times G^H \to G$ is equivalent to $G^H \to G^H$ and thus it always holds). Thus (H, H^G) is a duality pair.

Conversely, let (F, H) be a duality. We may clearly assume that F is a core (i.e. every homomorphism $F \to F$ is an automorphism) and further F is connected. Thus $F \times H \to F$ and $F \not\to F \times H$ (as $F \to F \times H$ would imply $F \to H$). We claim that there is no graph K satisfying $F \times H \to K \to F$ and $F \not\to K \not\to F \times H$: If $K \to F \not\to F \times H$ then the duality implies $K \to H$ and thus $K \to F \times H$ which contradicts our assumptions. This completes the proof of theorem. □

This Theorem 4.4 leads to yet another proof of Density Theorem 1.1:

THIRD PROOF OF THEOREM 1.1. By 4.2 (K_2, K_1) and (K_1, K_0) are the only duality pairs. Thus $(K_2 \times K_1)$ (which is equivalent to (K_1, K_2)) and $(K_0 \times K_1, K_1)$ (which is equivalent to (K_0, K_1) are the only gaps. □

Thus, modulo the above "arrow calculus" involved in the above proof of 4.4, the density theorem has been known even before it has been formulated.

Finally, let us remark that the above also shows that Theorem 3.1 gives all non-gap pairs for directed graphs: Let (T, H_T) be all duality pairs for oriented graphs (characterized by Theorem 4.3). Then $(T \times H_T, T)$ are exactly all gaps. This gives also yet another proof of Theorem 3.1.

References

[1] J. Bang-Jansen and P. Hell, *The effect of two cycles on the complexity of colourings by digraphs*, Discrete Applied Math. 26(1990), 1-23.

[2] J. Bang-Jensen, P. Hell and G. MacGillivray, *Hereditarily hard colouring problems*, submitted to J. Comput. Systems Science.

[3] D. Duffus, N. Sauer: Lattices arising in categorical investigations of Hedetniemi' s conjecture, Discrete Math. 152(1996), 125-139

[4] P.Erdos, *Graph Theory and Probability*, Canad. J. Math. 11(1959), 34-38

[5] R. Frucht, *Herstellung von Graphen mit vorgegebener abstrakter Gruppe*, Compos. Math. 6(1938), 239-250

[6] W. Gutjahr, *Graph colorings*, Ph. D. Thesis, Free University, 1991.

[7] W. Gutjahr, E. Welzl and G. Woeginger, *Polynomial graph colorings*, (1992).

[8] R.P.Häggkvist, P.Hell, D.J.Miller and V.Neuman Lara, *On Multiplicativite graphs and the product conjecture*, Combinatorica 8(1988), 63-74.

[9] Z. Hedrlín, A. Pultr, *Symmetric Relations (Undirected Graphs) with Given Semigroups*, Monatshefte f. Mathematik 69,4(1965), 318-322

[10] P. Hell and J. Nešetřil, *On the complexity of H-coloring*, J. Combin. Th. (B) 48(1990), 92-110.

[11] P. Hell, J. Nešetřil and X. Zhu, *Complexity of tree homomorphisms*,

[12] P. Hell, J. Nešetřil and X. Zhu, *Duality and Polynomial Testing of Tree Homomorphisms*, Trans. of the American Math. Society, 348 (1996), 1281-1297.

[13] P. Komárek, *Some new good characterizations of directed graphs*, Časopis Pěst. Mat. 51 (1984), 348-354.

[14] P. Komárek, *Good charactersizations of graphs*, PhD thesis, Charles University, Prague 1987

[15] L. Lovász, *Operations with structures*, Acta Math. Acad. Sci. Hungar.(1967), 321-328

[16] H.A. Maurer, J.H. Sudborough and E. Welzl, *On the complexity of the general coloring problem* Inform. and Control 51 (1981), 123-145.

[17] J. Nešetřil, *Theory of graphs* ,(In czech) SNTL (Praha), 1979.

[18] J. Nešetřil, *Amalgamation of Graphs and its Applications*, Ann. N.Y.Acad. Sci. 319(1979), 415-428

[19] J. Nešetřil, *Structure of Graph Homomorphisms*, Combinatorics, Probability and Computing

[20] J. Nešetřil, *Structure of Graph Homomorphisms I*.In: Graph Theory Notes of New York, New York Acad. Sci. 1998, pp.7-12.

[21] J. Nešetřil and A. Pultr, *On classes of relations and graphs determined by subobjects and factorobjects*, Discrete Math. 22 (1978), 287-300.

[22] J. Nešetřil and C. Tardif, *Density via Duality*, submitted.

[23] J. Nešetřil and C. Tardif *A new proof of the density of directed and undirected graphs*, submitted 1998.

[24] J. Nešetřil and C. Tardif *Duality Theorems for Finite Structures*, submitted.

[25] J. Nešetřil, X. Zhu, *Paths Homomorphisms*, Proc. Cambridge Phil. Soc.120 (1996), 207-220.

[26] C.Tardif, *Fractional multiples of graphs and the density of vertex transitive graphs*, to appear in J. Algebraic Combinatorics.

[27] E. Welzl, *Color Families are Dense*, J. Theoretical Comput. Sci. 17 (1982), 29-41.

[28] E. Welzl, *Symmetric graphs and interpretations*, J. Combin. Th.(B), 37(1984), 235-244.

Department of Applied Mathematics, Charles University, Malostranské nám. 25, 118 00 Prague, Czech Republic

E-mail address: nesetril@kam.ms.mff.cuni.cz

Department of Mathematics and Statistics, University of Regina, Regina, Saskatchewan, Canada, S4S 0A2

E-mail address: tardif@math.uregina.ca

DIMACS Series in Discrete Mathematics
and Theoretical Computer Science
Volume **49**, 1999

Spectra, Graphs and Proteins. Towards Understanding of Protein Folding.

Petr Pančoška, Vít Janota, and Jaroslav Nešetřil

Introduction

Proteins are molecular assemblies performing wide variety of life-essential functions, many of them known many still waiting to be understood [**1**]. Reflections of protein structure, which is the vehicle defining and enabling their functions, can be found under many expected and unexpected circumstances (see Figure 1). Mathematics was involved directly and indirectly in the discoveries of secrets of protein structure from very beginning. Mathematical methods of physics were instrumental for the structural interpretation of spectroscopic measurements and x-ray crystallographic data that forms core of the current knowledge base about the protein structure [**2**]. Mathematical chemistry formulates variety of challenging problems in modeling of protein conformation using energy minimization, molecular mechanics and dynamic computations, to mention only some [**3, 4, 5, 6, 7**]. Computational biology established itself as a powerful and rapidly expanding tool that is essential for quantitative and rational utilization of exponentially growing pool of genetic information [**8, 9**]. Discrete mathematics played in some sense supportive role in various areas of molecular sciences (graph encoding of the chemical structures being one example [**10**]). In this contribution, we would like to summarize some applications of discrete mathematics that might help to establish a "missing direct link" between the computational and molecular biology and physical and chemical experimental results measured on biomolecules. As the future of the discrete mathematics was summarized at the meeting, we can take advantage of such excellent overview of newest achievements in this area of mathematics and demonstrate their potential in applications in protein structural studies. Shortly, we want to rationalize the conjecture, that when it comes to applications to proteins, the roads of discrete mathematics are paved with gold.

Primary Structure and Periodicity in Strings

Accepting what was numerously demonstrated, namely that protein function is determined by its structure, the encoding of the details of the protein fold in renewable molecular media is essential feature needed for successful assembly of living systems. Only the first part of the structural encoding (sequence of nucleic acid base triplets defining the protein primary structure) is completely understood

1991 *Mathematics Subject Classification.* Primary 05C90.

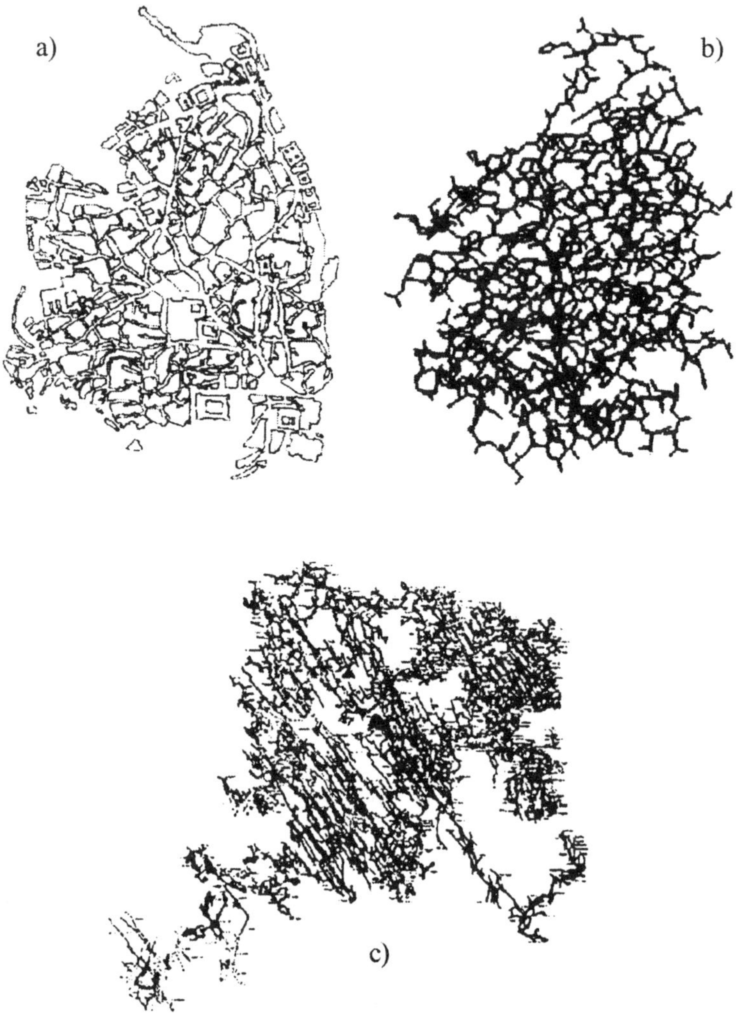

FIGURE 1. Comparison of a) two-dimensional rendering of city of Toledo (Spain) and b) concanavalin A x-ray determined structure. As an open problem, try to decide if the picture c) represents a protein structure or some human or natural artifact (see [**20**] for the solution).

[**11, 12**]. How the ordered linear string of "letters" from the "alphabet" of twenty amino acids is converted into three-dimensional native structure of an active enzyme is still an unsolved problem. Figure 2 summarizes some features of typical representatives of the molecular structure of this alphabet "letters".
Discrete mathematics can help discover any specific "syntax rules" in the protein primary structure. Although the direct analysis of the amino acid sequence is tempting, additional aspects of the protein structure should be accounted for in these analyses as they can alter the order and significance of the amino acid letters for the structural decoding of the sequence. The molecular tool for such an

FIGURE 2. Structures of amino acids glycine (G), alanine (A) and phenylalanine (F) and the molecular encoding of alanine by the nucleic acid base triplet guanine-cytosil-guanine (GCG).

alteration of the sequence order is protein secondary structure. Although its conventionally accepted role is to stabilize the three-dimensional shape of the protein, which is achieved by additional intra-chain through-space interactions (hydrogen bonds and weak forces), secondary structure also specifically positions the side chains of amino acids of its primary sequence substring relatively to each other in space. This positioning is important for interactions between different parts of the polypeptide chain that, in turn, define the overall protein fold.

Figure 3 shows idealized examples of structural consequences of amino acid periodicity for two regular conformations α-helix and β-strand. Periodic sequential polypeptides with sequences -(GGF)$_n$-, -(GF)$_n$- and -(GGGF)$_n$- were folded into helix and sheet secondary structures. For helical backbone conformation (see Figure 3a), the odd periodicity of F in the -(GGF)$_n$- sequence results in the right-handed helical arrangement of bulky F side-chains and even periodicity of F in -(GF)$_n$- sequence results in two left-handed and more extended helical strips of F side-chains on the segment surface. One of these strips can be eliminated using -(GGGF)$_n$- primary structure.

Considering the same sequences in sheet secondary structure (see Figure 3b), even

a)

b)

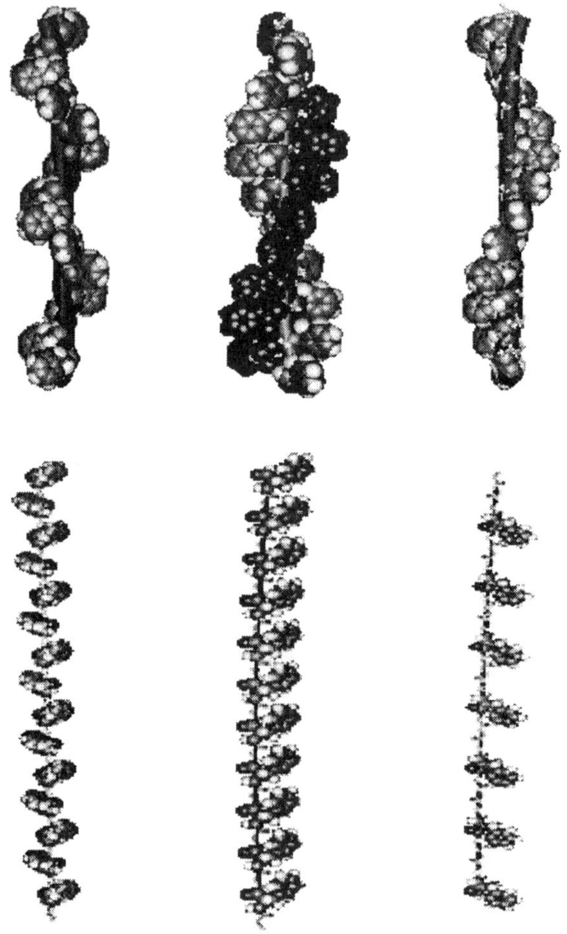

FIGURE 3. Periodic sequential polypeptides with amino acid sequences -(GGF)$_n$-, -(GF)$_n$- and -(GGGF)$_n$- folded into a) α-helix; b) β-sheet secondary structure.

periodicity of amino acids in the sequence results in uniform location of F side chains on one side of the ideal β-strand with variable "density" for -(GF)$_n$- and -(GGGF)$_n$-. Odd periodicity generates alternation of the F side chain location on both sides of the strand.

Example of amino acide sequence given in Figure 4 idealizes the sequence-defined structural location of two types of side chains (bulky F and small G) on secondary structure segments. Different relative position of amino acids in the primary structure string transforms into distinct relative spatial relation of respective side chains with different chemical properties on the secondary structure segment surface. Chirality of this arrangement remains related to the odd-even type (translational symmetry) of the primary structure periodicity. For helical secondary structure segments, the diagram of so called "helical wheel" (see Figure 5a) can be used for engineering of oriented segment surfaces by placing amino acids with selected side chain properties into proper positions of the primary sequence.

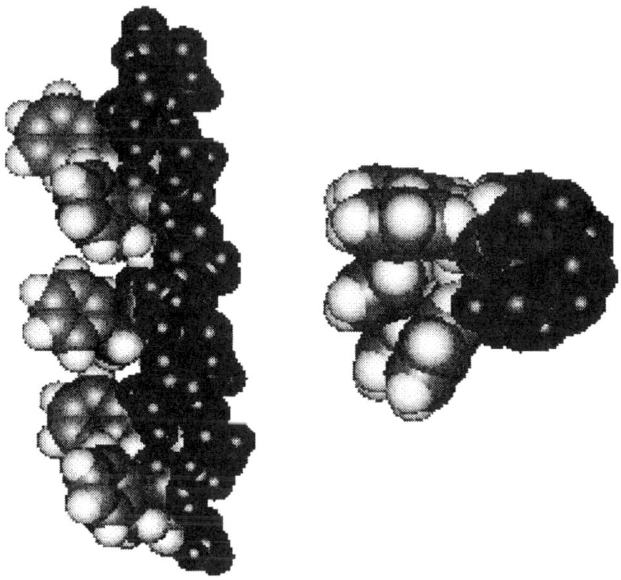

FIGURE 4. Side (left) and top (right) view of a secondary structure segment folded into α-helix conformation with amino acid sequence -GFGGFGGGFGGGFGGFGGG-. Here the helical wheel diagram was used to generate aromatic (phenylalanine) surface on one side of the segment and hydrogen (glycine) surface (black) on the opposite side of the segment.

Figure 5b represents the reduced wheel diagram that accounts for the fact that the most frequent length of helical segment in native proteins is about $10 - 11$ amino acids. This scheme shows that given "central" residue i in the sequence folded into a typical α-helix, the residues i+1 and i−1 together with their counterparts at $i\pm2$, $i\pm3$, $i\pm4$ and $i\pm5$ define the chemical properties of three different surface regions $\{i-5,\ i-4,\ i-1,\ i+2,\ i+3\}$, $\{i-4,\ i-3,\ i,\ i+3,\ i+4\}$ and $\{i-3,\ i-2,\ i+1,\ i+4,\ i+5\}$ of the segment.
Two different views of these surface regions are shown Figures 4 and 6. The first example also visualizes that (at least formally) one can consider these surfaces as superposition of portions of helical location of side chains shown in Figure 3 for sequential polypeptides with odd ($i\pm3$) and even ($i\pm4$) periodicity of their primary sequence properties.
The above analysis of the impact of the protein secondary structure on the transformation of the amino acid sequence into physically relevant substrings might be important for discrete mathematical methods for analysis of periodicity in strings, summarized and reviewed in another contribution of this volume by William F. Smyth.

Secondary Structure Segments and Eulerian Graphs

By removing all structural details other than assignment, position and length of secondary structure segments we can obtain representation of protein structure

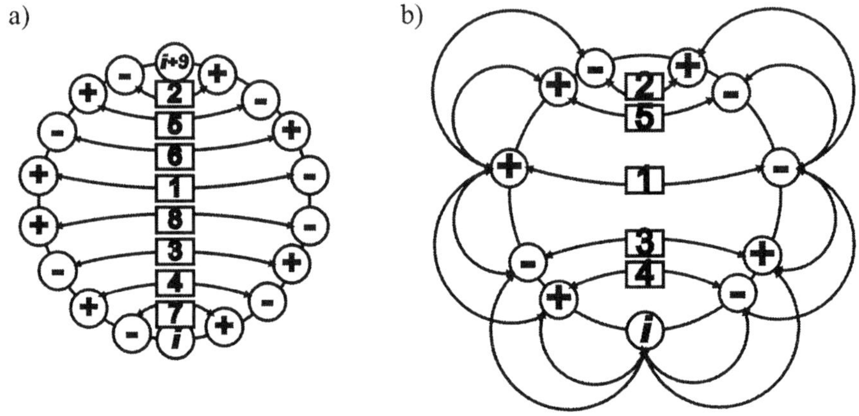

FIGURE 5. Helical wheel graphs. The large circle represents the surface of α-helix segment, small circles are locations of amino acid side chains. Numbering of the respective side chain locations is related to the residue i (bottom). Relative distance in the primary structure of a position $i \pm k$ from the residue i are shown by the numbers k in the square boxes and by the $+$ and $-$ sign in the position symbol. a) full helical wheel; b) reduced wheel for typical helical segment length of $10 - 11$.

that is practically interesting: it can be related to experimental spectroscopic measurements [**13, 14, 15**]. Construction of such representation which in principle allows to distinguish quantitatively structural differences between proteins with identical secondary structure composition distributed over the different number of secondary structure segments is shown for two hypothetic proteins in following scheme (schematic representation of layout of secondary structure segments of both proteins is given in Figure 7):

Protein A

C-E-C-H-C-H-C-E-C-E-C-E-C-H-C-E-C-E-C-E-C-E-C-E-C

C-ce-E-ec-C-ch-H-hc-C-ch-H-hc-C-ce-E-ec-C-ce-E-ec-C-ce-E-ec-C-ch-H-hc-
-C-ce-E-ec-C-ce-E-ec-C-ce-E-ec-C-ce-E-ec-C-ce-E-ec-C

$$\begin{matrix} H = 3, & he = 0, & hc = 3 \\ eh = 0, & E = 9, & ec = 9 \\ ch = 3, & ce = 9, & C = 13 \end{matrix} \Longrightarrow \begin{bmatrix} 3 & 0 & 3 \\ 0 & 9 & 9 \\ 3 & 9 & 13 \end{bmatrix}$$

Protein B

C-E-C-E-H-C-E-C-E-C-E-C-E-C

C-ce-E-ec-C-ce-E-eh-H-hc-C-ce-E-ec-C-ce-E-ec-C-ce-E-ec-C-ce-E-ec-C

$$\begin{matrix} H = 1, & he = 0, & hc = 1 \\ eh = 1, & E = 6, & ec = 5 \\ ch = 0, & ce = 6, & C = 7 \end{matrix} \Longrightarrow \begin{bmatrix} 1 & 0 & 1 \\ 1 & 6 & 5 \\ 0 & 6 & 7 \end{bmatrix}$$

Here H represents helices, E strands and C other secondary structure segments. Through the integral non-negative matrices shown in the scheme with elements defined by the numbers of regular secondary structure segments (diagonal, n_{ii}) and those of the oriented contact structures - head and tail sections at the boundaries of the regular secondary structure segments (off-diagonal, n_{ij}, $i \neq j$) - we can define

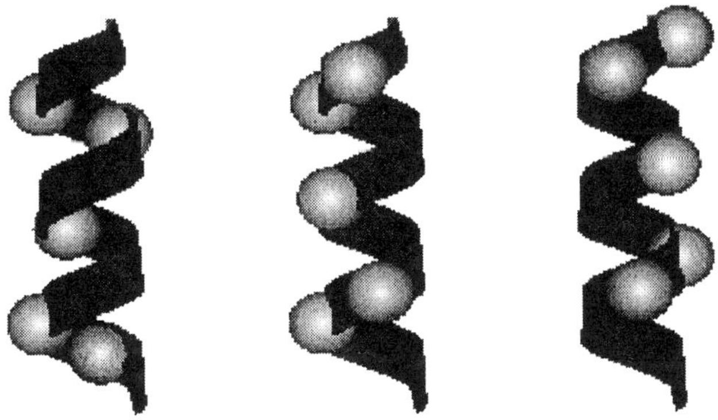

FIGURE 6. Location of side chains with primary structure positions $\{i{-}5,\ i{-}4,\ i{-}1,\ i{+}2,\ i{+}3\}$, $\{i{-}4,\ i{-}3,\ i,\ i{+}3,\ i{+}4\}$ and $\{i{-}3,\ i{-}2,\ i{+}1,\ i{+}4,\ i{+}5\}$ on the surface of an α-helical segment.

the adjacency matrices of the oriented (and necessarily Eulerian) graphs that are also shown in Figure 7. This representation of the segment topology is instrumental in the quantification of multiple protein structures that are compatible with any given matrix. Formally, such a multiplicity is a direct consequence of the reduction of structural information by the algorithm defining the above matrix structural descriptor. Factually, this structural multiplicity reflects the ambiguity of experimentally available information about the protein structure that was introduced deliberately to reflect the limited structural sensitivity of the physical phenomenon on which the experimental (spectroscopic in particular) measurements are based. Using modified Aardene-Ehrenfest theorem [**16**], the number of segment layouts for any matrix can be calculated as the number N_c of the non-equivalent Eulerian paths in the graph, defined by the $[n_{ij}]$ matrix:

$$N_c = \det \mid L^* \mid \frac{\prod_{i=1}^{n} (\mid n_{ii} \mid -1)!}{\prod_{i \neq j} (\mid n_{ij} \mid)!} \tag{1}$$

Here L^* is a matrix generated from $[n_{ij}]$ by sign change of the off-diagonal elements At the same time, all matrix descriptors $[n_{ij}]$ are non-negative matrices which links their properties to Perron-Frobenius theorem and opens the possibility to apply the theory of Markov chains for more detailed analyses [**17**].

In a related study of generalized 4x4 [HETC] matrix descriptors (from the original C category of above descriptors the new T (turn) category was separated) we have shown that there are two types of relations between their matrix elements n_{ij}:

a) Primary algebraic relations (sums of the off-diagonal elements in any given row or column should be equal to corresponding diagonal entries) that are the direct consequence of the Eulerian character of the corresponding graph reflecting the structural fact that in native proteins the segments are formed on a linear polypeptide chain.

b) Less obvious relationship was found between the number of contacts of secondary structure segments with those of other types in known protein structures.

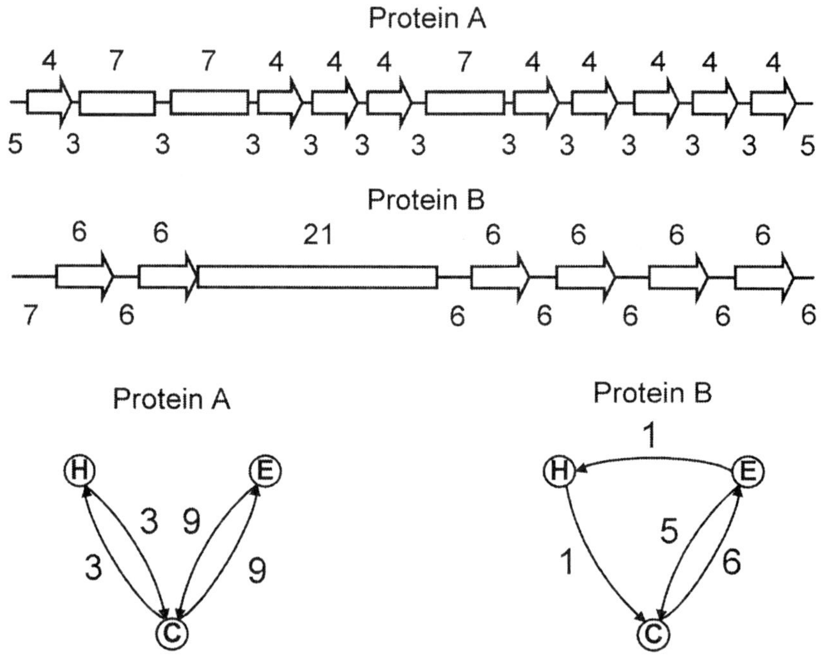

FIGURE 7. Example of quantification of layout of secondary structure segments. Top: schematic representations of layout of secondary structure segments. Rectangles=helices (H), arrows=strands (E), lines=other (C). Numbers above segments are their lengths in amino acid residues. Bottom: Corresponding Eulerian oriented graphs are shown with numbers given by the edge multiplicities.

Matrix elements n_{jk} can be expressed in terms of diagonal entries as $n_{jk} = n_{ii} + q$, $(j,k = i;\ q = 0, \pm 1, \pm 2,\ \ldots\ , \pm N)$.
Standard analysis provides the following expression for $P(N, q)$ describing the probability of occurrence of given q:

$$P(N,q) \approx \left(1 - \frac{|\,q\,|}{N_{\pm}}\right) \ln\left(\frac{N_{\pm}}{|\,q\,|}\right) \tag{2}$$

Figure 8 summarizes how function (2) matches the real q-distribution found in 492 representative x-ray protein structures.

Tertiary Protein Structure and Book Graphs

Next step in description of even higher and practically important aspect of protein structure using discrete mathematical objects requires classification of the natural protein folds into characteristic types. Figure 9 summarizes structures of nine representative protein folding types that cover typical features of all currently known proteins [**18**].
Quantitative encoding of these folding types represents a problem of its own. Variable protein structures (with exception of crude partitioning into all helix, all

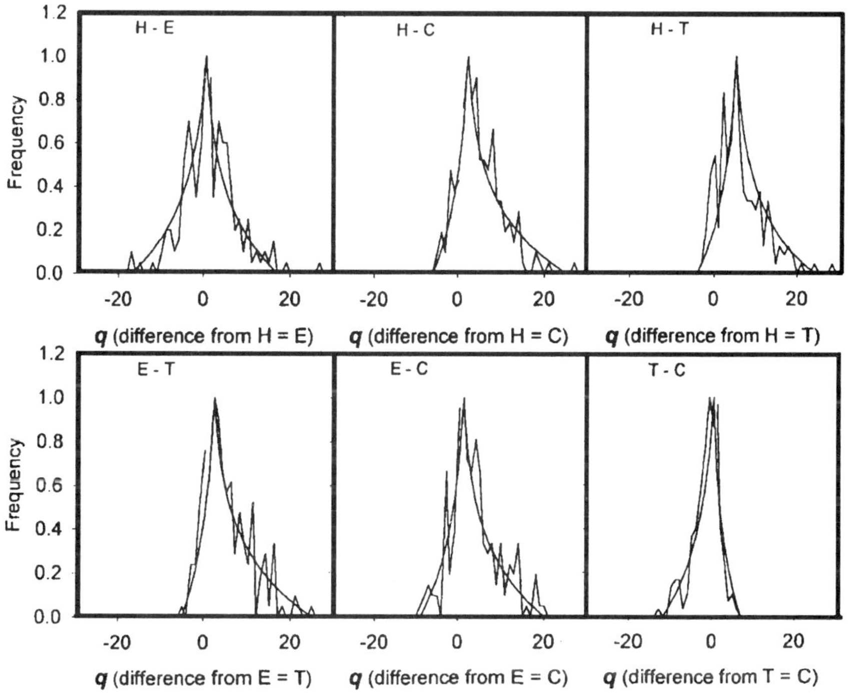

FIGURE 8. Comparison of distributions of q-values relating the numbers of secondary structure segments (H, E, T and C) in real proteins with the probabilistic function (2) (smooth lines).

sheet and mixed folding types) form necessarily overlapping categories. Based on the previous model of the protein secondary structure being organized in secondary structure segments, the transition to quantitative descriptor of the tertiary (three-dimensional) structure can be achieved by finding the (minimal) network of through-space contacts between the segments. Structurally, the contact is defined by the closest distance d_{ij} between the α-carbon atoms of the polypeptide backbone. These distances (calculated from Cartesian coordinates of C_α atoms provided by crystallographic data) are traditionally encoded in the distance matrix (see Figure 10). The set of these close distances can be conveniently reduced when only one - the closest - contact for segments i and j is retained for all $j \neq i$. This preserves the dominant network of distances that might be sufficient for complete description of the protein three-dimensional fold. Structurally, this reduction is shown for example of ribonuclease T on Figure 11a.

Accepting the above "structural" conjecture that this reduced network of inter-segment contacts is sufficient to define uniquely the three-dimensional structure of a protein fold (proof might be an interesting distance geometry problem) allows to encode the protein fold into a graph G_V shown in Figure 11b. Tests of graphs G_V for a representative set of 492 protein structures from a protein data bank revealed that with negligible exceptions, the page subgraphs of G_V are planar. This property connects this protein structure encoding to the enumeration of possible topologies of single stranded ribonucleic acid molecules [**8, 19**]. Enumeration uses

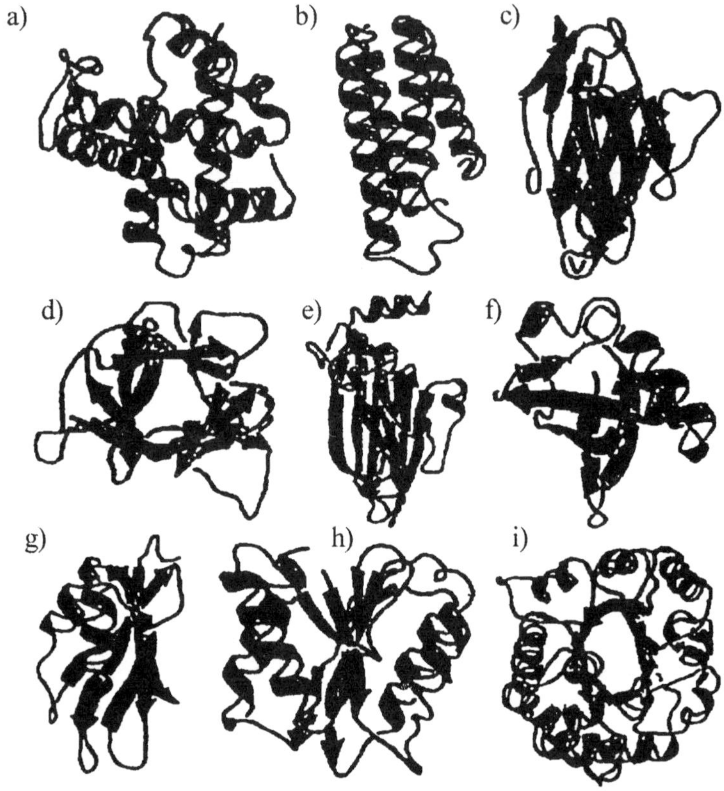

FIGURE 9. Representative examples of protein folding superfamilies: a) globin; b) α up-down; c) immunoglobulin; d) trefoil; e) jelly-roll; f) ubiquinon α/β roll; g) α-β sandwich; h) α/β doubly wound; i) triose phosphate isomerase (TIM) barrel.

the bijection between these subgraphs and trees (see Figure 12).
The number N_T of possible topologies is given for n secondary structure segments and k_i inter-segment contacts of a given type (k_i = number of edges at respective pages of G_V) is then

$$N_T = \prod_{i=1}^{6} \frac{1}{k_i} \binom{n-k_i}{k_{i+1}} \binom{n-k_i-1}{k_i-1} \tag{3}$$

For practical applications (for example for analysis of experimental structural information in optical spectra of proteins) the detailed information content of G_V is still not compatible with the global structural information obtainable from the experimental data. To match the structural details of the descriptor and the structural sensitivity of spectra, the invariant of G_V the six-component vector g_V defined by the numbers of edges at respective pages of G_V was found to be a convenient mathematical descriptor. Figure 13 shows that the transformation of protein tertiary structure into g_V conserves the basic information about the protein fold family as defined by types at Figure 9. These folding types are grouped in clusters if the vectors g_V are used as the generalized protein coordinates in cluster analysis.

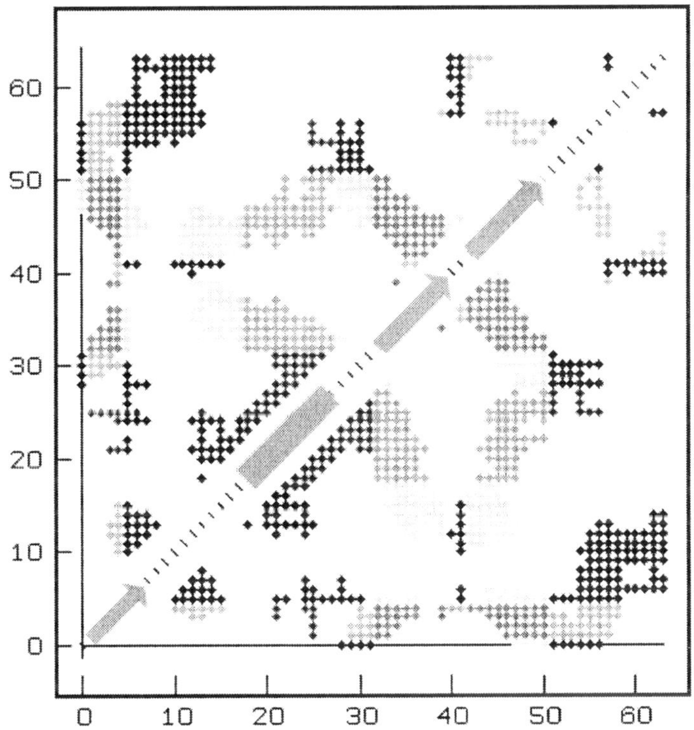

FIGURE 10. Rendering of distances d_{ij}<1.5 nm between α-carbon atoms of polypeptide backbone of a protein. Axes represent the index of amino acid residues i (horizontal) and j (vertical). The helical (square) and sheet (arrows) segments are indicated on the diagonal. Off-diagonal islands represent pairs of residues with d_{ij}<1.5 nm. Different shades of gray represent different types of segments in contact.

This encoding of the protein structure has another consequence. It can be used to identify various possible paths of unfolding of the native structure by transforming them into the tree-like scheme that shows some similarity to problems discussed at this conference by G. Breitwell (see Figure 14).

Secondary structure segments are not uniform in properties. There is a real gradient of physical properties given by the polypeptide molecular structure. In a more detailed model of folding, the relative orientation of "vectorial" segments should be taken into account. Simple example of such a scheme for protein consisting of three β-strands is given in Figure 15a and a more realistic example of one possible fold from a set of all folding paths is shown in Figure 15b for a protein with 7 strands and 8 connecting C-segments.

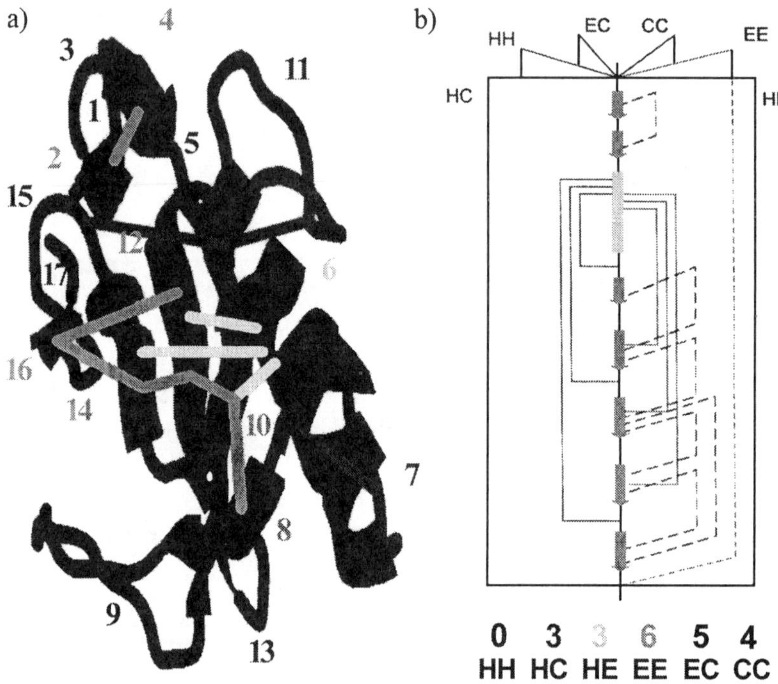

FIGURE 11. a) structural representation of the reduction of complete set of α-carbon distances $d_{ij}<1.5$ nm to the closest contacts between secondary structure segments for ribonuclease T1. The reduced set of distances is shown by gray bars (light gray H-E contacts, darker gray E-E contacts, dark gray C-C contacts). Numbers show order of the secondary structure segments along the amino acid sequence; b) graph representation of the above reduced set of close contacts. With the $d_{ij}<1.5$ nm limit, the book graph G_V is formed, with spine defined by the order of secondary structure segments along the sequence (path in the Eulerian graph defined by matrix $[n_{ij}]$). Pages contain edges representing close inter-segment contacts grouped according to the secondary structural type of vertices they connect.

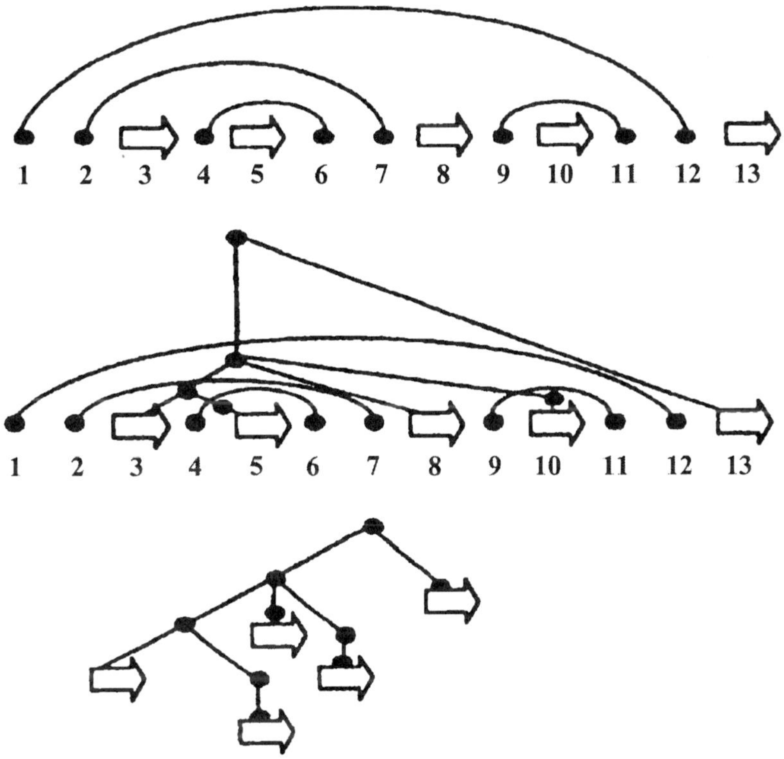

FIGURE 12. Correspondence between one possible topology of C-C contacts on the CC page of a graph G_V out of a set $S_{13,4}$ of possible topologies on the chain of 13 segments with 4 inter-segment contacts and a member $T_{5,3}$ of the set of linear trees (according to Waterman [**8, 19**]).

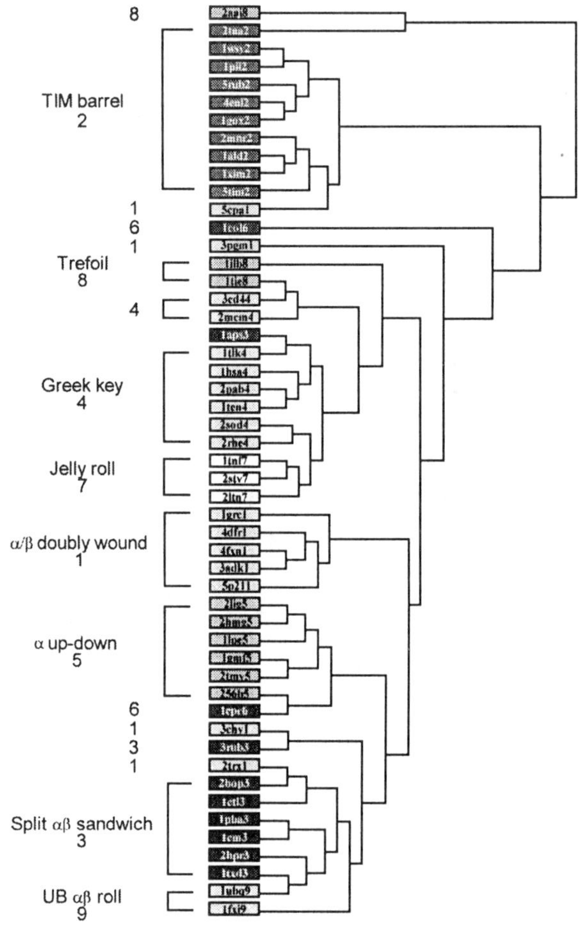

FIGURE 13. Dendrogram summarizing the cluster analysis results for proteins representing various tertiary fold families shown in Figure 9. Clustering was performed using Lance-Williams flexible algorithm with the vector g_V (invariant of the book graph G_V) as the protein coordinate.

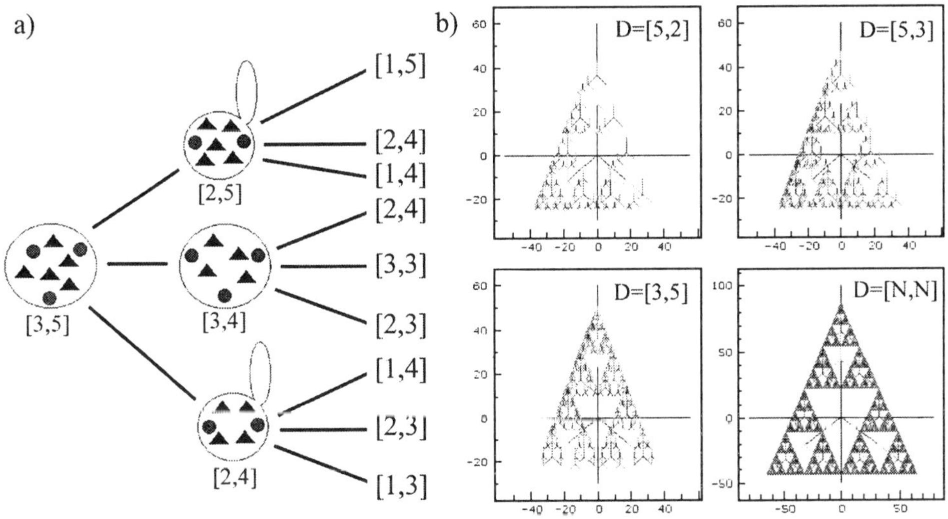

FIGURE 14. a) partial and b) complete model scheme of denaturation of a native protein structure that is based on the book graph invariant representation of the fold. For simplicity, only two types (e.g. H-H and E-E) of segment contacts are considered. In scheme (a) invariants g_V of the form [x,y] are used to characterize each state (vertices in a fractal-like scheme (b)). Simple model of structure unfolding is considered for generation of schemes (b), assuming that only one inter-segment contact of a given type is removed at any given instant of the unfolding.

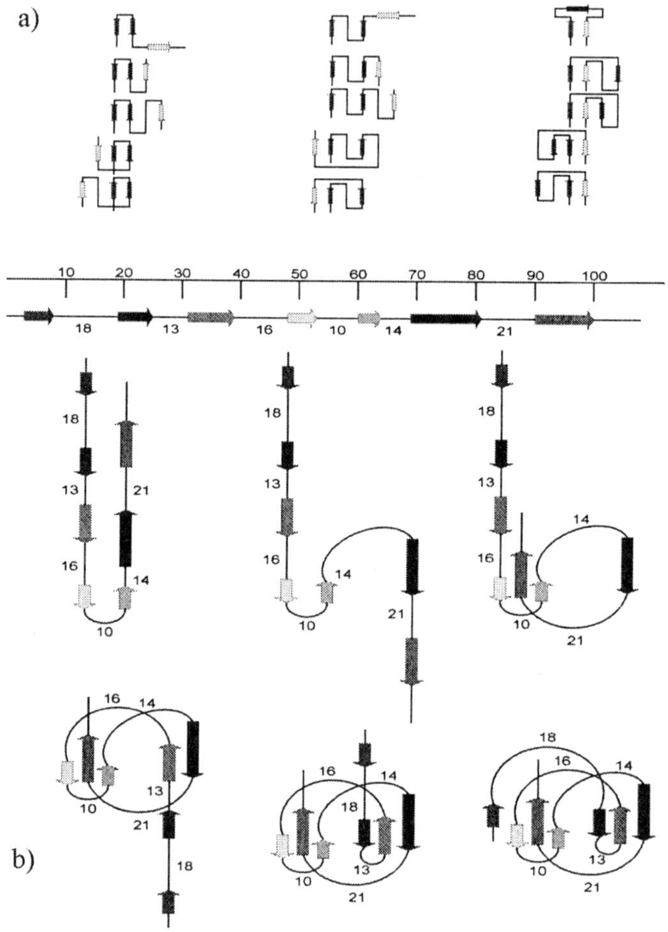

FIGURE 15. a) all possible orientations and contacts of a three-strand protein with completely connected fold (every strand is close to other strands); b) one possible pathway in folding of a protein chain with 7 strands into a native tertiary structure.

Conclusion

The primary goal of this contribution was to demonstrate several unconventional approaches to encoding of various aspects of protein structure into descriptors suitable to analysis by discrete mathematics. Eulerian and book graphs, non-negative matrices, trees, strings on finite set of amino acid symbols that are reorganized by secondary structural "reading frames", Markov chains, distance geometry problems, meaningful statements - all of these and many others can be tailored to represent biomolecular features. Known mathematical properties of these objects can help to solve important problems of protein structural studies. Ideally, the molecular applications should induce the formulation of new challenging mathematical problems. In this mutual exchange of inspiration we will find the proof of our original theorem, that for biomolecular research, the roads of discrete mathematics are paved with gold.

References

[1] C. B.RANDEN, J. TOOZE, (1991) *Introduction to protein structure*, Garland Publishing, New York & London.

[2] A. T. BRUNGER, M. NILGES, (1993) *Computational challenges for macromolecular structure determination by X-ray crystallography and solution NMR-spectroscopy*, Q. Rev. Biophys., 26, pp. 49-125.

[3] G. M. GRIPPEN, T. F. HAVEL, (1988) *Distance Geometry and Molecular Conformation*, John Wiley & Sons, New York.

[4] T. F. HAVEL, (1991) *An evaluation of computational strategies for use in the determination of protein structure from distance geometry constraints obtained by nuclear magnetic resonance*, Prog. Biophys. Mol. Biol., 56, pp. 43-78.

[5] J. B. SAXE, (1979) *Embeddability of weighted graphs in k-space is strongly NP-hard*, in Proc. 17th Allerton Conference in Communications, Control and Computing, Monticello, IL, pp. 480-489.

[6] H. A. SCHERAGA, (1992) *Predicting three-dimensional structures of oligopeptides*, in Reviews in Computational Chemistry, K. B. Lipkowitz and D. B. Boyd, eds., Vol. 3, VCH Publishers, New York, pp. 73-142.

[7] J. E. STRAUB, (1996) *Optimization techniques with applications to proteins*, in Recent Developments in Theoretical Studies of Proteins, R. Elber, ed., World Scientific, Singapore, pp. 137-196.

[8] M. S. WATERMAN, (1995) *Introduction to computational biology. Maps, sequences and genomes.*, Chapman & Hall, London.

[9] J. SETUBAL, J. MEIDANIS, (1997) *Introduction to computational molecular biology*, PWS Publishing Company, Boston.

[10] P. J. ARTYMIUK, H. M. GRINDLEY, E. M. MITCHELL, D. W. RICE, P. WILLETT, (1990) *Graph matching techniques for databases of three dimensional chemical structures*, in D. Bawden & E. M. Mitchell, eds., Chemical information systems: Beyond the structure diagram, Chichester: Ellis Horwood, pp. 50-62.

[11] R. TAMARIN, (1993) *Principles of genetics*, Wm. C. Brown, Dubuque, IA.

[12] J. D. WATSON, (1987) *Molecular biology of the gene*, Vol. 1-2, Benjamin Cummings, Redwood City, CA.

[13] P. PANČOŠKA, H. FABIAN, G. YODER, V. BAUMRUK, T.A. KEIDERLING, (1996) *Protein structural segments and their interconnections derived from optical spectra. Thermal unfolding of Ribonuclease T1 as an example.*, Biochemistry, 35, pp. 13094-13106.

[14] V. CABIAUX, K. A. OBERG, P. PANČOŠKA, T. WALZ, P. AGRE, A. ENGEL, (1997) *Comparison of AQP-I and Bacteriorhodopsin structures: A Fourier-transform infrared spectroscopy study.*, Biophys. J., 73, pp. 406-417.

[15] P. PANČOŠKA, E. BITTO, V. JANOTA, T. A. KEIDERLING, (1994) *Quantitative analysis of vibrational circular dichroism spectra of proteins. Problems and perspectives.*, Faraday Disc. 99, pp. 287-310.

[16] P. PANČOŠKA, V. JANOTA, J. NEŠETŘIL, T. A. KEIDERLING, (1998) *Novel matrix descriptor for spectroscopic determination of the connectivity of secondary structure segments in proteins. Analysis of general properties using graph theory.*, Analytical Biochem., in the press.

[17] E. SENETA, (1981) *Non-negative matrices and Markov chains*, Springer Series in Statistics, Springer-Verlag, New York, Heidelberg, Berlin.

[18] C. A. ORENGO, D. T. JONES, J. M. THORTON, (1994) *Protein superfamilies and domain superfolds*, Nature, 372, pp. 631-635.

[19] W. R. SCHMITT, M. S. WATERMAN, (1994) *Linear trees and RNA secondary structure*, Disc. Appl. Math., 51, pp. 317-323.

[20] Figure 1c is a map of Wind Cave (Wind Cave National Park, South Dakota, U.S.A.)

Department of Chemistry, University of Illinois at Chicago, 845 West Taylor Street, M/C 111, Chicago, Illinois 60607-7061 U.S.A.

Center for Discrete Mathematics, Theoretical Computer Science and Applications (DIMATIA), Charles University, Malostranské náměstí 25, 118 00 Praha 1, Czech Republic

Department of Chemical Physics, Charles University, Faculty of Mathematics and Physics, Ke Karlovu 3, 121 16 Praha 2, Czech Republic

E-mail address: pancoska@tigger.cc.uic.edu

Center for Discrete Mathematics, Theoretical Computer Science and Applications (DIMATIA), Charles University, Malostranské náměstí 25, 118 00 Praha 1, Czech Republic

E-mail address: janota@kam.ms.mff.cuni.cz

Center for Discrete Mathematics, Theoretical Computer Science and Applications (DIMATIA), Charles University, Malostranské náměstí 25, 118 00 Praha 1, Czech Republic

E-mail address: nesetril@kam.ms.mff.cuni.cz

DIMACS Series in Discrete Mathematics
and Theoretical Computer Science
Volume **49**, 1999

Meaningless Statements

Fred S. Roberts

ABSTRACT. Increasingly, discrete mathematics is influenced by connections with other fields. We give one example of a broad topic in the field of discrete mathematics that comes from the social sciences, and in particular from the theory of measurement that has been developed to put measurement, especially in the social sciences, on a firm mathematical foundation. The key concept is that of meaningful statement, a statement whose truth or falsity remains unchanged after admissible transformations of scales of measurement. We apply this concept to combinatorial optimization, graph coloring, scheduling, linear programming, 0-1 optimization, and multiperson games.

1. The Future of Discrete Mathematics

Increasingly, discrete mathematics is influenced by connections with other fields. In the mathematical sciences, it is influenced by connection with probability, geometry, algebra, analysis, topology, number theory, ... Outside the mathematical sciences, it is influenced by connection with biology, chemistry, physics, manufacturing, engineering, ... In this paper, I will give one example of a broad topic in the field of discrete mathematics that comes from the social sciences, but has been developed not only by economists and psychologists, but also by philosophers of science, physicists, logicians, and, of course, mathematicians. It is motivated by the theory of measurement that has been developed to put measurement, especially in the social sciences, on a firm mathematical foundation.

In this paper, I will give a brief overview of measurement theory and define a central concept of the theory, that of a meaningful statement. I will then apply the theory of meaningfulness to discrete mathematics in various ways. In particular, I will ask whether or not statements in combinatorial optimization, that a specific solution is optimal, are meaningful. I will talk about the meaningfulness of conclusions about the generalization of graph coloring called T-coloring. I will give a variety of examples of meaningful and meaningless conclusions about scheduling problems. This will lead to the question of when a greedy solution to an optimization problem is a meaningful optimal solution. I will talk in general about

1991 *Mathematics Subject Classification.* Primary 90.30; Secondary 05C, 05.55, 90C, 90.50, 90.70, 90.99, 92G.

The author thanks the U.S. National Science Foundation for its support under grants SBR97-09134 and INT96-05174 to Rutgers University.

meaningfulness of conclusions in linear programming and in 0-1 optimization. Finally, I will mention meaningfulness of conclusions about solutions for multiperson games.

The reader will see that questions of meaningfulness give rise to interesting questions in discrete mathematics. The detailed questions and answers are, however, perhaps not as important as the new way of thinking, and new types of questions, that applications such as those from the social sciences are bringing into discrete mathematics.

2. The Theory of Measurement

The theory of measurement was developed to place issues of measurement on a firm mathematical foundation. A primary motivation for the development of this theory was the need to understand, using measurement in the physical sciences as a paradigm, what it means to measure things in the social sciences, things like preference, value, loudness, etc. The ideas of the modern theory of measurement go back to 19th century work of Helmholtz [**10**] and others. The modern theory of "representational measurement" goes back to the foundational papers of Scott and Suppes [**36**] and Suppes and Zinnes [**42**], and is described in such books as [**12, 15, 41, 25, 28**].

A scale of measurement arises when we assign numbers to objects. Thus, we can think of a *scale* as a function f from some set A of objects into the real numbers, $f : A \to \Re$. A key idea is that of admissible transformation of scale. An *admissible transformation* sends one acceptable scale into another, for example changing from centigrade into Fahrenheit or from kilograms into pounds. In most cases, one can think of an admissible transformation ϕ as a function defined on the range of $f; \phi : f(A) \to \Re$. For instance, the admissible transformation changing degrees centigrade into degrees Fahrenheit is given by $\phi(x) = (9/5)x + 32$ and that for changing kilograms into pounds is given by $\phi(x) = (2.2)x$.

A variety of interesting classes of scales can be described by studying the associated class of admissible transformations. This idea was introduced by the social scientist S.S. Stevens in a series of papers (e.g., [**37, 38, 39**]). For instance, if we are allowed to change the unit and nothing else, then the class of admissible transformations is given by the functions of the form $\phi(x) = \alpha x, \alpha > 0$. A scale with this as its class of admissible transformations is called a *ratio scale.* Ratio scales arise when we are measuring mass, temperature on the Kelvin scale when there is an absolute zero, and time intervals such as minutes, hours, seconds. If we are allowed to change both the unit and the zero point, then the class of admissible transformations is given by the functions of the form $\phi(x) = \alpha x + \beta, \alpha > 0$. A scale with this as its class of admissible transformations is called an *interval scale.* Interval scales arise in the measurement of temperature (as for example when we change from centigrade to Fahrenheit) and in measurement of time in the calendar sense (as in the statement that this is the year 1997). If only order matters, then the class of admissible transformations consists of all strictly increasing functions ϕ, i.e., functions ϕ so that $x \geq y \Leftrightarrow \phi(x) \geq \phi(y)$. In this case, we say we have an *ordinal scale.* Ordinal scales arise when we have gradations, as in grades of leather or lumber or wool, or in measures of hardness of minerals. They often arise when we measure preferences using utilities.

The key idea in this paper is that of meaningfulness. The definition I will use is due to Suppes [**40**] and Suppes and Zinnes [**42**]. We shall say that a statement involving numerical scales is *meaningful* if its truth or falsity is unchanged after any (or all) of the scales is transformed by an admissible transformation. (As an aside, we note that there are situations, called *irregular*, where one may not be able to obtain one acceptable scale from another by an admissible transformation. In this case, we have to modify the definition to say: A statement involving numerical scales is *meaningful* if its truth or falsity is unchanged after any (or all) of the scales is replaced by another acceptable scale. We shall disregard this complication. It is discussed, however, in the paper [**35**]. Alternative formulations of the definition are discussed in many places in the literature, for instance in [**7, 15**]. See [**33**] for other references.) To illustrate the definition, consider the statement "The piece of yellow chalk is three times as long as the piece of white chalk." Is this a meaningful statement? Length is measured on a ratio scale (we only change units, as for example from centimeters to inches). Hence, the admissible transformations are of the form $\phi(x) = \alpha x, \alpha > 0$. The statement in question is the statement

$$f(a) = 3f(b).$$

We want this to hold if and only if

$$\phi(f(a)) = 3\phi(f(b)).$$

But this latter statement becomes

$$\alpha f(a) = 3\alpha f(b),$$

which is equivalent to the original since $\alpha > 0$. Thus, the original statement is meaningful. In general, for ratio scales, it is meaningful to compare ratios: $f(a)/f(b) > f(c)/f(d)$. For interval scales, it is meaningful to compare intervals: $f(a) - f(b) > f(c) - f(d)$. For ordinal scales, it is meaningful to compare size: $f(a) > f(b)$.

Consider another example, the statement: "The mouse weighs 1000 times what the elephant weighs." This is meaningful as with the last example, since weight is measured on a ratio scale. However, it is presumably false. Meaningfulness is different from truth. It is concerned with the kinds of assertions it makes sense to make, in particular the kinds of assertions that are not accidents of the particular choice of scale, i.e., of units, zero points, etc. If the truth of a conclusion can depend upon some arbitrary choices such as about units or zero points, we would probably not want to put much weight behind that conclusion. The notion of meaningfulness has a long history in the theory of measurement, and is closely related to the concept of invariance that has played such a central role in the history of mathematics. It has found applications in the study of average performance, importance ratings, indices of consumer confidence, psychophysical scaling, blockmodeling of social structure, structural modeling in decisionmaking, analysis of matching experiments, statistical analysis, performance analysis, and elsewhere. Some of these applications are investigated in the book [**28**] and the papers [**33, 29**]. In this paper, I would like to discuss its relevance to problems of discrete mathematics, in particular problems having to do with optimization.

3. Meaningfulness of Conclusions from Combinatorial Optimization

A classical problem of combinatorial optimization is the shortest path problem. Consider a network in which there are three vertices, x, a, and y, and there are arcs from x to a with weight 2, a to y with weight 4, and x to y with weight 40. Then the shortest path from x to y is the path x, a, y. Is this conclusion meaningful? What if weights are measured on an interval scale? This could happen if the weight on an arc measures the temperature encountered while traveling along that arc and we are seeking a route that minimizes the sum of temperatures encountered. Consider the admissible transformation $\phi(x) = 2x + 70$. Now the path x, y has smaller weight (150) than the path x, a, y (weight 152). The original conclusion was meaningless. It is, however, meaningful if weights are measured on a ratio scale. This kind of observation was first noted in the paper [**30**]. The shortest path problem can be formulated as a linear programming problem, and so our example shows that the conclusion that z is an optimal solution in a linear programming problem can be meaningless if the cost parameters are measured on an interval scale.

Consider next the minimum spanning tree problem: Given a weighted graph, what is the spanning tree with a total sum of weights as small as possible? If weights are given on an interval scale, is the conclusion that a given spanning tree is minimum a meaningful conclusion? The answer turns out to be yes. This is a corollary of the result that Kruskal's algorithm (the greedy algorithm) (see [**13**] and [**19**]) gives a minimum spanning tree. In this algorithm, we order edges in increasing order of weight and examine edges in this order. We include an edge if it does not form a cycle with edges previously included. We stop when all vertices are included. Since this algorithm depends only on the order of the weights, the optimal solution remains unchanged under the scale transformations associated with interval scales, i.e., the transformations of the form $\phi(x) = \alpha x + \beta, \alpha > 0$. In fact, it remains unchanged under strictly increasing transformations, and so the conclusion of optimality is meaningful even for ordinal scales.

Note that the analysis of meaningfulness is something like sensitivity analysis. We are interested in the sensitivity (or invariance) of conclusions under certain kinds of changes. However, the classes of perturbations of data in the study of meaningfulness are quite different than those studied in sensitivity analysis. While sensitivity analysis deals mainly with small perturbations of data, meaningfulness deals with admissible transformations that preserve certain relationships among "similar" types of data, so whether the change in problem parameters is small or large is irrelevant as long as a relationship in question remains unchanged.

4. T-Coloring

The following problem arises from questions of channel assignment in communications. Let $G = (V, E)$ be a graph and let T be a set of nonnegative integers. Find an assignment f of a positive integer $f(x)$ to each vertex x in G so that

$$x, y \in E \Rightarrow |f(x) - f(y)| \notin T.$$

Such an assignment f is called a T-*coloring* of graph G. In channel assignments, V is a set of transmitters, there is an edge between two transmitters if they conflict, $f(x)$ is the channel assigned to transmitter x, and T is a set of disallowed separations between channels assigned to conflicting transmitters. If $T = \{0\}$, then the T-coloring is an ordinary graph coloring. We shall always assume that 0 belongs to

T, for otherwise the trivial assignment in which each transmitter gets the same channel is a T-coloring. If the elements of T are nonnegative integers and the values of $f(x)$ are positive integers, as we shall assume in what follows, then the case $T = \{0, 1\}$ means that conflicting transmitters get not only different channels, but non-adjacent channels. The notion of T-coloring was introduced by Hale [**9**] and by Cozzens and Roberts [**5**], and has been studied extensively. For instance, several Ph.D. theses have been written about these colorings; see [**3, 14, 26, 43, 45**] . See [**31, 32**] for surveys of the subject.

Consider a triangle with vertices a, b, c and a T-coloring with $T = \{0, 1, 4, 5\}$. T-coloring the vertices in a greedy manner using the vertex order a, b, c, always choosing for a given vertex the lowest channel that does not violate the T-coloring rules, we get the assignment $f(a) = 1, f(b) = 3, f(c) = 9$. However, there is a "better" solution: $f(a) = 1, f(b) = 4, f(c) = 7$. The second solution is better in the sense that $\max_{x,y \in V} |f(x) - f(y)|$ is smaller with the second assignment. The quantity $\max_{x,y \in V} |f(x) - f(y)|$ is called the *span* of f and the minimum span over all T-colorings f is called the *optimal span (optimal T-span)* of G. It is natural to ask the question: for what graphs and what sets T does the greedy algorithm give the optimal span? Unfortunately, this is a difficult question, and it is not yet completely solved even for complete graphs. There is a considerable literature devoted to this subject (see summaries in some of the theses and papers cited above). What if the numbers in T change by a scale transformation? Is the answer to the above question meaningful? Consider the case where elements of T are measured on a ratio scale. Then even for complete graphs, the conclusion that the greedy algorithm obtains the optimal span is meaningless. Take the previous example of the triangle, but consider the transformation $\phi(x) = 2x$. Then T becomes the set $\alpha T = \{0, 2, 8, 10\}$ and now the greedy algorithm gives the assignment $f(a) = 1, f(b) = 2, f(c) = 5$. This has an optimal span.

(As an aside, we note that it is not as easy to give an example of a complete graph K_n and a set T and an integer α so that the greedy algorithm obtains the optimal span for K_n with T, but not with αT. One interesting example due to Cozzens and Roberts [**6**] is K_6 with $T = \{0, 1, 2, 3, 4\} \cup \{8\} \cup \{10, 11, 12\} \cup \{31, 32\}$. Then the greedy T-coloring gives an optimal span for K_6, but does not do so for αT when $\alpha = 2$.)

THEOREM 4.1 (Cozzens and Roberts [**6**]). *For fixed T, the conclusion that the greedy algorithm obtains the optimal span for* all *complete graphs is meaningful if the elements of T are measured on a ratio scale.*

For example, the greedy algorithm obtains the optimal span for all complete graphs if $T = \{0, \alpha, 2\alpha, \ldots, r\alpha\}$ for any $\alpha = 1, 2, 3, \ldots$. This result and the related analysis allowed for the discovery of new examples of sets T for which the greedy algorithm gave optimal spans for all complete graphs. Thus, considerations of meaningfulness had a side benefit of leading to results of independent interest in graph theory.

Note that limiting the set T to integers could be problematical in the interpretation of this example. As Pierre Hansen has pointed out (personal communication), suppose $T = \{0, 1\}$ and frequencies are measured in kilohertz (kHz). Then frequencies at distance 1 kHZ are not allowed on conflicting transmitters. Neither, of course, are frequencies at distance 1/2 kHz However, if we take $\alpha = 1000$, then the new units are Hertz (Hz) and now frequencies at distance 500 Hz $=$ 1/2 kHz are allowed on conflicting channels. This changes the problem.

Analyses of the type we have made in this section might be relevant to other types of generalizations of ordinary graph colorings such as set colorings (see survey in [**31**]).

5. Scheduling with Earliness and Tardiness Penalties and Priorities

Discrete mathematics plays a very important role in the analysis of scheduling problems. Here we shall ask the question: Is it meaningful to say that a given schedule is an optimal solution to a scheduling problem. In particular, we shall note that for certain kinds of scheduling problems that involve scales of measurement, this conclusion can be meaningless. The particular type of scheduling problem we shall consider arose from the Air Mobility Command of the United States Air Force, and it can be characterized as a scheduling problem with earliness/tardiness penalties and priorities. Suppose that n items (equipment, people) are to be transported from an origin to a destination. Item i has a *desired arrival time* d_i. For simplicity, we assume this is a positive integer. Transportation time will be assumed to be constant. For any given arrival time, there is available a certain *capacity* c for transportation (number of seats, cargo capacity, etc.) (Note that we are assuming constant capacity.) A *schedule* S is an assignment of a positive integer arrival time C_i to each item i subject to the constraint that the number of C_i that can equal a given integer is at most c. Penalties are applied for tardy arrivals and perhaps for early arrivals. These depend upon the desired and scheduled arrival times of the different items and also upon the priority or status of the different items. The priority of item i is denoted by w_i. The interest in scheduling problems where penalties are applied to early arrivals as well as late arrivals is closely tied to the concept of "just-in-time" production, the goal being to have the right amount of materials of the right quality at the right time in the right place to produce the right quantity of items demanded by the next step of the production.

In the transportation scheduling problem as we have described it, a *penalty function* depends upon the vectors $\mathbf{C}(S)=(C_1, C_2, \ldots, C_n)=\mathbf{C}$, $\mathbf{d}=(d_1, d_2, \ldots, d_n)$, and $\mathbf{w} = (w_1, w_2, \ldots, w_n)$. We denote the penalty function by an expression such as $F(\mathbf{C}, \mathbf{w}, \mathbf{d})$. It is natural to consider the case where F is *summable* in the sense that

$$F(\mathbf{C}, \mathbf{w}, \mathbf{d}) = \sum_{i=1}^{n} g(C_i, w_i, d_i).$$

Summable penalty functions are called *separable* if g can be expressed as

$$g(C, w, d) = \begin{cases} h_t(w)f(C, d) & \text{if } C \geq d \\ h_e(w)f(C, d) & \text{otherwise} \end{cases}$$

for functions h_t and h_e defined on the positive reals and $f(C, d)$ defined on ordered pairs of natural numbers. A summable, separable penalty function is called *symmetric* if $h_t = h_e$. A simple example of a summable, separable, symmetric penalty function is the function

$$F(\mathbf{C}, \mathbf{w}, \mathbf{d}) = \sum_{i=1}^{n} w_i |C_i - d_i|.$$

Here, we are using the weighted sum of the deviations between desired and scheduled arrival times, where the weighting factor is the priority. The scheduling problem we shall study is to find a schedule that minimizes the penalty. There is a large literature concerned with this problem. For example, if $h_e(w) = h_t(w)$ for all w and $f(C,d) = |C-d|$ and $c = 1$, Garey, Tarjan and Wilfong [**8**] showed that the problem is NP-complete if transportation times can vary, but solvable in polynomial time if they are constant as in our problem. Some survey papers that deal with penalty functions similar to the ones we have described are [**1, 2, 11**].

Suppose that priorities are measured on some scale. It then makes sense to ask whether or not the conclusion that schedule S is optimal is a meaningful conclusion. To give a simple example, suppose that priorities are measured on a ratio scale. We are interested in the assertion A:

$$F(\mathbf{C}, \mathbf{w}, \mathbf{d}) \leq F(\mathbf{D}, \mathbf{w}, \mathbf{d})$$

for all schedules $\mathbf{D}$. And we ask, if α is a positive number, does this assertion imply the assertion B:

$$F(\mathbf{C}, \alpha\mathbf{w}, \mathbf{d}) \leq \mathbf{F}(\mathbf{D}, \alpha\mathbf{w}, \mathbf{d})$$

for all schedules $\mathbf{D}$. Here, $\alpha\mathbf{w}$ means the vector $(\alpha w_1, \alpha w_2, \ldots, \alpha w_n)$. Suppose that F is summable and separable and that

$$h_e(\alpha w) = K(\alpha)h_e(w) \text{ for } \alpha > 0, K(\alpha) > 0. \tag{5.1}$$

Then assertion A implies assertion B, so, as noted in [**16**], it is meaningful to conclude that schedule $\mathbf{D}$ is optimal. Equation 5.1 arises in measurement theory in various contexts, and arose originally in psychophysics. See [**34**].

Consider a different example, also noted in [**16**]. Let F be a summable, separable, symmetric objective function defined by taking $c = 1, f(C,d) = |C-d|$, and $h_e(w) = h_t(w) = 2^{w-1}$ for all w. If $\mathbf{w} = (1,2,2,2)$ and $\mathbf{d} = (1,2,2,2)$, then consider $\mathbf{C} = (1,2,3,4)$. We have $F(\mathbf{C}, \mathbf{w}, \mathbf{d}) = 6$ and it is easy to see that $\mathbf{C}$ is optimal. However, after a change of scale from $\mathbf{w}$ to $\alpha\mathbf{w} = 2\mathbf{w}$, we find that $F(\mathbf{C}, 2\mathbf{w}, \mathbf{d}) = 24$ while $F(\mathbf{D}, 2\mathbf{w}, \mathbf{d}) = 22$ for the schedule defined by $\mathbf{D} = (4,1,2,3)$. Thus, $\mathbf{C}$ is no longer optimal after a change of unit. As noted in [**16**], the priority functions h_e and h_t used in this example might be somewhat unrealistic. However, they illustrate the point. If we use the same penalty function except to change h_e and h_t to be linear, then the conclusion of optimality of a schedule is meaningful under ratio scales. However, this turns out not to be the case under interval scales, as is also noted in [**16**].

A special case of the scheduling problem we have considered is the case of constant desired arrival times (also known as the problem of common due dates). Suppose that the penalty function is summable, separable, and symmetric. Suppose it is also *w-increasing* in the sense that both h_e and h_t are strictly increasing in w and *$|C-d|$-increasing* in the sense that $f(C,d) = A(|C-d|)$ for A an increasing function. Mahadev, Pekeč, and Roberts [**16**] show that in this case, a schedule is optimal if and only if it is attainable by a greedy algorithm. Moreover, any strictly increasing transformation of scale values from w_i to $\phi(w_i)$ gives the same greedy solution. The following conclusion results:

THEOREM 5.1 (Mahadev, Pekeč, and Roberts [**16**]). *Suppose that the objective function is summable, separable, symmetric, w-increasing, and $|C-d|$-increasing. Then the statement that $\mathbf{C}$ is an optimal solution for the constant desired arrival* *time problem is meaningful if priorities are measured on an ordinal, interval, or ratio scale.*

Perhaps the simplest case of non-constant desired arrival times is that where $\mathbf{d} = (d, d, \ldots, d, k)$, where $k \neq d$. We shall call this the (d,k)-*problem*. This problem is very specialized, but it leads to some interesting mathematical questions. In the case of ratio scale priority measurement, if we use the objective function defined by

$$F(\mathbf{C}, \mathbf{w}, \mathbf{d}) = \sum_{i=1}^{n} w_i |C_i - d_i|,$$

then the conclusion of optimality is meaningful for a (d, k) problem. This follows since Equation 5.1 holds. However, the conclusion of optimality fails if we have interval scale priority measurement. This is the case if $c = 1, \mathbf{w} = (\mathbf{9}, \mathbf{9}, \mathbf{9}, \mathbf{1}), \mathbf{d} = (\mathbf{2}, \mathbf{2}, \mathbf{2}, \mathbf{1})$. Then, as pointed out in [**16**], $\mathbf{C} = (2, 1, 3, 4)$ is an optimal schedule, but after the admissible transformation $\phi(w) = \alpha w + \beta$ with $\alpha = 1/8, \beta = 7/8$ is applied, the schedule $\mathbf{D} = (2, 3, 4, 1)$ has a lower penalty function than $\mathbf{C}$. This leads to the question of determining conditions under which the conclusion of optimality for a (d, k)-problem is meaningful under interval scale priority measurement.

THEOREM 5.2 (Mahadev, Pekeč, and Roberts [**16**]). *Suppose that the objective function is summable, separable, symmetric, and w-increasing, and $f(C, d) = K|C - d| + L$ for $K > 0$. Suppose that $k > d$ or $k < d$ and $d \geq \lceil (n+c)/2c \rceil$. Then the statement that C is an optimal solution for the (d, k)-problem is meaningful if priorities are measured on an ordinal, interval, or ratio scale.*

The results we have just described leave a variety of open problems. For example, little is known about meaningfulness even in the case of scheduling problems where $\mathbf{d} = (d, d, \ldots, d, k, k)$. We have also left open the meaningfulness for the (d, k)–problem when $k < d$ and $d < \lceil (n + c)/2c \rceil$ Little work has been done on the case of nonconstant transportation times. It remains an open question to find scheduling situations where the conclusion of optimality is meaningful for interval scales but not for ordinal scales.

6. Single Machine Scheduling

A problem analogous to the scheduling problem considered in the previous section arises when we study single machine scheduling. Consider the problem of scheduling n jobs on a single machine in which each job has a specified due date or completion time and a penalty is applied for a completion time different from the desired one. In many practical problems, a penalty is only applied for tardy completions; while more generally, a penalty is applied to both early and tardy completions, perhaps in a different way. Because penalties can be applied to early completions, we allow the machine to lie idle and we schedule without preemption, i.e., we do not allow a job to be interrupted once it is started. Among the penalties studied in the literature are those that involve weighting factors that weight the deviations from desired completion times.

Typically weights are measured using some scale of measurement, and we can ask about the effect on the solution to a scheduling problem if we make admissible changes of scale. Mahadev, Pekeč, and Roberts [**17**] show that in some cases such changes can transform an optimal solution into a non-optimal one, and systematically describe those situations when this anomaly occurs.

To formulate the problem precisely, let us suppose we have n jobs in which the ith job has *processing time* $p_i > 0$ and *desired completion time* $d_i \geq 0$. At most one job can be carried out at a time and there is no preemption. A *schedule* S (*feasible schedule* S) assigns a *completion time* C_i to each job i, with time starting at 0 and each job i getting processing time p_i. As in the previous section we let $\mathbf{C}(S) = (C_1, C_2, \ldots, C_n)$ and we also use the vectors $\mathbf{d} = (d_1, d_2, \ldots, d_n)$ and $\mathbf{p} = (p_1, p_2, \ldots, p_n)$. Since no job may start before time 0, we have $C_i \geq p_i$, for all i. We consider two special cases, those with *common due dates*, i.e., where $d_i = d$ for all i, and those with *common processing times*, i.e., where $p_i = p$ for all i. The first case is abbreviated **cdd**, with **ncdd** standing for the situation where we are not in case cdd. The second case is abbreviated **cpt**, with **ncpt** standing for the situation where we are not in case cpt.

Typically, we have an *objective function* $F(S) = F(d_1, d_2, \ldots, d_n, C_1, C_2, \ldots, C_n)$ and we seek a (feasible) schedule S that minimizes $F(S)$. Note that the objective function does not explicitly depend upon the processing times. These play a role in determining whether or not a schedule is feasible. We only consider objective functions that depend upon the deviations between desired and actual completion times, and specifically we define the *earliness* of the i-th job $E_i(S) = E_i = \max\{0, d_i - C_i\}$ and the *tardiness* of the i-th job $T_i(S) = T_i = \max\{0, C_i - d_i\}$. One objective function arises if we simply consider a weighted sum of earliness and tardiness, obtaining

$$F(S) = F_{sumE|T}(S) = \sum_{i=1}^{n} \alpha_i E_i + \sum_{i=1}^{n} \beta_i T_i,$$

where α_i and β_i are positive weights. If only tardiness is penalized, we consider the function

$$F(S) = F_{sumT}(S) = \sum_{i=1}^{n} \beta_i T_i.$$

In objective functions $F_{sumE|T}(S)$ and $F_{sumT}(S)$, we minimize the sum of the deviations. An alternative is to minimize the maximum deviation; we then have the objective functions

$$F(S) = F_{maxE|T} = \max\{\alpha_1 E_1, \alpha_2 E_2, \ldots, \alpha_n E_n, \beta_1 T_1, \beta_2 T_2, \ldots, \beta_n T_n\}$$

and

$$F(S) = F_{maxT} = \max\{\beta_1 T_1, \beta_2 T_2, \ldots, \beta_n T_n\}.$$

A variety of other objective functions have been studied in the literature.

In the case where $\alpha_i = \beta_i$ for all i, the objective function has symmetric weights and we say that we are in the case of *symmetry*; we abbreviate this case as **s**. If we are not in case s, we say we are in case **ns**. If $\beta_i = \beta$ for all i and if $\alpha_i = \alpha$ for all i when we are considering earliness penalties, then we say that we have *common weights* and abbreviate this case as **cw**. If we do not have common weights, we say we are in case **ncw**. In [**17**], Mahadev, Pekeč, and Roberts systematically study the four penalty functions $F_{sumE|T}$, F_{sumT}, $F_{maxE|T}$, and F_{maxT} under all combinations of the four assumptions cdd or ncdd, cpt or ncpt, s or ns, and cw or ncw. They show that if weights are measured on a ratio scale, then conclusions about optimality of a schedule are meaningful under the four objective functions we have defined. However, this is not always true for interval scales or ordinal scales. The results of [**17**] can be summarized both for interval and ordinal scales as follows: in the case of common weights, if weights are symmetric or if we only apply penalties for tardiness, then conclusions about optimality are meaningful for the four objective functions considered. In other cases, these conclusions can be meaningless (in the sense that there exist weights such that there is no optimal schedule for which the conclusion of optimality is meaningful), except in the cases ncw, cdd, cpt, s for objective function $F_{sumE|T}$ and ncw, cdd, cpt, ns for objective function F_{sumT}. The case of common due dates and non-common weights for objective function F_{maxT} is different than all other cases considered. In this case, there always exists an optimal schedule (given by the largest weight order) for which the conclusion of optimality is meaningful. However, there can be other optimal schedules for which the conclusion of optimality is meaningless.

The above results are mathematically very straightforward. They are more interesting for the points they make than for the mathematical issues involved. However, one surprising result is that the results about the meaningfulness of the conclusion of optimality are the same for interval and ordinal scales in all the cases that were considered in [**17**]. It is not hard to give examples of problems in combinatorial optimization where the conclusion of optimality is meaningful for interval scales but not for ordinal scales. (For instance, this is true for the traveling salesman problem where edge weights represent data measured on an interval scale.) It would be interesting to characterize scheduling problems for which the fact that the conclusion of optimality is meaningful for any choice of weights measured on an interval scale implies that the conclusion of optimality is meaningful for any choice of weights measured on an ordinal scale. This is a topic that will require further research.

The work of [**17**] leaves room for considerably more analysis. For one thing, it considers only single machines, rather than multiple machines. Second, the discussion is limited to four specific objective functions. A variety of other objective functions are studied in the literature (see for example the survey paper [**2**]) and it would be interesting to extend the analysis to some of these other objective functions as well.

7. When is a Greedy Solution a Meaningful Optimal Solution?

We have observed in Section 5 that in scheduling under ordinal scales, a "greedy" solution is optimal under certain conditions and this conclusion is meaningful. We would like to ask: Under a general setting, for ordinal scales, when is the greedy solution to an optimization problem a meaningful optimal solution? We give a general formulation due to Pekeč [**20, 21**]. In this section, we concentrate on the special case where weights are positive.

We use the notation $\Re^n_+$ for the set of vectors $\mathbf{x} \in \Re^n$ for which $x_i \geq 0$ for all i. Suppose $X = X_1 \times X_2 \times \ldots \times X_n$ and X_i is a metric space with a metric d_i. Consider an objective function $P : X \times \Re^n_+ \to \Re$. Let $F \subseteq X$ correspond to the set of feasible solutions. The problem is to minimize $P(\mathbf{x}, \mathbf{w})$ over $\mathbf{x} \in F$. Pekeč makes the following assumptions.

Axiom 1. For every i, there is a unique $a_i \in X_i$ so that for all $\mathbf{x} \in X, \mathbf{w} \in \Re^n_+$,

$$P(x_1, \ldots, x_{i-1}, a_i, x_{i+1}, \ldots, x_n; \mathbf{w}) \leq \mathbf{P}(\mathbf{x}, \mathbf{w}).$$

In this axiom a_i is the "desirable value" for the i-th variable.

Axiom 2. If $x_j = y_j$ for $j \neq i$, then

$$P(\mathbf{x}; \mathbf{w}) \leq P(\mathbf{y}; \mathbf{w}) \Leftrightarrow d_i(x_i, a_i) \leq d_i(y_i, a_i)$$

This axiom says that P is increasing as the i-th variable moves away from the desirable value.

Axiom 3 (w-unboundedness). If $x_j = y_j$ for $j \neq i$, then

$$\lim_{w_i \to \infty} |P(\mathbf{x}; \mathbf{w}) - P(\mathbf{y}; \mathbf{w})| = \infty.$$

This axiom says that there can be an unbounded change in the value of P for any change of the value of the i-th variable.

Axiom 4 (Controllability). If $x_j = y_j$ for $j \neq i$, then there is $M_i(\mathbf{x}, \mathbf{y})$ so that for all $\mathbf{w} \in \Re^n_+$,

$$\lim_{w_i \to 0} |P(\mathbf{x}; \mathbf{w}) - P(\mathbf{y}; \mathbf{w})| < M_i$$

This axiom says that moving a variable with a small weight cannot result in big changes in the value of the objective function.

These four axioms are satisfied by a linear objective function $P(\mathbf{x}; \mathbf{w}) = \mathbf{w}^T \mathbf{x} = \sum w_i x_i$ and by a quadratic objective function $P(\mathbf{x}; \mathbf{w}) = \sum w_i w_j x_i x_j$.

The w_i could be measured on scales. The decision as to whether or not $\mathbf{x} \in F$ could also depend upon scales. We shall only allow the former here. Moreover, we shall limit ourselves to the case where the w_i are measured on an ordinal scale.

THEOREM 7.1 (Pekeč [**20, 21**]). *Consider an optimization problem* $\min_{\mathbf{x} \in F} P(\mathbf{x}; \mathbf{w})$, *where F is a closed set and P satisfies Axioms 1 to 4. If $w_1, w_2, \ldots, w_n$ are measured on an ordinal scale, then every meaningful optimal solution is a "greedy" solution.*

Let us add two additional axioms.

Axiom 5. For $k = 1, 2, \ldots, n$, $c \geq 0$, the equation $d_k(y_k, a_k) = c$ has at most one solution $y_k \in X_k$.

Axiom 6. For all $\mathbf{x} \in F$, the function $f_{\mathbf{x}} : \Re^n_+ \to \Re$ defined by $f_{\mathbf{x}}(\mathbf{w}) = P(\mathbf{x}; \mathbf{w})$ is continuous.

THEOREM 7.2 (Pekeč [**20, 21**]). *Under Axioms 1 to 6, the following are equivalent:*

(a): *For any choice of $w_1, w_2, \ldots, w_n$ measured on an ordinal scale, any optimal solution to the problem* $\min_{\mathbf{x} \in F} P(\mathbf{x}; \mathbf{w})$ *is meaningful.*

(b): *For any such choice of $w_1, w_2, \ldots, w_n$, any optimal solution to the problem* $\min_{\mathbf{x} \in F} P(\mathbf{x}; \mathbf{w})$ *is greedy.*

These results have a negative implication for optimization under ordinal scales. Suppose there is a problem with a choice of weights so that the greedy algorithm doesn't give an optimal solution, for example the Traveling Salesman Problem. If w_i are measured on an ordinal scale, then there cannot be a meaningful optimal solution. In other words, here, it doesn't make sense to ask for an optimal solution!

It should be noted that Theorem 7.2 gives a characterization of problems for which only the greedy algorithm provides a solution, by characterizing the structure of the set of transformations of parameters of the objective function that leave the conclusion of optimality invariant. Problems solvable by a greedy algorithm are well understood in the case of optimizing a linear objective function over the set of 0-1 vectors–they are characterized by the matroid structure of the set of feasible solutions. (See for example [**19**]). However, the characterization in Theorem 7.2 applies in a much more general set-up. This indicates that it might be worth studying algorithms by way of understanding classes of transformations of parameters of the objective function under which the conclusion of optimality is invariant. For example, the least one should expect from an efficient optimization or approximation algorithm is that the same class of transformations that does not affect the conclusion of optimality should also not affect the solution obtained by the algorithm in question. Some observations along these lines are presented in [**20**].

It should be noted that the hypothesis that the function P is defined on $X \times \Re^n_+$ is needed for the proof of Theorem 7.2. It would be interesting to try to obtain a similar result if P is defined on $X \times \Re^n$.

8. Linear Programming

Theorem 7.2 suggests that for many problems, it will not make sense to ask for an optimal solution if we measure parameters of the objective function on ordinal scales. In this section, we consider interval and ratio scales. We concentrate on linear programming, and present just a few of the results of Pekeč [**20, 24**]. We have already seen in our discussion of the shortest path problem in Section 3 that if we have interval scales, the conclusion of optimality can be meaningless. Let us now consider a general linear programming problem:

$$\begin{aligned} &\max P(\mathbf{x};\mathbf{w}) = \mathbf{w}^T\mathbf{x} \\ &\text{subject to:} \\ &\mathbf{x} \in F := \{\mathbf{x} \in \Re^n : \mathbf{x} \geq 0, A\mathbf{x} = \mathbf{b}\} \end{aligned} \tag{8.1}$$

We limit ourselves to the situation where the w_i are measured on some scale, and assume that the a_{ij} and b_j are fixed. The situation when they are measured on scales also needs to be considered. A trivial observation is the following. Let $\mathbf{x}^*$ be an optimal solution to the linear programming problem 8.1. If $w_1, w_2, \ldots, w_n$ are measured on a ratio scale, then the conclusion that $\mathbf{x}^*$ is an optimal solution to 8.1 is meaningful. The following straightforward theorem handles the interval scale case.

THEOREM 8.1 (Pekeč [**20, 24**]). *Let $\mathbf{x}^*$ be an optimal solution to the linear programming problem 8.1. If $w_1, w_2, \ldots, w_n$ are measured on an interval scale, then the conclusion that $\mathbf{x}^*$ is an optimal solution to 8.1 is meaningful if and only if $x_1 + x_2 + \ldots + x_n$ is a constant on F.*

Analyzing meaningfulness of the conclusion of optimality when the set of feasible solutions depends on some problem parameters might not be a useful thing to do given that even a small change of, say, one coefficient in the constraint matrix A, can result in a change of all extremal points of the set of feasible solutions. However, there is a one to one correspondence between all extremal points and all basis matrices B, i.e., all possible sets of basic variables. (For definitions of basic variables and related concepts see any basic LP book, e.g., [**4**]). One can ask whether the set of basic variables of an optimal solution is invariant under certain transformations of problem parameters. In other words, one can ask whether the conclusion that the set S of basic variables defines an optimal solution is meaningful if, say, all coefficients in the i-th row of the constraint matrix A are measured on a ratio scale. In [**20, 24**] several problems of this type are analyzed. The following is a sample result.

THEOREM 8.2 (Pekeč [**20, 24**]). *Let B be an optimal basis for problem 8.1. If for every row i of the constraint matrix A, coefficients of the i-th row are measured on a common ratio scale, then the statement that the basic variables of B define an optimal basis for problem 8.1 is meaningful if and only if B^{-1} has no negative entries.*

The linear programming problems are the only problems where the meaningfulness of the conclusion of optimality, when the set of feasible solutions depends on the choice of the scale of measurement, has been studied. It would be interesting to see a similar analysis for problems where the structure of the set of feasible solutions is dependent on the problem parameters in a more complicated way (than simple linear inequalities). It is not even clear how this analysis would extend to semi-definite programming problems (given the absence of the notion of basic variable, basis, etc.).

9. 0-1 Optimization

Consider the linear 0-1 optimization problem

$$\max_{\mathbf{x}\in\mathcal{H}} \mathbf{w}^T\mathbf{x} \tag{9.1}$$

where $\mathcal{H} \subseteq \{0,1\}^n$. More generally, consider the problem

$$\max_{\mathbf{x}\in\mathcal{H}} P(\mathbf{x};\mathbf{w}) \tag{9.2}$$

where $\mathcal{H} \subseteq \{0,1\}^n$, $P : \{0,1\}^n \times \Re^n \rightarrow \Re$. A vector $\mathbf{x} \in \{0,1\}^n$ can be thought of as the incidence vector of a set $S \subseteq \{1,\dots,n\}$. Hence, we can reformulate 9.2 as

$$\max_{S\in\mathcal{H}} f_S(\mathbf{w}) \tag{9.3}$$

where $\mathcal{H}$ is a collection of subsets of $\{1,\dots,n\}$ and $f_S : \Re^n \rightarrow \Re$, in other words, $\mathcal{H}$ is a hypergraph. We ask: For what objective functions is the conclusion of optimality for a problem like 9.2 (or 9.3) meaningful? If w_i are measured on a ratio scale, the surprising result is that, under certain assumptions, the conclusion is meaningful only if the objective function is as in 9.1, i,.e., only if we have a linear 0-1 optimization problem.

Suppose that we have ratio scales. Then, meaningfulness means that for all $\alpha > 0$,

$$P(\mathbf{x}^*;\mathbf{w}) = max_{\mathbf{x}\in\mathcal{H}}P(\mathbf{x};\mathbf{w}) \Longleftrightarrow P(\mathbf{x}^*;\alpha\mathbf{w}) = max_{\mathbf{x}\in\mathcal{H}}P(\mathbf{x};\alpha\mathbf{w}).$$

We make the following assumptions.

Axiom L (Locality). For every $S \subset \{1,\dots,n\}$, and for all $j \notin S$, $\frac{\partial f_S}{\partial w_j} = 0$.

This axiom says that changing the weight w_j corresponding to any element $j \notin S$ will not change the value of f_S.

Axiom N (Normality). For all i, $f_i(\mathbf{w}) = w_i$.

This axiom is not a serious restriction. One can always introduce new weights $w_i' = f_i(w_i)$.

Axiom C (Completeness). For all nonempty $S \subset \{1, \ldots, n\}$, $f_S(\Re^n) = \Re$.

This says that every number in $\Re$ is obtained for some vector in $\Re^n$.

Axiom S (Separability). For all f_S and all i, there is $g_i : \Re \to \Re$ such that $g_i \in C'(\Re)$ and $\frac{\partial f_s}{\partial w_i}(\mathbf{w}) = g_i(w_i)$.

This says that the rate of change of $f_S(\mathbf{w})$ with respect to changing w_i should depend only on w_i and not on values of $w_j, j \neq i$, and the dependence should be "smooth."

THEOREM 9.1 (Pekeč [**22**]). *Consider an optimization problem* $\max_{\mathbf{x} \in \mathcal{H}} P(\mathbf{x}; \mathbf{w})$, *and suppose* $w_1, w_2, \ldots, w_n$ *are measured on a ratio scale. If Axioms L, N, C, and S hold, then every optimal solution is meaningful if and only if every* f_S *is linear, i.e., of the form*

$$f_S(\mathbf{w}) = \sum_{i \in S} c_{S,i} w_i.$$

Consider the example where

$$P(\mathbf{x}; \mathbf{w}) = (x_1 {w_1}^p + \ldots + x_n {w_n}^p)^{1/p}.$$

This function satisfies Axioms L, N, and C, but not Axiom S. Note that it does satisfy meaningfulness under ratio scales.

As we noted in Section 6, it is not hard to give examples of problems in combinatorial optimization where the conclusion of optimality is meaningful for interval scales but not for ordinal scales. It would be interesting to characterize 0-1 optimization problems for which the fact that the conclusion of optimality is meaningful for any choice of w_i measured on an interval scale implies the same for any choice of w_i measured on an ordinal scale.

10. Multi-person Games

In game theory, the study of games with many players and presented in characteristic function form has many different solution concepts and many different interesting applications, going back to the classic book by von Neumann and Morgenstern [**44**]. We use the terminology of multi-person game theory from [**27**] and [**18**]. Let I be a set of players and consider *coalitions* or subsets of I. A game in *characteristic function form* is defined by a function $v : 2^I \backslash \{\emptyset\} \to \Re$. Sometimes, the function is defined on $\emptyset$ also, and then we assume $v(\emptyset) = 0$; however, we shall not take this approach here. Moreover, sometimes the condition

$$v(S \cup T) \geq v(S) + v(T) \text{ if } S \cap T = \emptyset,$$

known as *superadditivity*, is added to the definition of characteristic function, but we shall not make this assumption. The number $v(S)$ is thought of as the value that coalition S can guarantee itself. We shall study various solution concepts for games, and ask what happens to a solution if v is measured on some scale and we have a change of scale.

Solution concepts for games in characteristic function form are defined in terms of vectors $\mathbf{x} = (x_1, x_2, \ldots, x_n) \in \Re^n$, where $n = |I|$ and x_i represents the "payoff" to player i. The solution concepts we shall mention are called the core, the stable set, the Shapley value, the Banzhaf value, and the nucleolus, and all are widely used in game theory. We first consider the idea of normalization. Suppose that $D \subseteq \Re^n$. If there is $\alpha > 0$ so that for all $\mathbf{x} \in D, \sum_{i=1}^n \alpha x_i = 1$, we say that $\alpha D = \{\alpha \mathbf{x} : \mathbf{x} \in D\}$ is the *normalization* of D. We cannot expect a solution to a game to stay fixed after an admissible transformation of scale, but we can expect the normalization of the solution to stay fixed, i.e., the relative distribution of the payoffs to stay the same. We present some results of Pekeč [**20, 23**]. We first consider the case of ratio scales.

THEOREM 10.1 (Pekeč [**20, 23**]). *Let $D \subseteq \{\mathbf{x} \in \Re^n : \sum_{i=1}^n x_i = 1\}$. If v defines a ratio scale, then it is meaningful to say that D is the normalization of the core of v. The same conclusion holds for stable set, Shapley value, Banzhaf value, and nucleolus.*

The case of interval scales is more complicated. Consider the case where $I = \{1,2\}, v(\{1\}) = 2, v(\{2\}) = 3, v(\{1,2\}) = 6$. It is well known (see [**27**]) that the core can be computed as

$$C(v) = \{\mathbf{x} : \sum_{i=1}^n x_i = v(I), \text{ and for all } S \subseteq I, \sum_{i \in S} x_i \geq v(S)\}.$$

In our example,

$$C(v) = \{(x_1, x_2) : x_1 \geq 2, x_2 \geq 3, x_1 + x_2 = 6\}.$$

The normalization of $C(v)$ is

$$D = \{(x_1, x_2) : 1/3 \leq x_1 \leq 1/2, x_1 + x_2 = 1\}.$$

Let $v' = v - 1$, i.e., $v' = \alpha v + \beta, \alpha = 1, \beta = -1$. Then $v'(\{1\}) = 1, v'(\{2\}) = 2, v'(\{1,2\}) = 5$ and

$$C(v') = \{(x_1, x_2) : x_1 \geq 1, x_2 \geq 2, x_1 + x_2 = 5\}.$$

The normalization of $C(v')$ is

$$D' = \{(x_1, x_2) : 1/5 \leq x_1 \leq 3/5, x_1 + x_2 = 1\}.$$

Thus, the statement that D is the normalization of the core is meaningless. So can be the statement that $\mathbf{x}$ is in the normalization of the core. The result is even stronger:

THEOREM 10.2 (Pekeč [**20, 23**]). *If a game v is defined on an interval scale and $D \subseteq \{\mathbf{x} \in \Re^n : \sum_{i=1}^n x_i = 1\}$, then the statement that D is the normalization of the core of v is meaningless. Moreover, if $\sum_{i=1}^n x_i = 1$, then the statement that $\mathbf{x}$ is in the normalization of the core of v is meaningless.*

There are similar results in [**20**] and [**23**] for stable sets.

Let $f(v)$ be the Shapley value, the Banzhaf value, or the nucleolus of game v. The normalization of $f(v)$ is the vector $[\sum_{i=1}^n f(v)_i]^{-1} f(v)$.

THEOREM 10.3 (Pekeč [**20, 23**]). *Let v be defined on an interval scale and let* $\mathbf{y} \neq (\frac{1}{n}, \frac{1}{n}, \ldots, \frac{1}{n})$, $\mathbf{y} \neq 0$. *Then if the statement that* $\mathbf{y}$ *is the normalization of the Shapley value is meaningful, the statement is false.*

In other words, except for 0 and symmetric vectors, if we can find an admissible transformation such that the vector is the normalization of the Shapley value, then we can find another admissible transformation such that it isn't. Similar results hold for the Banzhaf value and for the nucleolus.To illustrate the theorem, consider the example given prior to Theorem 10.2. Here, the Shapley value is given by (5/2, 7/2). Its normalization is (5/12, 7/12). If $v' = v - 1$, then the Shapley value becomes (2, 3) and its normalization is (2/5, 3/5). The statement that the normalization of the Shapley value is (5/12, 7/12) is meaningless.

In the case of ordinal scales, there is the following impossibility result:

THEOREM 10.4 (Pekeč [**23**]). *There is no solution concept* $f : \Re^{2^n-1} \to \Re^n$ *that attains infinitely many values in* $\Re^n$ *and such that for every ordinal scale n-person game there exists* $x \in \Re^n$ *so that the statement "*$x = f(v)$*" is meaningful.*

Put another way, if we exclude certain solution concepts that don't depend on game values, such as "player one takes everything," then for any solution concept f one can think of, we can construct an ordinal scale game so that "$f(v) = x$" is meaningless *for every* x.

It would be interesting to perform meaningfulness analyses for non-cooperative games, for instance to analyze the Nash equilibrium and its refinements from the point of view of measurement theory.

11. Closing

It is hard to predict what future trends and problems will stimulate the development of new mathematical methods. It is also hard to predict which methods of mathematics, in particular discrete mathematics, will turn out to be most useful, both theoretically and in applications to other areas. What is safe to say is that discrete mathematics, with its many connections to other fields, both mathematical and non-mathematical, is thriving.

Acknowledgements

The author thanks Aleksandar Pekeč for his helpful comments. He thanks Jan Kratochvíl and Jarik Nešetřil for their hospitality and extends his heartiest congratulations to DIMATIA on the occasion of the launching of the new center.

References

[1] Abdul-Razaq, T.S., Potts, C.N., and van Wassenhove, L.N., "A Survey of Algorithms for the Single Machine Total Weighted Tardiness Scheduling Problem," *Discrete Appl. Math.*, **24** (1990), 235-253.

[2] Baker, K.R., and Scudder, G.W., "Sequencing with Earliness and Tardiness Penalties: A Review," *Operations Research*, **38** (1990), 22-36.

[3] Bonias, I., *T-colorings of Complete Graphs*, Ph.D. Thesis, Department of Mathematics, Northeastern University, Boston, MA, 1991.

[4] Chvátal, V., *Linear Programming*, Freeman, 1983.

[5] Cozzens, M.B., and Roberts, F.S., "*T*-colorings of Graphs and the Channel Assignment Problem," *Congressus Numerantium*, **35** (1982), 191-208.

[6] Cozzens, M.B., and Roberts, F.S., "Greedy Algorithms for T-colorings of Complete Graphs and the Meaningfulness of Conclusions about Them," *J. Comb. Inform. Syst. Sci.*, **16** (1991), 286-299.

[7] Falmagne, J.-C., and Narens, L., "Scales and Meaningfulness of Quantitative Laws," *Synthese*, **55** (1983), 287-325.

[8] Garey, M.R., Tarjan, R.E., and Wilfong, G.T., "One-Processor Scheduling with Symmetric Earliness and Tardiness Penalties," *Mathematics of Operations Research*, **13** (1988), 330-348.

[9] Hale, W.K., "Frequency Assignment: Theory and Applications," *Proc. IEEE* **68** (1980), 1497-1514.

[10] Helmholtz, H.V., "Zählen und Messen," in *Philosophische Aufsätze,*" Fues's Verlag, Leipzig, 1887, pp. 17-52. Translated by C.L. Bryan, "Counting and Measuring," Van Nostrand, Princeton, NJ, 1930.

[11] Koulamas, C., "The Total Tardiness Problem: Review and Extensions," *Operations Research* **42** (1994), 1025-1041.

[12] Krantz, D.H., Luce, R.D., Suppes, P., and Tversky, A., *Foundations of Measurement*, Vol. I, Academic Press, New York, 1971.

[13] Kruskal, J.B., "On the Shortest Spanning Tree of a Graph and the Traveling Salesman Problem," *Proc. Amer. Math. Soc.*, **7** (1956), 48-50.

[14] Liu, D.D., *Graph Homomorphisms and the Channel Assignment Problem*, Ph.D. Thesis, Department of Mathematics, University of South Carolina, Columbia, SC, 1991.

[15] Luce, R.D., Krantz, D.H., Suppes, P., and Tversky, A., *Foundations of Measurement*, Vol. III, Academic Press, New York, 1990.

[16] Mahadev, N.V.R., Pekeč, A., and Roberts, F.S., "Effects of Change of Scale on Optimality in a Scheduling Model with Priorities and Earliness/Tardiness Penalties," *Mathl. Comput. Modelling*, **25** (1997), 9-22.

[17] Mahadev, N.V.R., Pekeč, A., and Roberts, F.S., "On the Meaningfulness of Optimal Solutions to Scheduling Problems: Can an Optimal Solution be Nonoptimal?" *Operations Research*, **46** (1998), in press.

[18] Owen, G., *Game Theory*, 3rd. ed., Academic Press, 1995.

[19] Papadimitriou, C.H., and Steiglitz, K., *Combinatorial Optimization: Algorithms and Complexity*, Prentice-Hall, Englewood Cliffs, NJ, 1982.

[20] Pekeč, A., *Limitations on Conclusions from Combinatorial Optimization Models*, Ph.D. Thesis, Department of Mathematics, Rutgers University, New Brunswick, NJ, 1996.

[21] Pekeč, A., "Optimization under Ordinal Scales: When is a Greedy Solution Optimal?" *Mathematical Methods of Operations Research*, **46** (1997), 229-239.

[22] Pekeč, A., "Hypergraph Optimization Problems: Why is the Objective Function Linear?" preprint, Center for Basic Research in Computer Science (BRICS), submitted for publication.

[23] Pekeč, A., "Meaningful and Meaningless Solutions for Cooperative N-person Games," preprint, Center for Basic Research in Computer Science (BRICS), submitted for publication.

[24] Pekeč, A., "Scalings in Linear Programming: Necessary and Sufficient Conditions for Invariance," preprint, Center for Basic Research in Computer Science (BRICS), submitted for publication.

[25] Pfanzagl, J., *Theory of Measurement*, Wiley, New York, 1968.

[26] Raychaudhuri, A., *Intersection Assignments, T-coloring, and Powers of Graphs*, Ph.D. Thesis, Department of Mathematics, Rutgers University, New Brunswick, NJ, 1985.

[27] Roberts, F.S., *Discrete Mathematical Models, with Applications to Social, Biological, and Environmental Problems*, Prentice-Hall, Englewood Cliffs, NJ, 1976.

[28] Roberts, F.S., *Measurement Theory, with Applications to Decisionmaking, Utility and the Social Sciences*, Addison-Wesley, Reading, MA, 1979.

[29] Roberts, F.S., "Applications of the Theory of Meaningfulness to Psychology," *J. Math. Psychol.*, **29** (1985), 311-332.

[30] Roberts, F.S., "Meaningfulness of Conclusions from Combinatorial Optimization," *Discrete Appl. Math.*, **29** (1990), 221-241

[31] Roberts, F.S., "From Garbage to Rainbows: Generalizations of Graph Coloring and their Applications," in Y. Alavi, G. Chartrand, O.R. Oellermann, and A.J. Schwenk (eds.), *Graph Theory, Combinatorics, and Applications*, Vol. 2, Wiley, New York, 1991, pp. 1031-1052.

[32] Roberts, F.S., "T-colorings of Graphs: Recent Results and Open Problems," *Discrete Math.*, **93** (1991), 229-245.

[33] Roberts, F.S., "Limitations on Conclusions Using Scales of Measurement," in S.M. Pollock, M.H. Rothkopf, and A. Barnett (eds.), *Operations Research and the Public Sector*, Vol. 6 in *Handbooks in Operations Research and Management Science*, North-Holland, Amsterdam, 1994, pp. 621-671.

[34] Roberts, F.S., "A Functional Equation that Arises in Problems of Scheduling with Priorities and Lateness/Earliness Penalties," *Mathl. Comput. Modelling*, **21** (1995), 77-83.

[35] Roberts, F.S., and Franke, C.H., "On the Theory of Uniqueness in Measurement," *J. Math. Psychol.*, **14** (1976), 211-218.

[36] Scott, D., and Suppes, P., "Foundational Aspects of Theories of Measurement," *J. Symbolic Logic*, **23** (1958), 113-128.

[37] Stevens, S.S., "On the Theory of Scales of Measurement," *Science*, **103** (1946), 677-680.

[38] Stevens, S.S., "Mathematics, Measurement, and Psychophysics," in S.S. Stevens (ed.), *Handbook of Experimental Psychology*, Wiley, New York, 1951, pp. 1-49.

[39] Stevens, S.S., "Measurement, Psychophysics, and Utility," in C.W. Churchman and P. Ratoosh (eds.), *Measurement: Definitions and Theories*, Wiley, New York, 1959, pp. 18-63.

[40] Suppes, P., "Measurement, Empirical Meaningfulness and Three-valued Logic," in C.W. Churchman and P. Ratoosh (eds.), *Measurement: Definitions and Theories*, Wiley, New York, 1959, pp. 129-143.

[41] Suppes, P, Krantz, D.H., Luce, R.D., and Tversky, A., *Foundations of Measurement*, Vol. II, Academic Press, New York, 1989.

[42] Suppes, P., and Zinnes, J., "Basic Measurement Theory" in R.D. Luce, R.R. Bush, and E. Galanter (eds.), *Handbook of Mathematical Psychology*, Vol. 1, Wiley, New York, 1963, pp. 1-76.

[43] Tesman, B., *T-colorings, List T-colorings, and Set T-colorings of Graphs*, Ph.D. Thesis, Department of Mathematics, Rutgers University, New Brunswick, NJ, 1989.

[44] von Neumann, J., and Morgenstern, O., *Theory of Games and Economic Behavior*, Princeton University Press, Princeton, NJ, 1944, 1947, 1953.

[45] Wang, D., *The Channel Assignment Problem and Closed Neighborhood Containment Graphs*, Ph.D. Thesis, Department of Mathematics, Northeastern University, Boston, MA, 1985.

DIMACS, Rutgers University, Piscataway, NJ, USA

DIMACS Series in Discrete Mathematics
and Theoretical Computer Science
Volume **49**, 1999

Graceful matchings in finite fields, the Factor-Difference sets of integers and integers of the form $a^2 + kb^2$ *†

Moshe Rosenfeld
Pacific Lutheran University
Tacoma, WA. 98744
moishe@math.washington.edu

1 Introduction

The DIMATIA-DIMACS joint conference in Štiřín was my first conference in Central Europe where Paul Erdős was not physically present, yet his spirit, his mathematical gospel and his legacy were very evident throughout the conference. The conference's theme was "The Future of Discrete Mathematics". No doubt Paul's work and problems will continue to play a dominant role in the future of discrete mathematics. We shall also see a steadily increasing trend of interaction between discrete mathematics and other classical areas of science in general, and pure mathematics in particular. The future will bring an ever increasing use of tools from various mathematical disciplines in combinatorics, a trend that is now in full swing. From this respect, "The Future of Discrete Mathematics" is already happening.

In this note, I'd like to discuss some problems from combinatorial number theory. At one point or another, I was fortunate to have had the opportunity to share them with Paul and enjoy his unique, deep insights, suggestions and comments. In the first section I'll discuss a matching problem that arose from a proof of the matchability of $GF^*(q)$. This proof uses a deep theorem in Number Theory by A. Weil. The first "pathfinders" to apply Weil's

*1991 *Mathematics Subject Classifcation 11B75*

†Parts of this work were done during visits of the author to Prague. This has been enabled by the Czech-US Science and Technology Research grant No. 94 051.

theorem in combinatorics were Ron Graham and Joel Spencer [3]. For an interesting account of the use of this tool and other tools from mathematics in combinatorics see the paper by Noga Alon [1].
In the next two sections I'll discuss some problems in combinatorial number theory. They include problems related to the Factor Difference Sets of integers, (joint work with Paul Erdős [2]), sequences of integers of the form $\{a^2 + kb^2\}$ and properties of consecutive integers of this form.

2 Graceful matchings

Let $\{A, B\}$ be a partition of $\{1 \dots 2n\}$ into two n-sets. We say that A and B are *gracefully matched* if one can sequence the two sets $A = \{a_1, \dots, a_n\}$ and $B = \{b_1, \dots, b_n\}$ so that all differences $\pm(a_i - b_i) \bmod 2n + 1$ are distinct. For example, the partition $A = \{1, 4, 5\}$ and $B = \{6, 3, 2\}$ is a *graceful match* while the partition $A = \{1, 4\}$ $B = \{2, 3\}$ is not gracefully matchable. A computer test produced a graceful match for all partitions of $\{1, \dots, 2n\}$ for $n = 3, 4, 5, 6, 8, 9$. Each of $10,000$ randomly selected partitions of $\{1, \dots, 20\}$ turned out to be gracefully matchable. Out of the 1716 distinct partitions of $\{1, \dots, 14\}$ only 4 are not gracefully matchable. They are:

1 4 5 6 9 11 14
2 3 7 8 10 12 13

2 5 6 8 9 11 14
1 3 4 7 10 12 13

1 4 6 9 10 11 14
2 3 5 7 8 12 13

2 3 5 8 11 12 14
1 4 6 7 9 10 13

These observations led us to:

> **Conjecture 1:** All partitions of $\{1, \dots, 2n\}$ for $n \geq 8$ are gracefully matchable.

Noga Alon (private communication) suggested to extend the notion of graceful matchability to abelian groups of order $2n + 1$. In this setting, the partition of $GF^*(9)$ into $\{\mathcal{R}, \mathcal{NR}\}$ where $\mathcal{R}$ is the set of quadratic residues and $\mathcal{NR}$ the set of non-residues is not gracefully matchable. It will be quite surprising if all partitions of abelian groups of order $2n + 1, n > 4$ turn out to be gracefully matchable. In the remainder of this section we'll present evidence in support of the conjecture and raise some other related questions.

Let us first identify two transformations that transform matchable partitions $\{A, B\}$ into matchable partitions $\{A', B'\}$.

T. Take an arbitrary subset of A and interchange it with its matched subset in B.

M. Choose α which is relatively prime to $2n+1$ and let $A' = \alpha A$ and $B' = \alpha B$ (all arithmetic is done mod $2n+1$).

It is a simple matter to check that indeed both transformations transform matchable partitions into matchable partitions. It is easy to see that many partitions are gracefully matchable. Indeed if A is an n-subset of $\{1, \ldots, 2n\}$ and $A \cap -A = \emptyset$ then the map $f(a) = -a$ is easily seen to be a graceful match for the partition $\{A, -A\}$. Thus, at least 2^{n-1} of the $\frac{1}{2}\binom{2n}{n}$ partitions of $\{1, \ldots, 2n\}$ are gracefully matchable. Clearly, we can also perform repeatedly sequences of the transformations T and M and get many more matchable partitions.

We were led to the above conjecture after proving that the partition of $GF^*(q)$ into $\{\mathcal{R}, \mathcal{NR}\}$ ($\mathcal{R}$ the quadratic residues and $\mathcal{NR}$ the non residues in $GF^*(q)$) is gracefully matchable for all $q = 2^r(2k+1)+1 > 9,\ \ k \geq 1$. The proof, a joint effort of Jan Kratochvíl, Jaroslav Nešetřil, Moshe Rosenfeld and Tamás Szőnyi, uses Weil's theorem [6].
We would like to sketch the proof as it indicates that there are many ways to gracefully match the partition $\{\mathcal{R}, \mathcal{NR}\}$ of $GF^*(q)$ and hence support our conjecture. Recall that χ, the quadratic residue character, is defined by:

$$\chi(x) = \begin{cases} 0 & \text{for } x - 0 \\ +1 & \text{if } x = y^2 \text{ for some } y \in GF^*(q) \\ -1 & \text{otherwise.} \end{cases}$$

The idea behind the proof is to seek a simple match of $\mathcal{R}$ and $\mathcal{NR}$ and then modify it so it will be graceful. The simplest match is a map of the form $f(r) = xr$. This will be a match if $x \in \mathcal{NR}$. It is easy to check that this is a graceful match when $q = 3 \bmod 4$ and $x = -1$ but, it will never produce a graceful match when $q \equiv 1 \bmod 4$. So in this case, we try to break $\mathcal{R}$ into two equal parts, select distinct suitable multipliers for each part and obtain the graceful matching. To do this, let α be a primitive element in $GF(q)$ where $q = 2^r(2m+1)+1,\ m \geq 1$ and let:

a. $\mathcal{B} = \{x \in \mathcal{NR} \mid x - 1 \in \mathcal{R}\}$ and $\mathcal{C} = \{x \in \mathcal{NR} \mid x - 1 \in \mathcal{NR}\}$.

b. For $x \in GF^*(q)$, $x = \alpha^j$, define $\theta(x) = j \bmod 2^r$.

Note that θ is a homomorphism of $GF^*(q)$ into the cyclic group $Z_{2^r} = \{0, 1, \ldots, 2^r - 1\}$.

We partition $\mathcal{R}$ into $\{\mathcal{R}_0, \mathcal{R}_1\}$ where:
$\mathcal{R}_0 = \{x \in \mathcal{R} \mid \theta(x) < 2^{r-1}\}$ and $\mathcal{R}_1 = \{x \in \mathcal{R} \mid 2^{r-1} \le \theta(x) < 2^r\}$

It is now a simple matter to verify that if we can find two elements a_0 and a_1 having the following properties:

1. $a_1 \in \mathcal{B}$ and $a_0 \in \mathcal{C}$

2. $\theta(a_1) = \theta(a_0)$

then the map $x \to xa_i \mid x \in R_i$ will determine a graceful match of the partition $\{\mathcal{R}, \mathcal{NR}\}$.
Let $a_i = \alpha x^2$ and $a_{1-i} = \alpha^{2^r+1}x^2$, $i = 0, 1$. Note that this means that $\theta(a_i) = \theta(a_{1-i})$. We then apply Weil's theorem [6] which states that:

$$\sum \chi(g(x)) \le (\deg(g) - 1)\sqrt{q},$$

(where χ is the quadratic character, and $g(x)$ is not a constant multiple of the square of another polynomial) to $g(x) = (\alpha x^2 - 1)(\alpha^{2^r+1}x^2 - 1)$.
Clearly, since $q = 2^r(2k+1) + 1$, $k \ge 1$, $\alpha^{2^r+1} \ne \alpha$ and thus $g(x)$ is not a constant multiple of the square of another polynomial. Since $\deg g(x) = 4$, for $q > 9$ $\exists x$ for which $\chi(g(x)) = -1$ and hence $a_0 \in \mathcal{C}$ and $a_1 \in \mathcal{B}$.

We note that for large q's there will be many x's for which $\chi(g(x)) = -1$ and hence many selections for a_0 and a_1. It is possible that these matchings combined with the transformations T and M will be enough to prove that all partitions of $\{1, \ldots, 2n\}$ are matchable at least when $2n + 1$ is prime.

After completion of the proof we learned that Olof Hanner [5] proved by elementary methods that the partition $\{\mathcal{R}, \mathcal{NR}\}$ is gracefully matchable. Our proof though points to many distinct possible matchings and thus might be more helpful in attacking our initial conjecture.

We also proved that $|\mathcal{B}| = |\mathcal{C}|$ and using another theorem of Weil we proved that if $q = 2^r(2k+1)+1$ and $2k+1 >> 2^r$ then $\forall i$, $0 \le i < 2^r$, $\exists c \in \mathcal{C}$ such that $\theta(c) = i$. We believe that an even stronger result can be proved:

> **Conjecture 2:** $\exists q_0$ such that $\forall q \geq q_0$, $q = 2^r(2k+1)+1$ an odd prime power, $\forall i, 0 \leq i < 2^r$, $\exists c \in \mathcal{C}$ such that $\theta(c) = i$ and $\exists b \in \mathcal{B}$ such that $\theta(b) = i$.

If true, this of course will yield many more distinct ways to gracefully match $\mathcal{R}$ and $\mathcal{NR}$.

3 The Factor-Difference set of integers

Define $D(n)$, the *Factor-Difference set* of n by:
$D(n) = \{d | d = |a-b|, n = ab\} = \{d_0 < d_1 < \ldots < d_k\}$.
In [2] P. Erdös and I posed the following question:

> Is it true that for every positive integer k one can find integers $N_1 < N_2 < \cdots < N_k$ such that $|\bigcap_{i=1}^k D(N_i)| \geq k$.

This question is still open. The best example known to me is the following sets of 3 integers that share 4 differences:

$$\{420, 3780, 14940, 76860\} \subset D(6925500) \cap D(37901500) \cap D(108448956)$$
$$\{420, 3780, 61695, 154332\} \subset D(2778300) \cap D(862552800) \cap D(5400442044)$$

These examples were found by Barry Guiduli. While $d_0(n)$ can be as small as we please, (i.e $d_0(n^2) = 0$) we proved that $d_1(n) \geq 2\sqrt[4]{n}$. This enabled us to obtain a close expression for $d_0(n_i)$ for some sequences $\{n_i\}$. For instance, we proved that:
$d_0(a(a+1)(a+2)(a+3)(a+4)(a+5)(a+6)(a+7)) = 16a + 56$ for $a \geq 5$.
It is natural to try to extend this observation to $d_0(n!)$. Note that $d_0(3!) = 1$, $d_0(4!) = d_0(5!) = d_0(7!) = 2$. We ask:

> **Problem 3.1**: Is $liminf(d_0(n!)) \leq K$ for some constant K?

We also wondered how many differences d_i can be "close" to $\sqrt[4]{n}$. Note that if $d_i = c\sqrt[4]{n}$ then n must have a divisor which is very close to $\sqrt{n}$. We showed that $N_a = a(a+1)(a+2)(a+3)(a+4)(a+5)(a+6)(a+7)$ has at least 4 differnces $d_i \leq 16\sqrt[4]{N_a}$. Paul recalled that I. Rusza asked the following directly related question:

> **Problem 3.2**: (I. Rusza) Is there an absolute constant K, so that $\forall c$ the number of divisors of n between $\sqrt{n}$ and $\sqrt{n} + c\sqrt[4]{n}$ is at most K for $n \geq n_0(c)$?

4 Sequences of the form $a^2 + kb^2$ and properties of consecutive integers

Properties shared by consecutive integers provide a fertile ground for many attractive elementary problems. For instance, it has been conjectured [4] (problem **B18**) that there are infinitely many triples $n, n+1, n+2$ of consecutive products of two primes (i.e $\{33, 34, 35\}$, $\{85, 86, 87\}$ etc.). Since $pq \neq 0 \bmod 4$ we cannot have 4 consecutive products. Consider the sequences $N(k) = \{a_1 < a_2 < \ldots \mid a_i = a^2 + kb^2\}$. Since $a^2 + b^2 \neq 3 \bmod 4$ there are at most 3 consecutive integers in $N(1)$. It is a simple exercise to show that there are infinitely many consecutive triples of the form $a^2 + b^2$. Looking at $a^2 + kb^2 \bmod 8$ we see that $a^2 + kb^2 \neq 5 \bmod 8$ hence there are at most 5 consecutive integers of the form $a^2 + kb^2$ and for $N(k)$ to have 5 consecutive integers we must have $k \equiv 2 \bmod 8$ and also there must be 5 consecutive integers which are quadratic residues $\bmod k$. The first such case is $k = 2$. In 1992 I asked whether $N(2)$ contains infinitely many quintuples of consecutive integers. Erick Wong, a 16 years old student at Simon Fraser University, found a very clever construction of infinitely many such quintuples ([7]). He showed that one can find infinitely many integers m for which $\{18m^2, 18m^2+1, 18m^2+2, 18m^2+3, 18m^2+4\} \subset N(2)$. He also proved that for every integer d there are infinitely many indices i for which $d_{i+1} - d_i = d$ $(d_i, d_{i+1} \in N(2))$, that is every gap size is realized infinitely often in the sequence $N(2)$. These observations led us to the following two questions:

> **Problem 4.1:** Are there infinitely many integers k for which $N(k)$ contains 5 consecutive integers $n, n+1, n+2, n+3, n+4$?

For instance the quintuples starting with 707776 or with 1584160 are subsets of $N(34)$.

> **Problem 4.2:** If $N(k)$ contains a quintuple of consecutive integers, does it contain infinitely many quintuples?

Actually even the case of consecutive triples is not obvious. That is:

> **Problem 4.3:** For which k does $N(k)$ contain infinitely many triples of consecutive integers?

As noted above, $N(1)$ contains infinitely many triples of consecutive integers, but the simple proofs do not extend to other values of k.

References

[1] N. Alon: *Tools from higher algebra* in Handbook of Combinatorics, North Holland (1995)

[2] Erdős Paul and M. Rosenfeld: *The factor-difference set of integers*, Acta. Arith. LXXIX.4, 353-359 (1997)

[3] Graham, R.L. and J. H. Spencer: *A constructive solution to a tournament problem*, Canad. math. bull. 14, 45-48 (1971)

[4] Guy, K.R., *Unsolved Problems in Number Theory*, Springer-Verlag, 2-nd edition (1994)

[5] O. Hanner: *Construction of Balanced Howell Rotations for* $2(p^r + 1)$ *Partnerships*, J. Comb. Theory, Series A 33, 205-212 (1982)

[6] R. Lidl, H. Niederreiter: *Finite fields*, Enc. of Math. and its appl. 20,

[7] Wong, E.: *Integers of the form* $a^2 + 2b^2$ (unpublished, manuscript available from M. R.)

DIMACS Series in Discrete Mathematics
and Theoretical Computer Science
Volume **49**, 1999

HOW TO SOLVE A TURÁN TYPE EXTREMAL GRAPH PROBLEM? (LINEAR DECOMPOSITION)

Miklós Simonovits

Dedicated to the memory of PAUL ERDŐS

ABSTRACT. The main purpose of this paper is to show that for many forbidden graphs L, "selected at random", some old, almost forgotten asymptotic or quasi-asymptotic results provide easy, almost immediate (and often "exact") solutions of the corresponding Turán type extremal problems. In some sense, the Petersen graph will be the "pretext" to formulate our theorems.

We shall get the Turán extremal number of the Petersen graph as an easy consequence of our results:

- for $n > n_0$ the following graph $H_{n,2,3}$ is the (only) extremal graph for the Petersen graph $\mathbb{P}_{10}$: one fixes a Turán graph $T_{(n-2),2}$ on $n-2$ vertices and joins two further vertices x and y to each other and to all the vertices of $T_{(n-2),2}$.
- We shall prove good general asymptotics for wide classes of excluded graphs (Theorem 2.2). Our results also provide algorithmic solutions of the corresponding extremal graph problems.
- We shall also prove an analog of the Andrásfai-Erdős-Sós theorem for some generalized Petersen graphs (Theorem 4.2).

The "motivating" theorems will lead to some further, more general theorems as well.

1. INTRODUCTION

1.1. Turán type extremal problems. In a Turán type extremal problem a family $\mathcal{L}$ of – so called – sample graphs is fixed and we consider only graphs G_n on n vertices not containing subgraphs $L \in \mathcal{L}$, where a "subgraph" means a not necessarily "induced" one.

PROBLEM 1.1 (Turán type). Given a family $\mathcal{L}$ of forbidden graphs, what is the maximum number of edges a graph G_n of order n can have without containing any $L \in \mathcal{L}$?

1991 *Mathematics Subject Classification.* Primary 05C35.

Research partially supported by HU-NSF grants OTKA T026069 and T16389, by AKP and also by DIMACS, New Jersey.

The maximum number of edges such a graph can have will be denoted by $\text{ext}(n, \mathcal{L})$. Let $T_{n,p}$ be the Turán graph on n vertices and p classes, i.e. n vertices be partitioned into p classes $A_1, \dots, A_p$ as uniformly as possible, and two vertices be joined iff they belong to different classes. In 1940 Turán [**28**], (see also [**29, 30**]) proved his famous theorem on the extremal number of K_{p+1}, according to which $T_{n,p}$ is the (only) extremal graph:

THEOREM 1.2 (Turán Theorem). *(a) $T_{n,p}$ contains no K_{p+1} and*
(b) if $e(G_n) \geq e(T_{n,p})$ and $G_n \neq T_{n,p}$ then $G_n \supseteq K_{p+1}$.

It is known from the Erdős–Stone-Simonovits theorem [**16, 12**] that $\text{ext}(n, \mathcal{L})$ is asymptotically determined by

$$p(\mathcal{L}) = \min_{L \in \mathcal{L}} \chi(L) - 1.$$

Namely,

$$\text{ext}(n, \mathcal{L}) = \text{ext}(n, K_{p+1}) + o(n^2).$$

Clearly, $\text{ext}(n, \mathcal{L}) = o(n^2)$ iff $p(\mathcal{L}) = 1$, which case will be called **degenerate**. (For corresponding stronger structural results see [**8, 22**].) The remainder terms in the above theorem depend primarily on the **Decomposition Class** $\mathcal{M}$ of $\mathcal{L}$. To define the decomposition class we introduce some notation.

Given two graphs Z and W, their **product** $Z \otimes W$ (often called by others their join) is the graph obtained by taking their vertex disjoint copies and joining each vertex of Z to each vertex of W. Given the graphs $U_1, \dots, U_p$, with pairwise disjoint vertex-sets, their product $\prod U_i$ is the graph obtained by joining each vertex of U_i to each vertex of U_j for each $i \neq j$.

DEFINITION 1.3. Given a family $\mathcal{L}$ of forbidden graphs with $p = p(\mathcal{L}) = \min_{L \in \mathcal{L}} \chi(L) - 1$, we shall call its **Decomposition Class** the family $\mathcal{M}$ of graphs M for which there exists an $L_0 = L_0(M) \in \mathcal{L}$ and a t such that

$$L_0 \subseteq M \otimes K_{p-1}(t, \dots, t).$$

In other words, a graph M is in the decomposition class if putting[1] it into a class A_i of a large $T_{n,p}$, the resulting graph *will* contain a forbidden $L \in \mathcal{L}$. Clearly, $p(\mathcal{M}) = 1$, the extremal graph problem of the decomposition class is always degenerate.

Whether we can solve a Turán type extremal problem or not depends primarily on the corresponding decomposition class $\mathcal{M}$. In some sense all the Turán type extremal problems reduce to the solution of the extremal graph problem of $\mathcal{M}$:

(a) If there is a tree or a forest in $\mathcal{M}$, then we can solve the problem in all known cases where we can solve the extremal graph problem of $\mathcal{M}$. This is what can be called linear decomposition: this is equivalent with saying that

$$\text{ext}(n, \mathcal{L}) - \text{ext}(n, K_{p+1}) = O(n).$$

(b) I conjecture that

CONJECTURE 1.4. *If $\mathcal{M}$ contains no trees, (neither forests) then some (or all?) extremal graphs for $\mathcal{L}$ for $n \geq n_0(\mathcal{L})$ are obtained by taking p graphs $U_1, \dots, U_p$ of order $n/p + o(n)$ and joining each vertex of each U_i to each vertex of each U_j ($i \neq j$). Then we say that the extremal graphs are of* **product forms**.

[1] "putting" means selecting $v(M)$ vertices in this class and joining them so that the resulting subgraph be isomorphic to M.

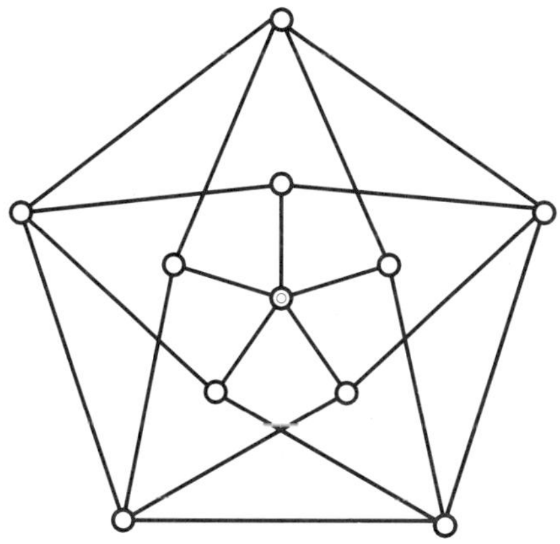

FIGURE 1:
Grötzsch Graph

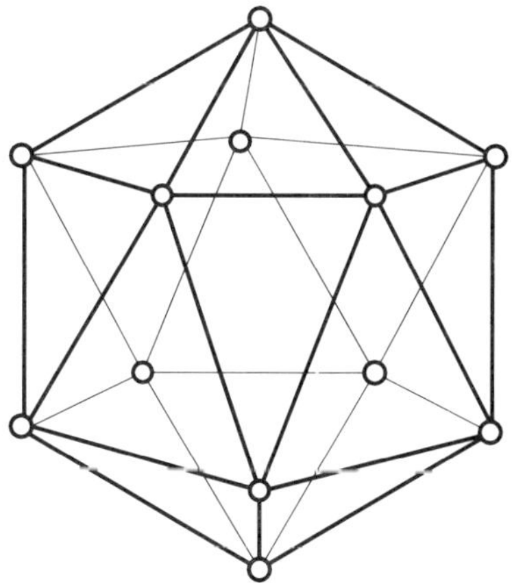

FIGURE 2:
Icosahedron graph

(We know that generally only $o(n^2)$ edges are missing between the classes and here the extra requirement is that **all** the edges between different classes must belong to the graph.)

One can easily see that – if Conjecture 1.4 holds for $\mathcal{L}$, then – U_i are also some extremal graphs for some degenerate families $\mathcal{M}_{i_n} \supseteq \mathcal{M}$. The details can be found in [**25**]. (See also our work with J. Griggs and G. Rubin Thomas, [**18**].)

Watch out! The main statement in (a) is not a theorem: it is an observation. Further, I think that Conjecture 1.4 is one of the most important, and most intriguing questions in this theory.

1.2. Some examples. Having proved his theorem, Turán also asked for the determination of $\text{ext}(n, L)$ for various other sample graphs L, among others, for the determination of $\text{ext}(n, P_k)$, or of $\text{ext}(n, L)$, if L is the graph of a platonic polyhedron: cube Q_8, octahedron O_6, icosahedron I_{12}, dodecahedron D_{20}. (The tetrahedron, K_4 was covered by his theorem.)

> *His aim was not so much to determine the extremal numbers for these particular sample graphs but to discover some new phenomena by solving the above special cases.*

(i) The question for P_k was answered by Erdős and Gallai [**11**].

(ii) Surprisingly enough, the cube Q_8 seems to be the most intractable. Erdős and I gave an upper bound $\text{ext}(n, Q_8) = O(n^{8/5})$, [**13**], which we conjectured to be sharp but we cannot prove this. As a matter of fact, there are no reasonable lower bounds known for this case. We cannot even prove that

$$\frac{\text{ext}(n, Q_8)}{n^{3/2}} \to \infty.$$

(iii) Erdős and I reduced the problem of the octahedron to the problem of C_4, [**14**]. We have proved that if S_n is extremal for the octahedron graph O_6, and n is sufficiently large, then $S_n = G_1 \otimes G_2$ where G_1 is extremal for C_4, G_2 is extremal for P_3 and $v(G_1) - v(G_2) = o(n)$.

(iv) The problems of the icosahedron and dodecahedron were solved by the author in [**23**] and [**24**]. To solve these problems, the author developed a special theory in [**24**] for the extremal graph problems where the decomposition class contains a path:

$$L \subseteq P_v \otimes K_{p-1}(v, \ldots, v) \text{ for some } L \in \mathcal{L}, \text{ for } v = v(L). \tag{1.1}$$

The solution of the dodecahedron and icosahedron problems helped to understand a lot about some kind of "non-degenerate" extremal graph problems. One could ask whether attacking some other special graphs, say the Grötzsch graph, or the Petersen graph, would that have lead to a completely different theory?

The surprising answer is that – as far as it can be judged – the theorem solving the problem of the Petersen graph would have led to *exactly* the same theory as the problem of the Dodecahedron graph. The problem of the Grötzsch graph was solved and is a subcase of some other results. Let us call an edge e of G color-critical if $\chi(G-e) < \chi(G)$. An old result of mine (for a special case proved first by Erdős) [**22, 23**]) asserts that

THEOREM 1.5 (Critical edge). *If $\chi(L) = p+1$ and L contains a color-critical edge, then $T_{n,p}$ is the (only) extremal for L, for $n > n_1$.*

This immediately implies Turán's theorem for large values of n and also solves the problem of the Grötzsch graph, since it is 4-chromatic and all its edges are critical. The icosahedron problem turned out to be much deeper and led to a more general theory. (The reason seems to be that there are almost extremal graphs for I_{12} which are fairly different from the extremal graph. In other words, the extremal graphs for I_{12} are less stable than for D_{20}.)

In this paper – among others – we shall determine the Turán and Zarankiewicz extremal numbers of the Petersen graph.[2] Further, we shall also prove an analog of the Andrásfai-Erdős-Sós theorem for Petersen graphs. These "motivating" theorems will lead to some more general theorems as well.

> *One aim of this paper is to illustrate on some new examples, how some old theorems or some old methods can be used to answer many new cases. New extremal problems can often immediately be solved by some old, almost forgotten general theorems.*

For the general theory of Extremal graph problems see e.g. [**3, 26**].

1.3. NOTATION. Here we also include some of the notations used above. We shall restrict our considerations to simple graphs, that is, to graphs without loops and multiple edges. Given a graph G, $e(G)$, $v(G)$ and $\chi(G)$ will denote the number of edges, vertices, and the chromatic number of G. We shall also use subscripts to indicate the number of vertices: $G_n, H_n, \ldots, S_n \ldots$ will always denote graphs of order n. Given a subset X of vertices in the vertex set $V(G)$ of the graph G, $e(X)$ will denote the number of edges both endvertices of which belong to X, and if Y is another set of vertices in $V(G)$, and $X \cap Y = \emptyset$, then $e(X, Y)$ will be the number of edges one endvertex of which is in X the other in Y. A graph of chromatic number $\leq p$ will be called p–colorable.

[2]The Zarankiewicz problem will be treated rather superficially.

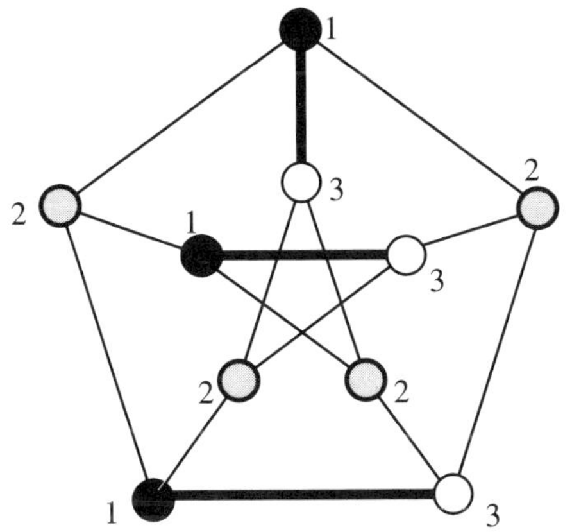

FIGURE 3:
Petersen graph

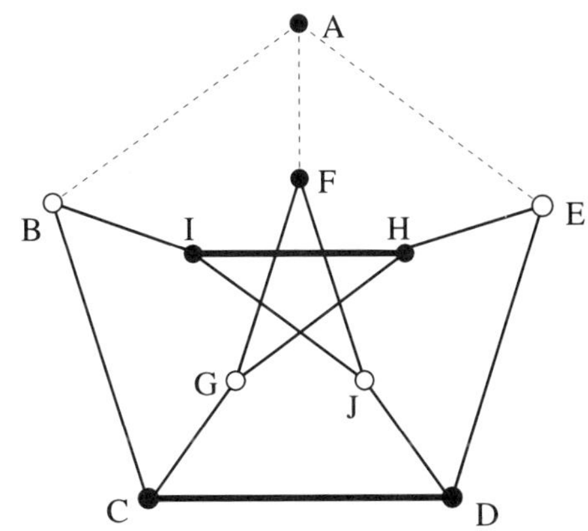

FIGURE 4:
Petersen Truncated

Special graphs. P_k denotes the path on k vertices. We denote the Petersen graph by $\mathbb{P}_{10}$ (while P_{10} is a path of 10 vertices!). $K_p(n_1, \dots, n_p)$ is the complete p-partite graph with n_i vertices in its ith class. $T_{n,p}$ is the Turán graph on n vertices and p classes, i.e. n vertices are partitioned into p classes $A_1, \dots, A_p$ as uniformly as possible, (the sizes of the classes differ by at most 1) and two vertices are joined iff they belong to different classes. $T_{n,p,k}$ is obtained from a $T_{n,p}$ by putting k new independent edges into one of its maximum size classes. (Of course, $k \le n/(2p)$ is assumed.) One of the most important special graphs of this paper is $H_{n,p,t} = K_{t-1} \otimes T_{n-t+1,p}$: the graph where $n-t+1$ vertices are partitioned into p classes $A_1, \dots, A_p$ as uniformly as possible and any two of these vertices are joined iff they belong to different classes. The remaining $t-1$ vertices have full degree $n-1$, i.e. they are joined to each other and to all the vertices in $\cup A_i$. Finally, $C_p[n]$ will denote the following "cyclical" graph: pn vertices are partitioned into p classes $A_1, \dots, A_p$, of size n each, and all the vertices of A_i are joined to all the vertices of A_{i-1} and A_{i+1}, where the indices are counted mod p. (In some places we shall misuse the notation, writing $C_p[\frac{1}{p}m]$ even if m is not divisible by p and meaning that m vertices are distributed in p classes as uniformly as possible. Which are the smaller or larger classes we do not care.)

— * * * —

Since the publication of [**23, 24**], many results have been proved which would follow (at least for $n > n_0(\mathcal{L})$) almost immediately from the theorems of [**23, 24**]. Here I shall prove – among others – four theorems in connection with the Petersen graph, (Theorems 1.6, 1.11, Lemma 3.2, Theorem 4.2) the first two of which are immediate corollaries of the theory built up in [**23, 24**].

THEOREM 1.6 (Petersen Extremal). *For $n > n_0$ $H_{n,2,3}$ is the (only) extremal graph for the Petersen graph $\mathbb{P}_{10}$.*

This theorem immediately follows from Theorem 2.2 of [**23**]:

THEOREM 1.7 ($H_{n,p,t}$-theorem). *(i) Let $L_1, \dots, L_\lambda$ be given graphs with $\min \chi(L_i) = p+1$. Assume that omitting any $t-1$ vertices of any L_i we obtain a graph of chromatic number $\geq p+1$, but L_1 can be colored in $p+1$ colors so that the subgraph of L_1 spanned by the first two colors is the union of t independent edges and (perhaps) of some isolated vertices. Then, for $n > n_0(L_1, \dots, L_\lambda)$, $H_{n,p,t}$ is the (only) extremal graph.*

(ii) Further, there exists a constant $C > 0$ such that if G_n contains no $L_i \in \mathcal{L}$ and

$$e(G_n) > e(H_{n,p,t}) - \frac{n}{p} + C,$$

then one can delete $t-1$ vertices of G_n so that the remaining G_{n-t+1} is p–colorable.

This theorem is strongly connected with Theorem 1.5. It is natural to ask if the uniqueness holds here as well or not:

OPEN PROBLEM 1.8. *Is there a family $\mathcal{L}$ of forbidden graphs for which for $n > n_0$ $H(n,p,t)$ is extremal but it is not the unique extremal graph?*

REMARK 1.9. The condition on L_1 is equivalent to that $L_1 \subseteq T_{m,p,t}$ for some m. One could also formulate this by saying that the decomposition class contains the graph consisting of t independent edges.

The meaning of (ii) is that the extremal structure is stable in some sense. To understand this stability better, we introduce the notion of chromatic properties, first only in its simplest form.

DEFINITION 1.10 ($\mathcal{B}_{p,t}$–property). We shall say that a graph has property $\mathcal{B}_{p,t}$ if one cannot delete $t-1$ vertices from it to get a p–colorable graph.

We shall not distinguish a property of graphs from the set of graphs having this property. If a graph $G \notin \mathcal{B}_{p,t}$ and $H \subseteq G$, then $H \notin \mathcal{B}_{p,t}$ either. To have such a property means that the graph is "big" in some sense, to not have means, that it is "small".

The assumptions of Theorem 1.7 include that each forbidden $L_i \in \mathcal{B}_{p,t}$ but $H_{n,p,t} \notin \mathcal{B}_{p,t}$. So trivially $H_{n,p,t}$ contains no forbidden L_i's. The meaning of Theorem 1.7 is that this property dominates the above extremal problem in the sense that the extremal graph is that very G_n which

(*) does not have the property $\mathcal{B}_{p,t}$, and
(**) has the maximum number of edges in $\overline{\mathcal{B}_{p,t}}$.

Theorem 1.7 immediately implies many known results, e.g. a result of Erdős and Gallai, [**11**] on independent edges, generalized by Erdős to independent triangles, [**5**] and by J. W. Moon [**21**] to independent K_{p+1}'s. Moon's theorem asserts that if L is the disjoint union of t vertex-disjoint K_{p+1}, then $H_{n,p,t}$ is the extremal graph for $n > n_0(p,t)$. Theorem 1.7 can also be applied to the dodecahedron graph: for D_{20}, $H_{n,2,6}$ is the extremal graph for $n > n_0$. (The reader can see 6 edges on Figure 6 the deletion of which yields a bipartite graph. On the other hand, it is not too difficult to check that the deletion of any 5 points leaves at least one odd cycle unchanged.) Applying Theorem 1.7 to $\mathbb{P}_{10}$ we get

THEOREM 1.11 (Petersen, Stability). *There exists a constant $C > 0$ such that if*

$$e(G_n) > e(H_{n,2,3}) - \frac{1}{2}n + C,$$

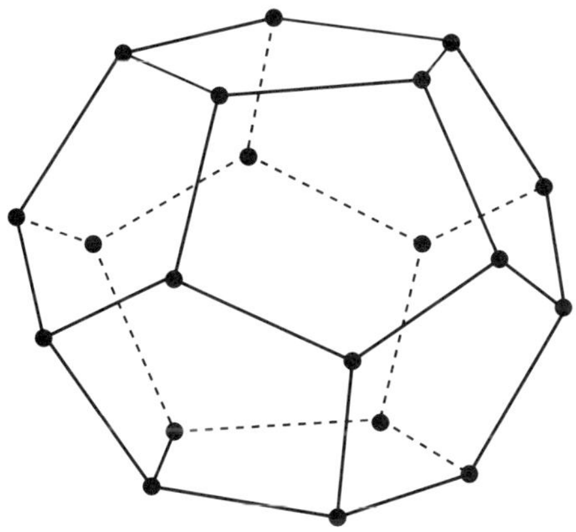

FIGURE 5:
Dodecahedron

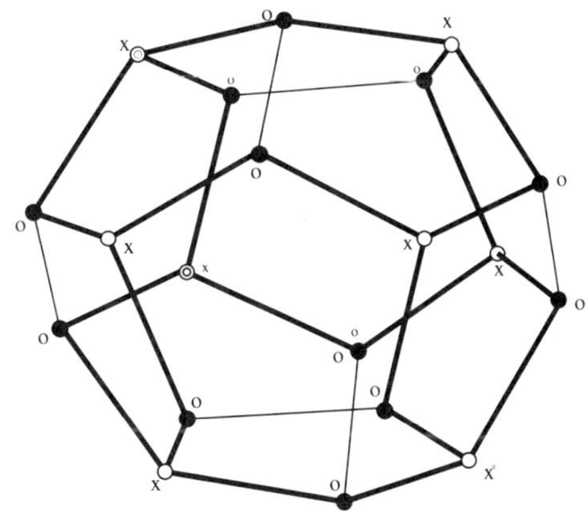

FIGURE 6:
Dodecahedron, 6 edges deleted, 2-colored

and $\mathbb{P}_{10} \not\subseteq G_n$, *then one can delete 2 vertices of* G_n *so that the resulting graph* G_{n-2} *be bipartite.*

Proof of Theorems 1.6 and 1.11 from Theorem 1.7. We show that $\mathbb{P}_{10}$ satisfies the conditions of Theorem 1.7 with $p = 2$ colors and $t = 3$ independent edges. Observe that

(a) the Petersen graph $\mathbb{P}_{10}$ is 3–chromatic (see Figure 1),

(b) one can color it in RED and BLUE so that the BLUE vertices are independent and the RED vertices span 3 independent edges: $\mathbb{P}_{10} \subseteq T_{n,2,3}$. (This can be seen on Figure 1 if we call colors 1 and 3 BLUE and 2 RED.

Further,

(c) deleting any 2 vertices of $\mathbb{P}_{10}$ we still have a 3-chromatic graph.

To prove property (c), observe that if we delete 2 vertices of $\mathbb{P}_{10}$, then (by symmetry) we may always assume that one of them is the A not seen on Figure 2. The remaining pentagons, $[BCDJI]$, $[CDEHG]$ and $[FGHIJ]$ cannot be represented by one vertex. ∎

Theorem 1.7/(ii) is sharp for $T_{n,p,t}$, but this does not prove e.g. that it is also sharp for the Petersen graph. In principle it could happen that, applying Theorem 1.7/(ii) the $-\frac{1}{2}n$ could be replaced by something much more negative. Construction 1.12 below shows that Theorem 1.11 is also sharp: Theorem 1.7 is always sharp when we apply it to a graph L not containing K_3.

CONSTRUCTION 1.12. *Let* $Z_{n,p,t}$ *be the graph obtained as follows. Take a Turán graph* $T_{n,p}$ *and fix* $t+1$ *vertices* $x_1, \dots, x_{t-2}$, y_1, y_2, y_3 *in its first class. Add the edges* y_1y_2, y_2y_3, *and* y_3y_1 *to the graph and also join each* x_i, $i = 1, \dots, t-1$ *to all the remaining vertices but to* y_1, y_2, y_3.

To prove the sharpness of Theorem 1.11 observe that

$$e(Z_{n,p,t}) = e(H_{n,p,t}) - \frac{n}{p} + O(1).$$

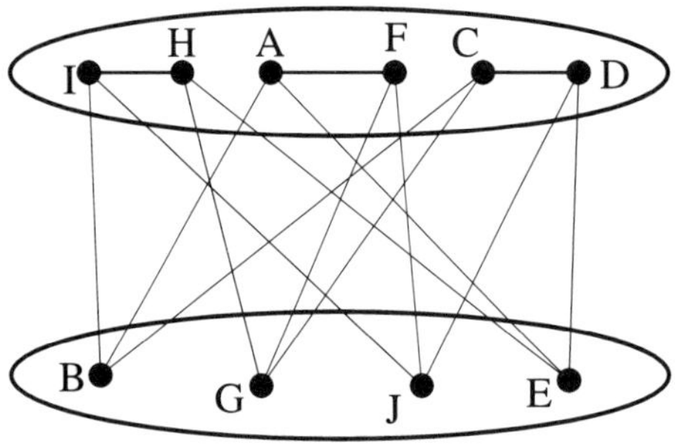

FIGURE 7:
Petersen embedded

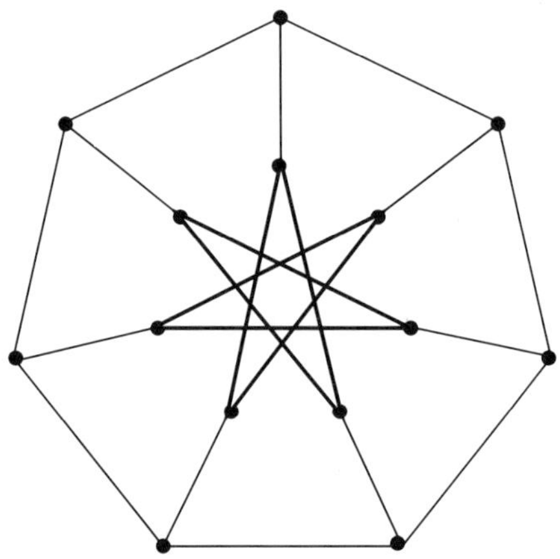

FIGURE 8:
Generalized Petersen Graph

Further, $Z_{n,p,t}$ has the following properties:

(a) if $K_3 \not\subseteq L \in \mathcal{B}_{p,t}$, then $L \not\subseteq Z_{n,p,t}$.

(b) $Z_{n,p,t} \in \mathcal{B}_{p,t}$.

Below we shall prove both (a) and (b) and they will obviously imply that Theorem 1.11 is sharp. To prove (a) observe that if $Z_{n,p,t}$ contained an L, then deleting e.g., y_1y_2 we would obtain a $Z_n^* \notin \mathcal{B}_{p,t}$ also containing L, but $L \in \mathcal{B}_{p,t}$, a contradiction. To prove (b) observe that to make $Z_{n,p,t}$ into a p-colorable graph by the deletion of some vertices one has to ruin *all* the $K_{p+1} \subseteq Z_{n,p,t}$. To achieve this, one has to delete at least 2 vertices of the triangle $y_1y_2y_3$ and all the vertices $x_1, \dots, x_{t-2}$, or at least cn other vertices. Indeed, each edge y_iy_j is contained in c_1n^{p-1} K_{p+1}'s and a vertex z forms at most $\binom{n}{p-2}$ K_{p+1} with e.g. y_1y_2. Therefore one has to delete either at least one of y_1, y_2, or at least c_2n other vertices. So we really have to delete at least two of $\{y_1, y_2, y_3\}$. Similarly, since each x_i is contained in c_3n^p copies of K_{p+1}, we have to delete all the x_i's as well, unless we are willing to delete c_4n further vertices. This completes the proof of the sharpness.

Properties of type $\mathcal{B}_{p,t}$ play fundamental role in some extremal graph problems (see e.g. [**23**]), above all, in cases where $H_{n,p,t}$ is the extremal graph. We can generalize the extremal graph problems, asking

PROBLEM 1.13. What is the maximum of $e(G_n)$ if G_n contains no $L \in \mathcal{L}$ and has property $\mathcal{B}_{p,t}$?

Assuming property $\mathcal{B}_{p,t}$ mostly decreases the extremal number: the extra condition *mostly* rules out $H_{n,p,t}$ and *mostly* in these cases there are no other extremal graphs for $\mathcal{L}$. However, this never changes the extremal number much, since we can always change the edges of a $H_{n,p,t}$ by rearranging and deleting $O(n)$ edges and increasing the chromatic number arbitrary high. The situation completely changes if we ask

PROBLEM 1.14. What is the maximum of the **minimum degree** a graph G_n can have if it contains no $L \in \mathcal{L}$ and has property $\mathcal{B}_{p,t}$?

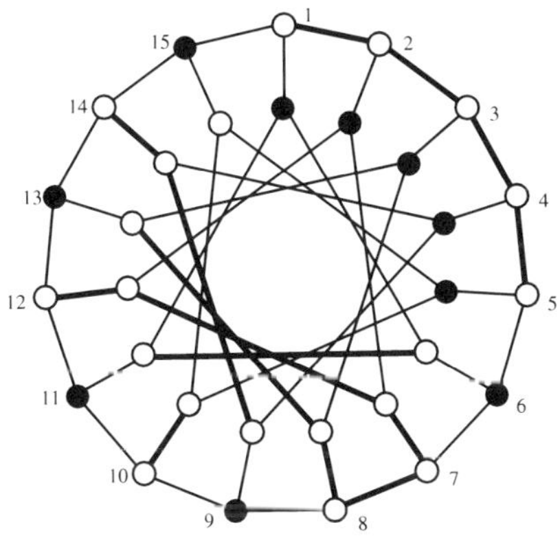

FIGURE 9:
Generalized Petersen,
$k = 15$, $\ell = 5$

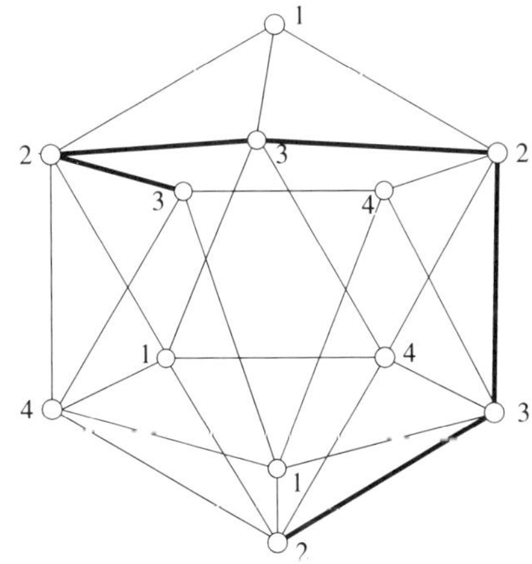

FIGURE 10:
Icosahedron and P_6 in it

We shall describe this situation in Section 4.

Above property $\mathcal{B}_{3,3}$ played an important role in our arguments, below we will go one step further to define the so called **Chromatic properties** or **Chromatic conditions**.

DEFINITION 1.15 (Symmetrical subgraphs). Let T_1 and T_2 be connected subgraphs of G. They are called symmetrical in G if either $T_1 = T_2$ or

(i) $V(T_1) \cap V(T_2) = \emptyset$, and

(ii) there are no edges joining T_1 to T_2, and

(iii) there exists an isomorphism $\omega : T_1 \to T_2$ such that for every $x \in T_1$, $u \in G - T_1 - T_2$, x is joined to u if and only if $\omega(x)$ is joined to u.

$T_1, \dots, T_k$ are symmetric if for every $1 \le i < j \le k$, T_i and T_j are symmetric.

The connectedness is assumed to rule out some trivial cases. We shall use this definition to state that under some conditions there are sequences of symmetrical extremal graphs: assuming the connectedness in this definition we get stronger statements.

DEFINITION 1.16 (Chromatic condition). A property $\mathcal{A}$ of graphs will be called chromatic condition if

(i) $G \in \mathcal{A}$ and $H \supset G$ implies $H \in \mathcal{A}$.

(ii) If $\rho = \rho(\mathcal{A})$ is a sufficiently large integer, then the following holds: if $T_1, \dots T_\rho$ are symmetric subgraphs of an $\mathcal{A}$-graph G, then $G - T_\rho$ is also an $\mathcal{A}$-graph.

(iii) there are graphs of property $\mathcal{A}$ and of arbitrary high girth.

Here (iii) is assumed only to rule out the uninteresting cases. Condition (ii) may seem to be artificial, however, this is the main point in the definition. It means that certain p-colorings of a part of a graph can automatically be extended to the whole graph, using the symmetry. E.g. if $\mathcal{A}$ is the family of at least q-chromatic graphs, then $\rho = 2$ can be taken.

EXAMPLE 1.17.

- *The property $\mathcal{B}_{p,t}$ that one cannot delete t vertices of G to get a p–colorable graph is one of the typical chromatic conditions. (Here $\rho = t + 2$ will do.)*
- *The property $\mathcal{B}_{p,t,h,\Delta}$ that one cannot delete $t \geq h$ vertices of G, at least h of which are of degree at most Δ, to get a p–colorable graph, is a more complicated one, but still a chromatic property.*

Given a chromatic condition $\mathcal{A}$, we may generalize our previous problem:

PROBLEM 1.18. What is the maximum number of edges a graph $G_n \in \mathcal{A}$ can have if it contains no $L \in \mathcal{L}$.?

Here mostly Definition 1.15(iii) ensures the existence of such graphs.

DEFINITION 1.19. Given a family $\mathcal{L}$ of graphs and a chromatic property $\mathcal{A}$, we call G_n $(\mathcal{L}, \mathcal{A})$–extremal if $G_n \in \mathcal{A}$, it contains no $L \in \mathcal{L}$ and has maximum number of edges under these conditions.

2. EXTREMAL GRAPH PROBLEMS WITH LINEAR DECOMPOSITION

In [**23**] I have given a fairly general theorem from which one can "almost algorithmically" solve many involved extremal graph problems. (The expression "almost algorithmically" will be explained in Remarks 2.3-2.4.)

As it is mentioned above, there I formulated a theorem (Theorem 2.7) which covered the case of the Dodecahedron graph and would have covered the case of the Petersen graph as well, assumed I would have considered it. To illustrate the applicability of Theorem 2.7, below we shall consider two different generalizations of the Petersen graph. The first one is the Kneser graph, the second one the cyclic generalization.

2.1. The Kneser graph. The Petersen graph is a special case of the Kneser graph.

DEFINITION 2.1. Given a set A of a elements and an integer $b < a/2$, define the Kneser graph $Z(a, b)$ as a graph on $\binom{a}{b}$ vertices, where the vertices are the b–tuples of A and two of these b-tuples are joined iff they are disjoint.

It is known and easy to see that $\mathbb{P}_{10} = Z(5, 2)$.

Obviously, the Kneser graphs $Z(a, b)$ can be colored in $p = a - 2b + 2$ colors as follows. For $a - 2b$ fixed elements $x_1, \dots, x_{a-2b}$ of A we fix $a - 2b$ colors, say $\psi_1, \dots, \psi_{a-2b}$. Whenever a b-tuple B contains x_j (a fixed element), color the corresponding vertex B of $Z(a, b)$ by ψ_j. If there are more than one such x_j's, use any of these colors. Now the vertices of $Z(a, b)$ not colored as yet correspond to the b-tuples of the $2b$–element set $M = A - \{x_1, \dots, x_{a-2b}\}$. Therefore they span a 1-factor of $Z(a, b)$ which can be colored with two further colors. This shows that

- $\chi(Z(a, b)) \leq a - 2b + 2$;
- One can color $Z(a, b)$ in $a - b + 1$ colors so that the first color-class spans a 1-factor, the others form independent sets of vertices.

It was a longstanding conjecture of Kneser, a deep theorem of Lovász [**20**] (proved slightly later in a simpler way by Bárány [**2**]) that the above upper bound is sharp:

$$\chi(Z(a, b)) = a - 2b + 2.$$

The next theorem is "kind of a generalization" of Theorem 1.7 and makes the solution of the extremal graph problem of any Kneser graph "almost" algorithmic.

THEOREM 2.2. *Let $\mathcal{L}$ be finite, $p(\mathcal{L}) = p$ but for some $L \in \mathcal{L}$*

$$L \subseteq T_{2pv,p,v} \text{ for } v := v(L), \tag{2.1}$$

then there exists an $n_0(\mathcal{L})$ such that, for $n > n_0$, there is an extremal graph U_n for $\mathcal{L}$, for which

(i) one can delete v vertices of U_n to get a $T_{n-v,p}$.

(ii) Furthermore, if $v := \max_{L\in\mathcal{L}} v(L)$, then one can also delete $m < v^2 \cdot 2^v$ vertices of U_n to get a $T_{n-m,p}$ where all the vertices of $x \in U_n - T_{n-m,p}$ are either joined to all the vertices of $T_{n-m,p}$ or to $p-1$ classes of $T_{n-m,p}$ completely but not at all to the remaining one. (This exceptional class may depend on x.)

REMARK 2.3. This is indeed an **algorithmic solution** of the extremal graph problem where $Z(a, b)$ is the excluded graph, or more generally, of any extremal problem of any finite $\mathcal{L}$ satisfying the conditions of Theorem 2.2. Indeed, in Theorem 2.2 the sequences (U_n) are completely given by the graphs $S_m := U_n - T_{n-m,p}$ and by the patterns according to which the vertices of S_m are joined to $T_{n-m,p}$. So we have to check only finitely many graph sequences (U_n), for finitely many S_m and finitely many ways of joining S_m to $T_{n-m,p}$ and decide if a graph $L_i \in \mathcal{L}$ is contained in U_n for large n, (i.e., the sequence is "good") or not (it is "bad"), Then we have to choose a "good" graph sequence (U_n) with the maximum number of edges: given two graph sequences, one can easily decide, which has more edges, though in principle this may depend on divisibility properties of n as well. Here we should be **careful** with the **infinite** families $\mathcal{L}$. If $\mathcal{L}$ is infinite, then we have to assume that there exists an **oracle** which tells us, if a sequence of graphs above is good or not ...

REMARK 2.4. The earlier results were weaker in the sense that they have not provided explicit (or implicit) upper bounds on the number of vertices to be deleted in Theorem 2.2(ii) and therefore having found the extremal graph one could not be sure if that is the extremal graph. (Those years we did not care so much about the algorithmic aspects.)

One could hope that the Kneser graphs satisfy the conditions of Theorem 1.7 and then one would get a much nicer looking result: $H_{n,p,s}$ would be the extremal graphs for $p = a - 2b + 1$ and $s := \frac{1}{2}\binom{2b}{b}$. However, generally this is not the case. As Z. Füredi pointed me out, sometimes one can delete much fewer vertices. The colorings of the Kneser graph can be much more complicated than the above one. A paper of P. Frankl and Z. Füredi [**17**] (disproving a conjecture of Erdős) gives a lot of information on this. Among others it contains the following construction. Let us consider e.g $Z(2b + 1, b)$. This is 3-chromatic. The above calculated $s = \frac{1}{2}\binom{2b}{b}$. If we split the base set into

$$X_1 \cup X_2 \text{ with } |X_1| = b + 1 \text{ and } |X_2| = b,$$

and color by RED those b-tuples which intersect X_1 in more than $\frac{1}{2}|X_1|$ elements, and color by BLUE those b-tuples which intersect X_2 in more than $\frac{1}{2}|X_2|$ elements, then the number of badly colored b-tuples is $\approx \frac{c}{\sqrt{b}}\binom{2b}{b} = o(s)$.

OPEN PROBLEM 2.5 (Kneser Coloring). *Which is the minimum $p(a, b)$ such that one can delete p vertices of the Kneser graph to get an $a - 2b + 1$-chromatic graph? And for edges?*

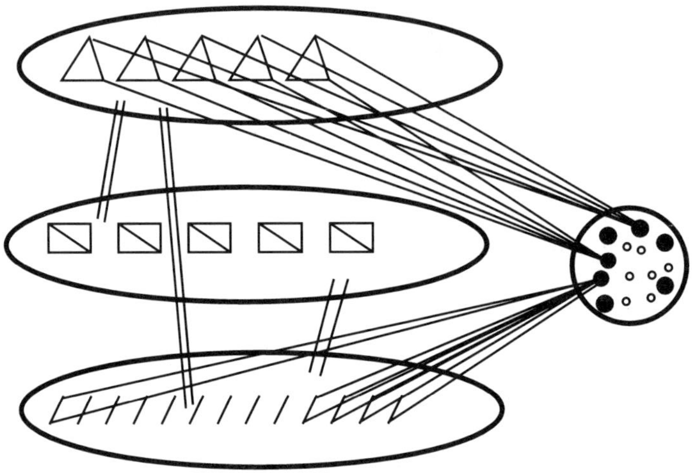

FIGURE 11:
Sequences of symmetrical graphs

Now we turn to the formulation of the general (old) theorem from which all *known* (?) Turán type extremal graph theorems with linear decomposition follow.

DEFINITION 2.6. (Family of symmetrical graphs) $\mathbb{D}(n,p,r)$ is the class of graphs G_n satisfying the following symmetry condition:

(i) It is possible to omit $\le r$ vertices of G_n so that the remaining graph G^* is a product (of graphs of almost equal order):

$$G^* = \prod_{\ell \le p} G^{m_\ell} \qquad \text{where} \qquad \left| m_\ell - \frac{n}{p} \right| \le r.$$

(ii) For every $\ell \le p$, there exist connected graphs $H_{p,j} \subseteq G^{m_\ell}$ and isomorphisms

$$\omega_{\ell,j} \ : \ H_{\ell,1} \to H_{\ell,j}$$

such that $H_{\ell,j} \quad (j = 1, \dots, k_\ell)$ are symmetric subgraphs of G_n, $v(H_{\ell,j}) \le r$ and

$$G^{m_\ell} = \sum_{j \le k_\ell} H_{\ell,j},$$

where the sum $\sum H_{\ell,j}$ is the vertex-disjoint union.

The graphs $H_{\ell,j}$ will be called the "blocks", the vertices in $G_n - G^*$ will be called "exceptional".

THEOREM 2.7 (Symmetrical extremal graphs, [**23**]). *Assume that a finite family $\mathcal{L}$ of forbidden graphs with $p = p(\mathcal{L})$, and a chromatic condition $\mathcal{A}$ are given. If for some $L \in \mathcal{L}$ and $v := v(L)$,*

$$L \subseteq P^v \otimes K_{p-1}(v, v, \dots, v), \tag{2.2}$$

then there exists a constant $r = r(L)$ such that, for every n, $\mathbb{D}(n,p,r)$ contains an extremal graph for $(\mathcal{L}, \mathcal{A})$. Furthermore, if for every $R > r$ there exists an n_R such that for $n > n_R$ $\mathbb{D}(n,p,R)$ contains only one extremal graph, then for sufficiently large n this is the only extremal graph. (Of course, the families $\mathbb{D}(n,p,R)$ form nested sequences, so this common graph sequence is the same.)

Now we turn to the proof of Theorem 2.2. The proof will be cut into two parts, the first being formulated in the following theorem.

THEOREM 2.8. *Assume that a finite family $\mathcal{L}$ of forbidden graphs with $p = p(\mathcal{L})$, and a chromatic condition $\mathcal{A}$ are given. If some $L \in \mathcal{L}$ is contained in a $T_{2pv,p,v}$, then – applying Theorem 2.7 –*

(i) all the blocks of $\mathbb{D}(n,p,r)$ will consist of isolated vertices: the product graph G^* will be a Turán graph $T_{n^*,p}$.

(ii) If there is no chromatic condition involved and s is the maximum integer for which $K_{p+1}(s,m,m,\ldots,m)$ contains no forbidden $L \in \mathcal{L}$ (even if m is very large), then

$$\text{ext}(n,\mathcal{L}) = \text{ext}(n,K_{p+1}) + \frac{s}{p}n + O(1). \tag{2.3}$$

(iii) Further, – still assuming the absence of the chromatic condition [3] – each vertex x of the "exceptional set" $U_n - G^*$ is joined either to all the vertices of this Turán graph or to all the vertices of $p-1$ classes and to none of the remaining class (but the remaining class may depend on x).

Proof. (a) Let I^s denote the set of s independent vertices and $H^*_{n,p,s+1} = I^s \otimes T_{n-s,p}$. By the definition of s, $H^*_{n,p,s+1}$ contains no forbidden subgraphs:

$$e(H^*_{n,p,s+1}) \le \text{ext}(n,\mathcal{L}).$$

This proves the lower bound in (2.3).

(b) We apply Theorem 2.7. Let U_n be extremal for $\mathcal{L}$ as described in Theorem 1.7. If a block contained an edge, then v blocks of that class would provide v independent edges, so U_n would contain a $T_{2pv,p,v}$, which was excluded. So the blocks are isolated vertices, the graph $U_n - S_m$ is a Turán graph. The second part of (i) follows from the definition of the Symmetrical Graph Sequences. Observe now that if there are ζ exceptional vertices joined to all the classes of this Turán graph and η vertices joined to fewer than $p-1$ classes, then

$$e(U_n) \le e(T_{n.p}) + \frac{\zeta - \eta}{p}n + O(1). \tag{2.4}$$

The number of "exceptional vertices" joined to all the classes of $T_{n^*,p}$ is at most s, by the definition of s. This proves the upper bound in (2.3) and also that the number of these vertices is *exactly* s. The only thing to be checked is that the "exceptional vertices" cannot be joined to fewer than $p-1$ classes. If we had an exceptional vertex y joined to at most $p-2$ classes, then

$$e(U_n) < \text{ext}(n,K_{p+1}) + \frac{s-1}{p}n + O(1) < e(H^*_{n,p,s+1})$$

would follow, a contradiction. ∎

The argument of (b) is not necessarily true if we extend the statement to extremal problems with chromatic conditions!

Proof of Theorem 2.2. Let U_n be an extremal graph for $\mathcal{L}$ described in Theorem 2.8. The classes of the Turán graph $T_{n^*,p}$ will be denoted by $A_1,\ldots,A_p$. We classify the vertices of $U_n - T_{n^*,p}$ as follows. W is formed by those s vertices which are joined to $T_{n^*,p}$ completely. $D_i \subseteq U_n - T_{n^*,p} - W$ is the class of vertices joined to $T_{n^*,p} - A_i$ completely, but not to A_i, $i = 1,\ldots,p$. By the assumption

[3] Or we could say that the chromatic condition contains all graphs.

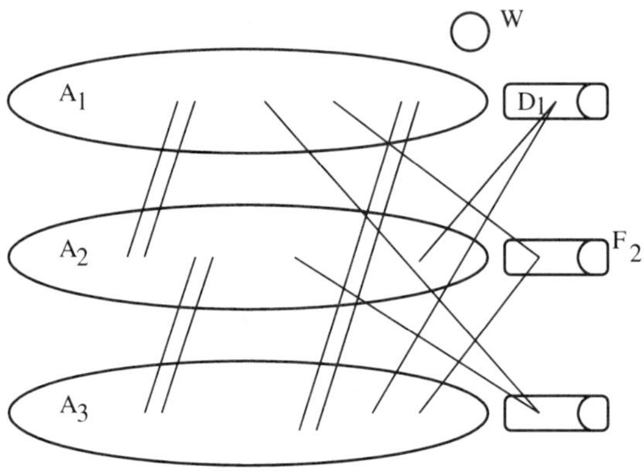

FIGURE 12:
The vertex-partition

on $\mathcal{L}$, $T_{2pv,p,t} \not\subset U_n$ and so there cannot be t independent edges between A_i and $W \cup D_i$. This and $t \le [v/(2p)]$ imply (i): the edges in $A_i \cup D_i$ can be represented by $2(t - |W|)$ vertices. So it is enough to delete $2p(t - |W|) + |W| \le v$ vertices to get a Turán graph.

To prove (ii) first we define the "horizontal"[4] and the "missing" edges. The edge xy is "horizontal" if x, y belong to the same $A_i \cup D_i$. A *non-edge* is a "missing edge" if x and y belong to different $A_i \cup D_i$'s. Let us fix in each D_i a maximal set of independent edges. The endvertices of these edges form the sets S_i. Put $F_i = D_i - S_i$. The vertices in F_i can be classified according to their connection to $(\cup S_i) \cup W$. There are at most 2^v classes. We show that each $x \in F_i$ is connected to all but at most v vertices of $V(U_n) - A_i - D_i$. (In other words, the "missing"-degree of x is at most v.) Indeed, if there were at least $v + 1$ "missing edges" incident to x, then adding these edges to U_n and deleting all the edges joining x to $(\cup S_i) \cup W$ we would get a U_n^* not containing forbidden subgraphs but having more than $e(U_n)$ edges, a contradiction.

Now, if there is a class $Q \subseteq F_i$ with $|Q| \ge v^2$ (in the above classification), then all the vertices $x \in Q$ are joined to all the vertices of $\cup_{i \ne j} F_j$. Otherwise there is a "missing edge" xy and we add xy to the graph U_n, getting a forbidden subgraph $L \subseteq U_n + xy$ and then we could replace x by an $x' \in Q$ joined to all the vertices in $N(x)$ and to y too: otherwise the "missing edge"-degree of y would be too large. Thus $L \subseteq U_n - x$, a contradiction again. This means that the vertices of Q are symmetric to each other.

Now, if we "increase" the size of Q by h (where h can be negative and positive as well), at the cost of changing the size of A_i in the other direction, then we get a family of graphs $U_{n,h}$ and $e(U_{n,h})$ will be a linear function of h. Since $U_{n,0}$ is extremal, and neither $U_{n,1}$, nor $U_{n,-1}$ contains forbidden subgraphs, $e(U_{n,h})$ must be constant. So we can replace U_n by $U_{n,-|Q|}$. This means that we can eliminate Q getting an other extremal graph. We can eliminate all the classes of size at least v^2. This completes our proof. ∎

[4]The "horizontal" edges will be used only later.

2.2. Cyclic generalization of the Petersen graphs.

DEFINITION 2.9. A generalized Petersen graph $\mathbb{P}_{k,\ell}$ is a 3-regular graph defined as follows: $x_1, \dots, x_k$ are the "outer" vertices, $y_1, \dots, y_\ell$ are the "inner" vertices, x_i is joined to y_i and $x_{i\pm 1}$, y_i is joined to x_i and $x_{i\pm\ell}$, (where the indices are counted mod k).

Observe that generally the "inner" and "outer" vertices behave differently, they play the same role only if $(k, \ell) = 1$: otherwise inside we have many disjoint cycles.

THEOREM 2.10. *A generalized cyclic Petersen graph $\mathbb{P}_{k,\ell}$ can be colored in RED and BLUE so that the RED vertices are independent and the BLUE vertices can be covered by a path.*

This implies that our theorems can be applied to $\mathbb{P}_{k,\ell}$ as well.

Proof of Theorem 2.10. There are many explicit colorings but here we shall be concise. We shall return to the detailed discussion of this topic in [**27**]. The cases $\ell = 1, 2$ can be treated separately and we skip them. ($\ell = 1$ is trivial.)

(a) Let us start with an observation. If we color the vertices of a 3-regular graph in RED and BLUE so that the RED vertices are independent and the BLUE ones form a forest, then we are home: any BLUE vertex having 3 BLUE neighbors can (one by one) be changed into RED. So we get the desired coloring: the RED vertices remain independent and the BLUE ones will have only BLUE degrees at most 2. Further, no BLUE cycle can arise. So the graph spanned by the BLUE vertices will be the union of paths.

(b) This gives an easy solution for ℓ even: color $y_1, \dots, y_\ell$ by RED, $x_1, \dots, x_\ell$ by BLUE, and then color the remaining y_i's by BLUE, the arc $x_{\ell+1}, x_{\ell+2}, \dots, x_k$ in RED-BLUE-RED- ... -RED. (This arc has now odd length.) We do not verify this simpler case, rather the next, slightly more involved one.

(c) If ℓ is odd, we may slightly change the above coloring: the outside path $x_{\ell+1}, x_{\ell+2}, \dots, x_k$ can be colored by a

RED-BLUE-BLUE-RED-BLUE-RED-BLUE ... -RED-BLUE-RED.

One can easily see that this works. (a) the RED points are independent. (b) One has to show that there is no cycle with all blue vertices. $y_1, \dots, y_\ell$ are RED and the BLUE path $x_1, \dots, x_\ell$ is separated by RED vertices from the remaining parts: so we may forget it. The only thing to be checked is that no BLUE cycle occurs (where a BLUE cycle means that all its vertices are BLUE). Indeed, if we delete the inside RED vertices, then the cycles inside are ruined. Call an edge BLUE if it joins BLUE vertices. There are no BLUE edges outside except $(x_{\ell+2}, x_{\ell+3}$, see Figure 5(a). and iw there were a BLUE cycle in the graph and we would walk around it, then arriving at any outside vertex but these two we are stuck. On the other hand, if we use the edge $(x_{\ell+2}, x_{\ell+3}$, then we otherwise must stay completely inside. Distinguishing the two cases $(k, \ell) = 1$ or not, we may easily cope with this case. ∎

CONJECTURE 2.11. *If L is a 3-regular graph of girth at least 5, then it can be two-colored in RED and BLUE so that all the RED vertices are independent and the BLUE vertices form a subgraph which can be covered by a path.*

(Maybe, this conjecture is not too difficult.)

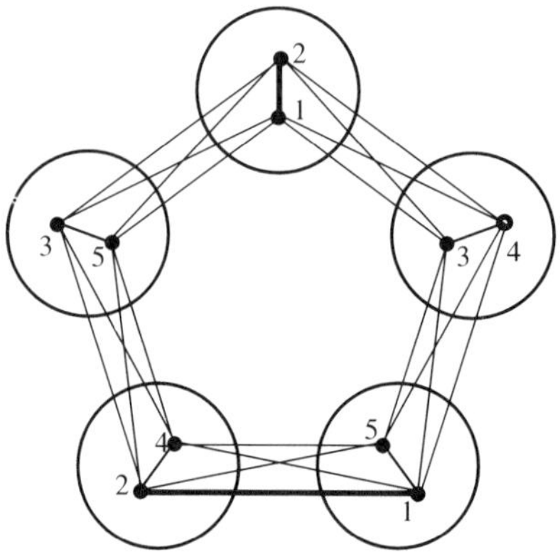

FIGURE 13:
The Łuczak graph

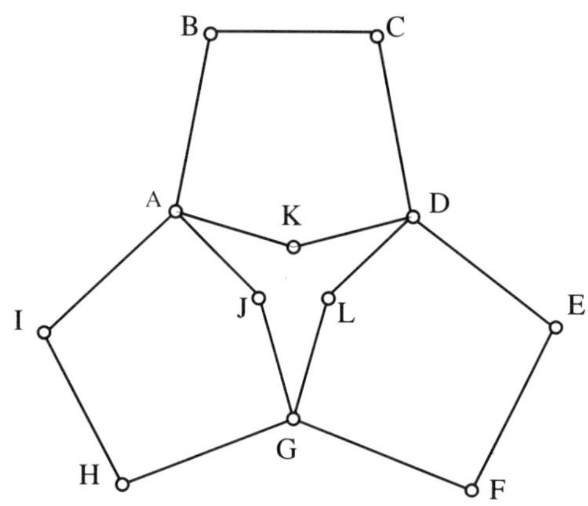

FIGURE 14:
Nešetřil Graph

2.3. The case of the Łuczak graph. To show the power of a method one can select certain cases "at random" and show that the method works in those cases. Below I shall discuss the case of two such graphs. The first one, L_{10} was used (as a counterexample for some graph entropy question) by Tomasz Łuczak, and is given in Figure 13. So I shall call it the Łuczak graph. The other graph, N_{12}, was used in a lecture by Jarik Nešetřil and below will be called Nešetřil graph. (These two graphs and the corresponding results can be generalized and we shall return to this in another paper [**27**]

PROBLEM 2.12. Determine $\text{ext}(n, L_{10})$. What are the extremal graphs?

This is really easy: One can see that L_{10} is 5-chromatic and that removing any vertex of L_{10} it remains 5-chromatic, but one can color it in 5 colors so that the first two colors span 2 independent edges (see Figure 13). Hence we can apply Theorem 1.7:

THEOREM 2.13. *For L_{10}, $H_{n,4,2}$ is the (only) extremal graph, for $n > n_0(L_{10})$.*

2.4. The Nešetřil graph. The Nešetřil graph consists of 3 pentagons $C_5(a_i, b_i, c_i, d_i, e_i)$ where the pentagons are glued together at the b_i's and e_i's cyclically: $b_1 = e_2$, $b_2 = e_3$, $b_3 = e_1$. Denote this graph by N_{12}. Relabel them as seen in Figure 14.

THEOREM 2.14. *There exists an $n_0(N_{12})$ such that for $n > n_0$, $H_{n,2,2}$ is the (only) extremal graph.*

REMARK 2.15. Theorem 2.14 does not follow from Theorem 1.7. Below we shall use the labeling of Figure 14.

(a) One can delete 3 independent edges, e.g. BC, EF, and IH to get a bipartite graph from N_{12} and the omission of 2 edges is obviously not enough.

(b) We could apply Theorem 1.7 if we could show that the omission of any 2 vertices leaves us with a 3-chromatic graph. The extremal graph would be $H_{n,3,2}$. However, this is not the case: the omission of A and L results a tree.

Proof of Theorem 2.14.

(a) $N_{12} \not\subseteq H_{n,2,2}$, since $N_{12} \in \mathcal{B}_{2,1}$ but $H_{n,2,2} \notin \mathcal{B}_{2,1}$.

(b) Apply Theorems 2.7 and 2.8 to an extremal graph U_n: there is a $T_{n^*,2} \subseteq U_n$ with $n^* = n - O(1)$. All the blocks are 1-vertex blocks and there can be at most one "exceptional" vertex $u \in S_n - T_{n^*,2}$ joined to $p = 2$ classes, otherwise, by Remark 2.15(b), $N_{12} \subseteq U_n$. (As a matter of fact, a $H_{\nu,2,3} \subseteq U_n$ for some $\nu \approx n/2$. So all the fixed graphs $L \notin \mathcal{B}_{2,2}$ can be built up in U_n, for $n > n_0(L)$.) If there is no exceptional vertex joined to both classes of $T_{n^*,2}$, then

$$e(U_n) \le e(T_{n,2}) + O(1) < e(H_{n,2,2}).$$

So there is exactly one vertex u joined to both classes. All the other "exceptional" vertices are joined to a class of $T_{n^*,2}$ completely, and to the other not at all. If there is a vertex $v \in S_n - T_{n^*,2}$ joined to (say) the second class of $T_{n^*,2}$ completely and to at least one vertex of the first class, then we can again find an $N_{12} \subseteq U_n$ (because N_{12} can be turned into a tree by deleting a vertex of degree 4 and an edge from the "opposite" pentagon. Now we may partition $V(U_n) - u$ into two classes according to whether they are joined to the first class of $T_{n^*,2}$ or the second one and the above observation yields that the vertices in these classes are independent. So $U_n \in \mathcal{B}_{2,1}$ and therefore either $e(U_n) < e(H_{n,2,2})$ or $U_n = H_{n,2,2}$. ∎

3. ZARANKIEWICZ TYPE PROBLEMS

Now we would like to answer the following question.

How large minimum degree ensures for a graph G_n a Petersen graph $\mathbb{P}_{10} \subseteq G_n$?

These type of questions are called Zarankiewicz type extremal problems.

PROBLEM 3.1 (Zarankiewicz type). Given a family $\mathcal{L}$ of forbidden graphs, what is the maximum of the minimum degree $\delta(G_n)$ a graph G_n of order n can have without containing any $L \in \mathcal{L}$?

Let us denote the maximum in such a problem by $\operatorname{dex}(n, \mathcal{L})$, assumed that $\mathcal{L}$ is a family of excluded subgraphs. It is obvious that

$$\operatorname{dex}(n, \mathcal{L}) \le \frac{2}{n} \cdot \operatorname{ext}(n, \mathcal{L}), \tag{3.1}$$

since any graph having

$$\delta(G_n) > \frac{2}{n} \cdot \operatorname{ext}(n, \mathcal{L})$$

has more than $\operatorname{ext}(n, \mathcal{L})$ edges and therefore must contain some of $L \in \mathcal{L}$.

On the other hand, in many cases some extremal graphs for the Turán type problem are almost regular. Then the difference between the Turán and Zarankiewicz type problems is negligible. If e.g. there exists an extremal graph sequence (Z_n) for the Turán problem in which the maximum degree and the minimum degree differ only by (at most) one, then

$$\operatorname{dex}(n, \mathcal{L}) = \left\lceil \frac{2}{n} \cdot \operatorname{ext}(n, \mathcal{L}) \right\rceil \tag{3.2}$$

In cases discussed by us the graph sequence $(H_{n,p,t})$ provided the extremal sequence, which (for $t > 1$) is far from being almost regular. Yet its minimum degree and average degree are near to each other. So

LEMMA 3.2. *If for $\mathcal{L}$ the extremal graph sequence is $(H_{n,p,t})$, then the solution of the Zarankiewicz problem is between*

$$\left(1-\frac{1}{p}\right)n \text{ and } \left(1-\frac{1}{p}\right)n+\frac{t-1}{p}.$$

This clarifies the situation in a weak asymptotic sense but the detailed discussion is more involved and postponed.

4. ANDRÁSFAI-ERDŐS-SÓS TYPE PROBLEMS

A completely new phenomenon occurs when we combine the Zarankiewicz type problems with chromatic condition type problems. Perhaps Andrásfai was the first to ask such questions:

PROBLEM 4.1 (Andrásfai). Determine the maximum of the minimum degree in a graph G_n under the condition

$$K_{p+1} \not\subseteq G_n \quad \text{and } \chi(G_n) > k.$$

For $k = p + 1$ this was solved by B. Andrásfai, P. Erdős and Vera T. Sós, [**1**]. In the simplest case, when we assume that G_n is triangle-free and non-bipartite, the pentagon like graph $H_n := C_5[n/5]$ shows that the minimum degree can be $\frac{2}{5}n - O(1)$. For the general case of K_{p+1} and $k = p + 1$ the extremal graph is $C_5[m/5] \otimes T_{n-m,p-2}$, where m is chosen so that the resulting graph be approximately regular. (This is obtained when $m = \frac{5n}{3p-1} + O(1)$.) P. Erdős and the author, [**15**] extended this result for the case where an arbitrary $p+1$-chromatic graph L with some color-critical edges is excluded. Of course, this does not cover the case of $\mathbb{P}_{10}$. To describe the case for the Petersen graph, we shall prove

THEOREM 4.2. *For every v (and $t \le v/2$) there exists a $K = K(v)$ such that if*

$$\delta(G_n) > \frac{2}{5}n + K$$

and $T_{v,2,t} \not\subset G_n$, then one can delete K vertices of G_n to get a bipartite graph.

REMARK 4.3.

(a) Theorem 4.2 is sharp, as shown by $C_5[\frac{1}{5}n]$. Clearly, $\delta(C_5[\frac{1}{5}n]) \ge \frac{2}{5}n - 2$ and $T_{v,2,t} \not\subset C_5[\frac{1}{5}n]$. Further, replacing $T_{v,2,t}$ by any graph $L \subseteq T_{v,2,t}$ we get the same sharpness if $K_3 \subseteq L$, since $C_5[\frac{1}{5}n]$ contains no K_3.

(b) Similarly, the theorem is sharp e.g. for $\mathbb{P}_{10}$: if one tries to embed $\mathbb{P}_{10}$ into $C_5[\frac{1}{5}n]$, then all the pentagons of $\mathbb{P}_{10}$ must "run around" in $C_5[\frac{1}{5}n]$. Therefore we may assume that (using the notations of Figure 2) $A \in A_1$, $B \in A_2$, $C \in A_3$, $D \in A_4$, $E \in A_5$. By symmetry we may assume that $F \in A_2$. Now we cannot put G anywhere: it must be in an A_i neighboring both A_3 and A_2, which is impossible.

(c) The theorem is *not sharp* if $\chi(L) = 3$ and $L \subseteq C_5[\mu]$ for some μ.

There are many deep, unsolved problems in this field but now we skip them. To prove Theorem 4.2 we shall use two results from extremal graph theory.

THEOREM 4.4 (Theorem 9 of [**4**]). *Let t be a natural number and let $c > 0$. Put $\ell(c) = \lceil \frac{c^{-1}-1}{2} \rceil$. Then there exists an $n_0 = n_0(t, c)$ such that if $n > n_0$ and $G_n \not\supseteq C_m[t]$ for $m = 3, 5, \ldots, 2\ell(c) + 1$, then G_n can be made bipartite by the omission of at most cn^2 edges.*

REMARK 4.5. For improved versions of the above theorem see a very recent result of Komlós, [**19**]. His theorem provides a much smaller $\ell(c)$, approximately the best one.

The other result we need is the following lemma.

LEMMA 4.6 (Lemma 4 of [**22**], p315). *Let M be a given positive integer and $c > 0$ be an arbitrary constant. Then there exist an M' and a $c' > 0$ such that if a set A of n elements contains M' subsets A_i each of which has at least cn elements, then there are M subsets*

$$A_{i_1}, \ldots, A_{i_M}$$

among them whose intersection has more than $c'n$ elements.

(Here M' is typically much bigger than M, c' much smaller than c. The lemma is contained in a lemma of Erdős, [**7**].)

Proof of Theorem 4.2. First we fix three constants:

$$\gamma := \frac{1}{100}, \qquad \eta := \gamma^2, \qquad \varepsilon := \gamma^4, \qquad \mu = \frac{100v^2}{\varepsilon}.$$

We shall consider a G_n not containing $T_{v,2,t}$. Our proof will have two main parts:

(A) when one can delete $< \varepsilon n^2$ edges from G_n to turn it into a bipartite graph and

(B) when one cannot do this.

In the first case we shall show that we can delete $O(1)$ vertices to make G_n bipartite. In the second one we shall apply Theorem 4.4 to get a $C_{2\ell+1}[M]$ for some not too large ℓ and very large M and then we shall deduce in (C) that $G_n \supseteq C_5[\mu]$. From this we shall deduce that the minimum degree is $< \frac{2}{5}n + K$.

(A) So first we assume that G_n can be turned into a bipartite graph by deleting $< \varepsilon n^2$ edges. Let us choose the vertex-partitioning $A \cup B = V(G_n)$ for which $e(A) + e(B)$ is the minimum possible. We shall prove that A contains $O(1)$ vertices so that deleting them we ruin all the edges in A. The same will hold for B. This will prove that we can omit $O(1)$ vertices to get a bipartite graph, as stated in the theorem.

(A1) By Lemma 4.6, there exists a constant h such that if $x_1y_1, \ldots, x_hy_h$ are h independent edges in A and $U_i = N(x_i) \cap N(y_i) \cap B$, $|U_i| > \varepsilon n$ for $i = 1, \ldots, h$, then we may choose v of these edges so that $|\cap U_i| > v$. Therefore $T_{2v,2,v} \subseteq G_n$ a contradiction.

(A2) Below we shall call *horizontal degree* the number of edges to the same class: For a given $x \in A$ we shall call $\alpha(x) := |N(x) \cap A|$ the horizontal degree and $\beta(x) := |N(x) \cap B|$. To show that

$$\frac{2}{5}n - 2\varepsilon n \leq |A|, |B| \leq \frac{3}{5}n + 2\varepsilon n, \tag{4.1}$$

first we assume that $|A| \geq |B|$. By the assumption that A and B contain only $\leq \varepsilon n^2$ edges, we get that the average of $\alpha(x)$ in A is at most $2\varepsilon n$. Therefore A contains an x for which $\beta(x) = d(x) - \alpha(x) \geq \frac{2}{5}n - 2\varepsilon n$:

$$|A| \geq |B| \geq \beta(x) \geq \frac{2}{5}n - 2\varepsilon n,$$

proving (4.1).

(A3) From now on $|A| \geq |B|$ will not be assumed. If $A' \subseteq A$ is the set of vertices x having horizontal degree $\alpha(x) < \eta n$, then A' contains at most h independent edges: each edge xy has endvertices joined to at least $\frac{2}{5}n - \eta n$ vertices from B. Therefore x and y have at least

$$\beta(x) + \beta(y) - |B| \geq \frac{4}{5}n - \frac{3}{5}n - 2\varepsilon n - 2\eta n > \frac{1}{10}n$$

common neighbors in B. So we may apply Lemma 4.6, as described in (A1) to see that the number of independent edges in A' is at most h, therefore they can be represented by at most $2h$ vertices.

(A4) Similarly, $|A - A'| < h$. Indeed, if $A'' := A - A'$ contains T points, [5] then we can select T independent edges $x_i y_i$ with

$$x_i \in A'', \qquad y_i \in A', \qquad \alpha(y_i) < 3\eta n.$$

Now, by the minimality condition on the partition,

$$\beta(x) \geq \alpha(x) \text{ and } \beta(x) \geq \frac{1}{2}d(x) \geq \frac{1}{5}n.$$

(A4/(i)) If $\beta(x_i) > \frac{1}{5}n + 10\eta n$, then

$$|N(x_i) \cap N(y_i) \cap B| > \frac{1}{5}n + 10\eta n + \frac{2}{5}n - \alpha(y_i) - |B| > (10\eta - 2\varepsilon)n.$$

This is enough to apply (A1).

(A4/(ii)) If $\beta(x_i) \leq \frac{1}{5}n + 10\eta n$, then both

$$\left|\alpha(x_i) - \frac{n}{5}\right| < 10\eta n, \text{ and } \left|\beta(x_i) - \frac{n}{5}\right| < 10\eta n.$$

If $|B| < \frac{3}{5}n - 6\eta n$, then the number of triangles on $x_i y_i$ is at least

$$|N(x_i) \cap N(y_i) \cap B| \geq \beta(x_i) + \beta(y_i) - |B| \geq \frac{1}{5}n + \frac{2}{5}n - 3\eta n - |B| > 3\eta n,$$

and we may apply (A1). If, on the other hand, $|B| \geq \frac{3}{5}n - 6\eta n$, then $|A| \leq \frac{2}{5}n + 6\eta n$. We estimate from below the number of K_3's of form $x_i yz$, where $z \in B \cap N(x_i)$ and $y \in N(x_i) \cap A$. Observe that each $z \in B \cap N(x_i)$ sends at least

$$\beta(z) + \alpha(x_i) - |A| \geq \frac{2}{5}n - \alpha(z) + \frac{1}{5}n - 10\eta n - \frac{2}{5}n - 6\eta n > \frac{n}{5} - 16\eta n - \alpha(z)$$

edges to $N(x_i) \cap A$. Further, $\sum \alpha(z) \leq e(B)$. So the number of triangles $x_i yz$ is at least

$$\sum_{z \in B \cap N(x_i)} |N(z) \cap N(x_i) \cap A| > \quad \alpha(x_i) \cdot \left(\tfrac{n}{5} - 16\eta n\right) - e(B)$$

$$> \left(\tfrac{n}{5} - 10\eta n\right)\left(\tfrac{n}{5} - 16\eta n\right) > \tfrac{n^2}{30}.$$

This means that (for fixed x_i) $x_i y$ is contained in at least $\frac{n^2}{30}/\alpha(x_i) > \frac{n}{20}$ triangles on the average (taken over $y \in A \cap N(x_i)$.) Therefore we may recursively choose the edges $x_i y_i$ contained in at least $\frac{1}{30}n$ triangles $i := 1, \ldots, T$, $T = o(n)$. Lemma 4.6 now provides us with a $T_{2v,2,v} \subseteq G_n$, unless $T \leq h$. So in this case we can again represent the above edges by at most h vertices $x_1, \ldots, x_T$.

(B) Next we assume that we cannot delete εn^2 edges of G_n to get a bipartite graph. By Theorem 4.4, for some $\ell \leq \frac{1}{\varepsilon}$ and for $M = \mu \cdot 10^{1/\varepsilon}$, G_n contains an $C_{2\ell+1}[M]$, assumed that n is sufficiently large.

[5] Here we assume e.g. that $T < \log n$ to deduce that $T < h$. If $|A - A'|$ is larger, then choose a subset of $[\log n]$ vertices.

Let us count the number of edges between $G_n - C_{2\ell+1}[M]$ and $C_{2\ell+1}[M]$. By the minimum degree condition,

$$e(G_n - C_{2\ell+1}[M], C_{2\ell+1}[M]) > \frac{2}{5}n \cdot (2\ell+1)M - O(M^2).$$

Each vertex $x \in G_n - C_{2\ell+1}[M]$ can be classified according to $N(x) \cap C_5[M]$. There are $O(2^{(2\ell+1)M}) = O_M(1)$ classes.

Let us say that an $x \in G_n - C_{2\ell+1}[M]$ is typically connected to a class A_i of $C_{2\ell+1}[M]$ if it is joined to at least $\frac{M}{10}$ vertices in this class and there are v other vertices of $G_n - C_{2\ell+1}[M]$ joined to the same vertices of $C_{2\ell+1}[M]$. A short calculation shows that if $2\ell+1 \geq 7$, then most vertices x are typically joined to at least 3 classes and these 3 classes must be pairwise non-neighbors. This implies that we have a $k < \ell$ for which $C_{2k+1}[M^*] \subseteq G_n$, with $M^* = [M/10]$. Iterating this step at most $[\frac{1}{\varepsilon}]$ times we get that a $C_5[\mu] \subseteq G_n$, for $\mu = 100v^2/\varepsilon$.

(C) So we assume that $G_n \supseteq C_5[\mu]$. Fix one such $C_5[\mu]$ and calculate the number of edges between $C_5[\mu]$ and $G_n - C_5[\mu]$:

$$e(C_5[\mu], G_n - C_5[\mu]) \geq 5\mu\delta(G_n) - 5\mu^2 > 2\mu n - 5\mu^2. \tag{4.2}$$

(C1) We change now the meaning of "typical connection": $x \notin C_5[\mu]$ is *typically* joined to a class A_i of $C_5[\mu]$ if $|N(x) \cap A_i| \geq v$ and there are v other vertices outside $C_5[\mu]$ also joined to each $y \in N(x) \cap C_5[\mu]$.

(*) If x is joined typically to A_i and A_{i+1}, then $T_{2v,2,v} \subseteq K_3(v,v,v) \subseteq G_n$ which is excluded.

We partition $V(G_n - C_5[\mu])$ into 7 classes:

(a) W is the set of those vertices which are joined typically to A_i and A_{i+1} (for some $i = 1, \ldots, 5$). By (*), $|W| < v \cdot 32^\mu = O(1)$.

(b) Z is the set of vertices "poorly" joined to $C_5[\mu]$: those vertices which are joined to at most $2\mu - 10v/\varepsilon$ vertices of $C_5[\mu]$. We will show that $|Z| < \frac{1}{2}\varepsilon n$.

(c) U_i $(i = 1, \ldots, 5)$ is the set of vertices not in Z, typically joined to A_{i-1} and A_{i+1}.

A vertex in U_i is joined to $C_5[\mu]$ by less then $2\mu + 3v$ edges. This, the bound on $|W|$, the definition of Z and (4.2) imply that $|Z| < \frac{3v}{10v/\varepsilon}n < \frac{1}{2}\varepsilon n$. So deleting all the edges represented by Z we deleted fewer than $\frac{1}{2}\varepsilon n^2$ edges.

Using the method of (A1) we see that U_i contains at most $O(1)$ independent edges, otherwise $T_{2v,2,v} \subseteq G_n$ and therefore $L \subseteq G_n$. So $e(U_i) = O(n)$, moreover, all the edges of U_i can be represented by $O(1)$ vertices.

Let us call a graph pentagon-like if it is contained in some $C_5[N]$. We have just proved that we can delete $T_1 := O(1)$ vertices of G_n to get a pentagon-like graph. But it is trivial that the minimum degree of a pentagon-like graph G_k is at most $\frac{2}{5}k$. So $\delta(G_n) \leq \frac{2}{5}n + T_1$. ∎

References

[1] B. Andrásfai, P. Erdős and V. T. Sós: On the connection between chromatic number maximal clique and minimal degree of a graph, Discrete Mathematics **8** (1974), 205-218, MR**49** #4831.

[2] I. Bárány: A short proof of Kneser's conjecture, Journal of Combinatorial Theory A**25**, (1978), 325-326.

[3] B. Bollobás: Extremal Graph Theory, Academic Press, London, (1978).

[4] B. Bollobás. P. Erdős, M. Simonovits and E. Szemerédi: Extremal graphs without large forbidden subgraphs, Advances in Graph Theory (ed. B. Bollobás), Annals of Discrete Mathematics 3 (1978), 29-41, North-Holland. MR**80a**# 05119.

[5] P. Erdős; Über ein Extremalproblem in der Graphentheorie, Archiv d. Math. **13**(1962), 222–227.

[6] P. Erdős: On the structure of linear graphs, Israel J. Math. **1** (1963), 156-160 MR**28** # 4533.

[7] P. Erdős: On extremal problems of graphs and generalized graphs, Israel J. Math. **2**(3) (1965), 183-190; MR**32**# 1134.

[8] P. Erdős: Some recent results on extremal problems in graph theory (Results), Theory of Graphs (International symposium, Rome, 1966), Gordon and Breach, New York and Dunod, Paris, 1967, pp. 117-123, MR 37, #2634.

[9] P. Erdős: On some new inequalities concerning extremal properties or graphs, Theory of graphs, Proc. Coll. Tihany (1966) (Eds. P. Erdős and G. Katona) Acad. Press, N.Y., (1968) pp77-81; MR**38**# 1026.

[10] P. Erdős: The Art of Counting, Selected Writings (in Combinatorics and Graph Theory, ed. J. Spencer), The MIT Press, Cambridge, Mass., (1973) MR**58**#27144

[11] P. Erdős and T. Gallai: On maximal paths and circuits of graphs, Acta Math. Acad. Sci. Hungar. **10** (1959), 337-356; MR**22**#5591.

[12] P. Erdős and M. Simonovits: A limit theorem in graph theory, Studia Sci. Math. Hungar. 1 (1966) 51-57. (Reprinted in [**10**], 1973, pp 194-200.) 51-57; MR**34**#5702.

[13] P. Erdős and M. Simonovits: Some extremal problems in graph theory, Combinatorial Theory and its applications, I. (ed. P. Erdős et al.), North-Holland, Amsterdam, (1970), pp. 377-390; MR**46**#84. Reprinted in the Art of Counting, MIT PRESS, 1973, pp 246-259).

[14] P. Erdős and M. Simonovits: An extremal graph problem, Acta Math. Acad. Sci. Hung. 22(3-4)(1971), 275-282. MR**45**#1790.

[15] P. Erdős and M. Simonovits: On a valence problem in graph theory, Discrete Mathematics 5 (1973), 323-334. MR**49**# 7175.

[16] P. Erdős and A. H. Stone: On the structure of linear graphs, Bull. Amer. Math. Soc. **52** (1946), 1089-1091. MR**8**# 333b.

[17] P. Frankl and Z. Füredi: Extremal problems concerning Kneser graphs. Journal of Combinatorial Theory. Series B **40**(3) (1986), 270–284. MR**87i**#05118

[18] J. Griggs, M. Simonovits, G. R. Thomas: Extremal graphs with bounded densities of small subgraphs, accepted in Journal of Graph Theory, 1997

[19] J. Komlós: Covering odd cycles, Combinatorica **17**(4) (1997), pp393–400.

[20] L. Lovász: Kneser's conjecture, chromatic number, and homotopy, Journal of Combinatorial Theory A**25**(3), (1978) 319-324.

[21] J. W. Moon: On independent complete subgraphs in a graph, Canad. J. Math. 20, (1968), 95-102; MR**36**#2529; also in: International Congress of Math. Moscow, (1966), vol 13.

[22] M. Simonovits: A method for solving extremal problems in graph theory, Theory of graphs, Proc. Coll. Tihany (1966) (Eds. P. Erdős and G. Katona) Acad. Press, N.Y., (1968) 279-319. MR**38**#2056.

[23] M. Simonovits: Extremal graph problems with symmetrical extremal graphs, additional chromatic conditions, Discrete Mathematics **7,** (1974), 349-376. MR**49**#2459.

[24] M. Simonovits: The extremal graph problem of the icosahedron, Journal of Combinatorial Theory 17B (1974) 69-79; MR**51**# 7945.

[25] M. Simonovits: Extremal graph problems and graph products, Studies in Pure Math. (dedicated to the memory of P. Turán) Akadémiai Kiadó+Birkhäuser Verlag (1982)

[26] M. Simonovits: Extremal Graph Theory, Selected Topics in Graph Theory, (ed. by Beineke and Wilson) Academic Press, London, New York, San Francisco, (1983) 161-200.

[27] M. Simonovits: On the extremal graph problem of the generalized Petersen graphs, (manuscript)

[28] P. Turán: On an extremal problem in graph theory, Matematikai Lapok, **48** (1941), 436-452 (in Hungarian), (see also [**29**], [**30**]). MR**8**# 284j

[29] P. Turán: On the theory of graphs, Colloq. Math. **3** (1954), 19-30, (see also [**30**]). MR**15**# 976b.

[30] Collected papers of Paul Turán: Akadémiai Kiadó, Budapest, 1989. Vol 1-3, (with comments of Simonovits on Turán's graph theorem).

Mathematical Institute of the Hungarian Academy of Sciences Budapest, Reáltanoda u 13-15, H–1053

E-mail address: miki@math-inst.hu

DIMACS Series in Discrete Mathematics
and Theoretical Computer Science
Volume **49**, 1999

Oriented list colouring of undirected graphs

Attila Sali and Gábor Simonyi

ABSTRACT. We introduce the choice number analogon of the oriented chromatic number of Sopena. It is shown that its minimum over all orientations of a graph can be arbitrarily large compared to the choice number of the underlying undirected graph. Investigating related problems we look at the minimum number of transitive triangles needed to cover all tournaments on n vertices. The choosability analogon of the dichromatic number of Erdős and Neumann-Lara is also considered briefly.

1. Introduction

Coloring the vertices of a graph so that no adjacent vertices receive the same color gives rise to the notion of chromatic number, one of the central notions of graph theory. Similar concepts for oriented or directed graphs have also been defined. (We make the following usual distinction between oriented and directed graphs: while a directed graph may have edges in both directions between the same pair of vertices, this is not allowed for an oriented graph.) The chromatic number of oriented graphs was introduced recently by Sopena [**11**] (cf. also [**7**], [**10**]). It is defined as follows. A legal coloring of an oriented graph is an assignment of colors to the vertices in such a way that adjacent vertices receive different colors and all edges between any two color classes go in the same direction. The oriented chromatic number $\chi^{\rightarrow}(G)$ is the minimal number of colors that is sufficient for a legal coloring of the oriented graph G. (In other words, it is the size of the smallest tournament T for which a homomorphism from G to T exists.)

Another recently well investigated graph parameter is $ch(G)$, the choice number of a(n undirected) graph G. (Cf. [**1**], [**6**].) Assign a list of k colors to each vertex of G and find a proper coloring of its vertices from the corresponding lists. The smallest k for which this is always possible (whatever the actual lists are) is the choice number of G.

In this paper we introduce the oriented choice number of an oriented graph in the obvious way (see details in next section) and the oriented choice number $och(G)$

1991 *Mathematics Subject Classification*. Primary 05C15, 05C20.

Research partially supported by the Hungarian Foundation for Scientific Research Grant (OTKA) No. T016389.

Research partially supported by the Hungarian Foundation for Scientific Research Grant (OTKA) Nos. F023442 and T016386.

of an *undirected graph* as the minimum of the former over all orientations of the edges of G. We will investigate the relation of $och(G)$ to other parameters of G. In particular, we will show that the gap between $och(G)$ and $ch(G)$ can be arbitrarily large while $och(G)$ is both upper and lower bounded by functions of $\tau(G)$, the minimal number of points needed to cover all edges.

Another notion of chromatic number for directed graphs, the so-called dichromatic number, is defined by Neumann-Lara, and the dichromatic number of an undirected graph by Erdős and Neumann-Lara, cf. [**9**], [**2**], [**3**]. In the last section we investigate the list coloring version of this concept. (A third chromatic number type invariant of directed graphs is given by Fachini and Körner [**5**]. This is of different nature, however, and we do not investigate it in this paper.)

The paper is organized as follows. The next section is devoted to the investigation of $och(G)$ and its relation to other graph parameters. Section 3 deals with miscellaneous problems connected to $och(G)$. In particular, we consider the question how many transitive triangles are needed to cover every tournament on n points. Section 4 deals with the list coloring analogon of the dichromatic number.

2. The oriented choice number

For an oriented graph Sopena [**11**] defines a proper coloring to be an assignment of colors to the nodes in such a way that in addition to the requirement that adjacent vertices get different colors we also have the following. It is not allowed that we have two (oriented) edges in such a way that the tail of one of them is colored the same color as the head of the other one while at the same time the tail of the second receives the same color as the head of the first. Now an oriented graph F is said to be k-choosable if for every possible assignment of lists of size k to its vertices there is a proper coloring of F where every vertex is assigned one of the colors of its list. The minimum k for which the oriented graph F is k-choosable will be denoted by $ch^{\rightarrow}(F)$. Our main concern will be the following graph parameter.

DEFINITION 2.1. The oriented choice number $och(G)$ of the undirected graph G is the smallest k for which there exists an orientation of the edges of G resulting in a k-choosable oriented graph.

In other words, $och(G)$ is the minimum of $ch^{\rightarrow}(\hat{G})$ for all possible oriented versions $\hat{G}$ of G.

It is obvious that $och(G) \geq ch(G)$. At first glance it is not clear, however, whether $och(G)$ is different from $ch(G)$. Indeed, if we would define the oriented chromatic number of an undirected graph in a similar manner (i.e., as the minimum value of the oriented chromatic number of the different orientations), we would only get a more complicated definition of the chromatic number itself.

First we give a simple example showing that $och(G) \neq ch(G)$ in general. Consider the chordless cycle C_6 on six vertices. Let the vertices be $v_1, ..., v_6$ in their cyclic order (that is $\{v_i, v_{i+1}\}$ is an edge for every i where addition is intended modulo 6.) It is easy to see that $ch(C_6) = 2$. We show $och(C_6) > 2$. First observe that if we do not orient the edges in an alternating way then an oriented path of length 2 appears that already has oriented chromatic number 3 and $ch^{\rightarrow}(G)$ is always at least as large as the oriented chromatic number of the same (oriented) graph. (This can be seen by assigning identical lists to the vertices.) So the only way $ch^{\rightarrow}(F)$ could be less than 3 for an oriented version F of C_6 is if every vertex of F is either a source or a sink. There is only one such orientation up to isomorphism so w.l.o.g. we may assume that the edges of F are $(v_1,v_2),(v_3,v_2),(v_3,v_4),(v_5,v_4),(v_5,v_6),(v_1,v_6)$. Now assign the following two-element lists to these vertices:

$$L(v_1)=L(v_4)=\{1,2\},\ L(v_2)=L(v_5)=\{2,3\},\ L(v_3)=L(v_6)=\{1,3\}$$

where $L(x)$ stands for the list assigned to vertex x.

CLAIM 2.2. One cannot find a proper coloring of the vertices of the above oriented graph F from the above lists.

PROOF. Indeed, if we color v_1 by 2 we are forced to color v_2 by 3, v_3 by 1 and v_4 by 2. Then coloring v_5 by neither of 2 or 3 is allowed. This means we should start coloring v_1 by 1 but then v_6, v_5 and v_4 are forced to be colored by 3, 2 and 1, respectively, and now v_3 cannot be colored properly. □

This example already shows that $och(G)=ch(G)$ does not hold in general. In fact, much more is true.

THEOREM 2.3. *For every positive integer k there is an integer $g(k)$ such that if a graph G contains at least $g(k)$ independent edges then $och(G)>k$.*

Before proving Theorem 2.3 let us formulate the following immediate

COROLLARY 2.4. *For every positive integer k there exists a graph G such that $ch(G)=2$ and $och(G)>k$.*

PROOF. Let G be the graph consisting of $g(k)$ independent edges, where $g(k)$ is the function in Theorem 2.3. This G has $ch(G)=2$ and its oriented choice number is larger than k by Theorem 2.3. □

We remark that if we want our graph to be connected then a path of $2g(k)$ vertices would also do in place of the graph in the proof of Corollary 2.4.

To prove Theorem 2.3 we recall the following result from [8] (cf. Theorem 10. on page 15) attributed to Erdős and Moser, and also Stearns.

THEOREM 2.5 (Erdős, Moser; Stearns). *For every n there exists a function $F(n)$ such that every tournament on at least $F(n)$ vertices contains a transitive subtournament on n vertices. For the smallest possible such $F(n)$ one has $2^{\frac{n-1}{2}}\le F(n)\le 2^n$.*

PROOF OF THEOREM 2.3. Fix some positive integer k. Consider a set of $F(2k)$ colors where $F(n)$ is the function in Theorem 2.5. The idea of the proof is that we will make it impossible that any tournament defined on $F(2k)$ colors as vertices would be consistent with a legal coloring from our lists. Let G be a graph with $m=2^{F(2k)}$ independent edges and fix any orientation of G. Let the independent edges be $e_1,\dots,e_m$. Consider all the possible $2^{F(2k)}$ tournaments on the $F(2k)$ colors, let them be called $T_1,\dots,T_m$. Now we give the lists assigned to those vertices of our graph that are the endpoints of the m independent edges. For giving the lists of the endpoints of e_i look at T_i and find a transitive subtournament of $2k$ points in it. Let the colors corresponding to this subtournament be $c_1,\dots,c_{2k}$ indexed according to their ordering in T_i. This means that $(c_j,c_r)\in E(T_i)$ iff $j<r$.

Denote the two endpoints of e_i by v_{i1} and v_{i2} and assume e_i is oriented towards the latter. Assign lists $L(v_{i1}) = \{c_{k+1}, c_{k+2}, \ldots, c_{2k}\}$ and $L(v_{i2}) = \{c_1, c_2, \ldots, c_k\}$ to v_{i1} and v_{i2}, respectively. Observe that any legal coloring of the endpoints of e_i will make it impossible that the coloring is consistent with (a homomorphism onto) T_i. Since every possible tournament on our colors is represented by some T_i this proves that no valid coloring exists. Thus $och(G) > k$. □

REMARK 2.6. The proof above does not really use that the T_i's contain large transitive subtournaments, only a bipartite subgraph of these is needed. Thus some smaller number replacing $F(2k)$ could be used in the argument.

Using the monotonicity of $och(G)$ (i.e., the fact that adding new edges to a graph $och(G)$ cannot decrease) Theorem 2.3 shows that $och(G)$ is linked to the matching number $\nu(G)$ of G in the sense that a lower bound on $och(G)$ can be formulated in terms of $\nu(G)$. The following observation shows that these two quantities are related, indeed, in the sense that an upper bound on $och(G)$ can also be given in terms of $\nu(G)$. Let $\tau(G)$ denote the minimum number of vertices covering all edges of G. It is obvious that $2\nu(G) \geq \tau(G)$ since the endpoints of edges in a maximum size matching cover every edge. Thus the following observation gives $och(G) \leq 2\nu(G) + 1$.

PROPOSITION 2.7. *For every graph G one has $och(G) \leq \tau(G) + 1$.*

PROOF. Consider G and a minimal set S of points covering all the edges. Orient all edges between a point in S and a point in $V(G) - S$ towards S and orient the rest of the edges arbitrarily. If we assign lists of $\tau(G) + 1$ colors to every node then we can color (in a greedy manner) every element of S first then delete the already used colors from every list in $V(G) - S$ and color the elements of $V(G) - S$ from the remaining lists. Since $|S| = \tau(G)$ (so at least one element remained on every list after coloring S) and $V(G) - S$ must be an independent set (otherwise some edge would not be covered) this is certainly possible thus proving the statement. □

REMARK 2.8. There are cases when $och(G) = \tau(G) + 1$, indeed. This happens for complete graphs and also for complete bipartite graphs with color classes of size $t = \tau(G)$ and t^t where already $ch(G) = \tau(G) + 1$. The latter can be seen by assigning disjoint t-element lists to the vertices of the smaller color class and all their transversals as lists to the vertices in the larger color class.

Theorem 2.3 is formulated in such a way that the existence of $g(k)$ is emphasized. It should be clear, however, that $g(k)$ is a well-defined function, namely, its value is the least integer m for which we have $och(G) = k$ if $G = mK_2$, i.e., the union of m independent edges. The proof of Theorem 2.3 and the statement of Proposition 2.7 give bounds on $g(k)$ but to determine the actual value seems to be an intriguing problem. We say a bit more about it later.

3. Covering tournaments

Let $C(k)$ mean the class of graphs G with $och(G) = k$ and let its superclass $C(k, r)$ be defined as follows. A graph G belongs to $C(k, r)$ if there exists an orientation of G such that for any assignment of k-element lists of colors *out of not more than r colors* a proper oriented coloring from the lists is possible.

Clearly $C(k,r) \subseteq C(k,r+1)$ and $C(k) = \cap_{r=k}^{\infty} C(k,r)$. It is also clear that $C(k,k)$ is identical to the class of k-chromatic graphs.

In the following we take a closer look on $C(2)$. A complete characterization of graphs belonging to this class seems rather tedious, therefore we make only some less ambitious remarks. It is clear that any graph in $C(2)$ should be bipartite, in fact, even its choice number should be 2. A characterization of 2-choosable graphs is given by Erdős, Rubin, and Taylor **[4]** as follows. They denote by $\Theta_{k,l,m}$ the graph consisting of three vertex disjoint paths of length k, l, m, respectively, between the same two distinguished endpoints. Their theorem states that deleting successively all degree one points of a 2-choosable graph the remaining part of each component must be (at most) a $\Theta_{2,2,2m}$, where m is a positive integer.
If a graph G has $och(G) = 2$ than it must be achieved by an orientation containing no two-edge oriented path. This means that for some 2-coloring of such a (necessarily 2-chromatic) graph all edges are oriented from one color class to the other. Thus for a bipartite graph with a given bipartition there is essentially only one orientation to be investigated when we want to decide whether its oriented choice number is 2 or larger. We will call this orientation the *alternating* orientation. (Cf. the example of C_6 in the previous section.)

We believe that $C(2) = C(2,4)$. (Note that it is not clear whether there exists for every k some finite r_k such that $C(k) = C(k, r_k)$.) One way to see the fact that $C(2) \neq C(2,3)$ is to realize that any graph consisting of independent edges only is in $C(2,3)$ but, for many edges, not in $C(2,4)$. We prove this next.

PROPOSITION 3.1. *Let F be a graph consisting of k independent edges. Then $F \in C(2,3)$. On the other hand, there exists k_0 for which $k \geq k_0$ implies $F \notin C(2,4)$.*

PROOF. We prove the first statement first. Assume we have three colors available, call them a, b, c. Consider the cyclic tournament K given by $a \to b \to c \to a$ on these colors. Now take the edges of F one by one. (Notice that the orientation is considered already fixed when the lists are assigned to the vertices.) If an edge has identical lists on its two endpoints then the two colors occurring in these lists can be put on the vertices in such a way that their order given by the orientation of the edge is consistent with K. If, on the other hand, the two lists are different then put a color on one of the endpoints that is not present in the other list. Now since K is cyclic, one of the colors in the other list can be used to color the other node properly. Since all edges of F can be colored this way, the first statement is proven.

Now we prove the second statement. Let T_0 denote the tournament on the available colors consistent with the proper coloring we are to find. Observe that the endpoints of an edge with list $\{a, b\}$ at the tail and list $\{b, c\}$ at the head cannot be properly colored if and only if the induced subtournament of T_0 on colors a, b, c is a transitive triangle with c being its source and a its sink. This means that an edge of the above type ensures that T_0 cannot contain this transitive triangle. Since any tournament on four points contains some transitive triangle, this way we can exclude every tournament from being a candidate for T_0 if we have enough edges to exclude at least one transitive triangle of each of them. This proves the second statement. □

The k_0 in the above theorem need not be very large, in fact its optimal value is at most 12. We will explain this in what follows. First we introduce a new notion inspired by the foregoing.

DEFINITION 3.2. Let T be a tournament on n labeled vertices. We say that a transitive triangle H covers T if it is on three vertices of T and those three vertices induce H in T.

REMARK 3.3. Note that Definition 3.2 requires more than H just being a subtournament. It is important that the vertices of H are labeled and it occurs in T on the triplet of vertices labeled the same.

It seems to be a question of independent interest what is the minimum number of transitive triangles that cover every possible tournament on n points. More precisely we mean the following. Let Q be a set of transitive triangles on the vertices $1, 2, \ldots, n$ such that every possible tournament on $\{1, \ldots, n\}$ contains at least one element of Q. Let $q(n)$ denote the minimum possible size of such a set Q. Our question is the value of $q(n)$. It is clear that $q(n)$ is not meaningful (or equals to infinity) for $n = 3$ since there are tournaments on three points not containing a transitive triangle. But $q(n)$ is well defined for all $n \geq 4$.

Two easy observations are as follows.
1) $q(n) \geq 9$ for all $n \geq 4$.
Reason: The number of n-vertex tournaments is exactly eight times the number of those any transitive triangle can cover. On the other hand, the two sets of tournaments covered by two different transitive triangles is disjoint iff the two triangles are on the same three vertices. Since the arcs in a given triple of vertices can be oriented cyclically one cannot cover all n-point tournaments by using triangles on only one triple of the points.

2) $q(n + 1) \leq q(n)$ for every n.
Reason: If we cover all tournaments on $\{1, \ldots, n\}$ then all tournaments on $\{1, \ldots, n+1\}$ are also covered.

It is somewhat frustrating that we do not know the exact value of $q(n)$ for every n. We believe it is 12 for all $n \geq 4$ but could not prove it for $n > 5$.

PROPOSITION 3.4. $q(4) = q(5) = 12$.

PROOF. By the second observation above it is enough to prove $q(4) \leq 12$ and $q(5) \geq 12$.

$q(4) \leq 12$: Consider all transitive triangles on $V_4 = \{1, 2, 3, 4\}$ in which 1 appears as a source or a sink. The number of such triangles is 12. Since either the out-degree or the in-degree of 1 is at least two in any tournament on V_4 this system of transitive triangles covers all possible tournaments on V_4.

$q(5) \geq 12$: Consider those tournaments on $V_5 = \{1, 2, 3, 4, 5\}$ that are the union of two oriented Hamiltonian cycles. The number of such tournaments equals the number of cyclic permutations of $1, \ldots, 5$, that is 24. One can easily check that any transitive triangle appears in at most (in fact, exactly) two of the above tournaments, thus at least 12 transitive triangles are needed to cover all of them. □

Proposition 3.4 shows that the k_0 of Proposition 3.1 need not be more than 12. (It does not necessarily give the optimum because of the possibility of edges with non-intersecting 2-element lists at their endpoints. More complicated tournament covering questions can be defined if we want to give an equivalent translation of the problem of determining the function $g(k)$ of the previous section. These problems, however seem to be more complicated than relevant therefore we do not discuss them.)

Let us consider the class of $C(2,3)$ once more. We sketch a way to characterize this class. We will call an edge an edge of type (xy, zt) if its tail is assigned the list of colors $\{x, y\}$ while its head is assigned the list of colors $\{z, t\}$. Considering $C(2,3)$ we have 3 colors available, let us call them a, b and c. Let G be a bipartite graph not in $C(2,3)$. Consider (one of) its alternating orientation(s) and an assignment of two element lists of the available colors to its vertices such that no proper coloring exists from these lists. (Such an assignment should exist if $G \notin C(2,3)$.) In the following we refer to our oriented graph with this assignment of color lists simply as G (or sometimes the assigned G to avoid confusion).

CLAIM 3.5. G must have an edge of each of the following six types: (ab, ac), (ac, ab), (ab, bc), (bc, ab), (ac, bc), (bc, ac).

PROOF. By our assumptions there is no tournament on $U = \{a, b, c\}$ to which G could be homomorphically mapped (in a way consistent also with the list assignment). Consider a transitive triangle on U. Let x be its source, z its sink, and y its middle point, where x, y, z is some permutation of the elements of U. One can easily check that the only obstacle of a homomorphism to this tournament can be an edge of type (yz, xy). This applies to all the six possible transitive triangles on U thus the six edges above must be present. □

Observe that once we have a bipartite graph with alternating orientation, then each of its points is either a source or a sink. Thus the Claim above implies that the assigned G must have at least six points: a source and a sink for each of the three possible lists $\{a, b\}$, $\{a, c\}$, $\{b, c\}$. In fact, we have the six edges given in the Claim and each of these six type points appear as an endpoint of two of them. These pairs of points of the same type can be identified or not, resulting in a graph of six edges on k points with $6 \leq k \leq 12$. This graph can now be investigated whether or not it is still possible to find a legal homomorphism of it onto a cyclically oriented triangle on the three colors. If, for example, we make all the possible identifications we arrive to C_6, our first example with $och(G) > ch(G)$. If fewer identifications are made then we may need additional edges to exclude homomorphisms to the cyclic triangular tournaments of 3 colors. A three edge path with edges of type (xy, yz), (xz, yz), (xz, xy) excludes a cyclic triangle and exchanging directions the other cyclic triangle can be excluded. Thus the two paths of length three, into which the C_6 above can be broken, also exclude all three vertex tournaments on the colors. If a cyclic triangle is not excluded by such a three edge path it can also be excluded by two connecting edges of type (xy, xz) and (xy, xy). Thus here we need some additional edge (to the previous six ones) which will have the same list on both of its endpoints. To exclude both cyclic triangles we either need two such two or three edge configurations or two can be glued together at a degree three point. (Thus we can exclude both cyclic triangles on a three edge star, for example.) This means that we have six edges to exclude the six transitive triangles and at most two additional edges (in appropriate locations) are needed to exclude

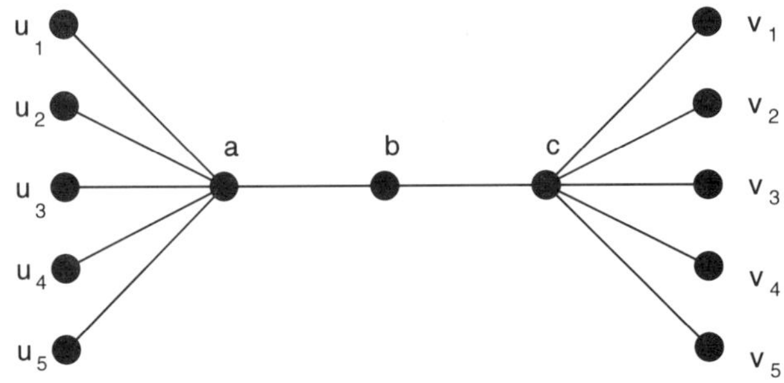

Figure 1.

the two cyclic triangles. Thus deciding whether a graph is in $C(2,3)$ needs only to decide whether certain configurations of at most eight edges appear in the graph or not. This is easy to check in polynomial time, say.

Finally in this section we give an example of a connected graph that belongs to $C(2,3)$ but not to $C(2,4)$.

Example Let T be the tree defined by $V(T) = \{a, b, c, u_1, \dots u_{m_1}, v_1, \dots v_{m_2}\}$ with $m_i \geq 3$ and $E(T) = \{\{a,b\},\{b,c\},\{a,u_i\},\{c,v_j\}\}$ where $i = 1, \dots, m_1$, $j = 1, \dots, m_2$. (In short: T is the union of two stars having a non-central vertex in common, see Figure 1.)
Consider the alternating orientation of T determined by (a,b) being an edge. We assign the following 2-element lists of colors x, y, z, t to the vertices. $L(a) = \{x,y\}$, $L(b) = \{x,z\}$, $L(c) = \{z,t\}$,$L(u_1) = \{x,t\}$, $L(u_2) = \{y,z\}$, $L(u_3) = \{y,t\}$,$L(v_1) = \{x,t\}$, $L(v_2) = \{y,z\}$, $L(v_3) = \{y,t\}$. Simple case checking shows that no proper oriented coloring from these lists is possible. On the other hand, if we have only three colors, than T can always be colored properly since only a and c are sources of the considered oriented version of T and we have seen that a graph not in $C(2,3)$ must have at least three sources.
We remark that there are infinitely many graphs in $C(2)$ having arbitrarily many sources and sinks in their alternating orientations. Such an infinite family of graphs can be constructed by putting arbitrarily large stars on two vertices of a $K_{2,3}$ that are not in the same color class. The reason of these graphs belonging to $C(2)$ is this. The given graphs have choice number 2 and one can easily check that any coloring that is proper as an ordinary coloring will also be legal as an oriented coloring for these graphs.

4. Dichoice number

In the previous sections we investigated the choice number analogon of the oriented chromatic number defined in [**11**]. In this section we do similarly with a different chromatic number type invariant of a directed graph defined by Neumann-Lara [**9**] and (its undirected version by) Erdős and Neumann-Lara, cf. [**2**] and also [**3**].
Let D be a directed graph. Its dichromatic number $d(D)$ is the smallest integer k for which the vertex set of D can be partitioned into k classes such that none of the classes contains an oriented cycle. For an undirected graph G its dichromatic number $d(G)$ is defined as the maximum of the dichromatic numbers of all oriented versions of G. Here we allow the orientation of each edge only in one direction;

observe however that the dichromatic number of a symmetrically directed graph (where each edge is present in both directions) is equal to the chromatic number of the underlying undirected graph. It follows that $d(G) \leq \chi(G)$ and it is remarkable that it is not known whether the gap between the two invariants can be arbitrarily large. More precisely, even that is not known whether there exists some k_0 such that $\chi(G) > k_0$ would imply $d(G) > 2$, cf. [**2**].

Here we introduce the choice number analogon $dch(G)$ of the undirected graph G and call it the dichoice number of G.

DEFINITION 4.1. Let G be an undirected graph. Its dichoice number $dch(G)$ is the smallest integer k satisfying the following conditions. For any assignment of k-element lists of colors for the vertices and any orientation of the edges of G there exists a coloring of the vertices from their lists such that no color class contains an oriented cycle.

It is clear that $dch(G)$ is upper bounded by $ch(G)$. For any tree T we have $dch(G) = 1$ so the dichoice number can even be smaller than the chromatic number. The main result in this section is that we show the existence of bipartite graphs with arbitrarily large dichoice number. Our construction is an extension of that of Erdős, Rubin, and Taylor [**4**] by which they showed the existence of bipartite graphs with arbitrarily large choice number.

THEOREM 4.2. *For any positive integer k there exists a graph G with $\chi(G) = 2$ and $dch(G) > k$.*

PROOF. Let $m = 2k\binom{2k-1}{k}$ and G be the complete bipartite graph $K_{m,m}$. We show that this G satisfies the requirements in the statement, that is its dichoice number is larger than k. To this end we give an orientation of its edges that has the property that we can assign k-element lists to the vertices for which no proper "dicoloring" exists from the lists. Let the color classes of G be called A and B. To define the orientation partition both A and B into $\binom{2k-1}{k}$parts of size $2k$ each. Within each such class label the vertices with $1, \ldots, 2k$. Now orient those edges of G from A to B that have different labels on their two endpoints and orient the rest from B to A.
Observe that vertices having the same label induce complete bipartite graphs isomorphic to $K_{\binom{2k-1}{k},\binom{2k-1}{k}}$. For these subgraphs of G we use the idea of Erdős, Rubin, and Taylor. Consider all possible k-element lists formed from a given set of $2k-1$ colors and assign each of these lists to exactly two vertices of each of the above mentioned subgraphs (i.e., those induced by vertices of identical labels), one in A and one in B, respectively.
Consider any coloring from the above lists. We show that there exists some color class containing an oriented cycle. By the argument in [**4**] for every i the graph induced by vertices labeled i must contain an edge whose two endpoints are colored by the same color. (This is because each $k-1$ element subset of the $2k-1$ colors are missing at some of these vertices thus at least k colors are used altogether both in A and B. But since there are only $2k-1$ colors they cannot all be different, thus some point in A got the same color as another one in B and they are adjacent.) Consider the edges with the same color on their endpoints for every i. This means $2k$ edges. Since we have only $2k-1$ colors there are two of these edges that have the same color on their altogether four endpoints. These two edges are oriented

from B to A since they both connect identically labeled vertices. These labels are different for the two edges, so the two other edges present among the four endpoints are oriented from A to B. This means that these vertices induce an oriented 4-cycle and the proof is complete. □

References

[1] N. Alon, Restricted colorings of graphs, *in:* "Surveys in Combinatorics, 1993, Proceedings, 14th British Combinatorial Conference"(K. Walker, ed.), 1–33, London Mathematical Society Lecture Notes, Vol. 187,Cambridge Univ. Press, Cambridge, UK, 1993.

[2] P. Erdős, Problems and results in number theory and graph theory, *Congress. Numer.* **27** (1979), 3–21.

[3] P. Erdős, J. Gimbel, and D. Kratsch, Some extremal results in cochromatic and dichromatic theory, *J. Graph Theory* **15** (1991) 579–585.

[4] P. Erdős, A. L. Rubin, and H. Taylor, Choosability in graphs, *Congr. Numer.* **26** (1980), 122–157.

[5] E. Fachini and J. Körner, Colour number, capacity and perfectness of directed graphs, to appear in *Graphs & Combinatorics.*

[6] T. R. Jensen and B. Toft, Graph Coloring Problems, Wiley, New York, 1995.

[7] A. V. Kostochka, E. Sopena, and X. Zhu, Acyclic and oriented chromatic numbers of graphs, *J. Graph Theory* **24** (1997) 331–340.

[8] J. W. Moon, *Topics on tournaments*, Holt, Rinehart and Winston, New York, 1968.

[9] V. Neumann-Lara, The dichromatic number of a digraph, *J. Combin. Theory Ser. B.* **33** (1982), no. 3., 265–270.

[10] A. Raspaud and E. Sopena, Good and semi-strong colorings of oriented planar graphs, *Information Processing Letters* **51** (1994) 171–174.

[11] E. Sopena, The chromatic number of oriented graphs, *J. Graph Theory* **25** (1997) 191–205.

Mathematical Institute of the HAS, Budapest, P.O.Box 127, H-1364 HUNGARY
E-mail address: sali@math-inst.hu

Mathematical Institute of the HAS, Budapest, P.O.Box 127, H-1364 HUNGARY
E-mail address: simonyi@math-inst.hu

DIMACS Series in Discrete Mathematics
and Theoretical Computer Science
Volume **49**, 1999

On the Limit Values of Probabilities for the First Order Properties of Graphs

Joel Spencer and Lubos Thoma

ABSTRACT. Consider the random graph $\mathcal{G}(n,p)$, where $p = p(n)$ is any threshold function satisfying $p(n) = \Theta(\ln n/n)$. We give a full characterization of the limit values of probabilities of $\mathcal{G}(n,p)$ having a property ψ, where ψ is any sentence of the first order theory of graphs.

CONTENTS

1. Introduction

Let $n \in \mathbb{N}$ and $0 \leq p = p(n) \leq 1$. The probability space $\mathcal{G}(n,p)$ is the usual probability space over the set of graphs on the vertex set $[n] = \{1, \ldots, n\}$ with the product measure $Prob(G) = p^{e_G}(1-p)^{\binom{n}{2}-e_G}$. The number of edges in a graph G is denoted, as usual, by e_G.

Let us define the first order language L of graphs. The language L consists of an alphabet containing a countable number of variable symbols $(x, y, z, \ldots)$ interpreted

1991 *Mathematics Subject Classification.* 05C80, 03C13.

Key words and phrases. Finite models, first order logic, random graph.

The second author gratefully acknowledges support as a DIMACS Postdoctoral Fellow. DIMACS is a cooperative project of Rutgers University, Princeton University, AT&T Research, Bellcore, and Bell Laboratories. DIMACS is an NSF Science and Technology Center, funded under contract STC-91-19999; and also receives support from the New Jersey Commission on Science and Technology.

The second author also gratefully acknowledges support as a member at the Institute for Advanced Study, Princeton NJ.

as vertices; logical symbols as Boolean connectives $\neg$, $\wedge$, $\vee$, quantifiers $\forall$, $\exists$, and brackets; the predicate of equality '=', and the adjacency '$\sim$' predicate, where $x \sim y$ is interpreted as $\{x, y\}$ being an edge. For $\psi \in L$, we write $Prob(\mathcal{G}(n,p) \models \psi)$ for the probability that a graph G in $\mathcal{G}(n,p)$ has the property ψ.

Our paper deals with the behaviour of $\mathcal{G}(n,p)$ for $p = p(n)$ being a threshold satisfying $p = \Theta(\ln n/n)$. Set

$$p_{k,l} = p_{k,l}(n;c) := \frac{\ln n}{kn} + \frac{l \ln\ln n}{kn} + \frac{c}{kn}, \tag{1}$$

where $l \geq k-1 \geq 0$ are integers and $c \in \mathbb{R}$. In their 1991 paper, T. Łuczak and the first author proved that the zero - one law for $\mathcal{G}(n,p), p = \Theta(\ln n/n)$, and the language L (cf. Theorem 5 in [ŁS 91] for the precise statement) holds except for p being "close" to $p_{k,l}$, in other words the probabilities $p_{k,l}$ are the only thresholds in the range $\Theta(\ln n/n)$.

As the general case is somewhat complicated we here describe our results when $k = 1$ and l is an arbitrary nonnegative integer. For a nonnegative integer s, let $\sigma_{l,s}$ be the following sentence of L : "there are *precisely* s vertices of degree *precisely* l". The number of vertices of degree l in $\mathcal{G}(n, p_{k,l})$ possesses asymptotically Poisson distribution with mean $e^{-c}/l!$. Set $q_{l,s} := e^{-e^{-c}/l!}(e^{-c}/l!)^s/s!$ which is the limiting probability of $\mathcal{G}(n, p_{k,l})$ satisfying $\sigma_{l,s}$.

Theorem 1. *Let $k = 1$, l be an arbitrary but fixed nonnegative integer, and $p = p_{k,l}(n;c)$. Let ψ be any sentence from the first order language of graphs L. Then $\mathcal{G}(n,p)$ satisfying ψ has a limiting probability (as $n \to \infty$) and the limit equals either a finite sum of the values $q_{l,s}$ defined above or one minus such a finite sum. [This includes zero and one with the null sum.] Furthermore, all such values are limiting probabilities of some first order sentence ψ. Further, the indices s of the finite sum are at most the quantifier rank of ψ.*

The general characterization of limit probabilities is given in Section 6. While we give our results for $p(n) = p_{k,l}(n;c)$ we note that an additive $o(n^{-1})$ term in (1) will not affect any of the limit probabilities. This can be shown by checking that it is so for each of the axioms of the almost sure theories as well as their extensions we develop. We omit the calculations.

We are inspired by the results of Lynch [L 92] on the limit probabilities for first order sentences over $\mathcal{G}(n,p)$ where $p = c/n$ or $p = cn^{-(u+1)/u}, u \in \mathbb{N}$. Lynch showed that all such limit probabilities could be expressed as functions of c that were generated from addition, subtraction, multiplication, division, and base e exponentiation. In one important sense our range $p = \Theta(\frac{\ln n}{n})$ is considerably simpler than the case $p = c/n$. Our limit probabilities can and will include hyperexponential terms, e.g. $e^{-e^{-c}}$, but they are never a tower of exponentials of any higher length. In this manner our range is similar to the other range studied by Lynch: $p = cn^{-(u+1)/u}, u \in \mathbb{N}$.

The paper is organized as follows. The next section contains a necessary background on Ehrenfeucht games and complete theories. Section 3 deals with a general procedure for computing the limit values of probabilities.

Sections 4 and 5 contain different cases of the proof of Theorem 1. In Section 4 we treat the classical $[k = 1, l = 0]$ Erdős - Rényi case. We give there the main ideas used later in the complete proof. The case $k = 1$ and $l \geq 0$ is dealt with in Section 5. We give all the details of the probabilistic as well as logical parts of the proof. The statement and proof of a generalization of Theorem 1 for $p_{k,l}(n;c)$, which is somewhat technical, is given in Section 6.

Our asymptotics are always for k, l fixed and n approaching infinity. Notation: For two functions $f(n)$ and $g(n)$, $f(n) = \Theta(g(n))$ means that there are two constants c_1 and c_2 such that $c_1 g(n) \leq f(n) \leq c_2 g(n)$ for n large enough.

2. Ehrenfeucht Games, Complete Theories, and Countable Models

In this section we introduce the tool used later in the proof - Ehrenfeucht games. Throughout this section, let L' denote a first order language with a finite number of predicates.

We start with the (standard) definition of *quantifier rank*, $qr(\psi)$, of a formula $\psi \in L'$. For $\psi \in L'$, the quantifier rank of ψ is defined recursively as follows

$$\begin{aligned} qr(\psi) &= 0, \quad \text{for } \psi \text{ atomic} \\ qr(\neg\psi) &= qr(\psi) \\ qr(\psi \vee \varphi) &= \max\{qr(\psi), qr(\varphi)\} \\ qr((\exists x)\psi) &= qr(\psi) + 1 \, . \end{aligned}$$

For an L'-model M we denote by $Th_t(L'; M)$ the set of all sentences of L' of quantifier rank at most t which hold for M, i.e.

$$Th_t(L'; M) = \{\psi \in L' \; ; \; M \models \psi \wedge qr(\psi) \leq t\} \, .$$

Ehrenfeucht games were introduced in [Eh 61]. The Ehrenfeucht game, $Ehr_t(M_1, M_2)$, is a t-round perfect information game between two players, played on two L'-models M_1 and M_2. In each round the first player (Spoiler) picks one of the two models M_1, M_2 and chooses one of its elements. The second player (Duplicator) must answer by selecting an element from the other model. For $i, 1 \leq i \leq t$, let a_i^1, a_i^2 denote the elements chosen by the players in the ith round of the game from M_1 and M_2, respectively. In the case the structures induced by $a_1^1, \ldots, a_t^1$ in M_1 and by $a_1^2, \ldots, a_t^2$ in M_2 are naturally isomorphic, the second player wins. We are going to use the following fact about Ehrenfeucht games (cf. [EF 95]).

Fact 2. *Let L' be a first order language with a finite number of predicates. Let M_1, M_2 be two L'-models and let t be a positive integer. The second player has a winning strategy for the game $Ehr_t(M_1, M_2)$ if and only if $Th_t(L'; M_1) = Th_t(L'; M_2)$.*

Two L'-models M_1, M_2 are called *elementarily equivalent* if and only if for every positive integer t . $Th_t(L'; M_1) = Th_t(L'; M_2)$. Let Σ be a theory. If Σ is not complete it has two extensions Σ_1, Σ_2 that have different truth value on some sentence ψ and by the Gödel Completeness Theorem there would be countable or finite models M_1, M_2 of those theories. Thus we obtain (cf. [BM 77]):

Fact 3. *Let Σ be a consistent first order theory which has no finite models. Then the following are equivalent.*

i. *Σ is complete.*
ii. *Any pair of countable models of Σ are elementarily equivalent.*
iii. *For any two countable models M_1, M_2 of Σ and any positive integer t, the second player has a winning strategy for the game $Ehr_t(M_1, M_2)$.*

Later we will be using the third condition, i.e. Ehrenfeucht games, to show a given theory is complete. We will need the following two facts about Ehrenfeucht games played on graphs. All graphs involved are considered as models of the first order language of graphs. The first fact describes a local sufficient condition for Duplicator to win the Ehrenfeucht game played on graphs.

Let r be a nonnegative integer and v a vertex of a graph. We define the rth (distance) neighborhood of v, denoted by $N(v;r)$, as the set of vertices of distance at most r from the vertex v.

Fact 4. *Let t be a positive integer. Let G_1 and G_2 be graphs satisfying the following property: for every $i = 1, \dots, t$, and for every choice of $w \in V(G_1)$ and $v_1, \dots, v_{i-1} \in V(G_2)$ there exists a vertex $v_i \in V(G_2)$ such that*

1. *the distance between v_i and v_α is at least 3^{t-i+1}, for every $\alpha = 1, \dots, i-1$, and*
2. *the 3^{t-i} neighborhood of w is isomorphic to the 3^{t-i} neighborhood of v_i, with an isomorphism mapping w into v_i.*

Further, assume the above condition holds with roles of G_1 and G_2 reversed. Then Duplicator has a winning strategy for $Ehr_t(G_1, G_2)$.

Proof. We will denote by $a_1^1, \dots, a_t^1$ and $a_1^2, \dots, a_t^2$ elements chosen in G_1 and G_2, respectively, by the players during the game $Ehr_t(G_1, G_2)$.

We are going to show by induction that Duplicator has a winning strategy during which he can maintain that for every $i = 1, \dots, t$, the neighborhoods $N(a_i^1; 3^{t-i})$ and $N(a_i^2; 3^{t-i})$ are isomorphic with the isomorphism mapping all a_j^1 to a_j^2 where $a_j^1 \in N(a_i^1; 3^{t-i}), j \leq i$. This implies that for every $i = 1, \dots, t$, the structures induced on the sets $\{a_1^1, \dots, a_i^1\}$ and $\{a_1^2, \dots, a_i^2\}$ are isomorphic.

For $i = 1$, if Spoiler chose a vertex a_1^1 in G_1, say, using the assumptions of the fact Duplicator can find a vertex a_1^2 in G_2 such that the neighborhoods $N(a_1^1; 3^{t-1})$ and $N(a_1^2; 3^{t-1})$ are isomorphic with the isomorphism mapping a_1^1 to a_1^2.

Assume the induction hypothesis to hold for all $j = 1, \dots, i-1$. Without loss of generality, suppose Spoiler chose its next move to be a vertex a_i^1 in G_1. In the case $a_i^1 \in \cup_{j=1}^{i-1} N(a_j^1; 2 \cdot 3^{t-i})$, Duplicator chooses a_i^2 according to the isomorphism between the neighborhoods $N(a_l^1; 3^{t-l})$ and $N(a_l^2; 3^{t-l})$ where the index l is such that $a_i^1 \in N(a_l^1; 2 \cdot 3^{t-i})$. As $N(a_i^1; 3^{t-i}) \subseteq N(a_l^1; 3^{t-l})$, we get that the neighborhoods $N(a_i^1; 3^{t-i})$ and $N(a_i^2; 3^{t-i})$ are isomorphic. Moreover, the isomorphisms of neighborhoods map a_j^1 to $a_j^2, 1 \leq j \leq i$. If $a_i^1 \notin \cup_{j=1}^{i-1} N(a_j^1; 2 \cdot 3^{t-i})$, Duplicator can choose a_i^2 according to the assumptions of the fact by setting $w = a_i^1$ and $v_1 = a_1^2, \dots, v_{i-1} = a_{i-1}^2$. Thus, the neighborhoods $N(a_i^1; 3^{t-i})$ and $N(a_i^2; 3^{t-i})$ are again isomorphic with the isomorphism mapping a_i^1 to a_i^2. Further, note that $N(a_i^1; 3^{t-i}) \bigcap (\cup_{j=1}^{i-1} N(a_j^1; 3^{t-i})) = \emptyset$ and $N(a_i^2; 3^{t-i}) \bigcap (\cup_{j=1}^{i-1} N(a_j^2; 3^{t-i})) = \emptyset$. This finishes the induction step. ▮4

Fact 5. (cf. Proposition 1.3.9 in [EF 95]) *Let t be a positive integer and let G_1, G_2, and H be graphs. If Duplicator has a winning strategy for the game $Ehr_t(G_1, G_2)$, then he has also a winning strategy for $Ehr_t(G_1 + H, G_2 + H)$ where '+' is the disjoint union of graphs.*

Proof. We need to show Duplicator has a winning stategy for the game $Ehr_t(G_1 + H, G_2 + H)$. If Spoiler chooses a vertex in the graph G_1 or G_2 then Duplicator uses his winning strategy from $Ehr_t(G_1, G_2)$. If Spoiler chooses a vertex from either copy of H, Duplicator selects the identical vertex in the other copy. ▮5

3. Completions of Almost Sure Theories and Limit Probabilities

In this section we describe a general procedure for characterizing the limit values of probabilities. The main theorem of this section is Theorem 8. We start with assumptions for this section.

Let L' be a (first or second order) language. For each $n \in \mathbb{N}$ we are given a probability space (Ω_n, μ_n), where Ω_n is a finite set of models of L' and μ_n is a probabilistic measure on Ω_n. [In the following sections we set (Ω_n, μ_n) to be $\mathcal{G}(n,p)$.] We assume all subsets of Ω_n are μ_n-measurable. For $\psi \in L'$ we write $\mu_n(\psi)$ for $\mu_n(\{M \in \Omega_n : M \models \psi\})$, the probability ψ holds in a "random" model.

A theory T is called the *almost sure theory* for the sequence (Ω_n, μ_n) if it consists of those sentences $\psi \in L'$ for which $\lim_{n\to\infty} \mu_n(\psi) = 1$.

Remark 6. Observe that any almost sure theory T is consistent. This holds due to the finiteness of proof, i.e. "compactness" . If ψ is a theorem of T, i.e. $T \vdash \psi$, then it follows from a finite number of axioms and therefore $\lim_{n\to\infty} \mu_n(\psi) = 1$. Hence, all theorems of T hold almost surely and therefore the sentence "false" is not in T.

Let $\sigma \in L'$ be a sentence with positive limit probability, i.e. $\lim_{n\to\infty} \mu_n(\sigma) > 0$, and let T' be the extension of T by the new sentence σ. We claim that the theory T' is consistent. Similarly as above, if ψ is a theorem of T', we get $\liminf_{n\to\infty} \mu_n(\psi) \geq \lim_{n\to\infty} \mu_n(\sigma) > 0$. Therefore, the sentence "false" cannot be in T'.

Definition 7. *Let T be a subtheory of the almost sure theory. Let I be a finite or countable index set. A family σ_s, $s \in I$, of sentences of L is called a* complete family of completions *if*

ι. *For all $s \in I$ the extension of T by adding the new sentence σ_s is a complete theory. [Henceforth we write $T(s)$ for this extended theory.]*

ιι. *For all $s \in I$*

$$a_s := \lim_{n\to\infty} \mu_n(\sigma_s)$$

exists and $0 < a_s < 1$.

ιιι. *The σ_s are mutually exclusive in the sense that $T \vdash (\sigma_s \to \neg\sigma_{s'})$ for all distinct $s, s' \in I$.*

ιιιι. $\sum_{s\in I} a_s = 1$.

For $\psi \in L'$, let $\mathfrak{M}[\psi] = \{s \in I; T(s) \vdash \psi\}$.

Theorem 8. *Given a complete family of completions $\sigma_s, s \in I$,*

$$\lim_{n\to\infty} \mu_n(\psi) = \sum_{s\in\mathfrak{M}[\psi]} a_s \tag{2}$$

for all $\psi \in L'$. In particular, the limiting probability always exists.

Remark 9. In general a complete family of completions implies convergence, i.e. that the limiting probabilities always exist. When combined with a characterization of the sets $\mathfrak{M}[\psi]$ a description of the limiting probabilities is achieved.

The assumptions of Definition 7 imply that the theory $\cap_{s\in I} T(s)$ is the almost sure theory for the sequence (Ω_n, μ_n).

Proof of Theorem 8. Let $\epsilon > 0$ be arbitrarily small. Pick $I_0 \subseteq I$ finite so that

$$\sum_{s \in I \setminus I_0} a_s < \epsilon$$

and set

$$MAIN := \vee_{s \in I_0} \sigma_s \ .$$

By $\iota\iota\iota$, $\mu_n(MAIN) = \sum_{s \in I_0} \mu_n(\sigma_s)$ so, as I_0 is finite, $\lim_{n\to\infty} \mu_n(MAIN) = \sum_{s \in I_0} a_s$. Hence by $\iota\iota\iota\iota$ we have $\lim_{n\to\infty} \mu_n(\neg MAIN) = \sum_{s \in I \setminus I_0} a_s$ and therefore for n sufficiently large $\mu_n(\neg MAIN) < \epsilon$. Set

$$\psi^* := \vee_{s \in I_0 \cap \mathfrak{M}[\psi]} \sigma_s$$

Then ψ, ψ^* are close in the sense that

$$(\psi \equiv \neg\psi^*) \Rightarrow \neg MAIN \bigvee \vee_{s \in I_0 \cap \mathfrak{M}[\psi]} (\sigma_s \wedge \neg\psi) \bigvee \vee_{s \in I_0 \setminus \mathfrak{M}[\psi]} (\sigma_s \wedge \psi) \ .$$

For $s \in I_0 \cap \mathfrak{M}[\psi]$, $\psi \in T(s)$ so $\mu_n(\sigma_s \wedge \neg\psi) \to 0$. For $s \in I_0, s \notin \mathfrak{M}[\psi]$, $\psi \notin T(s)$ so by completeness $\neg\psi \in T(s)$ and hence $\mu_n(\sigma_s \wedge \psi) \to 0$. As I_0 is finite for n sufficiently large

$$\mu_n(\psi \equiv \neg\psi^*) < \epsilon + \epsilon + \epsilon = 3\epsilon$$

so that

$$|\mu_n(\psi) - \mu_n(\psi^*)| < 3\epsilon \ .$$

But for n sufficiently large we also have

$$|\mu_n(\psi^*) - \sum_{s \in I_0 \cap \mathfrak{M}[\psi]} a_s| < \epsilon$$

so that

$$|\mu_n(\psi^*) - \sum_{s \in \mathfrak{M}[\psi]} a_s| < 2\epsilon$$

and thus

$$|\mu_n(\psi) - \sum_{s \in \mathfrak{M}[\psi]} a_s| < 5\epsilon$$

which, as ϵ was arbitrary, completes the proof. $\blacksquare_8$

4. The Erdős-Rényi Case

The classic work of Paul Erdős and Alfred Rényi [ER 60] contains many beautiful results. Surely one of the most striking is their description of the threshold for that most fundamental of graph concepts - connectivity. They showed that if

$$p = \frac{\ln n}{n} + \frac{c}{n} + o(\frac{1}{n}) \tag{3}$$

then

$$Prob(\mathcal{G}(n,p) \text{ connected}) = e^{-e^{-c}} + o(1) \ . \tag{4}$$

The precision of the result and the novelty of the double exponential have well stood the test of time.

In one sense this result is not directly pertinent to our current investigations as connectivity is not a first order property. But in their proof Erdős and Rényi actually examined the property of containing no isolated vertices, which we shall denote by σ_0 and can be written $(\forall x)(\exists y)(x \sim y)$ in the First Order world. They showed that σ_0 has the limit probability (4) and that connectivity and σ_0 have the same limit probability in this range of p.

We show that the double exponential is essentially the only kind of limit probability that can appear in the First Order world. Our results are given more formally and more generally in Sections 5 and 6, here we sketch the situation in this historically interesting case.

The first order sentences that hold almost surely for these p can be split into positive and negative statements. On the positive side:

- For every $m \geq 3$ and every $r \geq 1$: There are (at least) r cycles of length (precisely) m.
- For every $r \geq 1$ and every $d \geq 1$: There are (at least) r vertices of degree (precisely) d.

We think of small cycles and small degree vertices as special phenomenon. The above two schema say that they do occur, the final three schema say that they hold in isolation.

- For every $t, d \geq 1$: There are no vertices x_1, x_2 with degrees both at most d which are less than t apart.
- For every $t, d \geq 1$ and every $m \geq 3$: There is no vertex x of degree at most d which lies less than distance t from a cycle of length at most m.
- For every $t \geq 1$ and $m \geq 3$: There are no two cycles of length at most m which lie less than distance t apart.

Roughly speaking, the special phenomenon hold only polylogarithmically often and they are spread out in the graph so that locally one never sees two such phenomenon. Let T denote the theory given by these schema, which turns out to be the almost sure theory for p in this range. We prove this in the next section.

The countable models M of T are particularly nice. Because no two special phenomenon can be any finite distance apart any component of T can have only one of them. If a component C contains a vertex of finite degree d then it *must* be the infinite tree TR_d with all other vertices having infinite degree. Note TR_d is uniquely determined and note that M must have countably many copies of TR_d. Similarly if C contains a cycle of size m then it *must* be the unicyclic C_m generated from the cycle by generating a disjoint infinite rooted tree, all vertices of infinite degree, from each of the vertices of the cycle. Again C_m is uniquely determined and M must have countably many copies of C_m. Let TR_ω denote the infinite tree with all vertices of infinite degree. The model M may have countably many copies of TR_ω, it may have any finite number of copies of TR_ω, in particular it may have no copies of TR_ω.

What about isolated vertices? Their number cannot be specified in the almost sure theory T since this range of p is precisely the threshold function for having an isolated vertex. We let σ_s denote the sentence: " There exist (precisely) s isolated vertices", defined for all nonnegative integers s including (especially) zero.

We claim that for any s the theory $T(s) = T + \sigma_s$, i.e. T extended by the sentence σ_s, is complete. The countable models M_s of $T + \sigma_s$ are those models M of T that have precisely s isolated vertices. So $T(s)$ is not $\aleph_0$-categorical, as the models may or may not have copies of TR_ω. But it suffices to show that any two models M_s, M'_s of $T(s)$ are elementarily equivalent (i.e., have the same first order properties) for which it suffices to show that for any fixed t Duplicator can win the t-move Ehrenfeucht game on M_s, M'_s. To show that we are going to use Facts 4 and

5. First, we observe that due to Fact 5 it is enough to show that Duplicator can win the game $Ehr_t(\widehat{M}_s, \widehat{M}_s')$, where $\widehat{M}_s, \widehat{M}_s'$ denote the models M_s, M_s', respectively, after deleting the s isolated vertices. Let $i \in [t]$ and suppose vertices $w \in \widehat{M}_s$ and $v_1, \ldots, v_{i-1} \in \widehat{M}_s'$ are given. We show that there is a vertex $v_i \in \widehat{M}_s'$ satisfying the assumptions of Fact 4. Let U be the component of $\widehat{M}_s$ containing w. If U is not TR_ω, then there is an isomorphic copy of U in $\widehat{M}_s'$ such that this copy does not contain any of the vertices $v_1, \ldots, v_{i-1}$. This holds due to the fact that there are countably many of these components in M_s' and, thus, in $\widehat{M}_s'$. Hence, there is a vertex $v_i \in \widehat{M}_s'$ satisfying the assumptions of Fact 4. In the case U is a copy of TR_ω, we choose a copy U' of TR_1 in $\widehat{M}_s'$ such that U' does not contain any of the vertices $v_1, \ldots, v_{i-1}$. This is again possible due to the fact that M_s', and thus $\widehat{M}_s'$, contains a countable number of components isomorphic to TR_1. To satisfy the assumptions of Fact 4 we choose for v_i any vertex of U' which is at distance more than 3^t from the vertex of degree 1 in U'. The argument given above works also with the roles of $\widehat{M}_s$ and $\widehat{M}_s'$ reversed. Combining Facts 4 and 5 we get that Duplicator has a winning strategy for the game $Ehr_t(M_s, M_s')$.

Let $\overline{M}_s$ denote that model of $T + \sigma_s$ that has no copies of TR_ω, in some sense a minimal model. Then a sentence ϕ is a theorem of $T+\sigma_s$ if and only if ϕ holds in $\overline{M}_s$. Consider the t move Ehrenfeucht game on $\overline{M}_s, \overline{M}_{s'}$ with $s, s' \geq t$. Duplicator has a clear win since the only "nonisomorphism" that Spoiler can hope to exploit is in the number of isolated vertices but he has only t moves and every time he selects an isolated vertex Duplicator can likewise select an isolated vertex in the other model. Any given ϕ has some finite quantifier rank t and therefore ϕ holds in either all or none of the theories $T + \sigma_s$, $s \geq t$. Letting $\mathfrak{M}[\phi]$ denote the set of s for which $T + \sigma_s \vdash \phi$ we can rephrase this by saying $\mathfrak{M}[\phi]$ is either finite or cofinite.

What does all this say about the possible limit probabilities for a first order sentence ϕ? Let X be the number of isolated vertices in $\mathcal{G}(n,p)$. Then $\mathrm{Exp}[X] = n(1-p)^{n-1} \sim e^{-c}$ for p given by (3). Further it can be shown that X approaches the Poisson in distribution so that

$$Prob(\mathcal{G}(n,p) \models \sigma_s) = Prob(X = s) \to q_s := e^{-e^{-c}} \frac{(e^{-c})^s}{s!} \tag{5}$$

as $n \to \infty$. Note that $\sum_{s \geq 0} q_s = 1$, meaning roughly that the number of isolated vertices is almost surely bounded. This, and the careful epsilondeltamanship of Section 3, implies that any first order ϕ has limiting probability $\sum_{s \in \mathfrak{M}[\phi]} q_s$. Thus the possible limiting probabilities are finite sums of the q_s or one minus such a sum. Further, the σ_s are themselves first order sentences and for any finite subset S of nonnegative integers the sentence $\phi = \vee_{s \in S} \sigma_s$ does have limiting probability $\sum_{s \in S} q_s$ while $\neg\phi$ has limiting probability one minus that – therefore we have a *complete characterization* of the possible limiting probabilities of first order sentences ϕ.

5. The Case of $k = 1$ and $l \geq 0$.

In this section we prove Theorem 1. We introduce the almost sure theories $T_{1,l}$ and their completions $T_{1,l}(s)$. At the end of the section, the proof of Theorem 1 is given. Solely for the purpose of this section we set $p = p_{1,l} = (\ln n + l \, \ln\ln n + c)/n$.

5.1. Axioms of the almost sure theories $T_{1,l}$. We start with defining axioms of the almost sure theories $T_{1,l}$, where $l \geq 0$ is an integer. For two vertices v, w in a graph, we use $dist(v,w)$ to denote the distance of v and w.

For a graph H, let $\varphi(H)$ be the statement that there is a subgraph isomorphic to the graph H, i.e. $(\exists\, x_1, \ldots, x_{v_H})\ (\wedge_{\{i,j\}\in E(H)}(x_i \sim x_j))$. The theories $T_{1,l}$ have the following seven schema, $\Lambda^1(.) - \Lambda^7(.)$:

- For every graph H with $v_H < e_H : \Lambda^1(H) \equiv \neg\varphi(H)$.
- For every graph H which is a tree or unicyclic connected: $\Lambda^2(H) \equiv \varphi(H)$.
- For all integers $t \geq 1$ and $m \geq 3 : \Lambda^3(t,m) \equiv (\exists\, t$ distinct cycles $C_m^1, \ldots, C_m^t$ each on m vertices).
- For all integers $r, s, t \geq 1, s \geq r : \Lambda^4(r,s,t) \equiv \neg(\exists$ a pair of vertices $x, y)(deg(x) = r \wedge deg(y) = s \wedge dist(x,y) = t)$.
- For all integers $r, t \geq 1$ and $m \geq 3 : \Lambda^5(t,r,m) \equiv \neg(\exists x, deg(x) = r)$ ($\exists$ cycle C_m on m vertices) $(dist(x, C_m) = t)$.

and schema dealing with vertices of degree $r, r \geq 0$. The case $r = l$ is not included here as $p = p_{1,l}$ is the threshold function for the existence of vertices with degree equal to l.

- For all integers $r, 0 \leq r < l : \Lambda^6(r) \equiv \neg(\exists x)(deg(x) = r)$.
- For all integers $t \geq 1$ and $r > l : \Lambda^7(t,r) \equiv (\exists$ distinct vertices $x_1, \ldots, x_t)$ $(deg(x_1) = \ldots = deg(x_t) = r)$.

Note that the group of axioms $\Lambda^6(.)$ is void for $l = 0$, i.e. for the theory $T_{1,0}$. In the following lemma, we show that all the schema of $T_{1,l}$ have the limiting probability equal to 1. In fact Lemma 12, proved later, and the results of Section 3 imply that the theories $T_{1,l}$ are the almost sure theories for $p = p_{1,l}$.

Lemma 10. *For every $l \geq 0$, all the axioms of $T_{1,l}$ are first order sentences and all of them hold for $\mathcal{G}(n,p)$ with probability $1 - o(1)$.*

Proof. Clearly, all the axioms $\Lambda^1(.) - \Lambda^7(.)$ are expressible in the first order language of graphs.

(1) The fact that $\Lambda^1(H)$ holds almost surely was originally proved by P. Erdős and A. Rényi in [ER 60]. For a graph H with $v_H < e_H$, let Y_H be a random variable counting subgraphs of $\mathcal{G}(n,p)$ isomorphic to H. Using Markov inequality, we get

$$\begin{aligned} Prob(Y_H > 0) \leq \ \mathrm{Exp}[Y_H] &= \binom{n}{v_H} p^{e_H} = O(n^{v_H} p^{e_H}) \\ &= O\Big(\frac{(\ln n)^{e_H}}{n^{e_H - v_H}}\Big) = o(1)\ . \end{aligned}$$

Thus, $\lim_{n\to\infty} Prob(\mathcal{G}(n,p) \models \Lambda^1(H)) = \lim_{n\to\infty} Prob(Y_H = 0) = 1$.

(2) For a graph H, let $m(H) = \max_{F\subseteq H} e_F/v_F$. It was proved by Bollobás that the function $n^{-1/m(H)}$ is a threshold for $\mathcal{G}(n,p) \models \varphi(H)$. Recall that $m(TR_t) = 1 - \frac{1}{t}$ for a tree TR_t on t vertices and $m(U) = 1$ for a unicyclic connected graph U. Thus, $\lim_{n\to\infty} Prob(\mathcal{G}(n,p) \models \Lambda^2(H)) = 1$.

(3) Let Y_m be a random variable counting the cycles C_m in $\mathcal{G}(n,p)$. We get for the moments of the random variable Y_m

$$\mathrm{Exp}[Y_m] = \binom{n}{m} p^m = O((np)^m) = O(\ln^m n)$$

and

$$\begin{aligned}\mathrm{Var}[Y_m] &= \binom{n}{m}\binom{n-m}{m}p^{2m} + O\Big(\sum_{i=1}^{m}\binom{n}{i}\binom{n-i}{m-i}\binom{n-m}{m-i}p^{2m-i}\Big)\\ &- (\ \mathrm{Exp}[Y_m])^2\ .\end{aligned}$$

Thus, we have

$$\begin{aligned}\frac{\mathrm{Var}[Y_m]}{(\ \mathrm{Exp}[Y_m]-t)^2} &= (1+o(1))\cdot\frac{\mathrm{Var}[Y_m]}{(\ \mathrm{Exp}[Y_m])^2}\\ &= (1+o(1))\cdot\Big[O\Big(\sum_{i=1}^{m}\frac{\binom{n}{i}\binom{n-i}{m-i}\binom{n-m}{m-i}}{\binom{n}{m}^2}\cdot p^{-i}\Big)\\ &+ \Big(\frac{\binom{n}{m}\binom{n-m}{m}p^{2m}}{\binom{n}{m}^2p^{2m}}-1\Big)\Big]\\ &= O\Big(\sum_{i=1}^{m}\frac{1}{(np)^i}\Big)+o(1)\\ &= o(1)\ .\end{aligned}$$

We get via the Chebyshev inequality $Prob(Y_m < t) \leq\ \mathrm{Var}[Y_m]\ /\ (\ \mathrm{Exp}[Y_m]-t)^2 = o(1)$ and, hence, $Prob(\mathcal{G}(n,p)\models\Lambda^3(t,m)) = 1 - Prob(Y_m < t) = 1-o(1)$.

(4) Let Y be a random variable counting pairs of vertices x, y satisfying $deg(x) = r \wedge deg(y) = s \wedge dist(x,y) = t$. Computing the expectation of Y, one gets

$$\begin{aligned}\mathrm{Exp}[Y] &= O\Big(\binom{n}{2}\binom{n-2}{t-1}\binom{n-t-1}{r-1}\binom{n-t-r}{s-1}\\ &\times\ p^{s-1+r-1+t}(1-p)^{2(n-s-r-t+1)}\Big)\\ &= O\Big(n(np)^{s+r+t-2}e^{-2np}\Big)\\ &= o(1)\ .\end{aligned}$$

By Markov inequality, $Prob(Y > 0) \leq\ \mathrm{Exp}[Y] = o(1)$. Hence, $Prob(\mathcal{G}(n,p)\models\Lambda^4(r,s,t)) = 1-o(1)$.

(5) Let Y be a random variable counting pairs (x, C_m) where x is a vertex of degree r, C_m is a cycle on m vertices, and $dist(x, C_m) = t$. Computing the expectation of Y, we get

$$\begin{aligned}\mathrm{Exp}[Y] &= O\Big(n\binom{n-1}{r-1}p^{r-1}(1-p)^{n-t+1-m-r}\\ &\times\ \binom{n-r}{m}(m-1)!\cdot m\cdot p^m\binom{n-r-m}{t-1}p^t\Big)\\ &= O\Big((np)^{r+m+t-1}e^{-np}\Big)\\ &= o(1)\ .\end{aligned}$$

By Markov inequality, $Prob(Y > 0) \leq\ \mathrm{Exp}[Y] = o(1)$. Therefore, $\lim_{n\to\infty} Prob(\mathcal{G}(n,p)\models\Lambda^5(t,r,m)) = 1$.

(6) Let Y_r be a random variable counting the vertices of $\mathcal{G}(n,p)$ with degree equal to r. We are going to show $Prob(Y_r > 0) = o(1)$, for $r < l$, and $Prob(Y_r \geq$

$t) = 1 - o(1)$, for $r > l$. We start with computing the expectation of Y_r

$$\begin{aligned}\operatorname{Exp}[Y_r] &= n\binom{n-1}{r}p^r(1-p)^{n-r-1} \\ &= O\Big(n^{r+1}\frac{\ln^r n}{n^r}e^{-\ln n - l\ln\ln n}\Big) \\ &= O\Big(\ln^{r-l} n\Big)\,.\end{aligned} \tag{6}$$

In the case $r < l$, we get by Markov inequality

$$Prob(Y_r > 0) \leq \operatorname{Exp}[Y_r] = O(\ln^{r-l} n) = o(1)$$

and, thus, $Prob(\mathcal{G}(n,p) \models \Lambda^6(r)) = 1 - o(1)$.

For $r > l$, we are going to use Chebyschev inequality. We have $Prob(Y_r < t) \leq Prob(|Y_r - \operatorname{Exp}[Y_r]| \geq |\operatorname{Exp}[Y_r] - t|) \leq \operatorname{Var}[Y_r] \;/\; (\operatorname{Exp}[Y_r] - t)^2$. Thus, we need to bound $\operatorname{Var}[Y_r]$ from above. One gets for the second moment of the random variable Y_r

$$\begin{aligned}\operatorname{Var}[Y_r] &= \operatorname{Exp}[Y_r] + 2\operatorname{Exp}[\binom{Y_r}{2}] - (\operatorname{Exp}[Y_r])^2 \\ &= \operatorname{Exp}[Y_r] + 2\binom{n}{2} \\ &\times \Big[\sum_{i=0}^{r}\binom{n-2}{i}\binom{n-2-i}{r-i}\binom{n-r-2}{r-i}p^{2r}(1-p)^{2n-2r-2} \\ &+ \sum_{i=0}^{r-1}\binom{n-2}{i}\binom{n-2-i}{r-i-1}\binom{n-r-1}{r-i-1}p^{2r-1}(1-p)^{2n-2r-2}\Big] \\ &- (\operatorname{Exp}[Y_r])^2 \\ &= \operatorname{Exp}[Y_r] + \Big[2\binom{n}{2}\binom{n-2}{r}\binom{n-r-2}{r}p^{2r}(1-p)^{2n-2r-2} \\ &- (\operatorname{Exp}[Y_r])^2\Big] + 2\binom{n}{2}p^{2r}(1-p)^{2n-2r-2} \\ &\times \Big[\sum_{i=1}^{r}\binom{n-2}{i}\binom{n-2-i}{r-i}\binom{n-r-2}{r-i} \\ &+ \sum_{i=0}^{r-1}\binom{n-2}{i}\binom{n-2-i}{r-i-1}\binom{n-r-1}{r-i-1}p^{-1}\Big]\,.\end{aligned}$$

Hence,

$$\begin{aligned}
\frac{\mathrm{Var}[Y_r]}{(\mathrm{Exp}[Y_r]-t)^2} &= (1+o(1))\cdot\frac{\mathrm{Var}[Y_r]}{(\mathrm{Exp}[Y_r])^2} \\
&= (1+o(1))\Big[\frac{1}{\mathrm{Exp}[Y_r]} + \Big(\frac{2\binom{n}{2}\binom{n-2}{r}\binom{n-r-2}{r}}{n^2\binom{n-1}{r}^2} - 1\Big) \\
&+ \frac{2\binom{n}{2}}{n^2\binom{n-1}{r}^2}\cdot\Big(\sum_{i=1}^{r}\binom{n-2}{i}\binom{n-2-i}{r-i}\binom{n-r-2}{r-i} \\
&+ \sum_{i=0}^{r-1}\binom{n-2}{i}\binom{n-2-i}{r-i-1}\binom{n-r-1}{r-i-1}p^{-1}\Big)\Big] \\
&= (1+o(1))\Big[\frac{1}{\mathrm{Exp}[Y_r]} + o(1) \\
&+ \frac{1}{n^{2r}}\Big(O(n^{2r-1}) + O(\frac{n^{2r-1}}{\ln n})\Big)\Big] \\
&= o(1)\ .
\end{aligned}$$

Therefore, we get $Prob(\mathcal{G}(n,p) \models \Lambda^7(t,r)) = Prob(Y_r \geq t) = 1 - o(1)$. ∎10

5.2. Complete theories $T_{1,l}(s)$. For fixed integers $l, s \geq 0$, let $T_{1,l}(s)$ be the extension of the theory $T_{1,l}$ by the axiom $\sigma_{l,s}$. In this section, we are going to show that these theories are complete. To be able to employ the technique of Ehrenfeucht games, we need the following description of countable models for the theory $T_{1,l}(s)$. Note that due to the schema $\Lambda^7(.)$, the theories $T_{1,l}(s)$ do not possess finite models.

Lemma 11. *Let $l, s \geq 0$ be integers and let M be a countable model for the theory $T_{1,l}(s)$. Then M contains*

I. *in the case $l = 0$, precisely s finite components each of them being an isolated vertex,*
in the case $l > 0$, precisely s infinite trees, each with precisely one vertex of finite degree equal to l and with all other degrees infinite,

and every other component of M is one of the following three types

II. *an infinite tree with precisely one vertex of finite degree bigger than l and with all other degrees infinite,*
III. *an infinite tree with all vertices of infinite degree,*
IV. *an infinite unicyclic connected component with all vertices of infinite degree, i.e. an infinite component consisting of a finite cycle and infinite rooted trees from each vertex of the cycle having all degrees infinite. These infinite rooted trees are pairwise vertex disjoint.*

Moreover, for every integer $r > l$, M must contain an infinite number of trees of type II with a vertex of degree r and, for every (finite) cycle C of length m, the model M must contain an infinite number of components of type IV containing C. Furthermore, the number of component of type III can be either infinite or any nonnegative integer, including 0.

Proof. The structure of a countable model M of $T_{1,l}(s)$ is determined by the axioms $\Lambda^1(.), \ldots, \Lambda^7(.)$, and $\sigma_{l,s}$ of the theory.

The tree-like or unicyclic structure of components of M is implied by axioms $\Lambda^1(.)$ and $\Lambda^2(.)$. Moreover, axioms $\Lambda^4(.)$ and $\Lambda^5(.)$ imply that no two special phenomena - a vertex of a finite degree or a finite cycle - are in the same component of M.

The existence and number of components of type I is determined by the axiom $\sigma_{l,s}$. Finally, the existence and number of components of types II and IV is implied by axioms $\Lambda^3(.)$, $\Lambda^6(.)$, and $\Lambda^7(.)$. Note that the axioms of the theory $T_{1,l}(s)$ do not determine neither the number nor the existence of components of type III. $\blacksquare_{11}$

Lemma 12. *For all integers $l, s \geq 0$, the theories $T_{1,l}(s)$ are complete.*

Proof. Fix $l, s \geq 0$. In the proof we use T to denote the theory $T_{1,l}(s)$. Note that T is a consistent theory as it is an extension of the theory $T_{1,l}$ by the axiom $\sigma_{l,s}$ with nonzero limiting probability (cf. Remark 6 and Fact 15).

To show T is complete, we are going to use the condition $\iota\iota\iota$ of Fact 3. Thus the following claim already implies the completeness of T.

Claim 13. *For every two countable models M_1, M_2 of the theory T and for every positive integer t, the second player has a winning strategy for the Ehrenfeucht game $Ehr_t(M_1, M_2)$.*

Proof. The proof is based on Facts 4 and 5. When using Fact 4 we maintain the notation introduced there. Due to Fact 5 it is enough to show that Duplicator can win the game $Ehr_t(\widehat{M_1}, \widehat{M_2})$, where $\widehat{M_1}, \widehat{M_2}$ denote the models M_1, M_2, respectively, after deleting all the components of type I.

Let $i \in [t]$ and suppose, without loss of generality, vertices $w \in \widehat{M_1}$ and $v_1, \ldots, v_{i-1} \in \widehat{M_2}$ are given. We show that there is a vertex $v_i \in \widehat{M_2}$ satisfying the assumptions of Fact 4. Let U be the component of $\widehat{M_1}$ containing the vertex w. In the case the component U is of the type II or IV, we can always find a component U' of the same type, i.e. isomorphic, in $\widehat{M_2}$ such that none of the vertices $v_1, \ldots, v_{i-1}$ belongs to U'. Hence, we can choose the vertex v_i to satisfy the conditions of Fact 4. If U is a component of type III, let U' be a component of $\widehat{M_2}$ isomorphic to the tree TR_{l+1}, i.e. the tree with precisely one vertex of finite degree $l+1$ and all other degrees infinite, and such that U' does not contain any of $v_1, \ldots, v_{i-1}$. We choose for the vertex v_i any vertex of U' of distance more than 3^t from the vertex of finite degree $l+1$. The argument given above works also with the roles of $\widehat{M_1}$ and $\widehat{M_2}$ reversed. Thus Fact 4 implies that Duplicator has a winning strategy for $Ehr_t(\widehat{M_1}, \widehat{M_2})$ and our claim follows. $\blacksquare_{13}$

This also completes the proof of Lemma 12. $\blacksquare_{12}$

Remark 14. Fix $l, s \geq 0$. Note that $T_{1,l}(s)$ is not $\aleph_0$-categorical. However, the proof of Lemma 12 implies that there is a particular model $\overline{M}$ of $T_{1,l}(s)$ not containing any trees of type III (cf. Lemma 11) such that for every t and every model M of $T_{1,l}(s)$ one has

$$Th_t(L; \overline{M}) = Th_t(L; M) .$$

This $\overline{M}$ is in some sense a minimal model of $T_{1,l}(s)$. Later, we are going to refer to this "bar" notation.

To justify that $\sigma_{l,s}$ form a complete family of completions and, also, to determine the possible limit values, we need the following fact giving the limit probabilities $q_{l,s}$. The proof of it is standard and can be found for example in [B 85].

Fact 15. *For every $l \geq 0$, let $X_l(n)$ be a random variable counting the vertices of degree precisely l in $\mathcal{G}(n, p_{1,l})$. Then, for every integer $l \geq 0$, the variables $X_l(n)$ converge in distribution, as $n \to \infty$, to a random variable with the Poisson distribution and mean equal to $e^{-c}/l!$. Thus,*

$$\begin{aligned} q_{l,s} &:= \lim_{n\to\infty} Prob(\mathcal{G}(n, p_{1,l}) \models \sigma_{l,s}) \\ &= \lim_{n\to\infty} Prob(X_l(n) = s) \\ &= \frac{e^{-sc}}{(l!)^s s!} \cdot exp\Big(-\frac{e^{-c}}{l!}\Big) . \end{aligned}$$

5.3. Proof of Theorem 1. We can now quickly prove Theorem 1. The results of Section 3 give the statement of Theorem 1 except that we still need to characterize the sets $\mathfrak{M}[\psi]$. The following shows that all $\mathfrak{M}[\psi]$ are finite or cofinite.

Claim 16. *Let $t \geq 0$ be an integer. For all $s, s' \geq t$, let M_s be a model for $T_{1,l}(s)$ and $M_{s'}$ be a model for $T_{1,l}(s')$. Then, $Th_t(L; M_s) = Th_t(L; M_{s'})$.*

Proof. In view of Remark 14 after the proof of Lemma 12, instead of the models M_s and $M_{s'}$ we can equivalently use models $\overline{M}_s$ and $\overline{M}_{s'}$. Both of the models $\overline{M}_s$ and $\overline{M}_{s'}$ do not contain trees of type III. (cf. Lemma 11). Thus, after deleting all the components of type I (i.e. all isolated vertices in the case $l = 0$ or all infinite trees containing a vertex of degree l in the case $l > 0$) the remaining parts of the models $\overline{M}_s$ and $\overline{M}_{s'}$ are *isomorphic.* Due to the assumption that the number of the components of type I in both models is at least t, the second player can win $Ehr_t(\overline{M}_s, \overline{M}_{s'})$. Due to the structure of the models $\overline{M}_s$ and $\overline{M}_{s'}$ described above, Duplicator can win by using a rather straightforward strategy - he just precisely copies the moves of the Spoiler. Hence, the claim follows. $\blacksquare_{16}$

The claim implies the finite/co-finite part of the statement in Theorem 1. Moreover, for a given $\psi \in L$ we can give a bound on the size of $\mathfrak{M}[\psi]$ or of its complement, $\mathfrak{M}^C[\psi] = (\mathbb{N} \cup \{0\}) \setminus \mathfrak{M}[\psi]$, depending on which of these two sets is finite. To see that, let $qr(\psi) = h$. Recall that the theories $T_{1,l}(s)$ are complete, i.e. that either $T_{1,l}(s) \vdash \psi$ or $T_{1,l}(s) \vdash \neg\psi$. We need to distinguish the following two cases.

In the case $T_{1,l}(h) \vdash \neg\psi$, Claim 16 implies that for every $s \geq h$ one has $T_{1,l}(s) \vdash \neg\psi$. Thus, $\mathfrak{M}[\psi] \subseteq \{0, 1, \ldots, h-1\}$. In the other case, $T_{1,l}(h) \vdash \psi$, we define $\varphi \equiv \neg\psi$. Hence, $\mathfrak{M}^C[\psi] = \mathfrak{M}[\varphi] \subseteq \{0, 1, \ldots, h-1\}$.

To finish the proof of Theorem 1, i.e. to get the full characterization, we need to show witnessing sentences in the language L. For $S \subset \mathbb{N} \cup \{0\}$ finite, let $\psi_S \equiv \bigvee_{s\in S} \sigma_{l,s}$. In the case $S \subseteq \mathbb{N} \cup \{0\}$ is co-finite, we set $\psi_S \equiv \neg \bigvee_{s \in (\mathbb{N}\cup\{0\})\setminus S} \sigma_{l,s}$.

6. The General Case

In this section we give a generalization of Theorem 1 for $p = p_{k,l} = (\ln n + l \ln\ln n + c)/(kn)$, though omitting some of the more detailed calculations. The results are summarized in Theorem 22.

The section is organized similar to Section 5. We start with describing schema for almost sure theories $T_{k,l}$; then we describe the completions of these - $T_{k,l}(\mathbf{r})$. Finally, we give a description of sets $\mathfrak{M}[\psi]$, ψ being a first order sentence of graphs.

We begin with a special notation. Let $\mathbf{d} = (d_1, \ldots, d_{k'})$ with all $d_i \geq 0$ integers. Let TR be a tree on $\{1, \ldots, h\}$ with $h \geq k'$ and with all leaves of TR in $\{1, \ldots, k'\}$.

We say $(TR, \mathbf{d})$ has type (k', l') where

$$l' = h - 1 + \sum_{i=1}^{k'} d_i \ .$$

We call $(TR, \mathbf{d})$ a *low-degree tree* and, for $r \geq 1$, define the first order sentence expressing that there exist (at least) r such low degree trees

> $LDT(TR, \mathbf{d}, r)$: There exist distinct $x_1^{(u)}, \ldots, x_h^{(u)}$, $1 \leq u \leq r$, with $x_i^{(u)} \sim x_j^{(u)}$ if and only if $\{i, j\} \in E(TR)$ [i.e., the $x_i^{(u)}$'s form r copies of TR] and $x_i^{(u)}, 1 \leq i \leq k'$, is adjacent to precisely d_i vertices other than the $x_1^{(u)}, \ldots, x_h^{(u)}$.

In the case $r = 1$, we write $LDT(TR, \mathbf{d})$ for $LDT(TR, \mathbf{d}, 1)$.

Remark 17. $LDT(TR, \mathbf{d})$ gives the existence of k' low degree vertices in close proximity. When $l' = k' - 1$, the minimal possible value, we must have $h = k'$ and all $d_i = 0$ and then $LDT(TR, \mathbf{d})$ means there exists a component TR. When $h = k' = 1$ and $\mathbf{d} = (d_1)$ the sentence $LDT(TR, \mathbf{d})$ means there exists a vertex of degree precisely d_1. When, for example, $k' = 2, h = 3$, $TR = \{\{1, 3\}, \{2, 3\}\}$ and $\mathbf{d} = (5, 7)$ the sentence $LDT(TR, \mathbf{d})$ means there exist two vertices of (total) degrees 6 and 8 that are at distance two. Note that no condition is placed on the degree of the vertex between them.

The theories $T_{k,l}$ have the following common schema:

- There are no bicyclic subgraphs.
- For every $r \geq 1$ there are (at least) r copies of every tree and unicyclic graph.
- For every $d \geq 0$, $t \geq 0$, $m \geq 3$ there is no vertex of degree d at distance t from a vertex on a cycle of size m.

With k given the following common schema hold in the theories $T_{k,l}$ for all $l \geq k-1$:

- For $(TR, \mathbf{d})$ of type $(k+1, l')$, any l': $\neg LDT(TR, \mathbf{d})$.
- For $(TR, \mathbf{d})$ of type (k', l') with $0 < k' < k$, any l', and any $r : LDT(TR, \mathbf{d}, r)$.

With k, l fixed we add:

- For $(TR, \mathbf{d})$ of type (k, l') with $l' < l$: $\neg LDT(TR, \mathbf{d})$.
- For $(TR, \mathbf{d})$ of type (k, l') with $l' > l$, any r: $LDT(TR, \mathbf{d}, r)$.

We are going to show that all the axioms of the theories $T_{k,l}$ hold with probability $1 - o(1)$ which combined with Lemma 19 and the results of Section 3 implies that $T_{k,l}$ are almost sure theories. Identically as in Section 5, we get that every sentence in the first three schema holds with probability $1 - o(1)$.

Let $(TR, \mathbf{d})$ have type (k', l') and set $X = X_{(TR,\mathbf{d})}$ equal the number of choices of $x_1, \ldots, x_h$ having the $LDT(TR, \mathbf{d})$ property. Neglecting constants

$$\text{Exp}[X] = \Theta\left(n^h p^{h-1} \prod_{i=1}^{k'} \binom{n}{d_i} p^{d_i} (1-p)^n\right) = \Theta\left(n (\ln n)^{l'} e^{-pnk'}\right) .$$

Neglecting polylogarithmic terms we get $\text{Exp}[X] = n^{1-(k'/k)+o(1)}$. If $k' > k$ then $\text{Exp}[X] \to 0$ while if $k' < k$ then $\text{Exp}[X] \to \infty$. In the latter case a variance calculation, which is omitted, gives that almost surely the random variable X is close to $\text{Exp}[X]$ and, thus, $LDT(TR, \mathbf{d}, r)$ almost surely for any r. In the case $k' = k$ the powers of n cancel and $\text{Exp}[X] = \Theta((\ln n)^{l'-l})$. If $l' < l$ then $\text{Exp}[X] \to 0$ while if $l' > l$ then $\text{Exp}[X] \to \infty$. When $\text{Exp}[X] \to 0$ almost surely $\neg LDT(TR, \mathbf{d})$

while when $\text{Exp}[X] \to \infty$ a variance calculation, which we omit, shows $X \sim \text{Exp}[X]$ almost surely so that $LDT(TR, \mathbf{d}, r)$ almost surely for any r.

There are, critically, only a finite number of $(TR, \mathbf{d})$ of a given type (k, l). We call them *equivalent* if the sentences $LDT(TR, \mathbf{d})$ are tautologically equivalent. [For example, $k = h = 2$, TR an edge on $\{1, 2\}$, $\mathbf{d} = (5, 7)$ and $\mathbf{d} = (7, 5)$ are equivalent.] Let $\beta = \beta(k, l)$ be the number of equivalence classes and let $(TR_\gamma, \mathbf{d}_\gamma)$, $1 \le \gamma \le \beta$, be a list containing precisely one $(TR, \mathbf{d})$ from each equivalence class. Let $\mathbf{r} = (r_1, \dots, r_\beta)$ with $r_\gamma \ge 0$ arbitrary integers. Define

$$\sigma_{\mathbf{r}} := \wedge_\gamma \left[LDT(TR_\gamma, \mathbf{d}_\gamma, r_\gamma) \wedge \neg LDT(TR_\gamma, \mathbf{d}_\gamma, r_\gamma + 1)\right] .$$

(When $r_\gamma = 0$ the bracketed expression is to be replaced simply by the sentence $\neg LDT(TR_\gamma, \mathbf{d}_\gamma)$.)

We claim that the family of $\sigma_{\mathbf{r}}$ forms a complete family of completions in the sense of Section 3. Our index set will be

$$I = \{(r_1, \dots, r_\beta) : r_\gamma \ge 0 \text{ for all } 1 \le \gamma \le \beta\} .$$

Note that, for example, due to the second schema of $T_{k,l}$ the theories $T_{k,l}$ as well as $T_{k,l}(\mathbf{r})$ do not possess finite models.

Lemma 18. *Let k and $l, l \ge k - 1 \ge 0$, be integers and $\mathbf{r} \in I$. Every countable model M of the theory $T_{k,l}(\mathbf{r})$ contains*

A. *for every low degree tree $(TR_\gamma, \mathbf{d}_\gamma), \gamma = 1, \dots, \beta$, precisely r_γ components containing this tree. Moreover, if $l = k - 1$, all these componets are finite and consist of a single copy of the tree $(TR_\gamma, \mathbf{d}_\gamma)$. In the case $l > k - 1$, all these components are infinite with all vertices of infinite degree except vertices $x_1, \dots, x_k$ in the copy of the low degree tree $(TR_\gamma, \mathbf{d}_\gamma)$.*

Every other component of the model M is one of the following four types:

B. *an infinite tree containing a copy of a tree $(TR, \mathbf{d})$ of type (k, l'), $l' > l$, with all vertices of infinite degree except the vertices $x_1, \dots, x_k$ in the copy of $(TR, \mathbf{d})$,*
C. *a finite tree on k' vertices where $0 < k' < k$,*
D. *an infinite tree with all but at most $k - 1$ vertices of infinite degree,*
E. *an infinite unicyclic connected component with all vertices of infinite degree, i.e. an infinite component consisting of a finite cycle and infinite rooted trees from each vertex of the cycle having all degrees infinite. These infinite rooted trees are pairwise vertex disjoint.*

The model M must contain an infinite number of components of type B for every low degree tree $(TR, \mathbf{d})$ of type $(k, l'), l' > l$. Moreover, it must contain an infinite number of the finite components of type C for every tree $(TR, \mathbf{d})$ of type $(k', k' - 1), 0 < k' < k$. Also, for every integer $m \ge 3$ and a cycle on m vertices the model M must contain an infinite number of components of type E containing that cycle. Finally, the number of components of type D can be either infinite or any nonnegative number, including 0.

Proof. The axioms of the theory $T_{k,l}$ and the sentence $\sigma_{\mathbf{r}}$ determine the structure of the model M. The first schema of $T_{k,l}$ implies that every component of M has a tree-like or unicyclic structure. Moreover, using the third and fourth schema of $T_{k,l}$ we infere that there are at most k vertices of finite degree in a given component and that there is no component with both a finite cycle and a vertex of finite degree.

The existence and number of components of type A is given by the axiom $\sigma_{\mathbf{r}}$. Because there are at most k vertices of finite degree in each component of M, every copy of $(TR_\gamma, \mathbf{d}_\gamma), 1 \leq \gamma \leq \beta$, must lie in a separate component. If $l = k-1$, the low degree tree $(TR_\gamma, \mathbf{d}_\gamma)$ contains necessary k vertices and all $d_i, i = 1, \dots, k$, equal zero. Thus, every component containing a copy of $(TR_\gamma, \mathbf{d}_\gamma)$ consists of a single copy of the tree $(TR_\gamma, \mathbf{d}_\gamma)$. In the case $l > k-1$, there is either more than k vertices in $(TR_\gamma, \mathbf{d}_\gamma)$ and/or a vertex joined to some of the vertices $x_1, \dots, x_k$ of $(TR_\gamma, \mathbf{d}_\gamma)$. Thus, there must be a vertex, say v, which degree is not determined by the d_i's. As there cannot be more than k vertices of finite degree in a component, the vertex v must have an infinite degree. Hence, the component is infinite.

The last schema of $T_{k,l}$ implies that there is an infinite number of components of type B. Moreover, as $l' > l \geq k-1$, such a component must contain a vertex which does not have a degree determined by any of $d_i, i = 1, \dots, k$. Similarly as above, due to the fact that there cannot be $k+1$ vertices of finite degree in a component, we conclude that such a vertex must have an infinite degree. This implies that the component itself is infinite.

If a component of M contains $k', k' < k$ vertices of finite degree which form a low degree tree $(TR, \mathbf{d})$ of type $(k', k'-1)$, then TR has necessary k' vertices and all $d_i, i = 1, \dots, k'$, equal zero. Thus, such a component consists only of a single copy of the tree $(TR, \mathbf{d})$. Moreover, the fifth schema of $T_{k,l}$ implies that there is an infinite number of such components.

The existence and number of components containing none or $k', 0 < k' < k$, vertices of finite degree which in the latter case form a low degree tree of type $(k', l'), l' > k'-1$, is not determined by the theory $T_{k,l}(\mathbf{r})$.

The second schema of $T_{k,l}$ implies the infinite number of unicyclic components.

Finally, the axioms $\neg LDT(TR, \mathbf{d})$ for any low degree tree $(TR, \mathbf{d})$ of type $(k, l'), l' < l$, or $(k+1, l')$, any $l' \geq k$, and the nonexistence of bicyclic subgraphs imply that all the components of the model M must be one of the types A - E. $\blacksquare_{18}$

Lemma 19. *Let k and $l, l \geq k-1 \geq 0$, be integers. For every $\mathbf{r} \in I$, the theory $T_{k,l}(\mathbf{r}) := T_{k,l} + \sigma_{\mathbf{r}}$ is complete.*

Proof. The theory $T_{k,l}(\mathbf{r})$ is consistent by Remark 6. To show it is complete we are going to use the condition iii of Fact 3. We use a notation for components of countable models of the theory $T_{k,l}(\mathbf{r})$ introduced in Lemma 18.

Let M_1, M_2 be two countable models for $T_{k,l}(\mathbf{r})$ and consider the t-move Ehrenfeucht game $Ehr_t(M_1, M_2)$. To show Duplicator has a winning strategy for this game we are going to use Facts 4 and 5. First, we observe that due to Fact 5 it is enough to show that Duplicator can win the game $Ehr_t(\widehat{M_1}, \widehat{M_2})$, where $\widehat{M_1}, \widehat{M_2}$ denote the models M_1, M_2, respectively, after deleting all components of type A. Let $i \in [t]$ and suppose vertices $w \in \widehat{M_1}$ and $v_1, \dots, v_{i-1} \in \widehat{M_2}$ are given. We show that there is a vertex $v_i \in \widehat{M_2}$ satisfying the assumptions of Fact 4. Let U be the component of $\widehat{M_1}$ containing w. If U is a component of type B, C, or E, there must be a component U' in $\widehat{M_2}$ isomorphic to U such that none of the vertices $v_1, \dots, v_{i-1}$ belongs to U'. This holds due to the fact that M_2, and thus $\widehat{M_2}$, contains a countable number of such components. Hence, there is a vertex $v_i \in \widehat{M_2}$ satisfying the assumptions of Fact 4. Thus, suppose U is a component of type D, i.e. U is an infinite tree containing $k', 0 \leq k' < k$, vertices of finite degree. Set $a = 3^{t-i}$ in Fact 4. The a-neighborhood of the vertex $w, N(w; a)$, contains at most

k' vertices of finite degree, including the possibility of no such vertex. Consider a low degree tree $(TR', \mathbf{d}')$ which consists of the low degree tree generated by the vertices of finite degree in the neighborhood $N(w; a)$ plus additional vertices of finite degree to bring the total to k which are all at distance $a+1$ from w and all are given degree $l+1$. Giving them this high degree gives this low degree tree $(TR', \mathbf{d}')$ type (k, l'), where $l' \geq l+1$, just from a single one of the new vertices. Therefore, there must be a component U' (of type B) in $\widehat{M_2}$ containing the low degree tree $(TR', \mathbf{d}')$ and such that it does not contain any of the vertices $v_1, \ldots, v_{i-1}$. Further, there is a vertex $v_i \in U'$ corresponding to w such that the a-neighborhoods of v_i and w are isomorphic as required in Fact 4. The argument given above works also with the roles of $\widehat{M_1}$ and $\widehat{M_2}$ reversed. Thus, combining Facts 4 and 5 we get that Duplicator has a winning strategy for the game $Ehr_t(M_1, M_2)$.

By Fact 3 the theory $T_{k,l}(\mathbf{r})$ is complete. ∎19

Let M be a countable model of the theory $T_{k,l}(\mathbf{r})$. Based on the proof of Lemma 19, we can define the *minimal model* $\overline{M}$ as that model which we obtain by deleting all components of type D from M, i.e. the model which contains all the components of M of types A, B, C, and E. The proof of Lemma 19 implies that $Th_t(L; M) = Th_t(L; \overline{M})$ for every t and every countable model M of $T_{k,l}(\mathbf{r})$.

For $1 \leq \gamma \leq \beta$, let X_γ be the number of low degree trees of type $(TR_\gamma, \mathbf{d}_\gamma)$ in $\mathcal{G}(n, p_{k,l})$. We calculate $\operatorname{Exp}[X_\gamma]$ as before but watch the constants. The e^{-npk} term yields an additional e^{-c} factor so that

$$\mu_\gamma := \lim_{n\to\infty} \operatorname{Exp}[X_\gamma] = \lambda_\gamma e^{-c} ,$$

where λ_γ is a calculatable positive rational number that involves the $d_i!$ and the number of automorphisms of $(TR_\gamma, \mathbf{d}_\gamma)$ The values of λ_γ are described in full detail in Lemma 21.

Remark 20. Describing the λ_γ can be somewhat complex. For example, let $l = k-1$ so all $d_i = 0$. Then β is the number of isomorphism types of trees on k vertices. X_γ is the number of isolated trees of type TR_γ, and λ_γ involves the number of automorphisms of TR_γ. Note, however, that for k, l fixed we have a finite list of $(TR, \mathbf{d})$ to consider and therefore a finite and computable list of the λ_γ.

By the following lemma, the X_γ have a joint limit distribution as independent random variables having Poisson distribution with means μ_γ so that

$$a_{\mathbf{r}} := \lim_{n\to\infty} \mu_n(\sigma_{\mathbf{r}}) = \prod_{\gamma=1}^{\beta} e^{-\mu_\gamma} \frac{\mu_\gamma^{r_\gamma}}{r_\gamma!}$$

and so $\sum_{\mathbf{r}\in I} a_{\mathbf{r}} = 1$ and the $\sigma_{\mathbf{r}}, \mathbf{r} \in I$, form a complete set of completions.

Lemma 21. *Let $X_\gamma, 1 \leq \gamma \leq \beta$, be random variables as above counting the trees $(TR_\gamma, \mathbf{d}_\gamma)$ in $\mathcal{G}(n, p_{k,l})$. Then the multidimensional random variable $(X_1, \ldots, X_\beta)$ converges in probability to the multidimensional random variable consisting of β independent random variables with Poisson distribution and means, $1 \leq \gamma \leq \beta$,*

$$\mu_\gamma = \frac{1}{k^l \cdot aut(TR_\gamma) \cdot \prod_{i=1}^{k} d_i!} \cdot e^{-c} ,$$

where $aut(TR_\gamma)$ is the size of the automorphism group of the tree TR_γ and $\mathbf{d}_\gamma = (d_1, \ldots, d_k)$. In particular, for $\mathbf{r} \in I$

$$\lim_{n\to\infty} Prob((X_1, \ldots, X_\beta) = \mathbf{r}) = \prod_{\gamma=1}^{\beta} e^{-\mu_\gamma} \frac{\mu_\gamma^{r_\gamma}}{r_\gamma!}.$$

Proof. Recall that all the trees $(TR_\gamma, \mathbf{d}_\gamma)$ are of the type (k,l). To be precise, let us write $X_\gamma(n) = X_\gamma$ for the random variable counting the trees $(TR_\gamma, \mathbf{d}_\gamma)$ in $\mathcal{G}(n, p_{k,l})$. To show convergence to Poisson distribution it is enough to show convergence of moments.

First, let us compute the expectation of the random variable $X_\gamma(n)$. Let h_γ denote the number of vertices of TR_γ. We get

$$\begin{aligned}
\mathrm{Exp}[X_\gamma(n)] &= (1+o(1))\binom{n}{h_\gamma}\frac{h_\gamma!}{aut(TR_\gamma)}p^{h_\gamma-1}\prod_{i=1}^{k}\binom{n}{d_i}p^{d_i}(1-p)^n \\
&= (1+o(1))\frac{1}{aut(TR_\gamma)\cdot\prod_{i=1}^{k} d_i!}\cdot n(np)^l e^{-knp} \qquad (7)\\
&= (1+o(1))\frac{1}{k^l\cdot aut(TR_\gamma)\cdot\prod_{i=1}^{k} d_i!}\cdot e^{-c}.
\end{aligned}$$

Using (7), we get for the t-factorial moment $\mathrm{Exp}_t[X_\gamma(n)]$ of $X_\gamma(n)$

$$\begin{aligned}
\mathrm{Exp}_t[X_\gamma(n)] &= \mathrm{Exp}\prod_{\alpha=0}^{t-1}(X_\gamma(n)-\alpha) \\
&= \prod_{\alpha=0}^{t-1}\mathrm{Exp}[X_\gamma(n-\alpha h_\gamma)] \\
&= (1+o(1))\Big(\frac{1}{k^l\cdot aut(TR_\gamma)\cdot\prod_{i=1}^{k} d_i!}\cdot e^{-c}\Big)^t.
\end{aligned}$$

Thus, the random variable $X_\gamma(n)$ converges in probability ($n \to \infty$) to a random variable with Poisson distribution $Po(\mu_\gamma)$.

Further, we get for the joint factorial moments

$$\mathrm{Exp}\prod_{t=1}^{u}\prod_{\alpha_t=0}^{w_t-1}(X_{\gamma_t}(n)-\alpha_t) = (1+o(1))\prod_{t=1}^{u}\Big(\frac{1}{k^l\cdot aut(TR_\gamma)\cdot\prod_{i=1}^{k} d_i!}\cdot e^{-c}\Big)^{w_t},$$

$1 \le u \le \beta$. Thus, the random variable $(X_{\gamma_1}, \ldots, X_{\gamma_u})$ converges in probability ($n \to \infty$) to a multidimensional random variable $(Y_{\gamma_1}, \ldots, Y_{\gamma_u})$ with independent random variables Y_{γ_i} possessing Poisson distributions $Po(\mu_{\gamma_i}), 1 \le \gamma \le u$. ∎21

Now we determine the sets $\mathfrak{M}[\psi]$. When ψ has quantifier rank s and $\mathbf{r}, \mathbf{r}'$ differ only on coordinates on which both are at least s then we claim $\mathbf{r} \in \mathfrak{M}[\psi]$ if and only if $\mathbf{r}' \in \mathfrak{M}[\psi]$. This is because when both minimal countable models $\overline{M}_{\mathbf{r}}$ and $\overline{M}_{\mathbf{r}'}$ have s or more components of the same type Duplicator can play isomorphically on them throughout an s-move Ehrenfeucht game.

By an s-box on I let us mean the set of $\mathbf{r} \in I$ where for each coordinate r_γ we require either that that coordinate have a specific value, with that value less than s, or we require that the coordinate be at least s. Then $\mathfrak{M}[\psi]$ is the disjoint union of a finite number of h-boxes where $h = qr(\psi)$. But conversely, given a finite union of disjoint s-boxes $\cup_{j=1}^m B_j$, let $\mathbf{r}(j) = (r_1(j), \ldots, r_\beta(j)) \in B_j$ and define

$$\begin{aligned} \varphi_j \quad := \quad & \wedge_{\{\gamma : r_\gamma(j) < s\}} [LDT(TR_\gamma, \mathbf{d}_\gamma, r_\gamma(j)) \wedge \neg LDT(TR_\gamma, \mathbf{d}_\gamma, r_\gamma(j)+1)] \\ \bigwedge \quad & \wedge_{\{\gamma : r_\gamma(j) \geq s\}} LDT(TR_\gamma, \mathbf{d}_\gamma, s) \ , \end{aligned}$$

$j = 1, \ldots, m$. (When $r_\gamma(j) = 0$ then the expression in the brackets above should be replaced by $\neg LDT(TR_\gamma, \mathbf{d}_\gamma)$.) Then φ_j and, thus, $\varphi := \vee_{j=1}^m \varphi_j$ are themselves first order sentences. Further, $\mathfrak{M}[\varphi] = \cup_{j=1}^m \mathfrak{M}[\varphi_j] = \cup_{j=1}^m B_j$, i.e. any disjoint union of s-boxes is $\mathfrak{M}[\varphi]$ for some first order sentence φ. (We note that φ may have higher quantifier depth than s.) By Remark 9 this completes the complete characterization of the limit probabilities.

Let $f_\psi(c)$ denote the limit probability of ψ as a function of the real number c. The form of $f_\psi(c)$ is simple to describe if we relax the condition of a complete characterization. The basic sentence $LDT(TR_\gamma, \mathbf{d}_\gamma, i) \wedge \neg LDT(TR_\gamma, \mathbf{d}_\gamma, i+1)$ has a limiting probability f of the form $qe^{-\lambda e^{-c}} e^{-ci}$ where λ, q are positive rational numbers. The sentence $LDT(TR_\gamma, \mathbf{d}_\gamma, s)$ will have an f of the form one minus a finite sum of such terms. To conclude this section, we summarize the results in the following theorem.

Theorem 22. *Let k, l be integers such that $l \geq k-1 \geq 0$. Let ψ be any sentence of the first order language of graphs. Then, in general, the limiting probability $f_\psi(c)$ has the form of a finite sum and difference of rational numbers times finite products of terms of the form above, i.e. a linear combination with rational coefficients of terms of the form $e^{-c\lambda_1} e^{-\lambda_2 e^{-c}}$ where the λ_1, λ_2 are themselves rational numbers. [We include 1 as the null product here.]*

In particular, $f_\psi(c)$ always exists and is always an infinitely differentiable function of c.

References

[B 85] B. Bollobás, *Random graphs,* Academic Press, 1985.

[BM 77] J.L. Bell, M. Machover, *A Course in Mathematical Logic,* North-Holland Publishing Company, 1977.

[CK 90] C.C. Chang, H.J. Keisler, *Model theory,* Elsevier Science Publishers B.V., 1990.

[EF 95] H.-D. Ebbinghaus, J. Flum, *Finite model theory,* Springer-Verlag, 1995.

[Eh 61] A. Ehrenfeucht, *An application of games to the completeness problem for formalized theories,* Fundamentæ Math. 49 (1961), 129-141.

[ER 60] P. Erdős, A. Rényi, *On the evolution of random graphs,* Magyar Tud. Akad. Mat. Kutató Int. Közl 5 (1960), 17-61.

[L 92] J. F. Lynch, *Probabilities of Sentences about Very Sparse Random Graphs,* Random Struct. Alg. 3 (1992), 33 - 53.

[ŁS 91] T. Łuczak, J. Spencer, *When does the zero-one law hold ?,* J. Amer. Math. Soc. 4 (1991), 451 - 468.

COURANT INSTITUTE, NEW YORK UNIVERSITY, 251 MERCER ST., NEW YORK NY 10012
E-mail address: spencer@cs.nyu.edu

SCHOOL OF MATHEMATICS, INSTITUTE FOR ADVANCED STUDY, PRINCETON, NJ 08540
E-mail address: thoma@dimacs.rutgers.edu

DIMACS Series in Discrete Mathematics
and Theoretical Computer Science
Volume **49**, 1999

RAMSEY THEORY AND PARTIALLY ORDERED SETS

William T. Trotter

ABSTRACT. Over the past 15 years, Ramsey theoretic techniques and concepts have been applied with great success to partially ordered sets. In the last year alone, four new applications of Ramsey theory to posets have produced solutions to some challenging combinatorial problems. First, Kierstead and Trotter showed that dimension for interval orders can be characterized by a single ramsey trail by proving that interval orders of sufficiently large dimension contain all small interval orders as subposets. Second, Winkler and Trotter introduced a notion of Ramsey theory for probability spaces and used the resulting theroy to show that interval orders can have fractional dimension arbitrarily close to 4. Third, Felsner, Fishburn and Trotter developed an extension of the product Ramsey theorem to show that there exists a finite 3-dimensional poset which is not a sphere order. Fourth, Agnarsson, Felsner and Trotter combined Ramsey theoretic techniques with other combinatorial tools to determine an asymtotic formula for the maximum number of edges in a graph whose incidence poset has dimension at most 4. In this paper, we outline how these applications were developed. Full details will appear in individual journal articles. This article also includes a brief sketch of how the applications of Ramsey theoretic techniques to posets have evolved.

1. Introduction

In recent years, there has been rapid growth in research activity centered on combinatorial problems for partially ordered sets, evidenced in part by the new AMS subject classification 06A07: Combinatorics of Partially Ordered Sets. In this article, we explore connections between Ramsey theory and partially ordered sets—especially with the poset parameter called *dimension*.

In this introductory section, we present only those concepts and notations essential to the results discussed in this paper. For additional background material, the reader is referred to the survey articles [**33**], [**34**], [**35**] [**36**], the recent article by Brightwell [**4**] and the author's monograph on posets [**32**].

We consider a *partially ordered set* (or *poset*) $\mathbf{P} = (X, P)$ as a structure consisting of a set X and a reflexive, antisymmetric and transitive binary relation P on X. We call X the *ground set* of the poset $\mathbf{P}$, and we refer to P as a *partial order* on

1991 *Mathematics Subject Classification.* 06A07, 05C35.

Key words and phrases. Ramsey theory, partially ordered set, dimension.

Research supported in part by the Office of Naval Research and the Deutsche Forschungsgemeinschaft.

X. The notations $x \leq y$ in P, $y \geq x$ in P and $(x, y) \in P$ are used interchangeably, and the reference to the partial order P is often dropped when its definition is fixed throughout the discussion. We write $x < y$ in P and $y > x$ in P when $x \leq y$ in P and $x \neq y$. When $x, y \in X$, $(x, y) \notin P$ and $(y, x) \notin P$, we say x and y are *incomparable* and write $x \| y$ in P.

Although we are concerned primarily with *finite* posets, i.e., those posets with finite ground sets, we find it convenient to use the familiar notation $\mathbb{R}$, $\mathbb{Q}$, $\mathbb{Z}$ and $\mathbb{N}$ to denote respectively the reals, rationals, integers and positive integers equipped with the usual orders. We also use $\mathbb{R}_0$ to denote the set of positive real numbers. Note that these infinite posets are *total* orders; in each case, any two distinct points are comparable. Total orders are also called *linear* orders, or *chains*. For a positive integer n, we let $\mathbf{n}$ denote the n-element chain $0 < 1 < \cdots < n-1$, while $[n]$ denotes the n-element set $\{1, 2, \ldots, n\}$.

When $\mathbf{P} = (X, P)$ is a poset, a linear order L on X is called a *linear extension* of P when $x < y$ in L for all $x, y \in X$ with $x < y$ in P. A set $\mathcal{R}$ of linear extensions of P is called a *realizer* of $\mathbf{P}$ when $P = \cap\mathcal{R}$, i.e., for all x, y in X, $x < y$ in P if and only if $x < y$ in L, for every $L \in \mathcal{R}$. The minimum cardinality of a realizer of $\mathbf{P}$ is called the *dimension* of $\mathbf{P}$ and is denoted $\dim(\mathbf{P})$.

Given a poset $\mathbf{P} = (X, P)$, let $\operatorname{inc}(\mathbf{P}) = \{(x, y) \in X \times X : x \| y \text{ in } P\}$. Then a family $\mathcal{R}$ of linear extensions of P is a *realizer* of P if and only if for every $(x, y) \in \operatorname{inc}(\mathbf{P})$, there exists $L \in \mathcal{R}$ so that $x > y$ in L. Call a pair $(x, y) \in \operatorname{inc}(\mathbf{P})$ a *critical pair* if

1. $u < x$ in P implies $u < y$ in P, and
2. $v > y$ in P implies $v > x$ in P

for all $u, v \in X$. Then let $\operatorname{crit}(\mathbf{P})$ denote the set of all critical pairs. It follows that that a family $\mathcal{R}$ of linear extensions of P is a realizer of P if and only if for every $(x, y) \in \operatorname{crit}(\mathbf{P})$, there exists some $L \in \mathcal{R}$ so that $x > y$ in L.

We say that a linear extension L *reverses* the incomparable pair (x, y) when $x > y$ in L. Let $S \subset \operatorname{inc}(\mathbf{P})$. We say that L *reverses* S when $x > y$ in L, for every $(x, y) \in S$. We say that a family $\mathcal{R}$ of linear extensions of P *reverses* S if for every $(x, y) \in S$, there is some $L \in \mathcal{R}$ so that $x > y$ in L. So the dimension of P is just the least integer t for which there exists a family $\mathcal{R}$ of linear extensions of P which reverses $\operatorname{crit}(\mathcal{R})$.

2. Three examples of Posets with Large Dimension

In this section, we briefly discuss three well known examples of posets with large dimension. These examples will help readers who are new to the subject of partially ordered sets with concepts discussed in subsequent sections.

For integers $n \geq 3$, $k \geq 0$, define the *crown* $\mathbf{S}_n^k$ as the height 2 poset with $n + k$ minimal elements $a_1, a_2, \ldots, a_{n+k}$, $n + k$ maximal elements $b_1, b_2, \ldots, b_{n+k}$ and $a_i < b_j$, for $j = i + k + 1, i + k + 2, \ldots, i - 1$. In this definition, we interpret subscripts cyclically so that $n+k+1 = 1$, $n+k+2 = 2$, etc. The following formula for the dimension of crowns is derived in [**30**].

THEOREM 2.1. *Let $n \geq 3$ and $k \geq 0$ be integers. Then*

$$\dim(\mathbf{S}_n^k) = \left\lceil \frac{2(n+k)}{k+2} \right\rceil.$$

□

The critical pairs in the crown $\mathbf{S}_n^k$ are just the incomparable pairs (a_i, b_j), where a_i is a minimal element and bj is a maximal element. In [**30**], Trotter proves that no linear extension of $\mathbf{S}_n^k$ can reverse more than $(k+2)(k+1)/2$ critical pairs. Since there are $(n+k)(k+1)$ critical pairs altogether, the lower bound in Theorem 2.1 follows immediately. It takes a little more work to show that this bound is tight.

When $k = 0$, the crown $\mathbf{S}_n^0$ (also denoted $\mathbf{S}_n$), is called the *standard example* of an n-dimensional poset. To see that $\dim(\mathbf{S}_n) \leq n$, observe that there are n critical pairs, namely the pairs (a_i, b_i) for $i = 1, 2, \ldots, n$. Clearly, n linear extensions are enough to reverse them. Conversely, it is easy to see that no linear extension of $\mathbf{S}_n$ can reverse two or more linear extensions, so that the dimension of $\mathbf{S}_n^0$ is at least n.

Note that $\mathbf{S}_n$ is isomorphic to the set of 1-element and $(n-1)$-element subsets of $[n]$ ordered by inclusion. More generally, for integers k, r and n, with $1 \leq k < r \leq n-1$, let $\mathbf{P}(k, r; n)$ denote the poset consisting of all k-element and r-element subsets of $[n]$ ordered by inclusion. Also, let $\dim(k, r; n)$ denote the dimension of $\mathbf{P}(k, r; n)$. So $\mathbf{S}_n$ is isomorphic to $\mathbf{P}(1, n-1; n)$ and $\dim(1, n-1; n) = n$.

Our second example of a family of posets of large dimension is $\{\mathbf{P}(1, 2; n) : n \geq 3\}$. In this case, there are $n(n-1)(n-2)/2$ incomparable pairs; however, an easy exercise shows that a linear extension may reverse $n(n-1)(n-2)/6$ critical pairs. So the "pigeon hole" argument used for the first example shows only that $\dim(1, 2; n) \geq 3$. However, we claim that $\lim_{n\to\infty} \dim(1, 2; n) = \infty$, although the argument now requires some elementary Ramsey theory. Suppose to the contrary that there exists a positive integer $t \geq 3$ so that $\dim(1, 2; n) \leq t$, for every $n \geq 3$. We obtain a contradiction when n is sufficiently large. Let $\mathcal{R} = \{L_1, L_2, \ldots, L_t\}$ be a realizer of $\mathbf{P}(1, 2; n)$. For each 3-element subset $\{i < j < k\} \subseteq [n]$, consider the critical pair $(\{j\}, \{i, k\})$, and choose an integer $\alpha \in [t]$ so that L_j reverses it, i.e., $\{j\} > \{i, k\}$ in L_α. Then we have a coloring of the 3-element subsets of $[n]$ with t colors. If n is sufficiently large, then (by Ramsey's theorem) there exists a 4-element subset $H = \{i < j < k < l\} \subseteq [n]$ and an integer $\alpha \in [t]$ so that all 3-element subsets of H are mapped to α. This means that $\{j\} > \{i, k\} > \{k\} > \{j, l\} > \{j\}$ in L_α, which is a contradiction.

Each of the first two examples is a height 2 poset, so posets of bounded height can have arbitrarily large dimension. Our third example is different. In this family, large height is required for large dimension. For each $n \geq 3$, let $\mathbf{I}(n) = (I_n, P_n)$ denote the poset defined by setting I_n to be the family of all 2-element subsets of $[n]$ with $\{i, j\} < \{k, l\}$ in P_n when $1 \leq i < j < k < l \leq n$. Again, we claim that $\lim_{n\to\infty} \dim(I_n, P_n) = \infty$. Suppose to the contrary that $\dim(I_n, P_n) \leq t$, for all $n \geq 3$. We obtain a contradiction when n is large.

For each 3-element subset $\{i < j < k\} \subseteq [n]$, $(\{i, j\}, \{j, k\})$ is a critical pair, so if $\mathcal{R} = \{L_1, L_2, \ldots, L_t\}$ is a realizer of P_n, then we may choose $\alpha \in [t]$ so that $(\{i, j\}, \{j, k\})$ is reversed in L_α. This is a coloring of the 3-element subsets of $[n]$ with t colors, so that if n is sufficiently large, there exists a 4-element subset $H = \{i < j < k < l\}$ and an integer $\alpha \in [t]$ so that all 3-element subsets of H are mapped to α. This implies that $\{i, j\} > \{j, k\} > \{k, l\} > \{i, j\}$ in L_α, which is a contradiction.

This last example is drawn from the family of posets known as *interval orders* and we will have more to say about them in Sections 5, 6 and 7.

3. Early applications of Ramsey theory to posets

Given a finite set S and an integer k with $0 \le k \le |S|$, we denote the set of all k-element subsets of S by $\binom{S}{k}$. Given integers t and k and finite sets $S_1, S_2, \dots, S_t$, an element of $\binom{S_1}{k} \times \binom{S_2}{k} \times \cdots \times \binom{S_t}{k}$ is called a *grid* (also, a $\mathbf{k}^t$ grid), and the sets $S_1, S_2, \dots, S_t$ are called *factor sets* of the grid. Using the natural order, a set of n integers is just an n-element chain, so considered as a poset, $S_1 \times S_2 \times \cdots \times S_t$ is isomorphic to $\mathbf{n}_1 \times \mathbf{n}_2 \times \cdots \times \mathbf{n}_t$, where $n_i = |S_i|$ for $i = 1, 2, \dots, t$.

The following theorem, called the Product Ramsey Theorem and stated here in poset form, has been applied in several different settings to posets. We refer the reader to [**14**] for the proof.

THEOREM 3.1. *Given positive integers m, k, r and t, there exists an integer n_0 so that if $n \ge n_0$ and f is any map which assigns to each $\mathbf{k}^t$ grid of $\mathbf{n}^t$ a color from $[r]$, then there exists a subposet $\mathbf{P}$ isomorphic to $\boldsymbol{m}^t$ and a color $\alpha \in [r]$ so that $f(g) = \alpha$ for every $\mathbf{k}^t$ grid g from $\mathbf{P}$.* □

Although it was not originally stated in these terms, most likely the first application of the product Ramsey theorem to posets can be traced to the proof of the following theorem [**31**].

THEOREM 3.2. *Let $\mathbf{P} = (X, P)$ be a poset and let $A \subseteq X$ be an antichain with $X - A \neq \emptyset$. Then*

$$\dim \mathbf{P} \le 1 + 2\,\mathrm{width}(X - A, P(X - A)).$$

□

The inequality in Theorem 3.2 is quite straightforward, but it takes a non-trivial Ramsey theoretic argument to show that it is best possible. It is interesting to note that when the proof of Theorem 3.2 was first published, the use of Ramsey theory to prove that the inequality is best possible was not the main point. Instead as suggested by the title of the paper, it was the ensuing corollary: irreducible posets of large height exist. Some years later, explicit constructions for irreducible posets with large height would be given by Trotter and Ross [**37**], [**38**] and by Kelly [**16**].

Here is another application of Ramsey theory, one which deals with the concept of *ranking functions* as developed by Walker in [**41**]. For integers n and r, let $L(n, r)$ denote the set of all r-term non-decreasing sequences from $[n]$. We then let $\mathbf{L}(n, r) = \big(L(n, r), P(n, r)\big)$ be the subposet of $\mathbb{R}^r$ induced by $L(n, r)$. We view the elements of $L(n, r)$ as possible results of a series of races among n competitors in which ties are allowed. For example, the element $(5, 5, 3, 3, 2, 1)$ of $L(10, 6)$ represents an outcome for a single competitor of two fifth place finishes, two third place finishes, one second place finish, and one first place finish. Note that the notation does not specify the individual races in which these respective placings were achieved.

Now suppose prize money is assigned to the finishing positions so that smaller finishing positions (corresponding to better achievement) receive higher monetary awards. Then total prize money determines a ranking function among the competitors (again, ties are allowed). Walker [**41**] proposed to call a linear extension L of $P(n, r)$ *consistent* if there was a way to assign monetary awards so that if x wins more money than y, then $x > y$ in L. He then proved that $L(n, r)$ is then the intersection of all its consistent linear extensions. Furthermore, when $r = 2$, $\mathbf{L}(n, 2)$ is a 2-dimensional poset and Walker showed that it is also the intersection of 2 consistent linear extensions. Walker also showed that $P(4, 3)$ is the intersection of

3 consistent linear extensions and conjectured that $L(n,r)$ is always the intersection of r consistent linear extensions.

However, in [**25**], Rödl and Trotter used Ramsey theory to show that for every $t \geq 3$, there exists an integer n_0 so that if $n > n_0$, $P(n,3)$ is not the intersection of t or fewer consistent linear extensions—even though it is 3-dimensional and thus the intersection of 3 linear extensions.

An *interval* in a poset $\mathbf{P} = (X, P)$ is just a subposet of the form $[x,y] = \{z \in X : x \leq z \leq y\}$, where $x < y$. In [**26**], Scheinerman defined the *poset boxicity* of a graph $\mathbf{G} = (V, E)$ as the least t for which $\mathbf{G}$ is the intersection graph of a family of intervals in a t-dimensional poset. In [**39**], Trotter and West show that a graph on n vertices has poset boxicity at most $O(\log\log n)$. They also use Ramsey theory to show that there exist posets of arbitrarily large poset boxicity—although the lower bound grows considerably more slowly than the upper bound.

4. Computational Aspects

Mirroring the general flavor of Ramsey theory, there are instances in which the major emphasis is on whether a Ramsey theoretic result is true—and in such cases, it is rarely possible to make precise estimates as to how large the parameters must be. However, in other instances, the existence question is relatively straightforward, so researchers try to make a precise determination for the parameters (or at least a relatively accurate asymptotic estimate). Here are a few examples involving posets.

In [**23**], Nešetřil and Rödl show that for every positive integer h, if $\mathbf{P} = (X, P)$ is a poset of height h, then there exists a poset $\mathbf{Q} = (Y, Q)$ of height $2h - 1$ so that if the points of $\mathbf{Q}$ are assigned to two colors (say by a mapping to [2]), then there is a monochromatic subposet isomorphic to $\mathbf{P}$. The value $2h-1$ is clearly best possible. This work is closely related to Nešetřil and Rödl's well known construction of graphs (and hypergraphs) with large chromatic number and large girth [**22**].

In [**18**], Kierstead and Trotter study the dual problem. Given an integer w, find the least integer $f(w)$ so that if $\mathbf{P} = (X, P)$ is a poset of width w, then there exists a poset $\mathbf{Q} = (Y, Q)$ of width $f(w)$ so that if the points of $\mathbf{Q}$ are two colored, then there exists a monochromatic subposet isomorphic to $\mathbf{P}$. It is elementary to show that $2w - 1 \leq f(w) \leq w^2$, but Kierstead and Trotter prove that $f(w) > 2w - 1$. Subsequently, Kierstead [**17**] showed that $f(w) > 5w/2$, but it is still not known whether $f(w) = O(w)$.

Given a poset $\mathbf{P} = (X, P)$, a function $f : X \to X$ is called a *regression* if $f(x) \leq x$ for all $x \in X$. A regression is called a *choice function* if $f(x)$ is a minimal element for all $x \in X$. Given a regression f on a poset $\mathbf{P} = (X, P)$, a k-element chain $C = \{x_1 < x_2 < \cdots < x_k\}$ is *f-monotone* if $f(x_1) \leq f(x_2) \leq \cdots \leq f(x_k)$. Note that if f is a choice function, then the statement that C is a f-monotone chain just means that f is constant on C.

For a positive integer n, let $\mathbf{B}_0(n)$ denote the poset consisting of all non-empty subsets of $[n]$. In [**24**], Perry proved that for each $k \geq 1$, if $n \geq 2^{k-1}$, then any choice function f on $\mathbf{B}_0(n)$ is constant on a chain of cardinality k. Furthermore, this result is best possible.

In [**42**], West, Trotter, Peck and Schor prove that if w and k are positive integers, then any regression on a poset of width at most w having at least $(w+1)^{k-1}$ points has a f-monotone chain of cardinality k. Furthermore, this result is best possible.

For a positive integer $n \geq 2$, let $\mathbf{P}_n$ denote the set of closed intervals of $\mathbb{R}$ with integer endpoints from $[n]$, partially ordered by inclusion. Evidently, $\mathbf{P}_n$ is a 2-dimensional poset. In [**3**], Alon, Trotter and West study the problem of determining the largest integer $f(n)$ for which every regression on $\mathbf{P}_n$ has a monotone chain of cardinality $f(n)$. They show that

$$\log^*(n) - 2 \leq f(n) \leq \log^*(n).$$

5. Shift Graphs, Interval Orders and Layers

Given integers n and k with $1 \leq k < n$, we call an ordered pair (A, B) of k-element sets a (k, n)-*shift pair* if there exists a $(k+1)$-element subset $C = \{i_1 < i_2 < \cdots < i_{k+1}\} \subseteq [n]$ so that $A = \{i_1, i_2, \ldots, i_k\}$ and $B = \{i_2, i_3, \ldots, i_{k+1}\}$. We then define the (k, n)-*shift graph* $\mathbf{S}(k, n)$ as the graph whose vertex set consists of all k-element subsets of $[n]$ with a k-element set A adjacent to a k-element set B exactly when (A, B) is a (k, n)-shift pair. It is customary to call a $(2, n)$-shift graph just a *shift* graph; similarly, a $(3, n)$-shift graph is called a *double shift graph.*

Note that a $(1, n)$-shift graph is just a complete graph on n vertices, but for $k \geq 2$, $\mathbf{S}(k, n)$ is triangle-free. Now it is an immediate consequence of Ramsey's theorem that for any fixed $k \geq 1$, $\lim_{n\to\infty} \chi(\mathbf{S}(k, n) \to \infty$. However, as in the preceding section, relatively accurate estimates are known for how fast the chromatic number of $\mathbf{S}(k, n)$ is growing. As mentioned previously, $\chi(\mathbf{S}(1, n) = n$, and for $k = 2$, it is folklore that $\chi(\mathbf{S}(2, n) = \lceil \lg n \rceil$. For $k = 3$, the chromatic number of $\mathbf{S}(3, n)$ is just the least t for which there are n antichains in $\mathbf{2^t}$, the lattice of all subsets of $[t]$ ordered by inclusion (see [**32**] for additional details and a discussion for larger values of k). In turn, the problem of estimating the number of antichains in $\mathbf{2^t}$ is a classical problem known as *Dedekind's problem.* Although no closed form solution is known, there are relatively accurate estimates, e.g., see Kleitman and Markowsky [**20**], and these are enough to establish the following asymptotic formula:

$$\chi\big(\mathbf{S}(3, n)\big) = \lg\lg n + \big(1/2 + o(1)\big) \lg\lg\lg n.$$

A poset $\mathbf{P} = (X, P)$ is called an *interval order* if there exists a function F which assigns to each element $x \in X$ a closed interval $[l_x, r_x]$ of the real line $\mathbb{R}$ so that $x < y$ in P if and only if $r_x < l_y$ in $\mathbb{R}$. The poset $\mathbf{I}_n$ introduced in Section 2 is called the *canonical interval order.* Although posets of height 2 can have arbitrarily large dimension (crowns, for example), this is not true for interval orders. For a positive integer n, let $d(n)$ denote the maximum dimension of an interval order of height n. In [**12**], Füredi, Hajnal, Rödl and Trotter exploit the connection with double shift graphs to show that:

$$d(n) = \lg\lg n + \big(1/2 + o(1)\big) \lg\lg\lg n.$$

Combining techniques from Spencer [**29**] with the estimate for the chromatic number of double shift graphs, Trotter [**32**] showed that the same estimate holds for $\dim(1, 2; n)$.

$$\dim(1, 2; n) = \lg\lg n + \big(1/2 + o(1)\big) \lg\lg\lg n.$$

6. Ramsey Trails in Interval Orders

In each of the next four sections, we outline a recent application of Ramsey theory to posets. Our first example involves the concept of Ramsey trails. Suppose $\mathcal{C}$ is a class of finite structures with a well defined notion of substructure, which is denoted $\subset$. Also suppose that f is a monotonic function mapping $\mathcal{C}$ to $\mathbb{R}_0$, i.e., if $G \subset H$, then $f(G) \leq f(H)$. A sequence $\mathcal{T} = \{G_n : n \geq 1\}$ of structures is called a *Ramsey trail* if

1. $G_n \subset G_{n+1}$ for all $n \geq 1$, and
2. $\lim_{n\to\infty} f(G_n) = \infty$.

Now suppose that r is a positive integer and that $\mathcal{T}_i = \{G_{i,n} : n \geq 1\}$ is a Ramsey trail for each $i = 1, 2, \ldots, r$. We say that this family *characterizes* f if for every integer t, there exists an integer s, so that if G is any structure in $\mathcal{C}$ with $f(G) > s$, then there is an integer $i \in [r]$ and an integer $n \geq 1$ so that $G_{i,n} \subset G$ and $f(G_{i,n}) > t$. The least r for which such a family exists is then called the *Ramsey complexity* of the function f. For example, consider the class of all graphs and the function f which assigns to a graph G the number of vertices in the graph. Then it follows that the Ramsey complexity of f is 2. This is evidenced by two ramsey trails, the set of all independent graphs and the set of all complete graphs. On the other hand, the existence of graphs with large girth and large chromatic number is enough to show that chromatic number cannot be characterized by any finite number of Ramsey trails. So the Ramsey complexity of the function χ is infinite.

Nevertheless, it is an important topic in graph theory to identify classes of discrete structures and monotonic functions defined on them for which the Ramsey complexity is finite. It is of special interest to recognize when it is 1. For example, a well studied problem in graph theory is to investigate classes of graphs for which chromatic number can be bounded as a function of maximum clique size. Such classes are said to be *χ-bounded*. As just a single example, Gyárfás [**15**] has shown that the set of circle graphs (intersection graphs of chords of a circle) is χ-bounded. To date the best result on this subject is due to Kostochka and Kratochvíl [**21**] who showed that a circle graph with maximum clique size ω has chromatic number $O(2^\omega)$.

For posets in general, dimension cannot be characterized by any finite number of interval orders, so dimension has infinite Ramsey complexity. But for many years, it was believed that the Ramsey complexity of dimension is 1 for the class of interval orders, and that dimension for interval orders could be characterized by a single Ramsey trail, namely the family of canonical interval orders. This conjecture has recently been settled in the affirmative by Kierstead and Trotter [**19**]. Since every interval order is a subposet of a sufficiently large canonical interval order, their theorem has the following attractive reformulation.

THEOREM 6.1 *For every interval order* $\mathbf{P}$, *there exists an integer* t, *so that if* $\mathbf{Q}$ *is any interval order with dimension at least* t, *then* $\mathbf{P}$ *is isomorphic to a subposet of* $\mathbf{Q}$. □

7. Fractional Dimension for Interval Orders

Let $\mathbf{P} = (X, P)$ be a poset and let $\mathcal{F} = \{M_1, \ldots, M_t\}$ be a multiset of linear extensions of P. Brightwell and Scheinerman [**5**] call $\mathcal{F}$ a *k–fold realizer* of P if for each incomparable pair (x, y), there are at least k linear extensions in $\mathcal{F}$ which reverse the pair (x, y), i.e., $|\{i : 1 \leq i \leq t, x > y \text{ in } M_i\}| \geq k$. The *fractional dimension* of $\mathbf{P}$, denoted by $\text{fdim}(\mathbf{P})$, is then defined as the least real number $q \geq 1$ for which there exists a k–fold realizer $\mathcal{F} = \{M_1, \ldots, M_t\}$ of P so that $k/t \geq 1/q$ (it is easily verified that the least upper bound of such real numbers q is indeed attained). Using this terminology, the *dimension* of $\mathbf{P}$ is just the least t for which there exists a 1–fold realizer of P. It follows immediately that $\text{fdim}(\mathbf{P}) \leq \dim(\mathbf{P})$, for every poset $\mathbf{P}$.

The *maximum degree*, denoted $\Delta(\mathbf{P})$, of a poset $\mathbf{P} = (X, P)$ is just the maximum degree in the associated comparability graph, i.e., the maximum number of points comparable to any one point. The dimension of a poset is bounded in terms of its maximum degree. The following upper bound is due to Füredi and Kahn [**13**].

THEOREM 7.1. *If* $\mathbf{P} = (X, P)$ *is a poset and* $\Delta(\mathbf{P}) \leq k$, *then* $\dim(\mathbf{P}) \leq 50k \log^2 k$. □

The best lower bound to date is due to Erdös, Kierstead and Trotter [**7**].

THEOREM 7.2. *There exists an absolute constant* $\epsilon > 0$ *so that for each* $k \geq 1$, *there exists a poset* $\mathbf{P}$ *with* $\Delta(\mathbf{P}) = k$ *and* $\dim(\mathbf{P}) > \epsilon k \log k$. □

For fractional dimension, the corresponding problem is much cleaner. Brightwell and Scheinerman [**5**] proved that if $\mathbf{P}$ is a poset with $\Delta(\mathbf{P}) = k$, then $\text{fdim}(\mathbf{P}) \leq k+2$. They conjectured that this inequality could be improved to $\text{fdim}(\mathbf{P}) \leq k+1$. This was proved by Felsner and Trotter [**9**], and the argument yielded a much stronger conclusion, a result with much the same flavor as Brooks' theorem for graphs.

THEOREM 7.3. *Let* k *be a positive integer, and let* $\mathbf{P}$ *be any poset with* $\Delta(\mathbf{P}) = k$. *Then* $\text{fdim}(\mathbf{P}) \leq k + 1$. *Furthermore, if* $k \geq 2$, *then* $\text{fdim}(\mathbf{P}) < k + 1$ *unless one of the components of* $\mathbf{P}$ *is isomorphic to* $\mathbf{S}_{k+1}$, *the standard example of a poset of dimension* $k + 1$. □

Felsner and Trotter [**9**] derive several other inequalities for fractional dimension, and these lead to some challenging problems as to the relative tightness of inequalities similar to the one given in the preceding theorem.

Fractional dimension is also relatively well behaved on the class of interval orders. In [**5**], Brightwell and Scheinerman proved that the fractional dimension of an interval order was less than 4, and they conjectured that this result was best possible. In [**40**], Trotter and Winkler settled this conjecture in the positive.

THEOREM 7.4. *For every* $\epsilon > 0$, *there exists an interval order* $\mathbf{P}$ *with* $\text{fdim}(\mathbf{P}) > 4 - \epsilon$. □

The techniques and concepts introduced by Trotter and Winkler in [**40**] are likely to be more important than the theorem which motivated the work in the first place. Specifically, they ask what common patterns must appear in arbitrary probability spaces, provided that the space contain events corresponding to subsets of a sufficiently large finite set. This question must first be discretized, and this is accomplished by considering approximations.

8. Circle Orders and Sphere Orders

Given a partially ordered set (poset) $\mathbf{P} = (X, P)$, a function F which assigns to each $x \in X$ a set $F(x)$ is called an *inclusion representation* of $\mathbf{P}$ if $x \leq y$ in P if and only if $F(x) \subseteq F(y)$. Every poset has such a representation. For example, just take $F(x) = \{y \in X : y \leq x \text{ in } P\}$. We refer the reader to the recent paper by Fishburn and Trotter [**11**] for additional background material on geometric inclusion representations and an extensive bibliographic listing.

As is well known, the finite posets of dimension at most two are just those which have inclusion representations using closed intervals of the real line $\mathcal{R}$. Because a closed interval of $\mathbb{R}$ can also be considered as a sphere in $\mathbb{R}^1$, it is natural to ask which posets have inclusion representations using disks (circles) in $\mathbb{R}^2$. For historical reasons, these posets are called *circle orders*. Fishburn [**10**] showed that all interval orders are circle orders. Also, the so called *standard examples* of n-dimensional posets, the 1-element and $(n-1)$-element subsets of $\{1, 2, \ldots, n\}$, ordered by inclusion, are circle orders. So among the circle orders are *some* posets of arbitrarily large dimension.

Call a poset $\mathbf{P}$ a *sphere order* if there is some $d \geq 1$ for which it has an inclusion representation using spheres in $\mathbb{R}^d$. Using the "degrees of freedom" theorem of Alon and Scheinerman [**2**], it follows that not all posets of dimension $d+2$ have inclusion representations using spheres in $\mathbb{R}^d$. In particular, when $d = 2$, we conclude that there are 4-dimensional posets which are not circle orders.

In [**27**], Scheinerman and Wierman used Ramsey theory to show that the countably infinite 3-dimensional poset $\mathbb{Z}^3$ is not a circle order.

These results leave open the following question:

QUESTION 8.1. *Is every finite 3-dimensional poset a circle order?* □

A somewhat more general question was posed by Brightwell and Winkler in [**6**].

QUESTION 8.2. *Is every finite poset a sphere order?* □

Using Ramsey theoretic techniqes which extend the product Ramsey theorem, both Question 1 and Question 2 are settled by the following theorem of Felsner, Fishburn and Trotter [**8**].

THEOREM 8.3. *There exists an integer n_0 so that if $n > n_0$, the finite 3-dimensional poset $\mathbf{n}^3$ is not a sphere order.* □

The techniques developed in [**8**] are likely to have applications to other combinatorial problems, especially the use of ramsey theory to control error in approximations and the concept of *uniform* induced functions.

9. Extremal Problems for Posets

In [**1**], Agnarsson, Felsner and Trotter study a natural extremal problem for posets, a problem which in fact was motivated by questions in ring theory. With a finite graph $\mathbf{G} = (V, E)$, associate a partially ordered set $\mathbf{P} = (X, P)$ defined by setting $X = V \cup E$ and $x < e$ in P if and only if x is an endpoint of e in $\mathbf{G}$. This poset is called the *incidence poset* of $\mathbf{G}$, and the extremal problem investigated in [**1**] is then to determine the maximum number $\mathrm{M}(p, d)$ of edges in a graph on p nodes if its incidence poset has dimension at most d.

The starting point for this research is the following well known theorem of W. Schnyder [**28**].

THEOREM 9.1. *A graph $\mathbf{G}$ is planar if and only if the dimension of its incidence poset is at most* 3. □

As an immediate consequence of Schnyder's theorem, $\mathrm{M}(p, 3)$ is just the maximum number of edges in a planar graph on p vertices, so $\mathrm{M}(p, 3) = 3p - 6$ for all $p \geq 3$.

We can also determine the exact value of $M(p,2)$, since the incidence poset of a graph has dimension at most 2 if and only if it is either a path or a subgraph of a path. It follows that $M(p,2) = p - 1$, for all $p \geq 2$.

For $d \geq 4$, it is likely to be very difficult to determine $M(p,d)$ precisely, except when p is relatively small in comparison to d. For this reason, it seems more reasonable to concentrate on asymptotic results for fixed d with $p \to \infty$. For $d = 4$, Agnarsson, Felsner and Trotter provide the following formula.

THEOREM 9.2.

$$\lim_{p\to\infty} \frac{M(p,4)}{p^2} = \frac{3}{8}.$$

The proof of this theorem requires several powerful combinatorial tools, including the product Ramsey theorem, Turán's theorem and the Erdős/Stone theorem. We refer the reader to [**1**] for the proof and additional details on the connections with ring theory.

References

[1] G. Agnarsson, S. Felsner and W. T. Trotter, The maximum number of edges in a graph of bounded dimension, and ring theoretic consequences, *Discrete Math.*, to appear.

[2] N. Alon and E. R. Scheinerman, Degrees of freedom versus dimension for containment orders, *Order* **5** (1988), 11–16.

[3] N. Alon, W. T. Trotter and D. B. West, Regressions and monotone chains II: The poset of integer intervals, *Order* **4**, (1987), 217–223.

[4] G. R. Brightwell, Partial orders, in *Graph Connections: Relationships between graph theory and other areas of mathematics*, L.W. Beineke and R.J. Wilson, eds., Oxford University Press, Oxford (1997) pp52–69.

[5] G. R. Brightwell and E. R. Scheinerman, Fractional dimension of partial orders, *Order* **9** (1992), 139–158.

[6] G. R. Brightwell and P. M. Winkler, Sphere orders, *Order* **6** (1989), 235–240.

[7] P. Erdös, H. Kierstead and W. T. Trotter, The dimension of random ordered sets, *Random Structures and Algorithms* **2** (1991), 253–275.

[8] S. Felsner, P. C. Fishburn and W. T. Trotter, Finite Three dimensional posets which are not sphere orders, *Discrete Math.*, to appear.

[9] S. Felsner and W. T. Trotter, On the fractional dimension of partially ordered sets, *Discrete Math.* **136** (1994), 101–117.

[10] P. C. Fishburn, Interval orders and circle orders, *Order* **5** (1988), 225–234.

[11] P. C. Fishburn and W. T. Trotter, Geometric containment orders: A survey, preprint.

[12] Z. Füredi, P. Hajnal, V. Rödl and W. T. Trotter, Interval orders and shift graphs, in *Sets, Graphs and Numbers*, A. Hajnal and V. T. Sos, eds., Colloq. Math. Soc. Janos Bolyai **60** (1991), 297–313.

[13] Z. Füredi and J. Kahn, On the dimensions of ordered sets of bounded degree, *Order* **3** (1986) 17–20.

[14] R. L. Graham, B. L. Rothschild and J. H. Spencer, *Ramsey Theory*, 2nd Edition, J. H. Wiley, New York, 1990.

[15] A. Gyárfás, On the chromatic number of multiple interval graphs and overlap graphs, *Discrete Math.* **55** (1985), 161–166.

[16] D. Kelly, On the dimension of partially ordered sets, *Discrete Math.* **35** (1981), 135–156.

[17] H. Kierstead, personal communication.

[18] H. A. Kierstead and W. T. Trotter, A ramsey-theoretic problem for finite ordered sets, *Discrete Math.* **63** (1987), 217–223.

[19] H. A. Kierstead and W. T. Trotter, Interval orders and dimension, *Discrete Math.*, to appear.

[20] D. J. Kleitman and G. Markovsky, On Dedekind's problem: The number of isotone boolean functions, II, *Trans. Amer. Math. Soc.* **213** (1975), 373–390.

[21] A. Kostochka and J. Kratochvíl, Covering and coloring polygon-circle graphs, *Discrete Math.* **163** (1997) 299–305.

[22] J. Nešetřil and V. Rödl, A short proof of the existence of highly chromatic graphs without short cycles, *J. Comb. Theory B* **27** (1979), 225–227.

[23] J. Nešetřil and V. Rödl, Combinatorial partitions of finite posets and lattices—Ramsey lattices, *Algebra Universalis* **19** (1984) 106–119.

[24] R. L. Perry, Representatives of subsets, *J. Comb. Theory* **3** (1967) 302–304.

[25] V. Rödl and W. T. Trotter, A note on ranking functions, *Discrete Math.* **67** (1987), 307–309.

[26] E. R. Scheinerman, Intersection graphs and multiple intersection parameters of graphs, Ph. D. thesis, Princeton Univ. (1984).

[27] E. R. Scheinerman and J. C. Wierman, On circle containment orders, *Order* **4** (1988), 315–318.

[28] W. Schnyder, Planar graphs and poset dimension, *Order* **5** (1989), 323–343.

[29] J. Spencer, Minimal scrambling sets of simple orders, *Acta Math. Acad. Sci. Hungar.* **22**, (1971) 349–353.

[30] W. T. Trotter, Dimension of the crown $\mathbf{S}_n^k$, *Discrete Math.* **8** (1974), 85–103.

[31] W. T. Trotter, Irreducible posets with arbitrarily large height exist, *J. Comb. Theory A* **17** (1974), 337–344.

[32] W. T. Trotter, *Combinatorics and Partially Ordered Sets: Dimension Theory*, The Johns Hopkins University Press, Baltimore, Maryland, 1992.

[33] W. T. Trotter, Progress and new directions in dimension theory for finite partially ordered sets, in *Extremal Problems for Finite Sets*, P. Frankl et al., eds., Bolyai Soc. Math. Studies **3**, 1994, 457–477.

[34] W. T. Trotter, Partially ordered sets, in *Handbook of Combinatorics*, R. L. Graham, M. Grötschel, L. Lovász, eds., Elsevier, Amsterdam, Volume I (1995), 433–480.

[35] W. T. Trotter, Graphs and partially ordered sets, Congressus Numerantium **116** (1996), 253–278.

[36] W. T. Trotter, New perspectives on interval orders and interval graphs, in *Surveys in Combinatorics*, R. A. Bailey, ed., London Mathematical Society Lecture Note Series **241** (1997), 237–286.

[37] W. T. Trotter and J. Ross, Every t–irreducible partial order is a subposet of a $t+1$–irreducible partial order, *Annals of Discrete Math.* **17** (1983), 613–621.

[38] W. T. Trotter and J. Ross, For $t \geq 3$, every t–dimensional partial order can be embedded in a $t+1$–irreducible partial order, in *Finite and Infinite Sets*, A. Hajnal, L. Lovász, and V. T. Sös, eds., Colloq. Math. Soc. J. Bolyal **37** (1984), 711–732.

[39] W. T. Trotter and D. B. West, Poset boxicity of graphs, *Discrete Math.* **64** (1987), 105–107.

[40] W. T. Trotter and P. Winkler, Ramsey theory and sequences of random variables, *Combinatorics, Probability and Computing*, to appear.

[41] W. J. Walker, Ranking functions and axioms for linear orders, *Discrete Math.* **67** (1987) 299–306.

[42] D. B. West, W. T. Trotter, G. W. Peck and P. Schor, Regressions and monotone chains: A ramsey-type extremal problem for partial orders, *Combinatorica* **4** (1984), 117–119.

Department of Mathematics, Arizona State University, Tempe, Arizona 85287, U.S.A.
E-mail address: `trotter@ASU.edu`

DIMACS Series in Discrete Mathematics
and Theoretical Computer Science
Volume **49**, 1999

Generalizations of Davenport–Schinzel Sequences

Pavel Valtr

ABSTRACT. In this survey paper we describe results concerning two generalizations of Davenport–Schinzel sequences — generalized Davenport–Schinzel sequences and Davenport–Schinzel trees. We also describe applications of generalized Davenport–Schinzel sequences to Turán type problems for geometric graphs.

We say that two sequences of equal length are *isomorphic* if any of them can be changed to the other one by a bijective renaming of elements appearing in the sequence (e.g., the sequences $abcb$, $yxzx$ of length 4 are isomorphic). We say that a sequence is k-sparse if no element appears twice in any contiguous part of length at most k.

Let u be a sequence over k elements and let n be an integer. A (n, u)-*sequence* (or *generalized* (n, u) *Davenport–Schinzel sequence*) is defined as a k-sparse sequence over at most n elements containing no subsequence isomorphic to u. The problem of estimating the maximum length $f_u(n)$ of a (n, u)-sequence is the central problem concerning generalized Davenport–Schinzel sequences. We survey results related to this problem, and prove most of them.

We remark that the classical Davenport–Schinzel sequences are described by the special case when the "forbidden" sequence pattern u is an alternating sequence $abab\ldots$ of some fixed length s.

Another generalization of Davenport–Schinzel sequences, Davenport–Schinzel trees, deals with labeled trees, where each vertex is assigned a label. In several variants, a (n, u) Davenport–Schinzel tree can be defined as a (rooted) tree with vertices labeled by n labels in such a way that, besides some technical conditions, the sequence of labels along any path (from the root) contains no subsequence isomorphic to u. We survey all known results about the maximum size of a (n, u) Davenport–Schinzel tree.

1. Introduction

This survey paper describes results concerning two generalizations of Davenport–Schinzel sequences — generalized Davenport–Schinzel sequences and Davenport–Schinzel trees.

1.1. Organization of the Paper. The paper is organized as follows. Section 1 contains an introduction and some basic results on generalized Davenport–Schinzel sequences. Further results on generalized Davenport–Schinzel sequences are described in Section 2. In Section 3 we describe and prove auxiliary lemmas

---

1991 *Mathematics Subject Classification.* Primary 52C10, 68R15, 68U05; Secondary 05C78.

Research was supported by DIMACS Center, by Czech Republic Grant GAČR 201/1996/0194, and by Charles University grants No. 193/1996 and 194/1996.

which are then used in Sections 4 and 5 to obtain two strongest known results about the asymptotic behavior of the maximum length of generalized Davenport–Schinzel sequences. Section 6 contains a lower bound construction for (generalized) Davenport–Schinzel sequences. Section 7 contains applications to geometric graphs. Section 8 is an introduction to Davenport–Schinzel trees. Section 10 describes further results on Davenport–Schinzel trees, and Section 11 contains a lower bound construction for Davenport–Schinzel trees. In Section 12 we briefly mention further generalizations of Davenport–Schinzel sequences.

1.2. Davenport–Schinzel Sequences. Motivated by a problem involving linear differential equations, Davenport and Schinzel [**DS65**] asked what is the maximum length $N_s(n)$ of a sequence over n letters containing no letter on two consecutive positions and no alternating subsequence $abab\ldots$ of length s. Since then many other applications of the functions $N_s(n)$, mostly in computational and discrete geometry, have been discovered (see [**SA95**]). It is trivial that $N_3(n) = n$ and not difficult to show that $N_4(n) = 2n - 1$ for any $n \geq 1$. Improving previous upper bounds and establishing the first non-linear lower bound on $N_5(n)$ Hart and Sharir [**HS86**] proved that $N_5(n) = \Theta(n\alpha(n))$ (see also [**Ko88, WS88**] for a simpler proof of the lower bound), where $\alpha(n)$ is the inverse of the Ackermann function (see Paragraph 1.7 for its definition). Currently, the best known bounds on $N_5(n)$ are

$$\frac{1}{2}n\alpha(n) - 2n \leq N_5(n) \leq 2n\alpha(n) + O(n\sqrt{\alpha(n)}).$$

The lower bound follows from a construction of Wiernik and Sharir [**WS88**] (see also [**SA95**]), the upper bound is given in Klazar's paper [**Kl99b**] in this volume.

Improving previous bounds of Sharir [**Sh87, Sh88**], the following bounds were proved in [**ASS89**] (see also [**SA95**]):

$$N_6(n) = \Theta(n2^{\alpha(n)}),$$

$$N_s(n) \approx n2^{\alpha(n)^{s/2-c}}, \quad \text{for } s \geq 7,$$

(the bound on $N_s(n)$ is simplified here – see [**SA95**] for details).

1.3. Basic Definitions and Notation. By a sequence, we always mean a finite sequence (possibly of length 0). Let u be a (finite) sequence. Then $|u|$ denotes the length of u, $S(u)$ denotes the alphabet of u, i.e., the set of elements (letters) appearing in the sequence u, and $||u|| = |S(u)|$ denotes the size of $S(u)$. For example, the sequence $u = 125625$ (for simplicity, we do not separate terms in a sequence by commas) has length $|u| = 6$, its alphabet is $S(u) = \{1, 2, 5, 6\}$, and thus $||u|| = 4$. Elements of $S(u)$ are called *letters*. For a letter $a \in S(u)$, an occurrence of a in u is called an *a-occurrence (in u)*. A contiguous part of a sequence is called *interval*. If an interval begins and ends with the same letter a, then we call it *a-interval*. We say that u is *k-sparse*, if no interval in u of length at most k contains two occurrences of the same letter, i.e., $|I| = ||I||$ for every interval I in u with $|I| \leq k$. Two sequences $u = a_1a_2\ldots a_k$ and $v = b_1b_2\ldots b_k$ of equal length are said to be *isomorphic*, if there is a bijection Ψ between $S(u)$ and $S(v)$ such that $b_i = \Psi(a_i)$ for each $i = 1, \ldots, k$. We say that a sequence v is *u-free* if it has no subsequence isomorphic to u. Otherwise we say that v *contains* u. For example, the sequences 12322 and 41311 are isomorphic. The sequence $v = 12321$ is 222-free but it contains 233 (since the subsequence 122 of v is isomorphic to 233).

The sequence 1341234 is 3-sparse but not 4-sparse, since its initial interval 1341 contains two 1-occurrences.

A sequence $aa\dots a$ consisting of k a-occurrences is shortly denoted by a^k. In a natural way, we write $u = u_1u_2\dots u_k$, if u is a concatenation of sequences (and letters) $u_1, u_2, \dots, u_k$. For a sequence u, $\overline{u}$ denotes the sequence u written in the reversed order. We say that a sequence u is *symmetric*, if it is isomorphic to $\overline{u}$.

1.4. Generalized Davenport–Schinzel Sequences. For an integer n and a sequence u, a u-free $||u||$-sparse sequence over at most n elements (letters) is called (n, u)-*sequence*. In the first six sections we survey and mostly also prove results about the maximum length $f_u(n)$ of a (n, u)-sequence. The requirement on the (n, u)-sequence to be $||u||$-sparse is a natural technical condition, because an arbitrarily long $(||u|| - 1)$-sparse sequence $a_1a_2\dots a_{||u||-1}a_1a_2\dots a_{||u||-1}u_1u_2\dots$ formed by periodically repeating occurrences of $||u|| - 1$ letters is u-free. Trivially, if u_1, u_2 are isomorphic then $f_{u_1}(n) = f_{u_2}(n)$ for all n.

The problem of estimating the function $f_u(n)$ was proposed on the combinatorial seminar lead by Jiří Matoušek and Jarik Nešetřil at the Charles University in Prague in 1988/1989. Since then, more than 10 papers and one Ph.D. thesis have been written on this topic and on the closely related topic of Davenport–Schinzel trees considered in Sections 8–11.

We remark that in two cases we change the notation usually used in the previous works. In most papers, the function $f_u(n)$ is denoted by $\mathrm{Ex}(u, n)$. Our notation reflects the fact that, first of all, $f_u(n)$ is studied asymptotically as a function of n when u is fixed. We also use a new notation "k-sparse" instead of the older "k-regular". In this case we think that the new notation is more suitable and we also achieve a unification with the notation used for the Davenport–Schinzel trees.

CLAIM 1. *The function $f_u(n)$ is finite for each sequence u and each integer n.*

We prove Claim 1 in Section 3. An almost linear upper bound on $f_u(n)$ was given by Klazar [**Kl92**].

THEOREM 2 (Klazar [**Kl92**]). *For any sequence u, there is a constant $c = c(u)$ such that*

$$f_u(n) \le n2^{c(\alpha(n))^{|u|-4}},$$

where $\alpha(n)$ is the functional inverse to the Ackermann function.

The Ackermann function and the notion of functional inverse are defined in Paragraph 1.7. We remark that the term $2^{c(\alpha(n))^{|u|-4}}$ appearing in Theorem 2 is a function which grows extremely slowly to infinity. In particular, it grows to infinity slower than the functional inverse of any unbounded primitive recursive function (see [**Ro67, SA95**]).

PROPOSITION 3. *For any sequence u, the function $f_u(n)$ behaves asymptotically in one of the following three ways:*

(i) $f_u(n) = O(1)$,
(ii) $f_u(n) = \Theta(n)$, *or*
(iii) $\frac{f_u(n)}{n}$ *tends very slowly to infinity as n tends to infinity (it tends to infinity slower than the functional inverse of any unbounded primitive recursive function).*

Proposition 3 is derived from the weak super-additivity of $f_u(n)$ (Lemma 16) in Paragraph 3.2.

It is easy to see that (i) in Proposition 3 holds if and only if $|u| = ||u||$ (more precisely, $f_u(n) = \min\{|u|-1, n\}$ holds in this case). Otherwise (i.e., if $|u| > ||u||$), (ii) or (iii) holds. The upper bound in Theorem 2 and the results on Davenport–Schinzel sequences mentioned in Paragraph 1.2 indicate that the problem to distinguish between the cases (ii) and (iii) might be very difficult:

PROBLEM 1. *Characterize sequences u with $f_u(n) = O(n)$.*

Problem 1 initiated most of the work covered by this survey. We describe all what is known about Problem 1 in Sections 2 and 6, and prove most of these results in Sections 3–6.

1.5. Applications to Geometric Graphs. One of the main aims of this paper is to give a clear writeup of a result on generalized Davenport–Schinzel sequences having applications to geometric graphs. The result is described in Theorem 6 in Paragraph 2.2. The paper is organized so that the reader interested only in the proof of this result and in its geometric applications does not have to read the other parts of the paper. The self-contained proof of Theorem 6 is given in Paragraphs 1.3, 1.4, 2.1, and 2.2, and in Section 4. The applications are described in Section 7.

The reader interested only in the statement of Theorem 6 and in its applications may skip from here to Paragraph 2.2 and then to Section 7.

1.6. Davenport–Schinzel Trees. Another generalization of Davenport–Schinzel sequences, Davenport–Schinzel trees, deals with labeled trees, where each vertex is assigned a label. In several variants, a (n, u) Davenport–Schinzel tree can be defined as a (rooted) tree with vertices labeled by n labels in such a way that, besides some technical conditions, the sequence of labels along any path (from the root) contains no subsequence isomorphic to u. In Sections 8–11 we survey all known results about the maximum size of a (n, u) Davenport–Schinzel tree.

1.7. The Ackermann Function, Functional Inverse. The Ackermann function (see e.g. [**Ro67, SA95**]) is defined recursively as follows. First, we define auxiliary functions $A_k(n), k = 1, 2, \ldots,$ in the following way. For a function g, we set $g^{(s)}(n) = g(g(g(\ldots(g(n))\ldots)))$, where g is applied s times. With this notation we define, for each $n \geq 1$,

$$\begin{aligned} A_1(n) &= 2n, \\ A_k(n) &= A_{k-1}^{(n)}(1), \quad k \geq 2. \end{aligned}$$

Note that $A_k(1) = 2$ for all k, $A_2(n) = 2^n$, and $A_3(n)$ is the "tower" function. The functions $A_k(n)$ are gradually faster and faster with increasing index k. The *Ackermann function* is defined as the diagonal function

$$A(n) = A_n(n).$$

The Ackermann function grows faster than any of the functions $A_k(n)$ and, in fact, it grows faster than any primitive recursive function (see [**Ro67**]).

Given an unbounded function g from $\mathbb{N}$ into itself we define its *functional inverse* γ from $\mathbb{N}$ into itself by

$$\gamma(n) = \min\{i \geq 1 : g(i) \geq n\}.$$

The function $\alpha(n)$ appearing in the estimates in Paragraph 1.2 and in Theorems 2, 30, and 31 is the functional inverse of the Ackermann function, i.e.,

$$\alpha(n) = \min\{i \geq 1 : A(i) \geq n\}.$$

2. (Non-)Linearity of $f_u(n)$

2.1. Sequences u with $f_u(n) = O(n)$. Problem 1 was solved for all sequences u with $||u|| \leq 2$ in [**AKV92**]:

THEOREM 4 (two-letter theorem [**AKV92**]). *If u is a sequence with $||u|| \leq 2$ then $f_u(n) = O(n)$ if and only if u is ababa-free.*

It is known that $f_u(n) \neq O(n)$ holds for any sequence u containing *ababa* and also for some other sequences u (see Paragraph 2.4 for details). The following theorem gives the class of all sequences for which $f_u(n) = O(n)$ is known:

THEOREM 5 ([**AKV92, KV94**]). *Let $\mathcal{L}$ denote the class of all sequences u with $f_u(n) = O(n)$. Then:*

(i) *The sequence a^i lies in $\mathcal{L}$ for any $i \geq 1$ (a is a letter).*
(ii) *(monotone property) If v contains u (i.e., is not u-free) and $v \in \mathcal{L}$ then $u \in \mathcal{L}$.*
(iii) *(letter duplication) Let u_1, u_2 be two sequences, let a be a letter, and let $u_1a^2u_2 \in \mathcal{L}$. Then $u_1a^3u_2 \in \mathcal{L}$.*
(iv) *(first–letter and last–letter duplications) Let u be a sequence and let a be a letter. Then, if $ua \in \mathcal{L}$ then $ua^2 \in \mathcal{L}$, and if $au \in \mathcal{L}$ then $a^2u \in \mathcal{L}$.*
(v) *(letter insertions) Let $u = au_1a^2u_2$ be a sequence in $\mathcal{L}$, where u_1, u_2 are sequences (possibly empty), a, b are letters, and $b \notin S(u)$. Then the sequence $u' = bau_1ab^2au_2$ also lies in $\mathcal{L}$.*
(vi) *(sequence insertion) Let u_1, u_2, w be three sequences, let a be a letter, let $u = u_1a^2u_2$, and let $S(u) \cap S(w) = \emptyset$. Moreover, let u, w lie in $\mathcal{L}$. Then also u_1awau_2 lies in $\mathcal{L}$.*
(vii) *(symmetry) If $u \in \mathcal{L}$, then $\overline{u} \in \mathcal{L}$.*

Theorem 5 implies Theorem 4:

Proof of Theorem 4. Theorem 5(i), (iii), (iv), (v) gives $f_u(n) = O(n)$ for any sequence $u = a^ib^ja^kb^l$, where a, b are letters and $i, j, k, l \geq 0$ are integers (parts (i),(v) yield $ab^ja^2b^l \in \mathcal{L}$, parts (iii) and (iv) consequently give $u = a^ib^ja^kb^l \in \mathcal{L}$).

This gives one implication in Theorem 4, the other implication follows from Theorem 5(ii) and $f_{ababa}(n) \neq O(n)$ (recall that $f_{ababa}(n) = N_5(n) = \Theta(n\alpha(n))$). □

The proof of Theorem 5, based on [**AKV92, KV94**], is given in this paper. Parts (i) and (vii) are trivial and parts (ii)-(vi) of Theorem 5 are special cases of more general results: Part (ii) follows from Lemma 14, part (iii) from Lemma 11, part (iv) from Lemma 12, part (v) from Theorem 18, and part (vi) from Theorem 20.

2.2. The Sequences updownup. For any two positive integers k, p, we define a sequence $updownup(k,p)$ of length $3kp$ as the sequence $w_k = v_k\overline{v_k}v_k$, where $v_k = a_1^pa_2^p \ldots a_k^p$, $\overline{v_k} = a_k^pa_{k-1}^p \ldots a_1^p$, and $a_1, \ldots, a_k$ are k distinct letters. Theorem 5(v) (or the more general Theorem 18) has the following important corollary with applications to geometric graphs (see Section 7).

THEOREM 6. *For any two positive integers k, p,*

$$f_{\mathrm{updownup}(k,p)}(n) = O(n).$$

Proof. Theorem 6 easily follows from Theorem 5. For $k = 1$, the statement follows from Theorem 5(i). We further proceed by induction on k. Suppose that $k \geq 1$ and that

$$v_k \overline{v_k} v_k = \mathrm{updownup}(k, p) \in \mathcal{L}.$$

Then, by Theorem 5(v),(vii),

$$v_k a_{k+1}^2 \overline{v_k} v_k a_{k+1} \in \mathcal{L}.$$

Consequently, Theorem 5(iii),(iv) yields

$$v_k a_{k+1}^{2p} \overline{v_k} v_k a_{k+1}^p = \mathrm{updownup}(k+1, p) \in \mathcal{L}.$$

□

2.3. Long 2-sparse Sequences with $f_u(n) = O(n)$. There are arbitrarily long sequences u over a small number of letters satisfying $f_u(n) = O(n)$ (e.g., the sequences $u = a^i$, a^ib^i, $a^ib^ia^i$, $a^ib^ia^ib^i$). However, all such long sequences are highly "non-sparse". The following problem was raised in [**KV94**]:

PROBLEM 2. *Determine the asymptotic behavior of the function*

$$h(k) = \max\{|u| : u \textit{ is a 2-sparse sequence with } ||u|| \leq k \textit{ and } f_u(n) = O(n)\}.$$

By Theorem 5(ii) and by $\frac{f_{ababa}(n)}{n} \to \infty$, any sequence with $f_u(n) = O(n)$ is necessarily *ababa*-free. It follows that $h(k) \leq N_5(k) = O(k\alpha(k))$. The best known lower bound on $h(k)$ is $h(k) \geq 3k-2$. In particular, it is not known whether $h(k) = O(k)$. This illustrates how far are the present results from answering Problem 1.

The best known lower bound $h(k) \geq 3k - 2$ can be derived from Theorem 6. Theorem 6 implies $f_{u_k}(n) = O(n)$ for the 2-sparse sequence u_k of length $3k - 2$ obtained from the sequence updownup$(k, 1)$ by removing the first occurrence of a_k and the last occurrence of a_1. This gives the bound $h(k) \geq 3k - 2$. (Another sequence giving $h(k) \geq 3k - 2$ is the sequence u from Theorem 43).

2.4. Sequences u with $f_u(n) \neq O(n)$. All sequences u containing *ababa* satisfy $f_u(n) = \Omega(n\alpha(n))$ by Theorem 5(ii) and by $N_5(n) = f_{ababa}(n) = \Omega(n\alpha(n))$. However, an analogue of Theorem 4 does not hold for the sequences u with $||u|| \geq 4$ — it was noticed in [**KV94**] that a slight modification of the construction of Wiernik and Sharir [**WS88**] giving $f_{ababa}(n) = \Omega(n\alpha(n))$ gives $f_u(n) = \Omega(n\alpha(n))$ for the *ababa*-free sequence $u = abcbadadbcd$ (see Section 6, see also [**Kl93**] for details).

Klazar [**Kl99b**] also gives an *ababa*-free sequence u with $f_u(n) = \Omega(n2^{\alpha(n)})$. He conjectures that for any $c > 0$ there is a *ababa*-free sequence u with $f_u(n) = \Omega(n2^{\alpha(n)^c})$.

2.5. 2-Sparse Sequences over 3 Letters with $f_u(n) = O(n)$. The above results enable us to decide the validity of $f_u(n) = O(n)$ for most 2-sparse sequences over at most 3 letters. Up to isomorphism, Fig. 1 shows all 2-sparse *ababa*-free sequences over at most 3 letters (recall that $f_u(n) = O(n)$ may hold only if u is *ababa*-free). Sequences of equal length are placed in a row (those of length 6 in two rows), non-symmetric sequences are placed in the pairs $\{u, \overline{u}\}$ in the dotted frames (clearly $f_u(n) = f_{\overline{u}}(n)$ holds for all u, n). The large rectilinear region contains

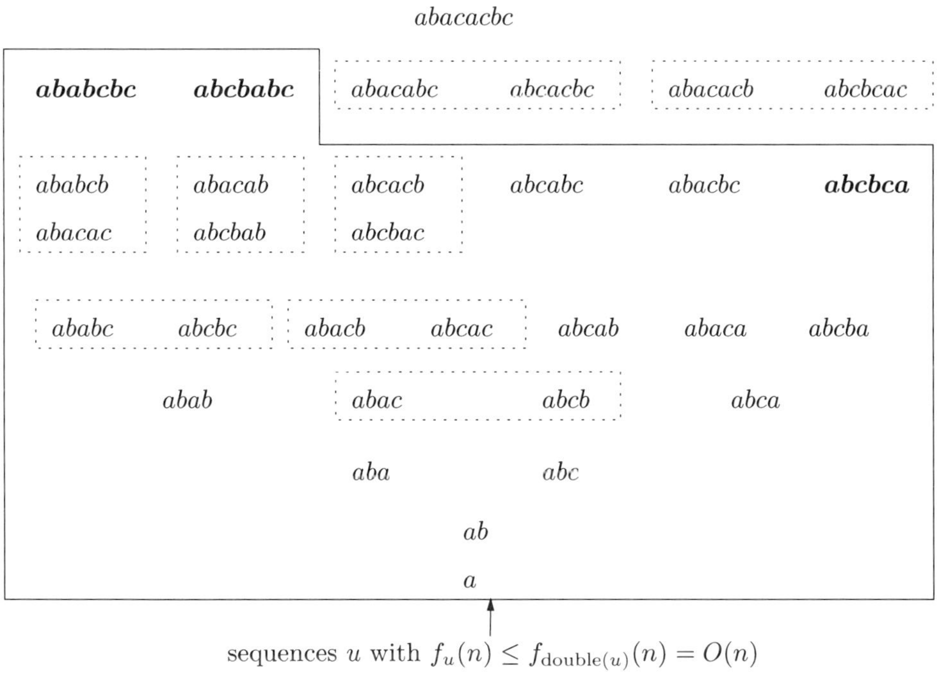

FIGURE 1. Table of *ababa*-free 2-sparse sequences over 3 letters. The rectilinear region contains sequences u with $f_u(n) \leq f_{\text{double}(u)}(n) = O(n)$. The validity of $f_u(n) = O(n)$ is not known for the remaining five sequences.

sequences u for which we can prove $f_u(n) = O(n)$. The validity of $f_u(n) = O(n)$ is unknown for the 5 sequences outside.

Interestingly, for all sequences inside the rectilinear region we can even prove that $f_{\text{double}(u)}(n) = O(n)$, where *double*$(u)$ is the sequence obtained from u by doubling each letter (e.g., if $u = aba$ then double$(u) = a^2b^2a^2$). We remark that, by Theorem 5(iii), $f_{\text{double}(u)}(n) = O(n)$ implies $f_{\text{double}(\text{double}(u))}(n) = O(n)$, $f_{\text{double}(\text{double}(\text{double}(u)))}(n) = O(n)$, etc. We now derive from Theorems 5 and 6 that indeed all sequences in the rectilinear region in Fig. 1 satisfy $f_{\text{double}(u)}(n) = O(n)$.

The rectilinear region contains three maximal sequences $t_1 = ababcbc$, $t_2 = abcbabc$, $t_3 = abcbca$ (typed in bold font in the figure). All other sequences are contained in t_1 or t_2 (e.g., the sequences *ababcb* and *abacbc* are contained in t_1 and the sequences *abacab*, *abcacb*, *abcabc* in t_2). Therefore, by Theorem 5(ii), it suffices to prove $f_{\text{double}(u)}(n) = O(n)$ for the three sequences $u = t_1, t_2, t_3$.

The sequence double(t_2) is contained in the sequence *updownup*$(3, 2)$ and therefore, by Theorem 6, $f_{\text{double}(t_2)}(n) = O(n)$.

We now apply Theorem 5(v) for the sequences $u = b^4c^2b^2c^2$ (which satisfies $f_u(n) = O(n)$ by Theorem 4), $u_1 = b$, and $u_2 = c^2b^2c^2$. We obtain $f_{u'}(n) = O(n)$ for the sequence $u' = ab^2a^2b^2c^2b^2c^2$. Theorem 5(iv) then gives $f_{\text{double}(t_1)}(n) = O(n)$.

Finally, $f_{\text{double}(t_3)}(n) = O(n)$ follows from $f_{a^4}(n) = O(n)$ and $f_{b^2c^2b^2c^2}(n) = O(n)$ by Theorem 5(vi).

3. Basic Tools and Properties of $f_u(n)$

In this section we survey (and also prove) basic tools and properties of $f_u(n)$ used in the study of generalized Davenport–Schinzel sequences. Most of them appeared first (explicitly or implicitly) in [**AKV92**] or in [**KV94**]. A new result is Proposition 3 which is derived from the weak super-additivity of $f_u(n)$ (Lemma 16) in Paragraph 3.2.

3.1. Tools and Properties Needed in Section 4. We say that a sequence u is *reducible*, if there is a non-trivial partition $u = u_1u_2$ of u into two intervals u_1 and u_2 such that $S(u_1) \cap S(u_2) = \emptyset$; otherwise we say that u is *irreducible*. If v is a (n, u)-sequence of the maximum length $f_u(n)$, then we say that v is (n, u)-*critical.*

First, we prove Claim 1 about the finiteness of $f_u(n)$ from the following easy observation.

OBSERVATION 7. *Let $a_1, \ldots, a_k$ be k distinct letters, let u be a sequence with $||u|| = k$, and let $v = v_1 \ldots v_{|u|}$ be a sequence partitioned into $|u|$ intervals v_i such that $\{a_1, \ldots, a_k\} \subseteq S(v_i)$ $(i = 1, 2, \ldots, |u|)$. Then v contains u.*

Proof. We take any sequence u' isomorphic to u such that $S(u') = \{a_1, \ldots, a_k\}$. We then find the sequence u' as a subsequence of v by appropriately choosing exactly one term in each interval v_i. □

Proof of Claim 1. We prove that $f_u(n) < ||u|| \cdot \binom{n}{||u||}|u|$. Let v be a (n, u)-sequence. Suppose to a contrary that the length of v is $||u|| \cdot \binom{n}{||u||}|u|$. We partition v into $\binom{n}{||u||}|u|$ intervals I_i of length $||u||$. Since v is $||u||$-sparse, each interval I_i satisfies $|I_i| = ||I_i|| = ||u||$. Thus, the alphabets $S(I_i)$ are $||u||$-element subsets of $S(v)$. Since $|S(v)| \leq n$ and there are $\binom{n}{||u||}|u|$ intervals I_i, there are $|u|$ intervals $I_{i_1}, \ldots, I_{i_{|u|}}$ with $S(I_{i_1}) = \cdots = S(I_{i_{|u|}})$. It follows from Observation 7 that the union of the intervals I_{i_j} contains u – a contradiction. □

The following lemma implies that, for any sequence u and for any integer $l \geq ||u||$, the maximum length of a u-free l-sparse sequence v with $||v|| \leq n$ is asymptotically $\Theta(f_u(n))$.

LEMMA 8 (sparsity lemma). *Let u be a sequence, and let $l \geq ||u||$. Then there exists a positive constant $\varepsilon = \varepsilon(u, l) > 0$ such that any u-free $||u||$-sparse sequence v has an l-sparse subsequence of length at least $\varepsilon|v|$.*

The proof of Lemma 8 relies on a simple and very useful greedy algorithm $A(l)$ introduced in [**AKV92**].

Algorithm $A(l)$. Let $l \geq 2$ be an integer. Given a sequence $u = a_1a_2 \ldots a_k$ of length k, the *algorithm* $A(l)$ finds an l-sparse subsequence $A(u, l)$ of u as follows. First, an auxiliary sequence B_0 is taken empty. Then, auxiliary l-sparse subsequences $B_1, B_2, \ldots, B_k$ of u are consecutively defined by

$$B_i = \begin{cases} B_{i-1}a_i, & \text{if } B_{i-1}a_i \text{ is } l\text{-sparse,} \\ B_{i-1}, & \text{otherwise.} \end{cases}$$

Finally, we set $A(u, l) := B_k$. (In other words, terms of u are considered one by one, and each of them is included into $A(u, l)$ if the l-sparsity of $A(u, l)$ is preserved.)

For example, the algorithms $A(2)$ and $A(3)$ applied on $u = 2121143223$ give the sequences $A(u,2) = 21214323$ and $A(u,3) = 21432$, respectively.

Clearly, the algorithm $A(2)$ finds the largest 2-sparse subsequence of the input sequence. It is worth mentioning that more generally, for any $l \geq 2$, the algorithm $A(l)$ finds an l-sparse subsequence whose length is at least $1/(l-1)$ of the length of the longest l-sparse subsequence.

Lemma 8 follows from the following property of Algorithm $A(l)$.

OBSERVATION 9. *If we apply the algorithm $A(l)$ on a $||u||$-sparse u-free sequence v, then the obtained sequence $A(v,l)$ has length at least $\frac{1}{f_u(l-1)+1}|v|$.*

Proof. Any interval in v of length $f_u(l-1)+1$ contains at least l distinct letters, and therefore it contains at least one letter of $A(v,l)$. Since the first term of v lies in $A(v,l)$, the observation follows. □

Proof of Lemma 8. Lemma 8 with $\varepsilon = \frac{1}{f_u(l-1)+1}$ follows immediately from Observation 9. □

LEMMA 10 (reduction lemma). *For any integer $l \geq 0$ and for any $\varepsilon \in (0,1)$, there is an $\varepsilon' > 0$ with the following property. If v is a $\lceil \frac{l}{\varepsilon} \rceil$-sparse sequence and v_1 is a subsequence of v of length at least $\varepsilon|v|$, then v_1 has an l-sparse subsequence of length at least $\varepsilon'|v|$.*

Though its proof is simple, Lemma 10 seems to be a very useful tool in the study of generalized Davenport–Schinzel sequences. Perhaps some generalizations, strengthenings, and/or modifications of Lemma 10 might lead to further progress in investigation of generalized Davenport–Schinzel sequences. In this survey, Lemma 10 is used in the proof of several results, some of which were previously proved by other, more specific and weaker, tools.

Proof of Lemma 10. Set $L = \lceil \frac{l}{\varepsilon} \rceil$. If $|v| \leq \frac{2l(l-1)}{\varepsilon}$ then the lemma trivially holds with any $\varepsilon' \leq \frac{\varepsilon}{2l(l-1)}$. Otherwise, we partition v into $\lceil |v|/L \rceil$ intervals of length at most L. At least $\frac{\varepsilon}{2lL}|v|$ of these intervals contain at least l terms of v_1 each, since otherwise

$$|v_1| < \frac{\varepsilon}{2lL}|v| \cdot L + \left\lceil \frac{|v|}{L} \right\rceil \cdot (l-1) < \frac{\varepsilon}{2l}|v| + (\frac{\varepsilon|v|}{l}+1) \cdot (l-1) = \varepsilon|v| + (l-1) - \frac{\varepsilon}{2l}|v| < \varepsilon|v|.$$

Since v is L-sparse, the algorithm $A(l)$ applied on v_1 selects at least one term from each such interval. Thus, $A(v_1,l)$ is an l-sparse subsequence of length at least $\frac{\varepsilon}{2lL}|v|$. □

LEMMA 11 (letter duplication). *Let u_1, u_2 be two sequences, and let a be a letter. Then*

$$f_{u_1a^3u_2}(n) = \Theta(f_{u_1a^2u_2}(n)).$$

Proof. Clearly, $f_{u_1a^3u_2}(n) \geq f_{u_1a^2u_2}(n)$. On the other hand, let v be a $(n, u_1a^3u_2)$-critical sequence. By the sparsity lemma (Lemma 8), v contains a $2||u_1a^2u_2||$-sparse subsequence v_0 of length $>> |v|$. Let v_1 be the sequence obtained from v_0 by removing every even (second, forth, etc.) occurrence of each letter of $S(v_0)$. Certainly, $|v_1| \geq \frac{|v_0|}{2}$. It is easy to see that v_1 is $u_1a^2u_2$-free (otherwise v_0 would contain $u_1a^3u_2$). By the reduction lemma (Lemma 10, with $l = ||u_1a^2u_2||, \varepsilon = 1/2$), v_1 contains a $||u_1a^2u_2||$-sparse subsequence v_2 of length $>> |v_0| >> |v| = f_{u_1a^3u_2}(n)$. Therefore,

$$f_{u_1a^2u_2}(n) \geq |v_2| = \Omega(f_{u_1a^3u_2}(n)).$$

□

LEMMA 12 (first–letter and last–letter duplications). *Let u be a sequence and let a be a letter. Then*

$$f_{ua^2}(n) = f_{ua}(n) + O(n)$$

and

$$f_{a^2u}(n) = f_{au}(n) + O(n).$$

Before proving Lemma 12, we make a simple observation.

OBSERVATION 13. *Let u be a sequence, and let $l = ||u|| > 1$. If w_1aw_2 is a u-free l-sparse sequence, where a is a letter, then w_1w_2 contains a l-sparse subsequence of length at least $|w_1w_2| - f_u(3l-4)$.*

Proof. If $|w_2| \leq f_u(3l-4)$, then w_1 is the required subsequence of w_1w_2. Now, let $|w_2| > f_u(3l-4)$. We write $w_2 = w_3w_4$, where $|w_3| = f_u(3l-4)+1$. We have $||w_3|| \geq 3l-3$. Let F be the set of letters appearing on the last $l-1$ positions in w_1 and on the first $l-1$ positions in w_4. Further, let w_3^* be the subsequence of w_3 consisting of the first occurrences of letters of $S(v_3) \setminus F$ in w_3. We have $|w_3^*| = ||w_3|| - |F| \geq (3l-3) - 2(l-1) = l-1$. It follows that the subsequence $w_1w_3^*w_4$ of w_1w_2 is l-sparse. Its length is at least $|w_1w_2| - (f_u(3l-4)+1)+(l-1) \geq |w_1w_2| - f_u(3l-4)$. The observation follows. □

Proof of Lemma 12. It suffices to prove the first estimate, the second estimate is analogous. Certainly, $f_{ua^2}(n) \geq f_{ua}(n)$. Now, let v be a (n, ua^2)-critical sequence. We obtain a (n, ua)-sequence of length $|v| - O(n)$ in n steps as follows. Let $S(v) = \{1, \ldots, n\}$. We set $v_0 = v$, and construct n (n, ua^2)-sequences $v_1, \ldots, v_n$ inductively as follows. Suppose that $i \in \{1, \ldots, n\}$ and that the sequence v_{i-1} has already been constructed. We write $v_{i-1} = w_1i_0w_2$, where i_0 is the last occurrence of i in v_{i-1}. By Observation 13, w_1w_2 contains a (n, ua^2)-subsequence of length at least $|v_{i-1}| - (f_u(3||u|| - 4) + 1)$. We take this sequence for v_i.

The (n, ua^2)-sequence v_n has length $|v| - O(n) = f_{ua^2}(n) - O(n)$. Moreover, v_n is ua-free, since otherwise the sequence $v_0 = v$ would not be ua^2-free. It follows that

$$f_{ua}(n) \geq |v_n| \geq f_{ua^2}(n) - O(n).$$

□

We remark that repeated applications of Lemma 11 give

$$f_{u_1a^iu_2}(n) = \Theta(f_{u_1a^2u_2}(n)),$$

for any fixed $i \geq 3$. Similarly, repeated applications of Lemma 12 give

$$f_{ua^i}(n) = f_{ua}(n) + O(n) \quad \text{and} \quad f_{a^iu}(n) = f_{au}(n) + O(n),$$

for any fixed $i \geq 2$.

3.2. Further Basic Tools and Properties. The following lemma is an easy consequence of the sparsity lemma:

LEMMA 14 (monotone property). *If u_1 contains u_2, then there is a constant $\varepsilon > 0$ such that, for any n,*

$$f_{u_1}(n) \geq \varepsilon \cdot f_{u_2}(n).$$

Proof. Let v be a (n, u_2)-critical sequence. By the sparsity lemma (Lemma 8), v contains a $||u_1||$-sparse subsequence of length at least $\varepsilon|v|$ for some $\varepsilon = \varepsilon(u_2, ||u_1||) > 0$. Thus, $f_{u_1}(n) \geq \varepsilon|v| = \varepsilon f_{u_2}(n)$. □

LEMMA 15 (additive property). *Let u_1, u_2 be two sequences with $S(u_1) \cap S(u_2) = \emptyset$. Then there are two positive constants $c_1, c_2 > 0$ such that, for any n,*

$$c_1(f_{u_1}(n) + f_{u_2}(n)) \leq f_{u_1u_2}(n) \leq c_2(f_{u_1}(n) + f_{u_2}(n) + 1).$$

Proof. The first inequality follows from the monotone property (Lemma 14). As of the second inequality, we show that, for any n,

$$f_{u_1u_2}(n) \leq (f_{u_1}(n) + 1) + 2||u_1u_2||f_{u_2}(n). \tag{1}$$

To see this, consider a (n, u_1u_2)-critical sequence v and suppose to a contrary that its length is bigger than the right-hand side in (1). Let $v = v_1v_2$ be a partition of v such that $|v_1| = f_{u_1}(n) + 1$ and $|v_2| > 2||u_1u_2||f_{u_2}(n)$. Then, v_1 has a subsequence $\tilde{u_1}$ isomorphic to u_1. Let v_2' be the sequence obtained from v_2 by removing all occurrences of letters of $S(\tilde{u_1})$. Since v_2 is $||u_1u_2||$-sparse and at most $||\tilde{u_1}|| = ||u_1|| = ||u_1u_2|| - ||u_2||$ terms were removed from any interval of length $||u_1u_2||$ in v_2, the sequence v_2' is $||u_2||$-sparse. It is also u_2-free, since otherwise v would contain u_1u_2. Therefore $|v_2'| \leq f_{u_2}(n) < \frac{|v_2|}{2||u_1u_2||}$. It follows that v_2 contains an interval of length $||u_1u_2||$ containing no term of v_2' — a contradiction with the $||u_1u_2||$-sparsity of v_2 and the choice of v_2'. This gives (1). The second inequality in the lemma immediately follows. □

We now prove that the function f_u is weakly super-additive unless $|u| = ||u||$, and then derive Proposition 3 from it.

LEMMA 16 (weak super-additivity). *If u is a sequence with $|u| > ||u||$, then there is a positive constant $\varepsilon = \varepsilon(u) > 0$ such that*

$$f_u(\sum_{i=1}^{r} n_i) \geq \varepsilon \sum_{i=1}^{r} f_u(n_i)$$

holds for any positive integers $r, n_1, n_2, \ldots, n_r$.

Proof. If u is irreducible and $|u| > 1$, then we take a (n_i, u)-critical sequence v_i for each $i = 1, 2, \ldots, r$, so that $S(v_i) \cap S(v_j) = \emptyset$ for $i \neq j$. Concatenation of these sequences yields a u-free $||u||$-sparse sequence $v = v_1v_2 \ldots v_r$ of length $|v| = \sum_{i=1}^{r} f_u(n_i)$ and with $||v|| \leq \sum_{i=1}^{r} n_i$. It follows that $f_u(\sum_{i=1}^{r} n_i) \geq \sum_{i=1}^{r} f_u(n_i)$ in this case.

We now prove the lemma by induction on the length of u. If u is irreducible then the lemma follows from above. Otherwise we may write $u = u_1u_2$, where $S(u_1) \cap S(u_2) = \emptyset$, and suppose that the lemma holds for u_1 (unless $|u_1| = ||u_1||$) and for u_2 (unless $|u_2| = ||u_2||$).

If $|u_1| > ||u_1||$ and $|u_2| > ||u_2||$, then, by the additive property (Lemma 15) and by the inductive hypothesis,

$$\begin{aligned} f_u(\sum_{i=1}^r n_i) &\geq c_1 \left(f_{u_1}(\sum_{i=1}^r n_i) + f_{u_2}(\sum_{i=1}^r n_i) \right) \\ &\geq c_1 \left(\varepsilon \sum_{i=1}^r f_{u_1}(n_i) + \varepsilon' \sum_{i=1}^r f_{u_2}(n_i) \right) \\ &\geq c_1 \min\{\varepsilon, \varepsilon'\} \sum_{i=1}^r (f_{u_1}(n_i) + f_{u_2}(n_i)) \\ &\geq c \sum_{i=1}^r (f_{u_1}(n_i) + f_{u_2}(n_i) + 1) \\ &\geq c \cdot \frac{1}{c_2} \sum_{i=1}^r f_u(n_i), \end{aligned}$$

where $c, c_1, c_2, \varepsilon, \varepsilon'$ are positive constants (independent of r, n_i).

If $|u_1| > ||u_1||$ and $|u_2| = ||u_2||$, then we may suppose that $|u_2| = 1$. Then, by the sparsity lemma (Lemma 8), $\varepsilon \cdot f_{u_1}(n) \leq f_u(n)$. On the other hand, it is not difficult to see that $f_u(n) \leq f_{u_1}(n) + ||u|| \leq (||u|| + 1) f_{u_1}(n)$. Thus,

$$\varepsilon \cdot f_{u_1}(n) \leq f_u(n) \leq (||u|| + 1) f_{u_1}(n),$$

and the weak super-additivity of $f_u(n)$ follows from the weak super-additivity of $f_{u_1}(n)$.

The case $|u_1| = ||u_1||$, $|u_2| > ||u_2||$ is analogous. □

Proof of Proposition 3. If $|u| = ||u||$ then $f_u(n) = \min\{|u| - 1, n\} = O(1)$. If $|u| > ||u||$, then the sequence $123\ldots n$ of length n is u-free and therefore $f_u(n) \geq n$. On the other hand, by Theorem 2, $\frac{f_u(n)}{n}$ is bounded from above by $2^{c(\alpha(n))^{|u|-4}}$, which tends to infinity slower than the functional inverse of any unbounded primitive recursive function (see [**Ro67**]). It remains to show that, for any sequence u, either $f_u(n) = O(n)$ or $\frac{f_u(n)}{n} \to \infty$.

Suppose that $f_u(n) \neq O(n)$ for some sequence u. Then, for any $M > 0$ there exists an n_M with $f_u(n_M) > M \cdot n_M$. By the weak super-additivity of $f_u(n)$ (Lemma 16),

$$f_u(n) \geq f_u\left(\left\lfloor \frac{n}{n_M} \right\rfloor n_M \right) \geq \varepsilon \left\lfloor \frac{n}{n_M} \right\rfloor f_u(n_M) > \varepsilon \frac{n}{2n_M} \cdot M \cdot n_M \geq \frac{\varepsilon}{2} Mn,$$

for any $n \geq n_M$. Since $\varepsilon = \varepsilon(u) > 0$ is a constant independent of M, we get $\frac{f_u(n)}{n} \to \infty$. □

The following lemma first appeared in [**Kl95**].

LEMMA 17 (weak sub-additivity). *For any sequence u and for any integer $k \geq 1$, there is a $c = c(u, k) > 0$ such that*

$$f_u(kn) \leq c f_u(n),$$

for any integer n.

Proof. Let v be a (kn,u)-critical sequence. By the sparsity lemma (Lemma 8), v has a $k||u||$-sparse subsequence v_0 of length at least $\varepsilon|v|$. Let $S(v_0)=S_1\cup S_2\cup\cdots\cup S_k$ be any partition of $S(v_0)$ into k sets of size at most n. By the pigeon-hole principle, letters of one of the sets S_i occur altogether at least $\frac{1}{k}|v_0|$ times in v_0. Let v_1 be the subsequence of v_0 consisting of these occurrences. Thus, $|v_1|\geq\frac{1}{k}|v_0|$ and $||v_1||\leq n$. By the reduction lemma (Lemma 10, with $l=||u||$ and $\varepsilon=\frac{1}{k}$), v_1 has a $||u||$-sparse subsequence of length at least $\varepsilon'|v_0|\geq\varepsilon'\varepsilon|v|=\varepsilon'\varepsilon f_u(kn)$. The lemma with $c=\frac{1}{\varepsilon'\varepsilon}$ follows. □

4. Letter Insertions

In this section we prove the following theorem, which generalizes Theorem 5(v).

THEOREM 18 (letter insertions). *Let $u=au_1a^2u_2$, $u'=bau_1ab^2au_2$, where u_1,u_2 are two sequences (possibly empty), a,b are letters, and $b\notin S(u)$. Then $f_{u'}(n)=O(f_u(n))$.*

(Analogously, if $u=u_1a^2u_2a$, $u'=u_1ab^2au_2ab$, and $b\notin S(u)$, then $f_{u'}(n)=O(f_u(n))$.)

Theorem 18 was proved in [**KV94**]. Here we give a simplified and modified version of the original proof. The following lemma is the key part of the proof of Theorem 18.

LEMMA 19 (density lemma). *For every sequence u and for every $p\geq 1$, there exists a constant $\varepsilon=\varepsilon(u,p)>0$ with the following property: Let v be a $||u||$-sparse u-free sequence, and let $<$ be any linear order on $S(v)$. Then v has a $||u||$-sparse subsequence $v^{(p)}$ of length at least $\varepsilon|v|$ such that, for any letter i and for any two different i-occurrences o_1, o_2 in $v^{(p)}$, at least p letters j with $j<i$ occur at least twice in the i-interval of v determined by o_1,o_2.*

Before proving Lemma 19, we derive Theorem 18 from it.

Proof of Theorem 18. Let v be a (n,u')-critical sequence. We linearly order $S(v)$ according to the order of the first occurrences of letters in v (i.e., $a<b$ if, in v, the first a-occurrence precedes the first b-occurrence). We apply the density lemma (Lemma 19) with $p=||u||+1$, and denote the obtained $||u||$-sparse sequence $v^{(p)}=v^{(||u||+1)}$ shortly by w.

We now prove that w is u-free. Suppose to the contrary that w contains u. Thus, w has a subsequence $\tilde{u}=\tilde{a}\tilde{u}_1\tilde{a}^2\tilde{u}_2$ isomorphic to u. By the choice of w, v has a subsequence $\tilde{a}\tilde{u}_1\tilde{a}\tilde{b}^2\tilde{a}\tilde{u}_2$ for at least $||u||+1=|S(\tilde{u})|+1$ letters $\tilde{b},\tilde{b}<\tilde{a}$. Choose $\tilde{b}$ such that $\tilde{b}\notin S(\tilde{u})$. Since $\tilde{b}<\tilde{a}$, at least one occurrence of $\tilde{b}$ precedes any occurrence of $\tilde{a}$ in v. Consequently, v has a subsequence $\tilde{b}\tilde{a}\tilde{u}_1\tilde{a}\tilde{b}^2\tilde{a}\tilde{u}_2$ isomorphic to u' – a contradiction. Thus, w is u-free.

From $|w|\geq\varepsilon|v|$ we get

$$f_u(n)\geq|w|\geq\varepsilon|v|=\varepsilon f_{u'}(n).$$

Consequently,

$$f_{u'}(n)=O(f_u(n)),$$

as required. □

Now, we prove the density lemma (Lemma 19).

Proof of Lemma 19. It suffices to prove the first part, the second part (in the brackets) then follows from the symmetry.

Suppose that $S(v) = \{1, 2, \dots, n\}$ and that $<$ coincides with the usual order of integers. By the sparsity lemma (Lemma 8), there is a $200||u||$-sparse subsequence v_0 of v with $|v_0| \geq \varepsilon_0|v|$. From now on, we consider the sequence v_0.

If $|v_0| < 399n$, then we may take $v^{(p)}$ as the subsequence of v formed by the first occurrences of letters of $S(v)$. Then the length of $v^{(p)}$ is $n > \frac{|v_0|}{399} \geq \frac{\varepsilon_0}{399}|v|$ and the lemma holds with any $\varepsilon \leq \frac{\varepsilon_0}{399}$. We further suppose that $|v_0| \geq 399n$.

Let $i \in S(v_0)$. All but at most 19 i-occurrences in v_0 can be covered by pairwise disjoint i-intervals, each containing exactly 20 i-occurrences. We fix such pairwise disjoint i-intervals and call them *i-blocks*. We partition the 20 i-occurrences in each i-block into 10 pairs of i-occurrences determining pairwise disjoint i-intervals further called *i-arcs*. An interval in v_0 is called *block* or *arc*, if it is an i-block or an i-arc for some $i \in S(v_0)$, respectively. Certainly, the sequence v_0 contains at least $\frac{|v_0|-19n}{20} \geq \frac{|v_0|-19\cdot\frac{|v_0|}{399}}{20} = \frac{|v_0|}{21}$ (and at most $\frac{|v_0|}{20}$) blocks. The number of arcs is 10 times bigger.

Now, we construct $n+1$ auxiliary sets $\mathcal{U}^{(0)}, \mathcal{U}^{(1)}, \dots \mathcal{U}^{(n)}$ of pairwise disjoint arcs in v_0. We start with $\mathcal{U}^{(0)} = \emptyset$. We now describe how to construct $\mathcal{U}^{(i)}$ from $\mathcal{U}^{(i-1)}$. We say that an i-block in v_0 is *good* if it intersects at least three arcs of $\mathcal{U}^{(i-1)}$, otherwise we say that it is *bad*. The set $\mathcal{U}^{(i)}$ is obtained from $\mathcal{U}^{(i-1)}$ in the following way: For each bad i-block B in v_0, remove from $\mathcal{U}^{(i-1)}$ the (at most two) arcs intersecting B and add to $\mathcal{U}^{(i-1)}$ the 10 i-arcs contained in B. The obtained set of arcs is taken for $\mathcal{U}^{(i)}$.

Now, consider the set $\mathcal{U} = \mathcal{U}^{(n)}$. Certainly, $\mathcal{U}$ is a set of pairwise disjoint arcs in v_0. Since v_0 is 2-sparse, the length of every arc is at least 3. Thus, $|\mathcal{U}| \leq \frac{|v_0|}{3}$. On the other hand, we can easily estimate the size of $\mathcal{U}$ from below by the number of bad blocks. If b is the number of bad blocks in v_0, then $|\mathcal{U}| \geq (10-2)b = 8b$, since to every bad block we can assign an addition of (its) 10 arcs to $\mathcal{U}$ and a removal of at most 2 arcs from $\mathcal{U}$. Thus, $8b \leq |\mathcal{U}| \leq \frac{|v_0|}{3}$, which implies $b \leq \frac{|v_0|}{24}$.

Since v_0 contains at least $\frac{|v_0|}{21}$ blocks, there are at least $\frac{|v_0|}{21} - \frac{|v_0|}{24} > \frac{|v_0|}{200}$ good blocks in v_0. Let v_1 be the subsequence of v_0 containing each term of v_0 which is the first term of a good block. Thus, the length of v_1 equals the number of good blocks in v_0, and is therefore at least $\frac{|v_0|}{200}$. We remark that if an i-block is good then it contains two occurrences of some letter j, $j < i$ (since it intersects at least three arcs of $\mathcal{U}^{(i-1)}$ and therefore contains at least one arc of $\mathcal{U}^{(i-1)}$). Thus, the lemma is partially satisfied by v_1 for $p = 1$: if i is a letter and o_1, o_2 are two occurrences of i in v_1, then the i-interval in v determined by o_1 and o_2 contains the good i-block starting with o_1 (or o_2), and therefore it contains two occurrences of some letter $j, j < i$.

However, the sequence v_1 does not have to be $||u||$-sparse. To achieve the $||u||$-sparsity, we apply the reduction lemma. By the reduction lemma (Lemma 10, with $l = ||u||$, $\varepsilon = \frac{1}{200}$), v_1 contains a $||u||$-sparse subsequence $v^{(1)}$ of length at least $\varepsilon_1|v_0| \geq \varepsilon_1\varepsilon_0|v|$. This completes the proof for $p = 1$.

For $p \geq 1$, we take $v^{(p)}$ as the subsequence of v obtained from v by p applications of the lemma with $p = 1$. (Or equivalently, we proceed by induction on p and set $v^{(p)} = (v^{(p-1)})^{(1)}$). Clearly, $v^{(p)}$ has the required length. It remains to verify the "nesting" property of $v^{(p)}$.

If i is a letter and o_1, o_2 are two occurrences of i in $v^{(p)}$, then some letter $j_0 < i$ occurs (at least) twice between o_1 and o_2 in $v^{(p-1)}$. Between these two occurrences of j_0, there are at least two occurrences of $p-1$ letters $j < j_0$ in v. These $p-1$ letters j and the letter j_0 occur each at least twice between o_1 and o_2 in v (see Fig. 2). □

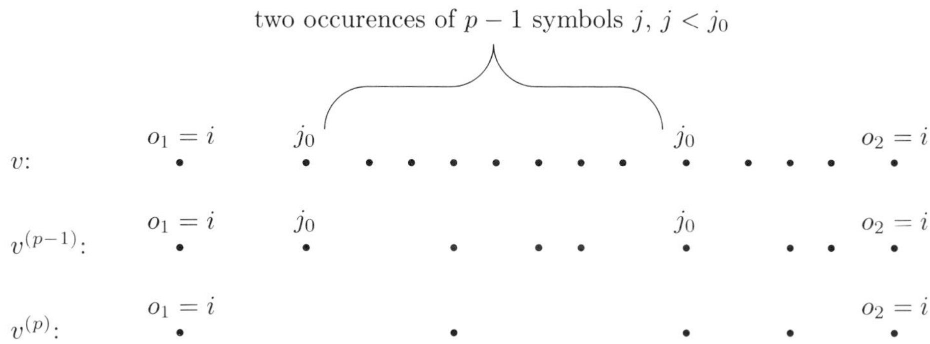

FIGURE 2. The second inductive step in the proof of Lemma 19.

5. Sequence Insertions

In this section we prove the following result of [**KV94**], which generalizes Theorem 5(vi).

THEOREM 20 (sequence insertion). *Let $u = u_1 a^2 u_2$, w be two sequences, where u_1, u_2 are two sequences, a is a letter, and $S(u) \cap S(w) = \emptyset$. Then*

$$f_{u_1 awau_2}(n) = O(f_w(f_u(n)) + f_u(n)).$$

In particular, if $f_u(n), f_w(n) = O(n)$, then also $f_{u_1 awau_2}(n) = O(n)$.

Proof. Denote $t = u_1 awau_2$, and set $M = f_t(2||u||) + 2||t||$ and $g(n) = M \cdot \max\{f_w(n), n\}$. The constant M and the function g are chosen so that the following two lemmas hold:

LEMMA 21. *Let s be a t-free $||t||$-sparse sequence of length $|s| = g(||s||)$. Then*

(i) *$||s|| > 2||u||$, and*

(ii) *for any set F of $||u||$ letters, there is a subsequence w_F of s isomorphic to w such that $S(w_F) \cap F = \emptyset$.*

LEMMA 22. *The function g is weakly super-additive, i.e., there is a positive constant $\varepsilon > 0$ such that*

$$g(\sum_{i=1}^{r} n_i) \geq \varepsilon \sum_{i=1}^{r} g(n_i)$$

holds for any positive integers $r, n_1, n_2, \ldots, n_r$.

Proof of Lemma 21. (i) We have $|s| = g(||s||) \geq M > f_t(2||u||)$ and thus $||s|| > 2||u||$.

(ii) Let s and F satisfy the assumptions. Then $|s| = g(||s||) \geq 2||t||f_w(||s||)$. Since s is $||t||$-sparse, each interval of length $||t||$ in s contains at most one occurrence of each letter of F. Thus, if we remove from s all occurrences of letters of F, then the obtained sequence is a $(||t||-|F| = ||t||-||u|| = ||w||)$-sparse sequence of length at least

$$\left\lfloor \frac{|s|}{||t||} \right\rfloor ||w|| \geq \left\lfloor \frac{2||t||f_w(||s||)}{||t||} \right\rfloor ||w|| > f_w(||s||).$$

Thus, the obtained sequence contains w, and (ii) follows. □

Proof of Lemma 22. If $|w| > ||w||$, then $f_w(n) \geq n$ and thus $g(n) = M \cdot f_w(n)$ for each n, and the statement follows from the weak super-additivity of $f_w(n)$ (Lemma 16).

If $|w| = ||w||$, then $f_w(n) \leq n$ and thus $g(n) = M \cdot n$ for each n, and the statement with $\varepsilon = 1$ follows. □

We may continue in the proof of Theorem 20. Let v be a (n,t)-sequence.

If $|v| > g(||v||)$, then v contains an initial nonempty interval v_1 with $|v_1| = g(||v_1||)$ (this is clearly true for the shortest nonempty initial interval v_1 of v satisfying $|v_1| \geq g(||v_1||)$). If $|v \setminus v_1| > g(||v \setminus v_1||)$, then $v \setminus v_1$ contains an initial interval v_2 with $|v_2| = g(||v_2||)$, etc. In this way we find a partition $v = v_1 v_2 \dots v_h v_0$ of v such that $|v_i| = g(||v_i||)$ for $i = 1, 2, \dots, h$ and $|v_0| \leq g(||v_0||)$.

Lemma 21(i) implies that $||v_i|| > 2||u||$, for $i = 1, 2, \dots, h$. We now show that it is possible to choose, for $i = 1, 2, \dots, h$, a subsequence v_i^* of v_i such that $|v_i^*| = ||v_i^*|| \geq \frac{||v_i||}{2}$ and that the two sequences

$$v_{\text{odd}} = v_1^* v_3^* \dots \qquad \text{and} \qquad v_{\text{even}} = v_2^* v_4^* \dots$$

are $||u||$-sparse. We can easily find the sequences v_i^* one by one for $i = 1, 2, \dots, h$ as follows. The sequence v_1^* is chosen from v_1 by picking exactly one occurrence of each letter of $S(v_1)$. Then, $|v_1^*| = ||v_1^*|| = ||v_1|| > \frac{||v_1||}{2}$. Analogously, we choose v_2^*. For $i > 2$, we suppose that v_{i-2}^* has already been chosen, and choose a subsequence v_i^* of v_i by picking one occurrence of each letter of $S(v_i)$ which does not occur among the last $||u|| - 1$ terms of v_{i-2}^*. In this way we maintain the $||u||$-sparsity of v_{odd} and v_{even}, and the length of v_i^* is at least $||v_i|| - (||u|| - 1) > \frac{||v_i||}{2}$ for each i. Now, it suffices to show that the sequences v_{odd}, v_{even} are u-free, since then the required estimate $|v| = O\left(f_w(f_u(n)) + f_u(n)\right)$ follows from the estimates

$$|v_0| \leq g(||v_0||) = O(f_w(n) + n), \quad n \leq f_u(n),$$

and

$$\begin{aligned}
|v_1 v_2 \dots v_h| &= \sum_{i=1}^{h} |v_i| \\
&= \sum_{i=1}^{h} g(||v_i||) \\
&\leq \sum_{i=1}^{h} g(2|v_i^*|) \\
&\overset{(*)}{=} O(g(2\sum_{i=1}^{h} |v_i^*|)) \\
&= O\left(g(2|v_{\text{odd}}| + 2|v_{\text{even}}|)\right) \\
&= O\left(g(4f_u(n))\right) \\
&\overset{(**)}{=} O\left(f_w(f_u(n)) + f_u(n)\right)
\end{aligned}$$

((*) follows from Lemma 22, (**) from the weak sub-additivity of f_w (Lemma 17)).

Suppose to a contrary that, say, v_{even} contains u. Thus, there is a subsequence $\tilde{u} = \tilde{u_1}\tilde{a}^2\tilde{u_2}$ of v_{even} isomorphic to u. By the construction, the two $\tilde{a}$-occurrences in $\tilde{u}$ (between $\tilde{u_1}$ and $\tilde{u_2}$) lie in distinct intervals $v_{2j}^*, v_{2k}^*, j < k$, of v_{even}. By Lemma 21(ii) (with $s = v_{2j+1}$ and $F = S(\tilde{u})$), the interval v_{2j+1} has a subsequence $\tilde{w}$ isomorphic to w such that $S(\tilde{w}) \cap S(\tilde{u}) = \emptyset$. Consequently, $\tilde{u_1}\tilde{a}\tilde{w}\tilde{a}\tilde{u_2}$ is a subsequence of v. Thus, v is not t-free – a contradiction. □

Remark. A more careful analysis of the length of $v_1 v_2 \dots v_h$ in the proof of Theorem 20 gives the estimate $f_{u_1awau_2}(n) = O(\frac{f_w(n)\cdot f_u(n)}{n} + f_u(n))$ which might look (but probably is not, due to Proposition 3 and the expected smoothness of the function $f_w(n)$) stronger than the estimate in Theorem 20.

6. A Lower Bound Construction for (Generalized) Davenport–Schinzel Sequences

Here we describe a construction of Wiernik and Sharir [**WS88**] (see also [**SA95**]) giving $N_5(n) = f_{ababa}(n) = \Omega(n\alpha(n))$. For any two positive integers k and m, we construct a sequence $\mathcal{S}(k,m)$ with the following five properties:

(C1) $\mathcal{S}(k,m)$ is *ababa*-free,

(C2) $c(k,m) := \frac{||\mathcal{S}(k,m)||}{m}$ is an integer,

(C3) first occurrences of letters of $S(\mathcal{S}(k,m))$ in $\mathcal{S}(k,m)$ lie in $c(k,m) = \frac{||\mathcal{S}(k,m)||}{m}$ pairwise disjoint intervals of length m, we call these intervals *fans*,

(C4) two adjacent terms in $\mathcal{S}(k,m)$ are equal if and only if the first of them is the last term in a fan,

(C5) each letter of $S(\mathcal{S}(k,m))$ appears exactly k times in $\mathcal{S}(k,m)$.

It follows from conditions (C1)–(C5) that $\frac{f_{ababa}(n)}{n}$ is unbounded, since the removal of the last term in each fan in $\mathcal{S}(k,m)$ gives a $(||\mathcal{S}(k,m)||, ababa)$-sequence of length at least $(k-1)||\mathcal{S}(k,m)||$. Thus, by Proposition 3, $\frac{f_{ababa}(n)}{n} \to \infty$. Actually, the sequences $\mathcal{S}(k,m)$ give $f_{ababa}(n) = \Omega(n\alpha(n))$ (see [**HS86, SA95**]).

The sequences $\mathcal{S}(k,m)$ are constructed by double induction on k and m as follows.

For $k = 1$ and any m, we take $\mathcal{S}(1, m) = 123 \dots m$ so that $\mathcal{S}(1, m)$ consists of one fan of length m.

For $k > 1$ and $m = 1$, we construct $\mathcal{S}(k, 1)$ from $\mathcal{S}(k-1, 2)$ by replacing each fan xy (of length 2) by an interval $xxyy$ (where the first x and the first y form one-term fans in $\mathcal{S}(k, 1)$).

For $k > 1$ and $m > 1$, we construct $\mathcal{S}(k, m)$ from the sequences $T = \mathcal{S}(k, m-1)$ and $V = \mathcal{S}(k-1, M)$, where $M = c(k, m-1) = \frac{||\mathcal{S}(k,m-1)||}{m-1}$ is the number of fans in T, as follows. Let $P = c(k-1, M)$ and let $F_1, F_2, \dots, F_P$ be the fans in V (listed from left to right). For each $i = 1, \dots, P$, we take a sequence T_i isomorphic to T such that the alphabets $S(V)$, $S(T_1), \dots, S(T_P)$ are pairwise disjoint. For $i = 1, \dots, P$, denote the terms in F_i by f_{ij} so that $F_i = f_{i1} f_{i2} \dots f_{iM}$, and let G_{i1}, $G_{i2}, \dots, G_{iM}$ be the fans in T_i (listed from left to right). Further, let T_i' be the sequence obtained from T_i by replacing G_{ij} by $G_{ij} f_{ij} f_{ij}$ for $j = 1, \dots, M$. Then $\mathcal{S}(k, m)$ is obtained from V by replacing F_i by T_i' for each $i = 1, \dots, P$. The intervals $G_{ij} f_{ij}$ are the fans in $\mathcal{S}(k, m)$. It is not hard to check that the construction obeys conditions (C1)–(C5). Moreover, the numbers $||\mathcal{S}(k, m)||$ grow analogously as the "Ackermann" numbers $A(k, m)$, and therefore the sequences $\mathcal{S}(k, m)$ give $N_5(n) = \Omega(n\alpha(n))$ (see [**HS86, SA95**] for details).

It was noticed in [**KV94**] that a slight modification of the above construction gives $f_u(n) = \Omega(n\alpha(n))$ also for the *ababa*-free sequence $u = abcbadadbcd$: Indeed, it is not hard to verify from the construction that the sequences $\mathcal{S}(k, m)$ are u-free. By the sparsity lemma (Lemma 8), each of them as an *ababa*-free (nearly 2-sparse) sequence contains a relatively long 4-sparse subsequence (see [**Kl93**] for details).

7. Applications to Geometric Graphs

7.1. Turán–Type Problems for Geometric Graphs. A *geometric graph* is a graph embedded in the plane so that the vertices are represented by distinct points in general position (i.e., no three lie on a line) and the edges by line segments connecting the corresponding pairs of points. We say that two edges in a geometric graph *cross*, if their relative interiors intersect.

Two of the basic Turán-type problems for geometric graphs are to estimate the maximum number $c_k(n)$ of edges in a geometric graph on n vertices with no k pairwise crossing edges and to estimate the maximum number $d_k(n)$ of edges in a geometric graph on n vertices with no k pairwise disjoint edges. For $k = 2$, geometric graphs with no pair of mutually crossing edges are (planar drawings of) planar graphs. Thus, by Euler's formula, $c_2(n) = 3n - 6$ (for $n \geq 3$), with equality attained for planar triangulations. It has been known for decades that $d_2(n) = n$ (e.g., see the book [**PA95**] for an elegant proof due to Perles) – an example showing $d_2(n) \geq n$ is shown in Fig. 3. It has been a challenging problem to decide for a fixed $k > 2$ whether or not $c_k(n) = O(n)$ and $d_k(n) = O(n)$.

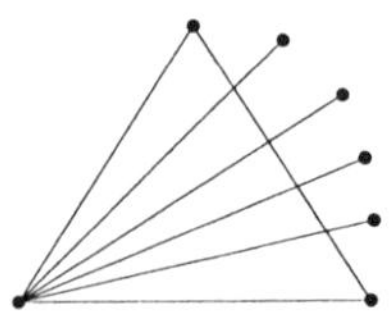

FIGURE 3. A geometric graph giving $d_2(n) \geq n$ (for $n = 7$).

The first proof of $d_k(n) = O(n)$ for any fixed $k > 2$ is due to Pach and Törőcsik [**PT94**]. They proved $d_k(n) \leq (k-1)^4 n$, which was later asymptotically improved to $d_k(n) \leq c_k \cdot k^3 n$ in [**TV98**] and further to $d_k(n) \leq c'_k \cdot k^2 n$ in [**To99**]. The best known lower bound on $d_k(n)$ (roughly $\frac{3}{2}kn$) can be found in [**TV98**].

Agarwal *et al.* [**AAPPS97**] proved that $c_3(n) = O(n)$, but the validity of $c_k(n) = O(n)$ for $k > 3$ is still open. The following best known upper bound on $c_k(n)$ for $k > 3$ relies on generalized Davenport–Schinzel sequences, namely on Theorem 6 about the sequences "updownup".

THEOREM 23. *For any fixed $k \geq 4$,*

$$c_k(n) = O(n \log n).$$

Theorem 23 was proved in [**V97, V98a**]. It improved the previous bounds $c_k(n) = O(n \log^{2k-4} n)$ ([**PSS96**]) and $c_k(n) = O(n \log^{2k-6} n)$ ([**AAPPS97**]).

A similar Turán-type problem deals with "parallel" edges. We say that two edges in a geometric graph are *parallel* if they are disjoint and the convex hull of their union is a convex quadrilateral. Equivalently, two edges are parallel if the line extension of each of them does not intersect the other edge (see Fig. 4 for a collection of pairwise parallel edges). Let $p_k(n)$ be the maximum number of edges

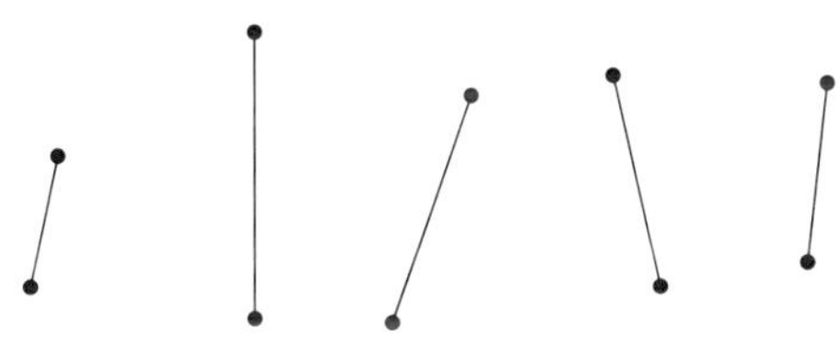

FIGURE 4. Five pairwise parallel edges.

in a geometric graph on n vertices with no k pairwise parallel edges. Davenport–Schinzel sequences were used to find the exact value of $p_2(n)$:

THEOREM 24 ([**KL98, V97**]). *For every $n \geq 4$,*

$$p_2(n) = 2n - 2.$$

Theorem 24 had been conjectured by Kupitz [**Ku79**] who gave a construction giving the lower bound $p_2(n) \geq 2n - 2$ (see Fig. 5).

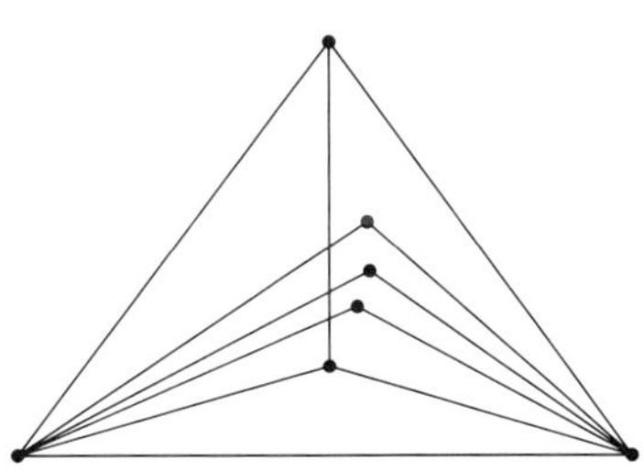

FIGURE 5. A geometric graph with n vertices and $2n-2$ edges such that no two edges are parallel.

Clearly, $p_k(n) \geq d_k(n)$ for all k and n, since any two parallel edges are disjoint. Using generalized Davenport–Schinzel sequences, a linear upper bound on $p_k(n)$ for any fixed k was given in [**V97**].

THEOREM 25 ([**V97**]). *For any fixed* $k \geq 2$,

$$p_k(n) = O(n).$$

In the rest of this section we prove Theorems 23–25. First we give the proof of Theorem 24 which relies on a simple result about cyclic Davenport–Schinzel sequences following from $f_{abab}(n) = 2n - 1$. Then we give the proof of Theorem 25 which relies on Theorem 6 about the sequences "updownup" and on Dilworth's theorem. Theorems 23 is then derived from Theorem 25 by a duality argument.

7.2. Geometric Graphs with no Pair of Parallel Edges. *Proof of Theorem 24.* The lower bound $p_2(n) \geq 2n-2$ follows from the construction of Kupitz shown in Fig. 5. We now give the proof of the upper bound $p_2(n) \leq 2n-2$ which is a simplified and modified version of the original proof from [**KL98, V97**] ([**KL98**] contains a proof of a slightly weaker upper bound $p_2(n) \leq 2n-1$, [**V97**] contains a refinement of the argument giving the tight bound $p_2(n) \leq 2n-2$).

Let $G = (V, E)$ be a geometric graph on $n > 3$ vertices with no pair of parallel edges. Suppose to the contrary that G has $2n-1$ edges. We take a directed graph $\overrightarrow{G}$ obtained from G by replacing each edge $ij \in E$ by a pair of conversely oriented edges $\overrightarrow{ij}$, $\overrightarrow{ji}$. The graph $\overrightarrow{G}$ has $4n-2$ directed edges. Let

$$D(G) = (e_0 e_1 \ldots e_{4n-3})$$

be the cyclic sequence of the edges of $\overrightarrow{G}$ listed in the counterclockwise order of their directions (a *cyclic sequence* is a sequence in which the first term and the last term are considered adjacent). From now on, we count all indices modulo $(4n-2)$. For each i, the edges e_i and $e_{i+(2n-1)}$ are equal up to their orientation.

Let $D'(G)$ be the cyclic sequence obtained from $D(G)$ by replacing each edge by its initial vertex, and let $S(G)$ be the longest 2-sparse cyclic subsequence of $D'(G)$. Clearly, the difference of the lengths of $D(G)$ and $S(G)$ is equal to the number of indices $i \in \{0, \ldots, 4n-3\}$ such that the edges e_i, e_{i+1} have the same initial vertex. However, if the edges e_i, e_{i+1} have the same initial vertex, then their "twins" $e_{i+(2n-1)}, e_{i+1+(2n-1)}$ do not. It follows that the length of $S(G)$ is at least half of the length of $D(G)$, i.e., at least $2n-1$. It is well-known (and easy to prove — see Theorem 58 in Section 12) that a 2-sparse *abab*-free cyclic sequence has length at most $2n-2$. Thus the sequence $S(G)$ has a subsequence *stst* for some $s \neq t$. It corresponds to four edges $\overrightarrow{ss'}, \overrightarrow{tt'}, \overrightarrow{ss''}, \overrightarrow{tt''}$, ordered in this way according to the counterclockwise order of their directions. Without loss of generality, we may suppose that the line $l = st$ is vertical and s lies below t.

Suppose first that the edges ss' and ss'' lie both on the same side of l (including the case when one of them coincides with the edge st). Without loss of generality, we may suppose that they lie to the right of l and that $\overrightarrow{ss''}$ has a larger slope than $\overrightarrow{ss'}$ (see Fig. 6). Then the edge $\overrightarrow{tt'}$ also lies to the right of l and its slope is larger that the slope of $\overrightarrow{ss'}$. It follows that ss' and tt' are parallel in this case — a contradiction. Similarly we obtain a contradiction if the edges tt' and tt'' lie on the same side of l.

Suppose now that exactly one of the edges ss' and ss'' lies to the right of l and also exactly one of the edges tt' and tt'' lies to the right of l. Without loss of

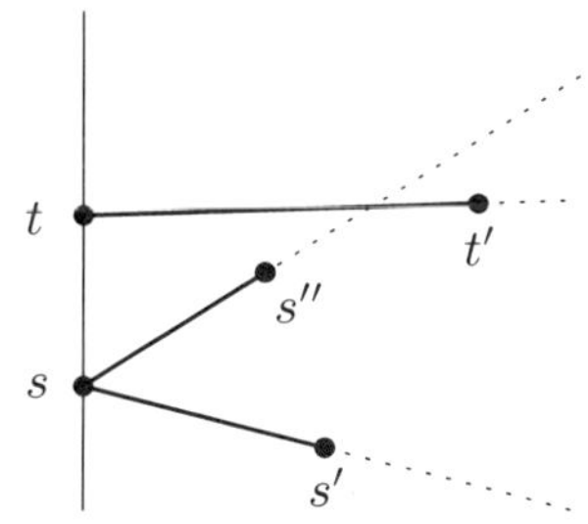

FIGURE 6. The edges ss' and tt' are parallel.

generality, ss' lies to the right and ss'' to the left of l. If tt' lies to the right of l, then ss' and tt' are parallel (see Fig. 7(a)) — a contradiction. If tt' lies to the left of l, then ss'' and tt' are parallel (see Fig. 7(b)) — a contradiction. □

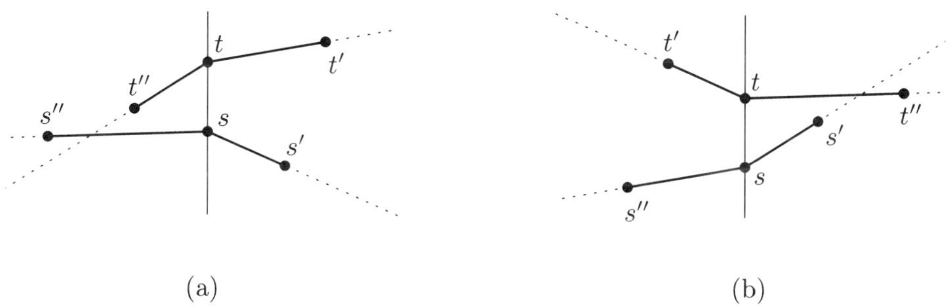

FIGURE 7. The graph contains two parallel edges.

7.3. Geometric Graphs with no k Pairwise Parallel Edges. In the proof of Theorem 25 (based on [**V97**]) we apply the following easy consequence of Dilworth's theorem [**Di50**].

THEOREM 26. *If the union of three partial orderings on a set I of size at least $(k-1)^3+1$ is a linear ordering on I, then at least one of the partial orderings contains a chain of size k.*

Proof. Let $\preceq_1$, $\preceq_2$, $\preceq_3$ be the three partial orderings on I. We suppose that neither $(I, \preceq_1)$ nor $(I, \preceq_2)$ contains a chain of size k. Then, we have to show that $(I, \preceq_3)$ contains a chain of size k. By Dilworth's theorem, $(I, \preceq_1)$ can be covered by at most $k-1$ antichains. By the pigeon–hole principle, one of these antichains has size at least $(k-1)^2+1$. We denote such an antichain by I_1. Analogously, $(I_1, \preceq_2)$ can be covered by at most $k-1$ antichains, and therefore has an antichain I_2 of size $(k-1)+1=k$. Since $\preceq_1 \cup \preceq_2 \cup \preceq_3$ is a linear ordering and I_2 is an antichain both in $(I, \preceq_1)$ and $(I, \preceq_2)$, it is a chain in $(I, \preceq_3)$ of size k. □

Proof of Theorem 25. Let $G=(V,E)$ be a geometric graph on n vertices and m edges with no k pairwise parallel edges. Similarly as in the proof of Theorem 24, we take a directed graph $\overrightarrow{G}$ obtained from G by replacing each edge $ij \in E$ by a pair of edges $\overrightarrow{ij}$, $\overrightarrow{ji}$. We order the edges of $\overrightarrow{G}$ in the conterclockwise order of their directions. We further split the obtained cyclic sequence into two sequences $D_1(G)$, $D_2(G)$ of equal length so that the directions of edges in $D_1(G)$ lie in $[0,\pi)$ and the directions of edges in $D_2(G)$ lie in $[\pi,2\pi)$. Further, let $P_i(G)$ be the cyclic sequence obtained from $D_i(G)$ by replacing each edge by its initial vertex ($i=1,2$).

We now show that at least one of the sequences P_1 and P_2 contains a relatively long l-sparse subsequence, where $l=(k-1)^3+1$. We partition the sequence $D_1(G)$ into $m/(4l^2)$ intervals of length $4l^2$ (suppose for simplicity that m is divisible by $4l^2$). Edges in every interval have altogether at least $2l$ distinct initial endpoints or at least $2l$ distinct terminal endpoints. Thus, for at least $m/(8l^2)$ intervals their edges have at least $2l$ distinct initial endpoints, or for at least $m/(8l^2)$ intervals their edges have at least $2l$ distinct terminal endpoints. In the first case, the sequence P_1 contains $m/(8l^2)$ pairwise disjoint intervals, each containing at least $2l$ distinct elements. In the second case, the sequence P_2 contains such $m/(8l^2)$ intervals. A simple greedy-type algorithm (e.g., the algorithm $A(l)$ introduced below Lemma 8) chooses at least l elements from each such interval so that the obtained subsequence S of P_1 or P_2 is l-sparse. Its length is at least $l\cdot m/(8l^2)=m/(8l)$. Now, it suffices to prove that P_1 and P_2 are updownup$(l,1)$-free, since then S is also updownup$(l,1)$-free and, by Theorem 6,

$$|E|=m\le 8l\cdot|S|\le 8l\cdot f_{\text{updownup}(l,1)}(n)=O(n).$$

Suppose to the contrary that P_1 (say) contains a subsequence $s_1s_2\dots s_{3l}$ isomorphic to updownup$(l,1)$. Thus, $s_1,s_2,\dots,s_l$ are pairwise different and, for $i=1,2,\dots,l$, $s_i=s_{(2l+1)-i}=s_{2l+i}$. For each i, we denote the edge of $D_1(G)$ corresponding to s_i by e_i. We obtain a contradiction by showing that k of the edges $e_1,e_2,\dots,e_{3l}$ are pairwise parallel.

For simplicity of notation, suppose that $s_i=i$ ($i=1,\dots,l$). Define three partial orderings $\preceq_1$, $\preceq_2$, $\preceq_3$ on the set $I=\{1,2,\dots,l\}$ as follows.

DEFINITION 27. *Let $i,j\in I$, and let $\operatorname{dir}(\overrightarrow{ij})$ denote the direction of the vector with initial vertex i and terminal vertex j. Then*

(i) $i\prec_1 j$, *if* $i<j$ *and* $\operatorname{dir}(\overrightarrow{ij})\in(\operatorname{dir}(\overrightarrow{e_l}),\pi]$,
(ii) $i\prec_2 j$, *if* $i<j$ *and* $\operatorname{dir}(\overrightarrow{ij})\in(\pi,\pi+\operatorname{dir}(\overrightarrow{e_{2l}})]$,
(iii) $i\prec_3 j$, *if* $i<j$ *and* $\operatorname{dir}(\overrightarrow{ij})\in(\pi+\operatorname{dir}(\overrightarrow{e_{2l}}),2\pi)\cup[0,\operatorname{dir}(\overrightarrow{e_l})]$.

Since the union of $\preceq_1$, $\preceq_2$, $\preceq_3$ is a linear ordering on I, Theorem 26 implies that one of the orderings $\preceq_1$, $\preceq_2$, $\preceq_3$ contains a chain $i_1,i_2,\dots,i_k$ of size k. We distinguish the corresponding three possible cases.

If $i_1\prec_1 i_2\prec_1\cdots\prec_1 i_k$, then the edges $e_{i_1},e_{i_2},\dots,e_{i_k}$ are pairwise parallel. Indeed, if $1\le j<j'\le k$, then the inequalities

$$0\le\operatorname{dir}(\overrightarrow{e_{i_j}})<\operatorname{dir}(\overrightarrow{e_{i_{j'}}})\,(\le\operatorname{dir}(\overrightarrow{e_l}))<\operatorname{dir}(\overrightarrow{i_ji_{j'}})\le\pi$$

show that the edges e_{i_j}, $e_{i_{j'}}$ are parallel (see Fig. 8).

Similarly, if $i_1\prec_2 i_2\prec_2\cdots\prec_2 i_k$ then the edges e_{2l+i_1}, e_{2l+i_2}, $\dots,e_{2l+i_k}$ are pairwise parallel, and if $i_1\prec_3 i_2\prec_3\cdots\prec_3 i_k$ then the edges $e_{(2l+1)-i_1}$, $e_{(2l+1)-i_2}$, $\dots,e_{(2l+1)-i_k}$ are pairwise parallel. □

7.4. Geometric Graphs with no k Pairwise Crossing Edges. In this paragraph, based on [**V97**], we derive Theorem 23 from Theorem 25.

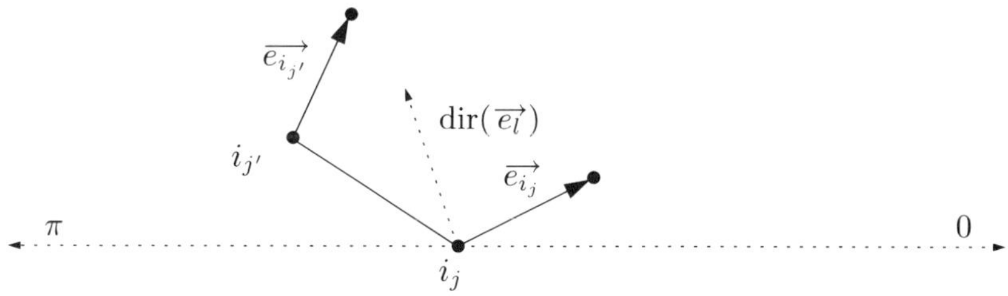

FIGURE 8. The edges $e_{i_j}, e_{i_{j'}}$ are parallel.

Let $G = (V, E)$ be a geometric graph on n vertices with no k pairwise crossing edges. Introduce Cartesian coordinate system so that the y-axis partitions V into two parts which are as equal as possible, thus the sets

$V^- = \{v \in V | \text{the } x\text{-coordinate of } v \text{ is negative}\}$,

$V^+ = \{v \in V | \text{the } x\text{-coordinate of } v \text{ is positive}\}$

have sizes

$$|V^-| = \lfloor n/2 \rfloor, \qquad |V^+| = \lceil n/2 \rceil .$$

Partition E into three subsets E^+, E^-, E' such that E^+ contains the edges with both endpoints in V^+, E^- contains the edges with both endpoints in V^-, and E' contains the edges crossing the y-axis.

Obviously,

$$|E^-| \le f_k(\lfloor n/2 \rfloor), \qquad |E^+| \le f_k(\lceil n/2 \rceil).$$

To obtain Theorem 14, it suffices to show that $|E'| = O(n)$, because then

$$f_k(n) \le f_k(\lfloor n/2 \rfloor) + f_k(\lceil n/2 \rceil) + O(n)$$

and, consequently,

$$f_k(n) = O(n \log n).$$

To bound the size of E', consider the mapping T given by $(x, y) \mapsto (\frac{1}{x}, \frac{y}{x})$. T is a projective transformation sending the y-axis to infinity. It is not hard to verify that two segments uv, $u'v'$ intersect if and only if the segments $T(u)t(v)$, $T(u')T(v')$ are parallel.

Consider the graph $\tilde{G}$ on the vertex set $\tilde{V} = \{T(v) | v \in V\}$ with two vertices $T(v), T(w)$ connected by an edge if and only if $\{v, w\} \in E'$. The graph $\tilde{G}$ contains no k pairwise parallel edges, since otherwise the corresponding k edges in E' would be pairwise crossing. By Theorem 25, $\tilde{G}$ contains $O(n)$ edges. Consequently, $|E'| = O(n)$ as required. □

8. Davenport–Schinzel Trees

Klazar [**Kl95**] further generalized the concept of Davenport-Schinzel sequences to labeled (colored) trees. He investigated the maximum size of a tree with each vertex labeled by one of n labels in such a way that, besides some technical conditions, the tree does not contain a subdivision of a given labeled path-pattern.

A *labeled tree* $\mathcal{T} = (T, l)$ is a tree $T = (V, E)$ with a labeling $l : V \to L$, which assigns a label to each vertex.

Let u be a sequence. We say that a labeled tree $\mathcal{T} = (T, l)$ is *$\hat{u}$-free* if, for every path $v_1 e_1 v_2 \ldots e_k v_k$ in $\mathcal{T}$, the sequence $l(v_1) l(v_2) \ldots l(v_k)$ is u-free. A *rooted labeled tree* is a labeled tree with one vertex specified as the *root*. We say that a rooted labeled tree $\mathcal{T} = (T, l)$ is *u-free* if, for every path $(\text{root} = v_1)e_1v_2 \dots e_kv_k$ in $\mathcal{T}$ starting in the root, the sequence $l(v_1)l(v_2)\dots l(v_k)$ is u-free.

We say that a labeled tree is *k-sparse*, if no two vertices at distance less than k are labeled by the same label. Thus, a labeled tree is 2-sparse, if no two adjacent vertices are labeled by the same label.

Let n be an integer and let u be a sequence. An *$(n, \hat{u})$-tree* is a $\hat{u}$-free $||u||$-sparse labeled tree whose vertices are labeled by at most n distinct labels. An *(n, u)-tree* is a u-free $||u||$-sparse rooted labeled tree whose vertices are labeled by at most n distinct labels.

We consider the problem of estimating the maximum size of a (n, u)-tree ($(n, \hat{u})$-tree, respectively) satisfying still some additional technical condition. An additional technical condition is needed to get reasonable tree analogues of Davenport–Schinzel sequences (e.g., if $u = abab$, an additional technical condition is needed to avoid arbitrarily large (n, u)-trees such as the star in Fig. 9.) It is possible to choose between

FIGURE 9. An arbitrarily large $(n, abab)$-tree (for $n \geq 2$) and an example of a spidery tree.

several natural technical conditions. We consider three different technical conditions which (somewhat surprisingly) lead to asymptotically equivalent extremal functions. Each of the three considered conditions restricts possible distributions of a label.

Let $\mathcal{T}$ be a labeled tree. For a label i, let $\mathcal{T}_i$ be the smallest (labeled) subtree of $\mathcal{T}$ containing all vertices labeled by i. We say that $\mathcal{T}$ is

- *skeletal* if, for every label i, all vertices in $\mathcal{T}_i$ of degree more than 2 are labeled by i,
- *weakly spidery* if $\mathcal{T}_i$ is a path for every label i,
- *spidery* if $\mathcal{T}$ is a labeled rooted tree and each $\mathcal{T}_i$ is a subpath of a path starting in the root (see Fig. 9).

A spidery tree is also called *spider* for short. Similarly, a spidery (n, u)-tree is also called *(n, u)-spider* for short.

We are interested in the following three functions:

$\tau_u(n) =:$ the maximum size (i.e., number of vertices) of a skeletal (n, u)-tree (unless $||u|| = 1$),

$\sigma'_u(n) =:$ the maximum size of a weakly spidery (n, u)-tree.

$\sigma_u(n) =:$ the maximum size of a spidery (n, u)-tree.

If $||u|| = 1$ then we define $\tau_u(n) = \sigma'_u(n)$ for technical reasons (the above definition would give $\tau_u(n) = +\infty$ for $|u| > 2, ||u|| = 1$). Here are unrooted analogues of the above functions:

$\hat{\tau}_u(n)$ =: the maximum size (i.e., number of vertices) of a skeletal $(n, \hat{u})$-tree (unless $||u|| = 1$),

$\hat{\sigma}'_u(n)$ =: the maximum size of a weakly spidery $(n, \hat{u})$-tree.

$\hat{\sigma}_u(n)$ =: the maximum size of a spidery $(n, \hat{u})$-tree.

If $||u|| = 1$ then we define $\hat{\tau}_u(n) = \hat{\sigma}'_u(n)$ for technical reasons (the above definition would give $\hat{\tau}_u(n) = +\infty$ for $|u| > 3, ||u|| = 1$).

The function $\hat{\tau}_u(n)$ (originally denoted by $Ex(u,n)^{\mathrm{T}}$) was introduced in [**Kl95**] and further investigated in [**Kl97, Kl99a, V98b**], while the function $\hat{\sigma}_u(n)$ is considered in [**V99a, V99b**]. The remaining four functions $\tau_u(n)$, $\sigma'_u(n)$, $\sigma_u(n)$, $\hat{\sigma}'_u(n)$ were introduced in [**V99a**], where it was shown that the functions $\hat{\tau}_u(n)$, $\hat{\sigma}'_u(n)$, $\hat{\sigma}_u(n)$ (and also the functions $\tau_u(n)$, $\sigma'_u(n)$, $\sigma_u(n)$) have the same asymptotic behavior for each u (see Theorem 29 below).

From the definitions, we have

$$f_u(n) \leq \sigma_u(n) \leq \sigma'_u(n) \leq \tau_u(n),$$
$$\hat{\sigma}_u(n) \leq \hat{\sigma}'_u(n) \leq \hat{\tau}_u(n),$$
$$\hat{\sigma}_u(n) \leq \sigma_u(n), \quad \hat{\tau}_u(n) \leq \tau_u(n), \quad \hat{\sigma}'_u(n) \leq \sigma'_u(n)$$

for each u and n. If u is symmetric, then also $f_u(n) \leq \hat{\sigma}_u(n)$.

CLAIM 28. *For any u and n, $\tau_u(n)$ is finite (and thus also the other five functions are finite).*

Proof of Claim 28. The maximum degree in a skeletal 2-sparse tree labeled by n labels is at most $2(n-1)$, since no vertex can have three neighbors labeled by the same label. The claim then follows from the finiteness of $f_u(n)$ (Claim 1), since the length of any path in a (n, u)-tree is bounded from above by $f_u(n)$. □

Somewhat surprisingly, for each u all the functions $\tau_u(n), \sigma'_u(n), \sigma_u(n)$ have the same asymptotic behavior:

THEOREM 29 (Valtr [**V99a**]). *For any sequence u,*

$$\tau_u(n) = \Theta(\sigma'_u(n)) = \Theta(\sigma_u(n)) \quad \text{and} \quad \hat{\tau}_u(n) = \Theta(\hat{\sigma}'_u(n)) = \Theta(\hat{\sigma}_u(n)).$$

Klazar proved an analogue of Theorem 2 for the labeled trees:

THEOREM 30 (Klazar [**Kl99a**]). *For any sequence u, there is a constant $c = c(u)$ such that*

$$\hat{\tau}_u(n) = O\left(n 2^{(\alpha(n))^c}\right).$$

Theorem 30 was improved in [**V99a**]:

THEOREM 31 (Valtr [**V99a**]). *For any sequence u, there is a constant $c = c(u)$ such that*

$$\hat{\sigma}_u(n) \leq \sigma_u(n) = O\left(n 2^{c(\alpha(n))^{|u|+||u||-5}}\right).$$

We remark that the proofs of Theorems 30 and 31 are derived from Theorem 2. The proof of Theorem 31 is quite simple, we sketch it in Section 9.

By Theorem 29, the bounds in Theorems 30 and 31 hold also for the functions $\sigma'_u(n)$, $\tau_u(n)$, $\hat{\sigma}'_u(n)$, $\hat{\tau}_u(n)$. Also other asymptotic results about any of the functions $\tau_u(n)$, $\sigma'_u(n)$, $\sigma_u(n)$ hold also for the other two functions. From now on, we will state them for the function $\sigma_u(n)$ (or for $\hat{\sigma}_u(n)$) which seems to be more natural and easier to investigate than the other, asymptotically equivalent, functions.

Using Theorem 30, we obtain an analogue of Proposition 3:

PROPOSITION 32. *For any sequence u, the function $\sigma_u(n)$ behaves asymptotically in one of the following three ways:*

(i) $\sigma_u(n) \leq 1$ *for all* n,

(ii) $\sigma_u(n) = \Theta(n)$, *or*

(iii) $\frac{\sigma_u(n)}{n}$ *tends very slowly to infinity as n tends to infinity (it tends to infinity slower than the functional inverse of any unbounded primitive recursive function).*

Proposition 32 is proved at the end of Paragraph 10.5. It is not difficult to see that case (i) in Proposition 32 holds only for sequences isomorphic to one of the three sequences ε (empty sequence), a, and ab, where a, b are letters. It is a more delicate (and open) question to distinguish between the cases (ii) and (iii) in Proposition 32.

We remark that a statement analogous to Proposition 32 holds for the function $\hat{\sigma}_u(n)$. In the light of Proposition 32, one could conjecture that the functions $f_u(n)$ and $\sigma_u(n)$ (or $\hat{\sigma}_u(n)$) have different asymptotic behavior only for some small class of exceptional sequences u. This turned out to be wrong, and in this survey we describe results showing a significant difference (within the bounds given by Propositions 3 and 32) between the functions $f_u(n)$ and $\sigma_u(n)$. One such result is the following one:

THEOREM 33 (two-letter theorem, [**V99b**]). (i) *If u is a sequence with $||u|| \leq 2$, then $\sigma_u(n) = O(n)$ if and only if u is ababa-free and aba^2b-free.*

(ii) *If u is a sequence with $||u|| \leq 2$, then $\hat{\sigma}_u(n) = O(n)$ if and only if u is ababa-free and ab^2a^2b-free.*

Comparison of Theorem 33 with the two-letter theorem for sequences (Theorem 4) gives

$$f_u(n) = O(n), \quad \hat{\sigma}_u(n) \neq O(n), \quad \sigma_u(n) \neq O(n)$$

for all sequences $u = a^i b^j a^k b^l, i, l \geq 1, j, k \geq 2$. Theorem 33 also shows that

$$\hat{\sigma}_u(n) = O(n), \quad \sigma_u(n) \neq O(n)$$

holds for any sequence $u = a^i b a^k b^l, i, l \geq 1, k \geq 2$.

There are also non-trivial results concerning the validity of $\sigma_u(n) = O(n)$ for sequences $u, ||u|| > 2$. We describe them in the following three sections. Here is an important rule allowing to derive an upper bound on $\hat{\sigma}_u(n)$ from the upper bounds on the functions $\sigma_v(n)$:

THEOREM 34 (derivation rule, [**V99b**]). *For any u,*

$$\hat{\sigma}_u(n) = O(\min\{\sigma_{\overline{v}}(n) + \sigma_w(n) | u = vw, S(v) \cap S(w) = \emptyset\}),$$

unless $|u| = ||u|| = 4$. (For the empty sequence ε and for each n, we take $\sigma_\varepsilon(n) = 0$.)

The bound in Theorem 34 is known to be asymptotically best possible for irreducible sequences and for sequences over at most two letters:

THEOREM 35 ([**V99b**]). *If u is irreducible, then*

$$\hat{\sigma}_u(n) = \Theta(\min\{\sigma_u(n), \sigma_{\overline{u}}(n)\}).$$

COROLLARY 36 ([**V99b**]). *If $||u|| \leq 2$, then*

$$\hat{\sigma}_u(n) = \Theta(\min\{\sigma_u(n), \sigma_{\overline{u}}(n)\}).$$

The following theorem determines, for example, the validity of $\sigma_{\text{double}(u)}(n) = O(n)$ and $\hat{\sigma}_{\text{double}(u)}(n) = O(n)$ for any sequence u.

THEOREM 37 ([**V99b**]). *If $u = a_1^{i_1} a_2^{i_2} \dots a_k^{i_k}$, where $a_1, a_2, \dots, a_k$ are letters (not necessarily distinct) and each i_j is bigger than 1, then*

(i) *$\sigma_u(n) = O(n)$ if and only if u is abab-free, and also*
(ii) *$\hat{\sigma}_u(n) = O(n)$ if and only if u is abab-free.*

9. Upper Bounds on $\sigma_u(n)$ Obtained from Upper Bounds on $f_u(n)$

Some upper bounds on $\sigma_u(n)$ can be obtained from the upper bounds on $f_{u'}(n)$, where u' is a certain extension of u. One of the results obtained in this way is Theorem 31. Two other results give good estimates on $\sigma_u(n)$ and $\hat{\sigma}_u(n)$ when u is an alternating sequence or a so-called "up-down sequence". Here is the basic lemma:

LEMMA 38 ([**V99a**]). *Let $u = wa_1$ be a sequence over an alphabet $S(u) = \{a_1, a_2, \dots, a_t\}$, $||u|| = t \geq 2$, obtained from a sequence w by adding a letter a_1 to its end. Define u' as the sequence $u' = wa_1a_2a_3 \dots a_t$ of length $|u| + ||u|| - 1$. Then*

$$\sigma_u(n) \leq f_{u'}(n).$$

Proof of Lemma 38 (sketch). Let $\mathcal{T}$ be a (n, u)-spider on $\sigma_u(n)$ vertices, and let l be the labeling on $\mathcal{T}$. Let $v_1, v_2, \dots, v_{|\mathcal{T}|}$ be the vertices of $\mathcal{T}$ ordered in the order in which they are visited during a fixed depth–first search starting from the root. Clearly, the sequence $l(v_1) \dots l(v_{|\mathcal{T}|})$ of length $|\mathcal{T}| = \sigma_u(n)$ is $||u||$-sparse. It is also u'-free (see [**V99a**]). The lemma follows. □

Proof of Theorem 31. Theorem 31 is an immediate consequence of Lemma 38 and Theorem 2. □

We now give the results about the alternating and "up-down" sequences.

By al(i) we denote an *alternating sequence* of length i (i.e., a sequence *ababa*... of length i, where a, b are two different letters).

THEOREM 39 ([**V99a**]). *For any i,*

$$N_i(n) = f_{\text{al}(i)}(n) \leq \hat{\sigma}_{\text{al}(i)}(n) \leq \sigma_{\text{al}(i)}(n) \leq f_{\text{al}(i+1)}(n) = N_{i+1}(n).$$

Proof. The first inequality is obvious, the last inequality follows from Lemma 38. □

For $k, i \geq 1$, we define updown(k, i) as the sequence

$$\text{updown}(k, i) = a_1^i a_2^i \dots a_{k-1}^i a_k^i a_k^i a_{k-1}^i \dots a_2^i a_1^i$$

of length $2ki$, where $a_1, \dots, a_k$ are k distinct letters and a_j^i denotes the sequence $a_j a_j \dots a_j$ of length i.

THEOREM 40 ([**V99a**]). *For any fixed $k, i \geq 1$,*

$$\hat{\sigma}_{\text{updown}(k,i)}(n) \leq \sigma_{\text{updown}(k,i)}(n) = O(n).$$

Proof. Let $u = \text{updown}(k,i)$, and let u' be as in Lemma 38. The sequence u' is a subsequence of the sequence $\text{updownup}(k,i) = a_1^i \dots a_k^i a_k^i \dots a_1^i a_1^i \dots a_k^i$ over the same alphabet. Thus, $f_{u'}(n) \le f_{\text{updownup}(k,i)}(n)$. By Theorem 6, $f_{\text{updownup}(k,i)}(n) = O(n)$ for any fixed k, i. Theorem 40 now follows from Lemma 38:

$$\sigma_u(n) \le f_{u'}(n) \le f_{\text{updownup}(k,i)}(n) = O(n).$$

□

We remark that the special case $k = 1$ in Theorem 40 was first shown in [**V98b**].

10. Further Results on Davenport–Schinzel Trees

In this section we describe results on the asymptotic behavior of $\sigma_u(n)$. Most general results from the previous sections bounding the function $f_u(n)$ hold also for the function $\sigma_u(n)$. However, the letter-insertions rule described in Theorem 5(v) and in Lemma 11 holds only in a somewhat weaker form, and this causes a significant difference in the asymptotic behavior of $\sigma_u(n)$ (or $\hat{\sigma}_u(n)$) and $f_u(n)$. Throughout the rest of the paper, $|\mathcal{T}|$ denotes the number of vertices in a spider $\mathcal{T}$.

10.1. (Non-)Linearity of $\boldsymbol{\sigma_u(n)}$. Here is an analogue of Theorem 5:

THEOREM 41. *Let $\mathcal{L}'$ denote the class of all sequences u with $\sigma_u(n) = O(n)$. Then:*

(i) *The sequence a^i lies in $\mathcal{L}'$ for any $i \ge 1$ (a is a letter).*
(ii) *(monotone property) If v contains u and $v \in \mathcal{L}'$ then $u \in \mathcal{L}'$.*
(iii) *(letter duplication) Let u_1, u_2 be two sequences, let a be a letter, and let $u_1a^2u_2 \in \mathcal{L}'$. Then $u_1a^3u_2 \in \mathcal{L}'$.*
(iv) *(first-letter and last-letter duplications) Let u be a sequence and let a be a letter. Then, if $ua \in \mathcal{L}'$ then $ua^2 \in \mathcal{L}'$, and if $au \in \mathcal{L}'$ then $a^2u \in \mathcal{L}'$.*
(v) *(letter insertions) Let $u = au_1a^2u_2$ be a sequence in $\mathcal{L}'$, where u_1, u_2 are sequences (possibly empty), a, b are letters, and $b \notin S(u)$. Then the sequence $u' = bau_1abau_2$ also lies in $\mathcal{L}'$.*
(vi) *(sequence insertion) Let u_1, u_2, w be three sequences, let a be a letter, let $u = u_1a^2u_2$, and let $S(u) \cap S(w) = \emptyset$. Moreover, let u, w lie in $\mathcal{L}'$. Then also u_1awau_2 lies in $\mathcal{L}'$.*

Theorem 41 is analogous to Theorem 5 with the only exceptions that an analogue of part (vii) does not hold and that only one letter b is inserted between the two a's in the sequence u' in part (v). However, these two little changes cause a significant (but necessary) difference. For many sequences u, Theorem 5 gives $f_u(n) = O(n)$ but Theorem 41 does not give $\sigma_u(n) = O(n)$ (or we even know that $\sigma_u(n) \ne O(n)$).

The proof of Theorem 41 is also analogous in many steps to the proof of Theorem 5 but it is more involved. It uses analogous tools and properties: sparsity and reduction lemmas, monotone and additive properties, weak super- and subadditivity. We formulate all of these tools and properties below and for all of them we at least indicate how to prove them. However, we do not give the proofs of the key parts (v) and (vi) of Theorem 41, the reader can find them in [**V99b**]. In Section 11 we describe a construction from [**V99b**] which gives a non-linear lower bound on $\sigma_u(n)$ ($\hat{\sigma}_u(n)$) for many relatively short sequences u:

THEOREM 42. *If u contains at least one of the four sequences $u = ababa$, aba^2b, $abcabc$, $abcacb$, then*

$$\sigma_u(n) = \Omega(n\alpha(n)).$$

THEOREM 43. *If u is a 2-sparse sequence with $|u| > 3||u|| - 2$, then*

$$\sigma_u(n) \geq \hat{\sigma}_u(n) = \Omega(n\alpha(n)).$$

Theorem 43 is proved in Section 11. On the other hand, it easily follows from Theorem 41(iv),(v) that there are arbitrarily long sequences u with $|u| = 3||u|| - 2$ satisfying $\sigma_u(n), \hat{\sigma}_u(n) = O(n)$:

THEOREM 44. *For any $k \geq 2$,*

$$\hat{\sigma}_u(n) \leq \sigma_u(n) = O(n),$$

where u is the sequence of length $3k - 2$ defined by $u = u_1u_2, \dots, u_{k-1}a_k$, $u_i = a_ia_{i+1}a_i$, $S(u) = \{a_1, \dots, a_k\}$.

Theorems 43 and 44 immediately give a solution to an analogue of Problem 2 for $\sigma_u(n)$ ($\hat{\sigma}_u(n)$):

COROLLARY 45. (i) *The maximum length of a 2-sparse sequence u over k letters with $\sigma_u(n) = O(n)$ is $3k - 2$.*

(ii) *The maximum length of a 2-sparse sequence u over k letters with $\hat{\sigma}_u(n) = O(n)$ is $3k - 2$.*

□

10.2. 2-Sparse Sequences over 3 Letters with $\sigma_u(n) = O(n)$. All known results about the asymptotic behavior of $\sigma_u(n)$ and $\hat{\sigma}_u(n)$ for 2-sparse sequences u with $||u|| \leq 3$ are shown in Fig. 10 and Fig. 11. They all follow from Theorem 41 and from the construction given in Section 11. See [**V99b**] for details.

Here are a few remarks:

By Fig. 10, the sequence $u = abcbab$ satisfies $\sigma_u(n) = O(n)$ and $\sigma_{\overline{u}}(n) \neq O(n)$ (by Theorem 33, this holds also for $u = ab^2ab$). This shows that there is no analogue of the symmetry property (Theorem 5(vii)) for the function $\sigma_u(n)$.

On the other hand, $\hat{\sigma}_u(n) = \hat{\sigma}_{\overline{u}}(n)$ for all u and n. Therefore, each pair $u, \overline{u}$ (in a dotted frame) is either fully inside or fully outside each of the rectilinear regions in Fig. 11.

10.3. Basic Tools. In this paragraph we formulate the (weaker) analogues of Lemmas 8-17. Four of them are proved in Paragraph 10.5, the remaining ones can be proved similarly as their sequence analogues in Section 3. The reader can find the proofs of all of them in [**V99b**]. In Paragraph 10.5 we also describe a tree analogue of the algorithm $A(l)$. In Paragraph 10.4 we formulate the tree analogues of Theorems 18 (letter insertions) and 20 (sequence insertion).

If $\mathcal{T}$ is a rooted tree and V is a set of vertices in $\mathcal{T}$ containing the root of $\mathcal{T}$, then the V*-reduction* of $\mathcal{T}$ is defined as the rooted tree on the vertex set V with the root in the root of $\mathcal{T}$ and with two vertices $v_1, v_2 \in V$ connected by an edge if and only if v_1, v_2 lie on a common path from the root in $\mathcal{T}$ and the subpath connecting them contains no other vertices of V (see Fig. 12). If $\mathcal{T}'$ is a V-reduction of $\mathcal{T}$ for some V, we say that $\mathcal{T}'$ is a *reduction* of $\mathcal{T}$. Obviously, any reduction of a spider is also a spider.

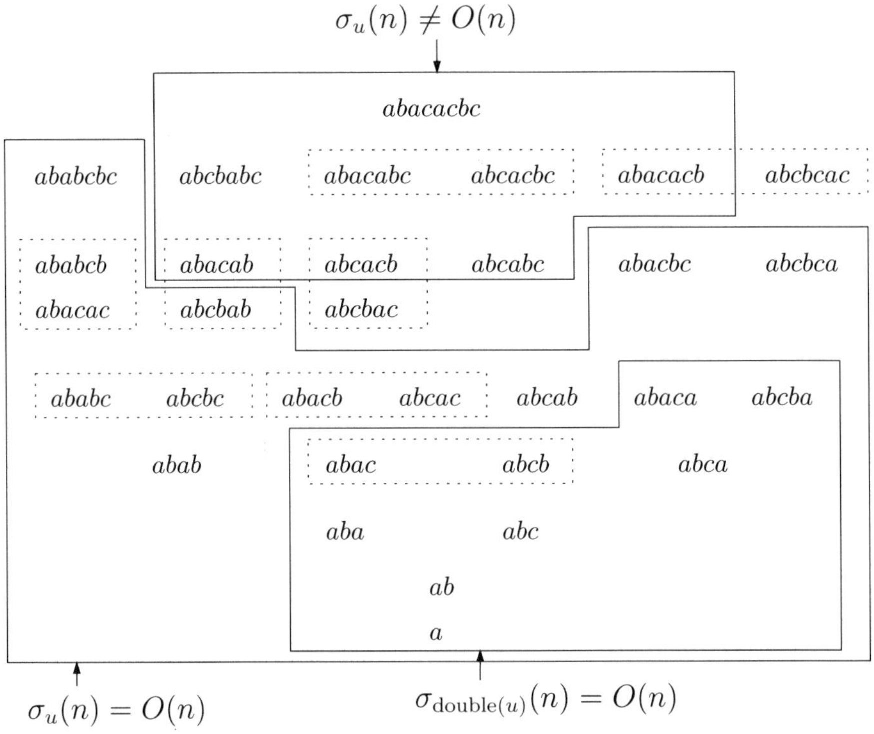

FIGURE 10. Table of *ababa*-free 2-sparse sequences over 3 letters analogous to Fig. 1. The three frames contain sequences u with $\sigma_u(n) = O(n)$, $\sigma_{\text{double}(u)}(n) = O(n)$, and $\sigma_u(n) \neq O(n)$, respectively. The validity of $\sigma_u(n) = O(n)$ is not known for the remaining two sequences.

LEMMA 46 (sparsity lemma). *Let u be a sequence, and let $l \geq ||u||$. Then there exists a positive constant $\varepsilon = \varepsilon(u, l) > 0$ such that for any u-free $||u||$-sparse spider $\mathcal{T}$ there is an l-sparse reduction of $\mathcal{T}$ with at least $\varepsilon|\mathcal{T}|$ vertices.*

LEMMA 47 (reduction lemma). *For any integer $l \geq 0$ and for any $\varepsilon \in (0,1)$, there is an $\varepsilon' > 0$ with the following property. Let $\mathcal{T}$ be a $\lceil l/\varepsilon \rceil$-sparse spider and let V_1 be a set of at least $\varepsilon|\mathcal{T}|$ vertices of $\mathcal{T}$. Then the V_1-reduction of $\mathcal{T}$ has an l-sparse reduction on at least $\varepsilon'|\mathcal{T}|$ vertices.*

LEMMA 48 (letter duplication). *Let u_1, u_2 be two sequences, and let a be a letter. Then*

$$\sigma_{u_1 a^3 u_2}(n) = \Theta(\sigma_{u_1 a^2 u_2}(n)).$$

LEMMA 49 (first–letter and last–letter duplications). *Let u be a sequence and let a be a letter. Then*

$$\sigma_{ua^2}(n) = \sigma_{ua}(n) + O(n)$$

and

$$\sigma_{a^2 u}(n) = \sigma_{au}(n) + O(n).$$

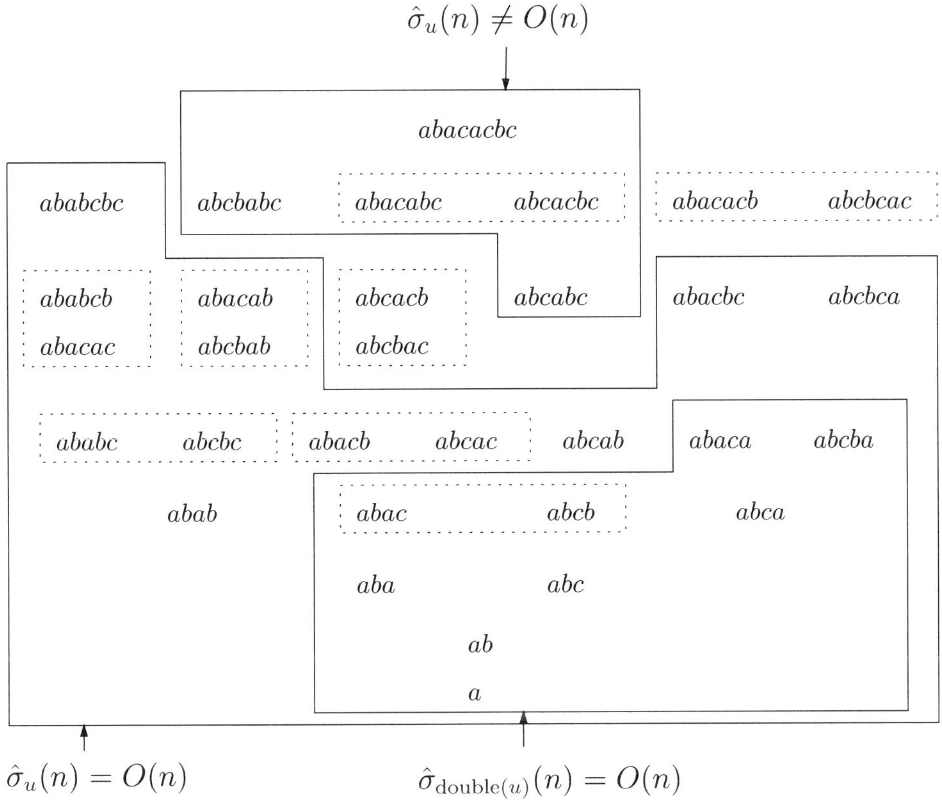

FIGURE 11. Table of *ababa*-free 2-sparse sequences. The three frames contain sequences u with $\hat{\sigma}_u(n) = O(n)$, $\hat{\sigma}_{\text{double}(u)}(n) = O(n)$, and $\hat{\sigma}_u(n) \neq O(n)$, respectively. The validity of $\hat{\sigma}_u(n) = O(n)$ is not known for the remaining two pairs of sequences.

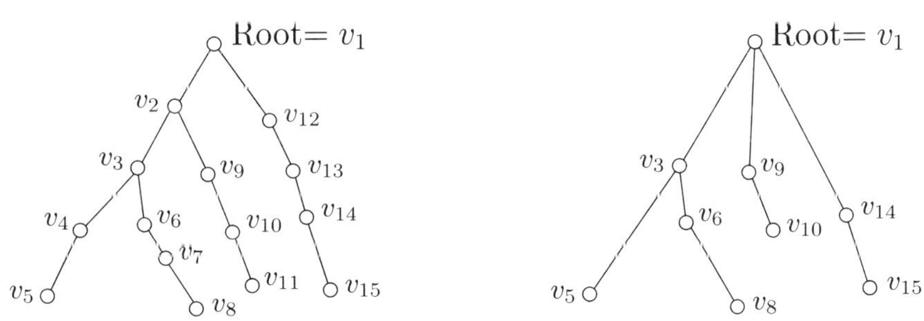

FIGURE 12. A rooted tree $\mathcal{T}$ and its V-reduction, where $V = \{v_1, v_3, v_5, v_6, v_8, v_9, v_{10}, v_{14}, v_{15}\}$.

LEMMA 50 (monotone property). *If u_1 contains u_2, then there is a constant $\varepsilon > 0$ such that, for any n,*

$$\sigma_{u_1}(n) \geq \varepsilon \cdot \sigma_{u_2}(n).$$

LEMMA 51 (additive property). *Let u_1, u_2 be two sequences with $S(u_1) \cap S(u_2) = \emptyset$. Then there are two positive constants $c_1, c_2 > 0$ such that, for any n,*

$$c_1(\sigma_{u_1}(n) + \sigma_{u_2}(n)) \leq \sigma_{u_1 u_2}(n) \leq c_2(\sigma_{u_1}(n) + \sigma_{u_2}(n) + n).$$

LEMMA 52 (weak super-additivity). *If u is a sequence with $|u| > \min\{||u||, 2\}$, then there is a positive constant $\varepsilon = \varepsilon(u) > 0$ such that*

$$\sigma_u(\sum_{i=1}^{r} n_i) \geq \varepsilon \sum_{i=1}^{r} \sigma_u(n_i)$$

holds for any positive integers $r, n_1, n_2, \ldots, n_r$.

LEMMA 53 (weak sub-additivity). (i) *For any sequence u and for any integer $k \geq 1$, there is a $c = c(u, k) > 0$ such that*

$$\sigma_u(kn) \leq c\sigma_u(n),$$

for any integer n.

(ii) *For any sequence u and for any integer $k \geq 1$, there is a $c' = c'(u, k) > 0$ such that*

$$\hat{\sigma}_u(kn) \leq c'\hat{\sigma}_u(n),$$

for any integer n.

10.4. Letter and Sequence Insertions. Here we formulate the tree versions of Theorems 18 and 20. They are proved in [**V99b**].

THEOREM 54 (letter insertions). *Let $u = au_1a^2u_2$, $u' = bau_1abau_2$, where u_1, u_2 are two sequences (possibly empty), a, b are letters, and $b \notin S(u)$. Then $\sigma_{u'}(n) = O(\sigma_u(n))$.*

THEOREM 55 (sequence insertion). *Let $u = u_1a^2u_2$, w be two sequences, where u_1, u_2 are two sequences, a is a letter, and $S(u) \cap S(w) = \emptyset$. Then*

$$\sigma_{u_1awau_2}(n) = O(\sigma_w(\sigma_u(n)) + \sigma_u(n)).$$

In particular, if $\sigma_u(n), \sigma_w(n) = O(n)$, then also $\sigma_{u_1awau_2}(n) = O(n)$.

We remark that, similarly as for the sequences, the proof of the letter-insertions theorem relies on a density lemma:

LEMMA 56 (density lemma). *For every sequence u and for every $p \geq 1$, there exists a constant $\varepsilon = \varepsilon(u, p) > 0$ with the following property: Let $\mathcal{T}$ be a $||u||$-sparse u-free spider, and let $<$ be any linear order on the set of labels used on $\mathcal{T}$. Then $\mathcal{T}$ has a $||u||$-sparse reduction $\mathcal{T}^{(p)}$ on at least $\varepsilon|\mathcal{T}|$ vertices such that, for any label i and for any two different vertices v_1, v_2 in $\mathcal{T}^{(p)}$ labeled by the label i, at least p labels j with $j < i$ occur on the path in $\mathcal{T}$ connecting v_1, v_2.*

10.5. Proofs of the Basic Tools. The proofs of Lemma 48 (letter duplication), Lemma 49 (first–letter and last–letter duplications), Lemma 50 (monotone property), and Lemma 53 (weak sub-additivity) are very similar to the proofs of their sequence analogues (Lemmas 11, 12, 14, 17); the reader can find them in [**V99b**].

In this paragraph we prove the other four lemmas described in the previous paragraph, and also derive Proposition 32 from one of them. All proofs are taken from [**V99b**], one of them is only sketched here.

First, we give an analogue of the algorithm $A(l)$ (introduced in Paragraph 3.1) for spiders, which, roughly speaking, simulates the algorithm $A(l)$ on the sequence of labels along each path from the root of a spider, and then makes the V-reduction on the obtained set of vertices V.

Let $k \geq 2$, and let $\mathcal{T} = (T, l)$ be a spider with root r. The *algorithm* $\mathcal{B}(k)$ finds an k-sparse reduction $\mathcal{B}(\mathcal{T}, k)$ of $\mathcal{T}$ as follows. The vertex set of $\mathcal{B}(\mathcal{T}, k)$ will be denoted by V. Let $r = v_1, v_2, \ldots, v_z$ be any path from r in $\mathcal{T}$, and let

$$C = l(v_1)l(v_2)\ldots l(v_z)$$

be the sequence of labels appearing along it. Further, let

$$A(C, k) = l(v_{i_1})l(v_{i_2})\ldots l(v_{i_t})$$

be the sequence obtained by running the algorithm $A(k)$ on the sequence C. Then V contains vertices $v_{i_1}v_{i_2}\ldots v_{i_t}$ and no other vertices of the path $v_1, \ldots, v_z$. Clearly, V is correctly defined. $\mathcal{B}(\mathcal{T}, k)$ is then the V-reduction of $\mathcal{T}$ (see Fig. 13).

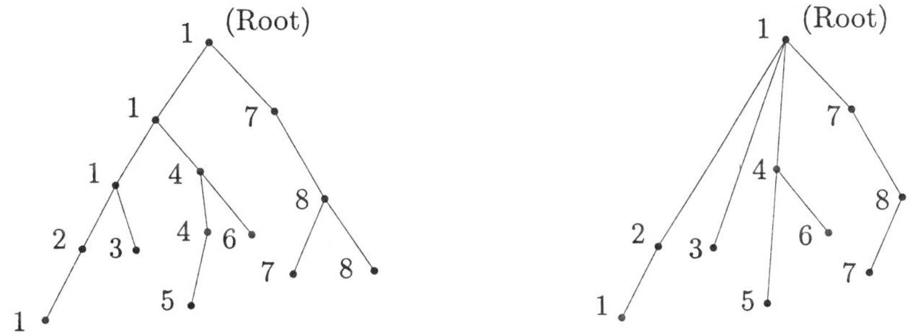

FIGURE 13. A spider $\mathcal{T}$ and the spider $\mathcal{B}(\mathcal{T}, 2)$ obtained from $\mathcal{T}$ by running the algorithm $\mathcal{B}(2)$.

OBSERVATION 57. (i) *$\mathcal{B}(\mathcal{T}, k)$ is a k-sparse spider.*
(ii) *If $\mathcal{T}$ is u-free, then any reduction of $\mathcal{T}$ (in particular, $\mathcal{B}(\mathcal{T}, k)$) is also u-free.*

Proof. (i) follows from the properties of the algorithm $A(k)$, (ii) is trivial. □

Proof of Lemma 46 (sparsity lemma). We obtain Lemma 46 by running the algorithm $\mathcal{B}(l)$ on $\mathcal{T}$. By Observation 57, the spider $\mathcal{B}(\mathcal{T}, l)$ is u-free and l-sparse, so it remains to show the required bound on the size of $\mathcal{B}(\mathcal{T}, l)$.

For a vertex v of $\mathcal{B}(\mathcal{T}, l)$, let $\mathcal{T}_v$ be the maximal subtree of $\mathcal{T}$ containing v and vertices "under" v not lying in the vertex set of $\mathcal{B}(\mathcal{T}, l)$ (a vertex x lies *under* v if v lies on the path from x to the root).

It follows from the construction that the spider $\mathcal{T}_v$ with the root in v is u-free and its vertices are labeled by at most $l - 1$ labels (namely, by the labels used on the first $l - 1$ vertices on the path in $\mathcal{B}(\mathcal{T}, l)$ from v to the root of $\mathcal{B}(\mathcal{T}, l)$). Thus, $\mathcal{T}_v$ contains at most $\sigma_u(l - 1)$ vertices. Since each vertex of $\mathcal{T}$ lies in exactly one tree $\mathcal{T}_v, v \in \mathcal{B}(\mathcal{T}, l)$, the lemma with $\varepsilon = \frac{1}{\sigma_u(l-1)}$ follows. □

Proof of Lemma 47 (reduction lemma). Let $\mathcal{T}$ be a $\lceil l/\varepsilon \rceil$-sparse spider, let V_1 be a set of at least $\varepsilon|\mathcal{T}|$ vertices of $\mathcal{T}$, and let $\mathcal{T}_1$ be the V_1-reduction of $\mathcal{T}$. We obtain the required reduction of $\mathcal{T}_1$ by running the algorithm $\mathcal{B}(l)$ on $\mathcal{T}_1$. Let $\mathcal{T}' = \mathcal{B}(\mathcal{T}_1, l)$

be the reduction obtained by the algorithm. $\mathcal{T}'$ is an l-sparse reduction of $\mathcal{T}_1$, so it remains to show that $\mathcal{T}'$ has at least $\varepsilon'|\mathcal{T}|$ vertices.

For a vertex $u \in V_1$, let $v(u)$ be the first vertex of $\mathcal{T}'$ on the path in $\mathcal{T}$ (or in $\mathcal{T}_1$) from u to the root (thus, if u lies in the reduction $\mathcal{T}'$ then $v(u) = u$). If the distance between u and $v(u)$ in $\mathcal{T}$ is less than $\lceil l/\varepsilon \rceil$, then we say that u is *of type I*; otherwise we say that u is *of type II*.

Let v be any vertex in $\mathcal{T}'$, and let U_v be the set of vertices $u \in V_1$ of type I with $v(u) = v$. By the $\lceil l/\varepsilon \rceil$-sparsity of $\mathcal{T}$, vertices of U_v are labeled by pairwise different labels. Since all vertices of U_v different from v are discarded by the algorithm $\mathcal{B}(l)$, they are labeled only by the (at most) $l-1$ labels used on the first (at most) $l-1$ vertices of the path in $\mathcal{T}'$ from v to the root. It follows that each U_v contains at most $1+(l-1) = l$ vertices. Consequently, there are at most $l \cdot |\mathcal{T}'|$ vertices of type I.

To estimate the number of vertices of type II, we use vertices of type II as colors. For each vertex $u \in V_1$ of type II, we use u as a color on the first $\lceil l/\varepsilon \rceil$ vertices on the path in $\mathcal{T}$ from u to the root. Thus, each vertex u of type II is used as a color on exactly $\lceil l/\varepsilon \rceil$ vertices (including u itself). On the other hand, we now show that any vertex x in $\mathcal{T}$ is colored by at most $l-1$ colors (vertices). Let $C(x)$ be the set of colors (vertices) of x. All vertices of $C(x)$ different from x lie "under" x at distance less than $\lceil l/\varepsilon \rceil$ from x. Therefore, since $\mathcal{T}$ is a $\lceil l/\varepsilon \rceil$-sparse spider, $C(x)$ contains no pair of vertices labeled by the same label. Since any vertex u in $C(x) \subseteq V_1$ is discarded by the algorithm $\mathcal{B}(l)$, it must be labeled by one of the labels used on the first (at most) $l-1$ vertices of the path in $\mathcal{T}'$ from $v(u) = v(x)$ to the root. It follows that the size of $C(x)$ is at most $l-1$. Thus, the number of vertices of type II is at most $(l-1)|\mathcal{T}|/\lceil l/\varepsilon \rceil$.

Altogether, $\mathcal{T}_1$ contains at most

$$l \cdot |\mathcal{T}'| + \frac{(l-1)|\mathcal{T}|}{\lceil l/\varepsilon \rceil}$$

vertices. Consequently,

$$|\mathcal{T}'| \geq \frac{1}{l}\left(|\mathcal{T}_1| - \frac{(l-1)|\mathcal{T}|}{\lceil l/\varepsilon \rceil}\right) \geq \frac{1}{l}\left(\varepsilon|\mathcal{T}| - \frac{\varepsilon(l-1)}{l}|\mathcal{T}|\right) = \frac{\varepsilon}{l^2}|\mathcal{T}|.$$

□

If $\mathcal{T}$ is a (n,u)-spider of the maximum size $\sigma_u(n)$, then we say that $\mathcal{T}$ is (n,u)-*critical*.

We now prove Lemma 52, and then derive Lemma 51 from it.

Proof of Lemma 52 (weak super-additivity). We choose a (n_i, u)-critical spider $\mathcal{T}_i$ for each $i = 1, 2, \ldots, r$, so that every label is used in at most one spider $\mathcal{T}_i$. We identify the roots of the spiders $\mathcal{T}_i$ and also their labels. This gives us a rooted skeletal u-free tree on $\sum \sigma_u(n_i) - (r-1)$ vertices labeled by at most $\sum n_i - (r-1) = n - (r-1)$ labels. We further add $r-1$ new vertices (leaves) connected to the root, and label them by pairwise distinct new labels. The obtained tree is a rooted skeletal u-free tree on $\sum \sigma_u(n_i)$ vertices labeled by at most n labels. Thus, $\sum \sigma_u(n_i) \leq \tau_u(n) = O(\sigma_u(n))$ (the second estimate follows from Theorem 29). □

Proof of Lemma 51 (additive property) (sketch). The first inequality follows from the monotone property (Lemma 50). For the second inequality, consider a (n, u_1u_2)-critical spider $\mathcal{T}$. Let $\mathcal{T}'$ be the maximum u_1-free subspider of $\mathcal{T}$ containing the root of $\mathcal{T}$. The size of $\mathcal{T}'$ is at most $\sigma_{u_1}(n)$. If we add to $\mathcal{T}'$ vertices of $\mathcal{T}$ connected by an edge to $\mathcal{T}'$, then the size of the obtained subspider $\mathcal{T}''$ is at most $\sigma_{u_1}(n)+n$. By the choice of $\mathcal{T}''$, any path in $\mathcal{T}$ connecting the root with a vertex outside $\mathcal{T}''$ contains u_1 on its intersection with $\mathcal{T}''$. The subgraph $\mathcal{S}$ of $\mathcal{T}$ induced by the vertices outside $\mathcal{T}''$ is a union of subspiders of $\mathcal{T}$. Each of them is "nearly" u_2-free: If some of these subspiders contains u_2, then $\mathcal{T}$ contains a sequence u_1u_2', where u_2' is isomorphic to u_2. However, some letter may occur both in u_1 and in u_2' — this complication might be excluded by taking certain relatively large reductions of each component in $\mathcal{S}$ (see [**V99b**] for details). These reductions are u_2-free and therefore, by the weak super-additivity of σ_{u_2} (Lemma 52), their overall number of vertices is at most $O(\sigma_{u_2}(n))$. A technical argument in [**V99b**] shows that $\mathcal{S}$ cannot be much larger, its size is at most $O(\sigma_{u_2}(n)+n)$. Thus,

$$\sigma_{u_1u_2}(n)=|\mathcal{T}|\le(\sigma_{u_1}(n)+n)+O(\sigma_{u_2}(n)+n)=O(\sigma_{u_1}(n)+\sigma_{u_2}(n)+n).$$

□

Proof of Proposition 32. If u is isomorphic to one of the three sequences ε (empty sequence), a, ab (where a,b are letters), then $\sigma_u(n)\le 1$. So suppose that u is isomorphic to none of the sequences ε, a, ab. Then the star in Fig. 14 gives $\sigma_u(n)\ge n$. On the other hand, by Theorem 2, $\frac{\sigma_u(n)}{n}$ is bounded from above by

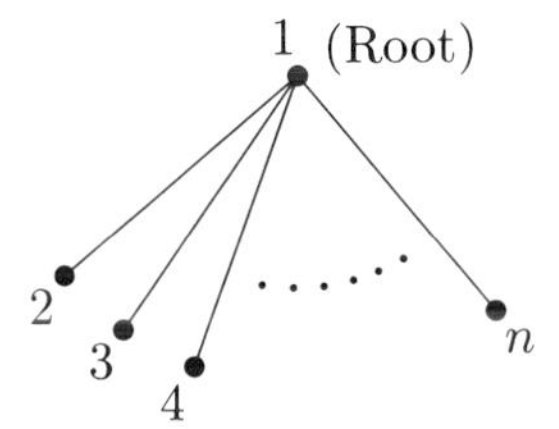

FIGURE 14. A labeled star giving lower bounds $\sigma_u(n)\ge n$ (unless $u\approx\varepsilon,a,ab$) and $\hat{\sigma}_u(n)\ge n$ (unless $u\approx\varepsilon,a,ab,abc$).

$2^{c(\alpha(n))^{|u|+||u||-5}}$, which tends to infinity slower than the functional inverse of any unbounded primitive recursive function (see [**Ro67**]). It remains to show that, for any sequence u, either $\sigma_u(n)=O(n)$ or $\frac{\sigma_u(n)}{n}\to\infty$.

Suppose that $\sigma_u(n)\ne O(n)$ for some sequence u. Then, for any $M>0$ there exists an n_M with $\sigma_u(n_M)>M\cdot n_M$. By the weak super-additivity of $\sigma_u(n)$ (Lemma 52),

$$\sigma_u(n)\ge\sigma_u\left(\left\lfloor\frac{n}{n_M}\right\rfloor n_M\right)\ge\varepsilon\left\lfloor\frac{n}{n_M}\right\rfloor\sigma_u(n_M)>\frac{\varepsilon n}{2n_M}\cdot M\cdot n_M=\frac{\varepsilon M}{2}n,$$

for any $n\ge n_M$ (where $\varepsilon>0$ is independent of n_M,n). We get $\frac{\sigma_u(n)}{n}\to\infty$. □

11. A Lower Bound Construction for Labeled Trees

We now describe a construction from [**V99b**] giving Theorems 42 and 43. For any two positive integers k and m, we construct a spider $\mathcal{T}(k,m)$ with the following five properties:

(C1) $\mathcal{T}(k,m)$ is $ababa$-, aba^2b-, $abcabc$-, and $abcacb$-free,

(C2) $c(k,m) := \frac{||\mathcal{T}(k,m)||}{m}$ is an integer,
(C3) for each label l appearing in $\mathcal{T}(k,m)$, there is a leaf $z(l)$ labeled by l; moreover, the leaves $z(l)$ lie in $c(k,m)$ fans, where *fan* is a set of m leaves having a common neighbor (father),
(C4) in each fan in $\mathcal{T}(k,m)$, there is a leaf labeled by the same label as its father; otherwise no two neighbors in $\mathcal{T}(k,m)$ are labeled by the same label,
(C5) $\frac{|\mathcal{T}(k,m)|}{||\mathcal{T}(k,m)||} \geq \frac{k}{2}$.

It follows from conditions (C1)–(C5) that $\frac{\sigma_{ababa}(n)}{n}$ and $\frac{\sigma_{aba^2b}(n)}{n}$ are unbounded, since the removal of all leaves in $\mathcal{T}(k,m)$ gives a $(||\mathcal{T}(k,m)||, ababa)$-spider (and a $(||\mathcal{T}(k,m)||, aba^2b)$-spider) of size at least $(\frac{k}{2}-1)||\mathcal{T}(k,m)||$. Since $\mathcal{S}(k,m)$ is also *abcabc*-free and *abcacb*-free, Lemma 46 now implies that also $\frac{\sigma_{abcabc}(n)}{n}$ and $\frac{\sigma_{abcacb}(n)}{n}$ are unbounded. Now, let u be any sequence containing at least one of the sequences *ababa*, aba^2b, *abcabc*, and *abcacb*. Then from above and by Proposition 32, $\frac{\sigma_u(n)}{n} \to \infty$. Actually, since the numbers $||\mathcal{T}(k,m)||$ grow analogously as the "Ackermann" numbers $A(k,m)$, the spiders $\mathcal{T}(k,m)$ give Theorem 42 (see [**V99b**] for details of the proof).

The spiders $\mathcal{T}(k,m)$ are constructed by double induction on k and m as follows.

For $k=1$ and any m, $\mathcal{T}(1,m)$ consists of the root and a single fan of m leaves connected to the root (see Fig. 15).

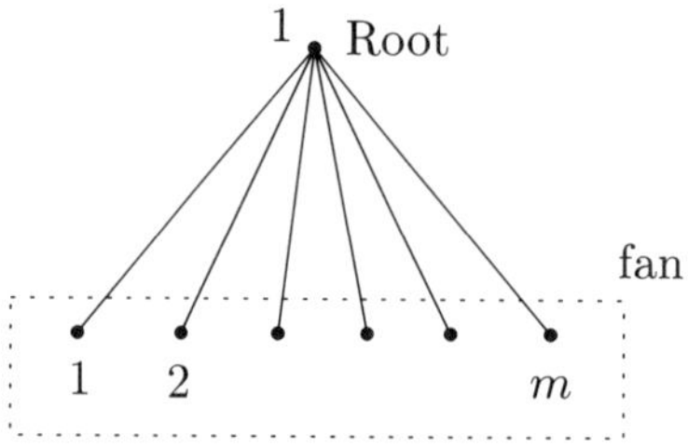

FIGURE 15. The spider $\mathcal{T}(1,m)$

For $k>1$ and $m=1$, we construct $\mathcal{T}(k,1)$ from $\mathcal{T}(k-1,2)$ by adding an edge and a vertex in each fan as shown in Fig. 16. We get two one-term fans in $\mathcal{T}(k,1)$ in place of each two-term fan in $\mathcal{T}(k-1,2)$.

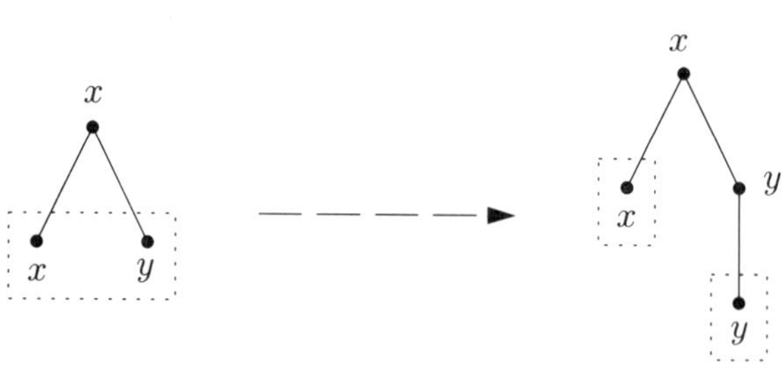

FIGURE 16. Getting from $\mathcal{T}(k-1,2)$ to $\mathcal{T}(k,1)$ (dotted rectangles denote fans).

For $k>1$ and $m>1$, we construct $\mathcal{T}(k,m)$ from the spiders $\mathcal{T}=\mathcal{T}(k,m-1)$ and $\mathcal{V}=\mathcal{S}(k-1,M)$, where $M=c(k,m-1)=\frac{||\mathcal{T}(k,m-1)||}{m-1}$ is the number of fans in $\mathcal{T}$, as follows. Let $P = c(k-1, M)$ and let $F_1, F_2, \ldots, F_P$ be the fans in $\mathcal{V}$ (listed in an arbitrary order). For each $i = 1, \ldots, P$, we take a spider $\mathcal{T}_i$ isomorphic to $\mathcal{T}$ such that no label appears in more than one of the spiders $\mathcal{V}, \mathcal{T}_1, \ldots, \mathcal{T}_P$. For $i = 1, \ldots, P$, we denote the leaves in F_i by $f_{i1}, f_{i2}, \ldots, f_{iM}$ (in an arbitrary order) and their common father by f_i. Further, we denote the root of $\mathcal{T}_i$ by r_i and the fans in $\mathcal{T}_i$ by $G_{i1}, G_{i2}, \ldots, G_{iM}$. For $j = 1, \ldots, M$, let g_{ij} be the common father of the leaves in G_{ij}.

The spider $\mathcal{T}(k, m)$ is obtained from the disjoint union of the spiders $\mathcal{V}, \mathcal{T}_1 \ldots, \mathcal{T}_P$ in the following four steps made for each $i = 1, \ldots, P$ (see Fig. 17):

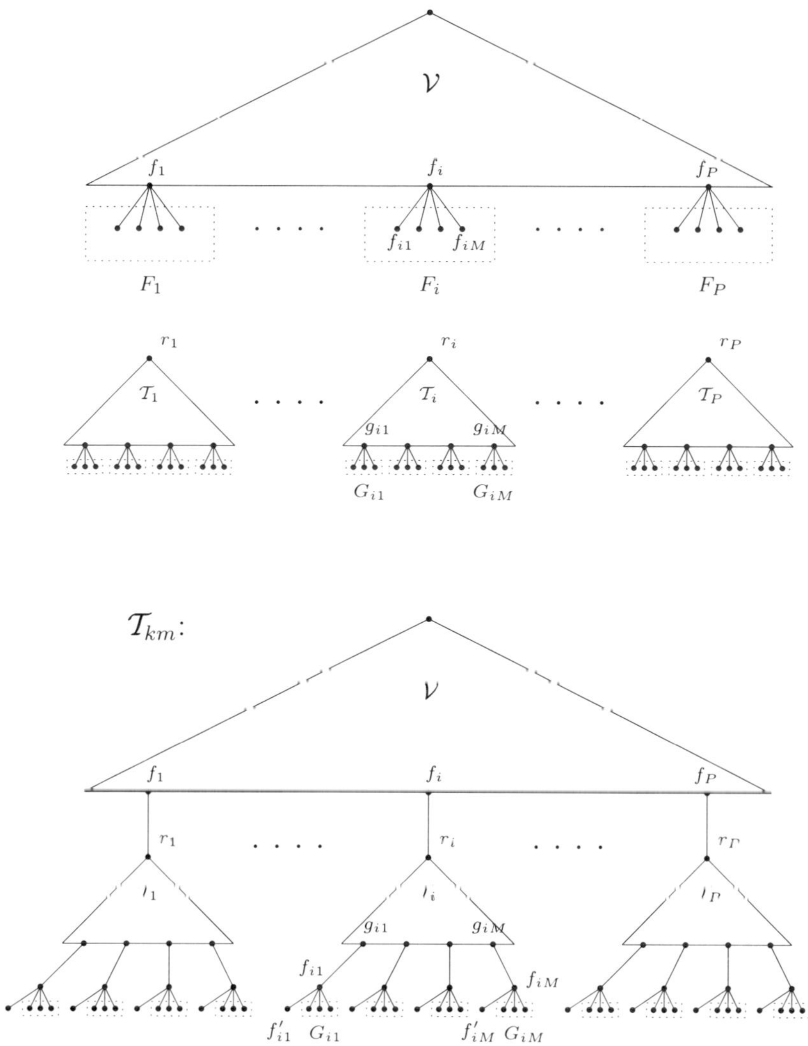

FIGURE 17. The inductive construction of $\mathcal{T}(k, m)$, $k > 1$, $m > 1$.

1. remove all edges connecting the leaves in the fans $F_i, G_{i1}, G_{i2}, \ldots, G_{iM}$ with their fathers,
2. connect f_i with r_i by an edge,

3. for each $j = 1, \ldots, M$, connect f_{ij} by an edge with the vertex g_{ij} and with all vertices of the fan G_{ij},
4. for each $j = 1, \ldots, M$, add a new vertex (leaf) f'_{ij} labeled by $l(f_{ij})$ and connect it by an edge to f_{ij}.

The fans in $\mathcal{T}(k,m)$ are $G_{ij} \cup \{f'_{ij}\}$, $i = 1, \ldots, P, j = 1, \ldots, M$. It is not too difficult to verify conditions (C1)–(C5).

It seems possible that $\sigma_u(n) = O(n)$ if and only if u is *ababa*-, *aba*^{2}*b*-, *abcabc*-, and *abcacb*-free (though the support for such a conjecture is quite weak at the moment):

PROBLEM 3. *Is there a ababa-, aba^{2}b-, abcabc-, and abcacb-free sequence u with $\sigma_u(n) \neq O(n)$?*

Our construction also gives Theorem 43:

Proof of Theorem 43. It suffices to prove that the above constructed spiders $\mathcal{T}(k,m)$ contain no 2-sparse sequence u with $|u| > 3||u|| - 2$. We proceed by induction on the construction. The statement is clearly true for the spiders $\mathcal{T}(1,m)$. If $\mathcal{T}(k,1)$ contains a 2-sparse sequence u, then already $\mathcal{T}(k-1,2)$ contains u. It remains to prove that the statement holds for $\mathcal{T}(k,m)$ with $k,m > 1$, provided it holds for $\mathcal{T} = \mathcal{T}(k,m-1)$ and for $\mathcal{V} = \mathcal{T}(k-1,M)$, where M is as in the construction of $\mathcal{T}(k,m)$. If a sequence u is contained in $\mathcal{T}(k,m)$, then it follows easily from the construction that $u = u_1u_2u_3$, where $S(u_1) \cap S(u_2) = \emptyset$ (and therefore $||u_1|| + ||u_2|| \leq ||u||$), $|u_3| \leq 2$, u_1 is contained in $\mathcal{V}$, and u_2 is contained in $\mathcal{T}$. Then, by the assumptions,

$$|u| = |u_1| + |u_2| + |u_3| \leq (3||u_1|| - 2) + (3||u_2|| - 2) + 2 \leq 3||u|| - 2.$$

□

12. Further Generalizations of Davenport–Schinzel Sequences

12.1. Cyclic (Generalized) Davenport–Schinzel Sequences. A *cyclic sequence* is a sequence in which the first and the last terms are considered adjacent. We say that a cyclic sequence is u-free, if it is *u-free* whenever we cut it into an ordinary sequence by a cut between any two adjacent terms. Analogously we define *k-sparse* cyclic sequences. A *cyclic (n,u)-sequence* is defined as a u-free $||u||$-sparse cyclic sequence over at most n letters. Let $c_u(n)$ be the maximum length of a cyclic (n,u)-sequence.

Clearly, $c_u(n) \leq f_u(n)$ for each u and n. For alternating sequences, we easily get the following relation to the function $N_s(n)$:

THEOREM 58. *For any two positive integers k and n,*

$$c_{\text{al}(2k)}(n) = f_{\text{al}(2k)}(n) - 1 \ (= N_{2k}(n) - 1).$$

(Recall that al(i) *denotes the alternating sequence ababa... of length i.)*

Proof. Let v be al$(2k)$-free 2-sparse cyclic sequence over n letters. Then the sequence obtained by doubling an arbitrary term in v and cutting v between the two adjacent equal terms is a al$(2k)$-free 2-sparse sequence over n letters. This gives $c_{\text{al}(2k)}(n) \leq f_{\text{al}(2k)}(n) - 1$.

On the other hand, if v is a al$(2k)$-free 2-sparse sequence over n letters, then the cyclic sequence v' obtained from v in a natural way is also al$(2k)$-free. It suffices to remove at most one term from v' to make it 2-sparse. It follows that $c_{\mathrm{al}(2k)}(n) \geq f_{\mathrm{al}(2k)}(n) - 1$. □

We remark that it is also not difficult to shown that

$$f_{\mathrm{al}(2k)}(n) - 1 \leq c_{\mathrm{al}(2k+1)}(n) \leq f_{\mathrm{al}(2k)}(n) + 2n.$$

12.2. Containment Problems for $\{0,1\}$-Matrices. Paper [**FP91**] contains a survey on containment problems for $\{0,1\}$-matrices which are often related to Davenport–Schinzel sequences. We chose one paper cited there, and briefly describe results obtained in it.

Let $M = (M_{ij})$ be an $m \times n$ matrix with all entries in $\{0,1\}$, and let $C = (C_{ij})$ be an $u \times v$ matrix with 1's and blanks as its entries. We say that M *contains* C if it contains an $u \times v$ submatrix C' such that $C_{ij} = 1$ implies $C'_{ij} = 1$ for all $1 \leq i \leq u$ and $1 \leq j \leq v$ (if C_{ij} is blank then C'_{ij} may be arbitrary).

For a matrix C and for two integers m and n, let $g(m,n;C)$ be the maximum number of 1's in an $m \times n$ matrix not containing C. An upper bound on $g(m,n;C)$ for various C is useful in solving some problems in combinatorial geometry and also in bounding the time and storage complexity of some geometric algorithms.

Füredi and Hajnal [**FH92**] systematically studied the asymptotic behavior of $g(n,n;C)$ for all $\{1,\mathrm{blank}\}$-matrices C containing exactly four 1's so that each row and column contains at least one 1. There are 37 such matrices C. Let's call them FH-matrices. Füredi and Hajnal [**FH92**] establish the asymptotic bahavior of $g(n,n;C)$ for 33 of them. It is interesting that various extremal-type methods are applied in [**FH92**], and also various bounds on $g(n,n;C)$ are obtained. For $C = \begin{pmatrix} 1 & 1 \\ 1 & 1 \end{pmatrix}$, $g(n,n,C) = \Theta(n^{3/2})$. For all other FH-matrices C, $g(n,n;C) = O(n \log n)$. In fact, $g(n,n,C) = O(n)$ holds for 28 of the 37 FH-matrices C, and for three other FH-matrices C no super-linear lower bound on $g(n,n,C)$ is known. For two FH-matrices C, $g(n,n;C) = \Theta(n \log n)$ is known. Interestingly, $g(n,n;C) = \Theta(n\alpha(n))$ is known for the two matrices

$$C = \begin{pmatrix} 1 & & 1 & \\ & 1 & & 1 \end{pmatrix} \quad \text{and} \quad C = \begin{pmatrix} 1 & & & \\ & & 1 & \\ & 1 & & 1 \end{pmatrix}$$

– the proofs rely on Davenport–Schinzel sequences.

Acknowledgments. I would like to thank Martin Klazar for valuable comments, suggestion of several simplifications, and finding many errors in an earlier version of the paper.

References

[AKV92] R. Adamec, M. Klazar and P. Valtr, Generalized Davenport–Schinzel sequences with linear upper bound, *Discrete Math.* **108** (1992), 219–229.

[AAPPS97] P.K. Agarwal, B. Aronov, J. Pach, R. Pollack, and M. Sharir, Quasi-planar graphs have a linear number of edges, *Combinatorica* **17** (1997), 1–7.

[ASS89] P. K. Agarwal, M. Sharir and P. Shor, Sharp upper and lower bounds on the lengths of general Davenport–Schinzel sequences, *J. Comb. Th., Ser. A*, **52** (1989), 228–274.

[Di50] R. Dilworth, A decomposition theorem for partially ordered sets, *Annals of Mathematics* **51** (1950), 161-166.

[DS65] H. Davenport and A. Schinzel, A combinatorial problem connected with differential equations, *Amer. J. Math.* **87** (1965), 684–694.

[FH92] Z. Füredi and P. Hajnal, Davenport-Schinzel theory of matrices, *Discrete Mathematics* **103** (1992), 233–251.

[FP91] Z. Füredi and J. Pach, Traces of finite sets: extremal problems and geometric applications, in: *Extremal Problems for Finite Sets*, Proc. Colloq., Visegrád, Hungary, 1991. (P. Frankl et al., Eds.) *Bolyai Society Mathematical Studies* **3** (1994), 251–282.

[HS86] S. Hart and M. Sharir, Nonlinearity of Davenport–Schinzel sequences and of generalized path compression schemes, *Combinatorica* **6** (1986), 151–177.

[KL98] M. Katchalski and H. Last, On geometric graphs with no two edges in convex position, *Discrete Comput. Geom.* **19** (1998), 399–404.

[Kl92] M. Klazar, A general upper bound in extremal theory of sequences, *Comment. Math. Univ. Carolinae* **33** (1992), 737–747.

[Kl93] M. Klazar, Two results on a partial ordering of finite sequences, *Comment. Math. Univ. Carolinae* **34** (1993), 697–705.

[Kl94] M. Klazar, A linear upper bound in the extremal theory of sequences, *J. Comb. Th., Ser. A*, **68** (1994), 454–464.

[Kl95] M. Klazar, *Combinatorial aspects of Davenport–Schinzel sequences*, doctoral thesis, Charles University, Prague, 1995.

[Kl97] M. Klazar, Combinatorial aspects of Davenport–Schinzel sequences, *Discrete Math.* **165/166** (1997), 431–445.

[Kl99a] M. Klazar, Extremal problems for labeled trees and Davenport–Schinzel sequences, to appear in *Discrete Math.* **197/198** (1999).

[Kl99b] M. Klazar, On the maximum lengths of Davenport–Schinzel sequences, *this volume.*

[KV94] M. Klazar and P. Valtr, Generalized Davenport–Schinzel sequences, *Combinatorica* **14** (1994), 463–476.

[Ko88] P. Komjáth, A simplified construction of nonlinear Davenport–Schinzel sequences, *J. Comb. Th., Ser. A,* **49** (1988), 262–267.

[Ku79] Y. Kupitz, *Extremal Problems in Combinatorial Geometry*, Aarhus Univ. Lecture Notes Series **53** (1979).

[PA95] J. Pach and P.K. Agarwal, *Combinatorial Geometry*, Wiley Interscience, New York (1995).

[PF94] J. Pach and Z. Füredi, Traces of finite sets: extremal problems and geometric applications, *in* Extremal Problems for Finite Sets, Proc. Colloq., Visegrád, Hungary, 1991. (P. Frankl et al., Eds.) *Bolyai Society Mathematical Studies* **3** (1994), 251–282.

[PSS96] J. Pach, F. Shahrokhi, and M. Szegedy, Applications of the crossing number, *Algorithmica* **16** (1996), 111–117.

[PT94] J. Pach and J. Törőcsik, Some geometric applications of Dilworth's theorem, *Discrete Comput. Geom.* **12** (1994), 1–7.

[Ro67] H. Rogers, *Theory of recursive functions and effective computability*, McGraw-Hill, New York (1967).

[Sh87] M. Sharir, Almost linear upper bounds on the length of general Davenport–Schinzel sequences, *Combinatorica* **7** (1987) 131–143.

[Sh88] M. Sharir, Improved lower bounds on the length of Davenport–Schinzel sequences, *Combinatorica* **8** (1988), 117–124.

[SA95] M. Sharir and P. K. Agarwal, *Davenport–Schinzel sequences and their geometric applications*, Cambridge University Press, Cambridge, 1995.

[Sz74] E. Szemerédi, On a problem by Davenport and Schinzel, *Acta Arithm.* **25** (1974), 213–224.

[To99] G. Tóth, manuscript in preparation.

[TV98] G. Tóth and P. Valtr, Geometric graphs with few disjoint edges, Proc. of Symp. Comput. Geom. 1998, also to appear in *Discrete Comput. Geom.*

[V97] P. Valtr, Graph drawings with no k pairwise crossing edges, in: G. DiBattista (ed.), *Graph Drawing '97*, Rome, September 1997, pp. 205–218.

[V98a] P. Valtr, On geometric graphs with no k pairwise parallel edges, *Discrete Comput. Geom.* **19** (1998), 461–469.

[V98b] P. Valtr, On an extremal problem for labeled trees, to appear in *Europ. J. Comb.* **19** (1998).

[V99a] P. Valtr, Three asymptotically equivalent definitions of Davenport–Schinzel trees, manuscript.

[V99b] P. Valtr, Davenport–Schinzel trees, manuscript in preparation.

[WS88] A. Wiernik and M. Sharir, Planar realization of nonlinear Davenport-Schinzel sequences by segments, *Discrete Comput. Geom.* **3** (1988), 15–47.

DEPARTMENT OF APPLIED MATHEMATICS, CHARLES UNIVERSITY, MALOSTRANSKÉ NÁM. 25, 118 00 PRAHA 1, CZECH REPUBLIC

E-mail address: valtr@kam.ms.mff.cuni.cz

DIMACS CENTER, RUTGERS UNIVERSITY, P.O.BOX 1179, PISCATAWAY, NJ 08855, U.S.A.